THE MICROSCOPE AND MICROSCOPICAL TECHNOLOGY.

A TEXT-BOOK FOR PHYSICIANS AND STUDENTS.

BY

DR. HEINRICH FREY,

PROFESSOR OF MEDICINE IN ZURICH, SWITZERLAND.

TRANSLATED FROM THE GERMAN AND EDITED BY

GEORGE R. CUTTER, M.D.,

CLINICAL ASSISTANT TO THE NEW YORK EYE AND EAR INFIRMARY.

ILLUSTRATED BY 343 ENGRAVINGS ON WOOD, AND CONTAINING THE PRICE-LISTS OF THE PRINCIPAL MICROSCOPE-MAKERS OF EUROPE AND AMERICA.

FROM THE FOURTH AND LAST GERMAN EDITION.

NEW YORK:
WILLIAM WOOD & CO.,
27 GREAT JONES STREET.
1872.

POOLE & MACLAUCHLAN, PRINTERS,
205-213 *East Twelfth St.*,
NEW YORK.

CONTENTS.

TRANSLATOR'S PREFACE.

Any attempt on my part, by way of introduction or commendation of Professor Frey's work, must, I feel, be altogether misplaced and unnecessary. Although the treatise has been but a few years before the public, four large editions have already been issued, and a copy is always found on the table of those microscopists who are able to read it in the original.

I have been induced to undertake the arduous labor of preparing a translation of the work by the hope that it might stimulate and facilitate the study of this important branch of science in this country.

An apology may be thought necessary for the style of the translation,—in having followed the German so literally. The nature of the subject, however, involving as it does such very minute descriptions, and the frequent repetition of the same terms, added to the impossibility of doing justice in any other way to the author's condensed style, have necessitated a rigid adherence to the original text.

The few additions which have been made are enclosed within brackets.

George R. Cutter, M.D.,
No. 228 East 12th St., New York.

October, 1872.

INTRODUCTION.

"To endeavor to discover new methods of investigation appears to me to be one of the most important duties of every observer. To communicate these to his pupils must be the desire of every teacher of any branch of natural science."—(L. Beale. *How to Work with the Microscope*, p. 3.)

WITHIN the last ten years the Microscope, that instrument which has conquered a new world of minuteness for natural science, has become widely known. Already a considerable number of these instruments are yearly issued from the large and celebrated establishments of Europe; and not less noticeable is the number of those which are constructed and sold by less renowned opticians. The opinion is now admitted that the microscope is quite as indispensable for the scientific, as the stethoscope and pleximeter for the practical requirements of the physician.

Through Schwann's classical work we have learned that the human body is formed, in all its parts, from cells and their derivatives, and that the cell is the ultimate organized unity of animal life. As in the province of anatomy we cannot understand the structure of any portion of the body without this little microscopical foundation-stone, even so little do we succeed in comprehending the physiological action, if we disregard the isolated action of these ultimate organized unities. The united action of an organ is only the result of all the single actions of cells, of the "elementary organisms," as they were afterwards called. Thus Histology has become an indispensable member in the series of anatomico-physiological sciences.

Health and disease appear, in the ingenuous view of man, to be separated by a wide abyss; an opinion which, in the domain of science, is drawn like a red thread through so many nosological systems of former days. The recognition of the contrary is properly greeted as a great advance in physiological opinion. The processes which take place in the diseased body are actually, for us, but modifications of those which occur in the normal; the same physiological laws obtain in the one case as in the other; and those also which, in a material regard, occur in the diseased body, the metamorphosis, separation, and new formation of its elements, obey the same laws of cell-life which we recognize in the normal organism. The eminent significance of pathological histology requires no wider discussion, nor is it necessary further to recommend the instrument by means of which histology is chiefly created.

Microscopy is, however, as some of our readers will have already experienced in their first attempts, delicate work. How many a student, how many a physician, impelled by the great value of such studies, has procured a microscope only to perceive, to his great dissatisfaction, how little he is qualified to use it. Here, as in all departments of human efficacy, a period of apprenticeship is necessary,—an arduous season of sowing before attempting to reap the plenteous harvest.

The microscope is a delicate implement; like other complicated instruments, its use must be learned. The faculty of seeing with it must likewise be acquired, which also requires some perseverance, if we would attain to the *accurate* vision which is here indispensable.

The art to observe and investigate requires the employment and knowledge of many small, and therefore, at first, apparently unimportant accessories. The time is past when it was thought possible to fathom the finer textural relations of a piece of fresh tissue by picking, and, perhaps, the assistance of pressure and a little acetic acid. Modern chemistry, to which medicine is extremely indebted, has also furnished the microscopist with a series of the most important accessories. Thus, now-a-days, knives and needles, the syringe, the scales, and many other artifices are employed in the investigation of the tissues of the body.

We shall readily comprehend, from what has been said, that our so industrious epoch in a microscopical regard, among many sound investigations, also annually produces over-hasty performances which show how little their authors have learned to overcome the most elementary difficulties.

This remark, however, is not written to discourage; it should, on the contrary, only indicate that the most complete familiarity with the instrument and with the entire *technique* should be the indispensable preliminary of every microscopical investigation.

Though that school which offers the practical instruction of a teacher is always the best, it is not permitted to every one to walk in this path to learning. Here written directions find their place, and, provided they are judicious and practical, may afford a sufficient compensation, and make a microscopical observer out of the beginner.

The literature of the microscope is already voluminous. Admirable and copious works exist in the German, Hollandish, and English languages, such as those of Mohl, Harting, and Carpenter. But in the German there is a great deficiency of concise works especially adapted to the practical wants of the physician, as we have only the obsolete work of Vogel. L. Beale has produced two able text-books for the English.

May our little work serve as a guide for students and physicians till the time, at least, when a better pen shall produce a better substitute.

That we premise the mechanism of the instrument and the use of its several parts is obviously necessary; for a knowledge of the implements must always precede the labor which they are to perform. That we limit ourselves, in this section, to that which is most important and indispensable, and touch but lightly the difficult, and in all its parts by no means definitely settled, optical theory, requires no further justification. Another portion of our work treats of the various methods of investigation at present in use. A third part, finally, completes the directions for investigating the various tissues and parts of the body in a normal and pathological condition. Possibly, in the department of pathology we may have been too concise for a portion of our readers. The investigation of sputum, pus, urinary sediment, and tumors usually occupies a much larger space in books on this subject. But, true to our maxim, that the most accurate knowledge of the normal relations should precede every investigation of their pathological condition, we endeavor, first of all, to make the former clear and then join the latter supplementarily. As every pathological new formation repeats, more or less, the type of the normal structure, so are the methods of investigating diseased tissues and portions of the body almost the same.

With regard to the literature of the microscope, we would particularly mention the following works:—

J. Vogel, Anleitung zum Gebrauche des Mikroskops, Leipzig, 1841. —H. v. Mohl, Mikrographie, Tübingen, 1846.—C. Robin, Du microskope et des injections, Paris, 1871.—P. Harting, Das Mikroskop, 2. deutsche Originalausgabe, besorgt von Theile, 3 Bde., Braunschweig, 1866.—W. Carpenter, The Microscope, 4th Edition, London, 1868.—L. Beale, How to Work with the Microscope, 4th Edition, London, 1867, and, The Microscope in its application to practical Medicine, 3d Edition, London, 1866.—H. Schacht, Das Mikroskop, 3. Auflage, Berlin, 1862.—C. Nägeli und S. Schwendener, Das Mikroskop, Leipzig, 1867.—L. Dippel, Das Mikroskop und seine Anwendung, Braunschweig, Bd. 1, 1867, und Bd. 2, Abth. 1, 1869.

Section First.

THEORY OF THE MICROSCOPE.

THAT wonderful organ, the human eye, has often been compared to the camera obscura, and the comparison is, indeed, an excellent one. As the collective lens projects an inverted diminished image at the background of the latter apparatus, which is received on the ground-glass plate, so the collective refracting media of the eye produce the same inverted diminished image at its profundity, which is received on the retina.

Probably all of our readers are aware that the dimensions which an object appears to the eye to possess, depend upon the size of the angle of vision; an angle which one receives by combining, through straight lines, the corresponding terminal points of the object, and of the image received in the eye.

A glance at fig. 1 will render this intelligible. The curved

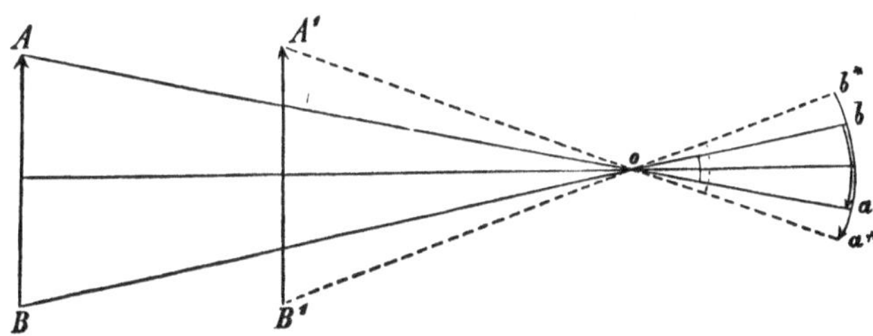

Fig. 1. Visual angle and apparent magnitude of the object.

line at *b a* represents the image, projected upon the fundus oculi, of the arrow A B, placed before the organ of vision; *a* is united by a line to A, *b* by a second one to B. Thus arises the visual angle A *o* B = *b o a*. All bodies whose terminal points touch the lines A *a* and B *b*, appear to the eye to be of the same size. A needle held close before the eye may, under these circumstances, have the same apparent dimensions as a long pole which stands in the distance. When the arrow is brought nearer to the eye, to A′ B′ for instance, it projects the image *b** *a**, causing the angle of vision A′ *o* B′; the arrow, however, appears to be larger. If the angle of vision falls below a certain size, the object ceases to be visible. A thick wire, for example, if considerably removed from our eye, is no longer perceived. If we bring the wire nearer and nearer, whereby the visual angle is also increased, it appears at first as a fine thread, then with increasing diameter. We therefore instinctively examine small objects at a certain proximity.

But a continued approach also finds its limit at last; the wire, which was still distinctly seen, becomes indistinct, and finally, having been brought quite close to the eye, becomes entirely invisible.

On what does this last condition depend?

It is known that the image of an object projected by a collective lens alters its position according as the object is removed from or brought nearer to the lens. In the first case the image approaches the lens, in the latter, it recedes farther behind it. Now, as the human eye acts in the same manner as a lens, and accurate vision only occurs when the rays of light, coming from any point of an object, are so refracted as to be again united at the retina, so, in reality, a distinct image can only be possible at a certain distance. But daily observation teaches us something further; we see remote and near objects, one after another, with equal accuracy. The eye must, therefore, possess a correcting apparatus, in order to adjust its refracting media for proximate and remote bodies; it accommodates, as the physiologists say.

This process of accommodation is, however, disregarding individual variations, limited. The image of an object, brought more and more near to the eye, falls at last behind the retina. In our fig. 2, the arrow placed at A would give a distinct image, the rays of light diverging from a point *p* being united at a point *r* on the retina.

Should the arrow, however, be brought so near as B to the organ of vision, this union can no longer take place at the retina. The rays of light proceeding from p^* come together farther behind this membrane, at r^*.

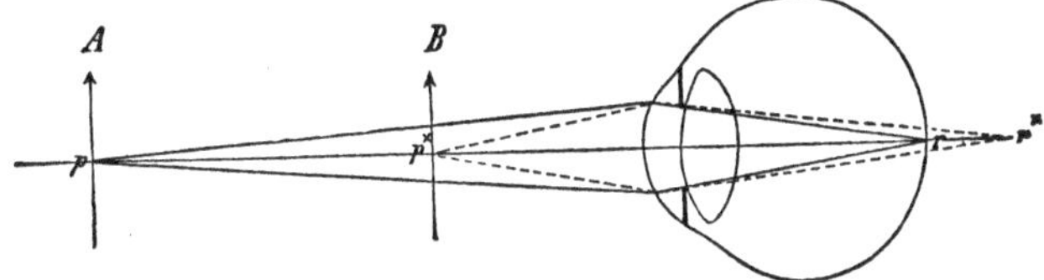

Fig. 2. Position of an object, and union of the rays which proceed from it in the eye.

Very small objects, therefore, when brought too near to the human eye become invisible; here, as we shall soon see, other accessories are necessary.

The distance from the eye at which objects of medium size can be most sharply discerned, is called the mean distance of vision. This is usually considered to be from 8 to 10 inches, or 25 centimetres, for the normal eye. The closest proximity at which an object is still visible is called the near point. Near-sighted eyes permit of a few inches nearer approach, far-sighted ones find their limit sooner; the first refract more, the latter less strongly.

Such a small body may, however, readily be made visible by placing a convex lens between it and the eye. The reason is easily understood. The point placed at O, fig. 3, produces its

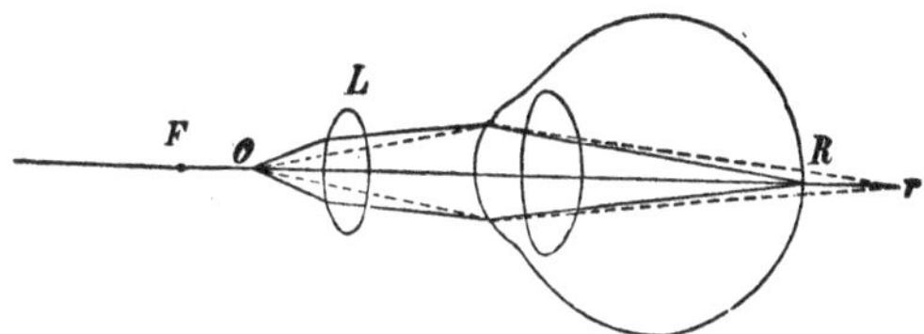

Fig. 3. Action of a convex lens when an object is brought near the eye.

image at r, and is therefore no longer perceptible to the eye. If we now place the lens L, whose focus is at F, between, the rays of light receive the direction indicated by the dotted lines, are less divergent on reaching the eye, and are, consequently, united on the retina at R. Thus a distinct image is formed.

It will also be observed, that by the use of such a lens the image thus received is enlarged.

Now whence does this arise?

Let us suppose that the object is placed at A B, fig. 4, and that a convex lens is brought between it and the eye. The cone of rays which proceeds from any point of the arrow, for example from A, projects its rays A *b*, A C, A *c* on to the lens, and these, with the exception of the rays A C, are refracted by the

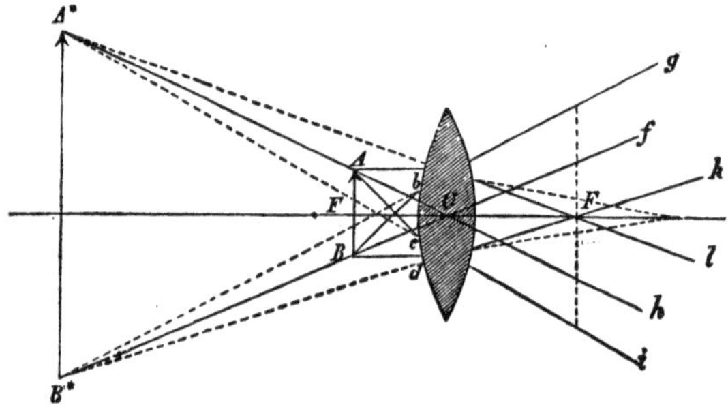

Fig. 4. An object magnified by a convex lens.

lens in the direction *b l* and *c i;* receiving a slight divergence, as if they proceeded from the more distant point A*, they arrive at the eye and are united on the retina. The same is repeated for the cone of rays B, etc.; thus an inverted image of the arrow is formed in the eye. The object appears to the organ of vision to be placed, not at A B, but at A* B*, and therefore enlarged. As a proof that the image caused by a convex lens is always apparently more remote than the object itself, look at a piece of paper through the lens, and attempt to touch the edge of it with the point of a needle. The needle will invariably be directed some distance beneath the paper.

Such convex lenses are usually called loups, so long as their magnifying power does not exceed 15 or 20 diameters, and so long as during their use they can be guided by the human hand. When the magnifying power of such lenses is greater, so that a stand is necessary to support them during their use, the combination is called a simple microscope. The impossibility of making a sharp line of demarcation between the two kinds of instruments is self-evident, as weak convex lenses are also fastened to a stand, and numerous so-called loup-supporters are also employed (fig. 5).

Fig. 5. Nachet's simple loup-stand.

There are many varieties of loups, but we must refer to more detailed works for their description. Their value and application in natural philosophy is also too well known to render it necessary for us to say more on the subject. A good loup, magnifying 10 or 15 times, is indispensable.

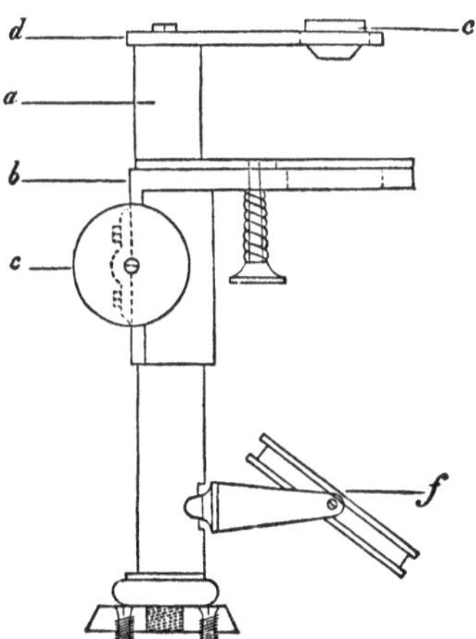

Fig. 6. Plössl's simple microscope.

The simple microscope of Plössl, of Vienna, is seen in fig. 6. A metal bar (*a*) supports, at half its height, a horizontal plate which is bored through its centre, the stage (*b*) of the microscope. This can be elevated or depressed by means of the rack (*c*). The movable mirror (*f*) placed beneath, serves to reflect the light on to the object to be examined, which is placed on the stage. If, instead of transmitted light, it be desired to examine the object by reflected light, after the manner of our usual vision, the mirror is thrown out of action, or an opaque plate is placed on the stage. The horizontal arm (*d*), on the upper extremity of the stand, carries the magnifying glass, the lens (*e*). It can be removed from the opening in the arm and replaced by another.

The simple microscope of Nachet, of Paris (fig. 7), likewise has a convenient form. The movement is made by a rack, which elevates or lowers the lens, in contradistinction to the

Fig. 7. Nachet's simple microscope.

stand of Plössl, in which the stage moves up and down. Two additional plates on the stage, bent downwards, serve as supports for the hands during the manipulation. Both instruments have clamps on the stage, for the purpose of fixing the object.

The simple microscope is now-a-days indispensable to the natural philosopher, as an instrument for dissecting. It is, however, no longer, or but little, employed for scientific investigations.

When a tube is placed over the magnifying-glass of the simple microscope, and the object is placed somewhat beyond the focus of the lens, an enlarged, inverted image of the same is projected within the tube. In fig. 8 we can readily perceive this relation. If we connect the lens L with a funnel, whose diameter extends from e^* to d^*, we may receive the image on a ground glass plate.

When this aerial image is again enlarged by means of a convex lens, we have the compound dioptrical microscope. The difference between these two instruments consists in this, that

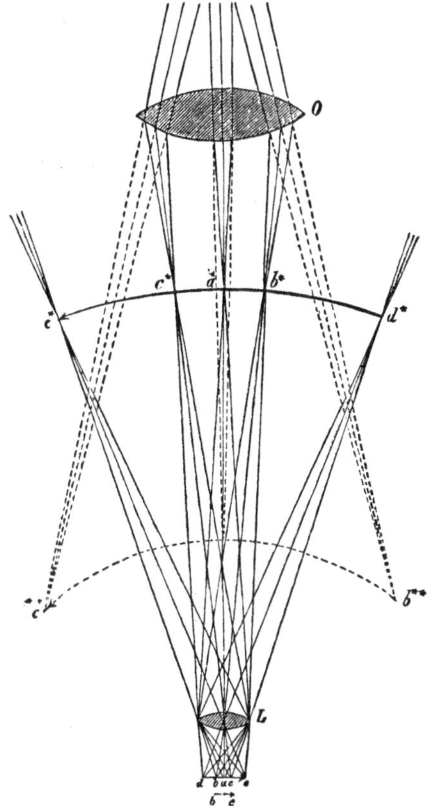

Fig. 8. The compound microscope in simplified form.

in the simple microscope the object itself, in the compound, on the contrary, the enlarged image of the object is seen. Our fig. 8 may represent the compound microscope in its simplest form. The united cone of rays c^* a^* b^* diverge from the elevation e^* d^* to the upper lens, from whence, after their refraction, they arrive, under a slight divergence, at the human eye. At the same time, however, we find that the cone of rays which proceed from the terminal points *d* and *e* of the arrow, are united at d^* and e^*, they do not arrive at the upper lens. We perceive, therefore, in our example, only the length *b–c* of the arrow. A smaller arrow, circumscribed in this dimension (see lower part of fig. 8), would, on the contrary, be distinctly seen. The dotted lines, which pass to c^{**} and b^{**}, the prolongation of the rays refracted by the upper lens, give, at the same time, the apparent size under which the arrow *b c* is seen.

An explanation of the image c^* a^* b^* of the arrow is necessary, in one other regard; it appears curved, while the arrow itself is straight. If we hold it as established that the point of union of a cone of rays falls farther behind the lens from a near object than it does from one lying more remote; and if we remember that *b* and *d*, *c* and *e* are situated farther from the optical middle point than *a*, then we can readily understand why the surface of the image is curved.

The knowledge of magnifying-glasses and the art of grinding them already existed in antiquity and the early middle ages. The invention of the compound microscope occurred, on the contrary, at a considerably later epoch. There is no longer any doubt that an humble Hollandish spectacle grinder, Zacharias Janssen, of Middelburg, probably about the year 1590, produced the first instrument of this kind. Other authorities mention, without sufficient foundation, the Netherlander, Cornelius Drebbel, Galilei and another Italian, Fontana, as the discoverers. Harting, with his usual carefulness, investigated this question of the invention several years ago.

The oldest compound microscopes were, however, very incomplete instruments, and were encumbered with great optical deficiencies. These imperfections rendered themselves sufficiently perceptible with weaker powers, and attained, with somewhat stronger glasses, such a rapid increase as to render the whole combination nearly useless.

In order to understand this, we must recall to our minds a few well-known optical principles.

The term angle of aperture of a lens denotes the angle which is formed by the focus and the two terminal points of the diameter of the lens. Thus, *g f h* is the angle of aperture of our fig. 9. Only so long as this angle remains small, do the peripheral and central rays actually reunite in a point, as we have thus far, for the sake of greater simplicity, assumed. When the

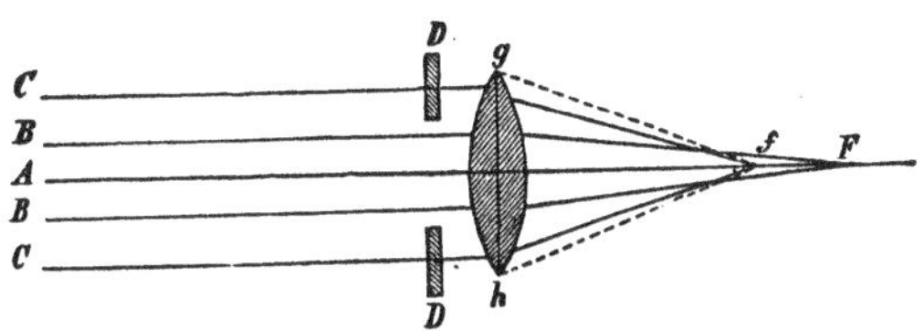

Fig. 9. Spherical Aberration.

angle of aperture is greater, only those rays of light (BB) which are nearly parallel to the axis (A) of the lens, and which pass through its central portions, are united at the focus F, while those rays (CC) which pass nearer the periphery of the lens are more strongly refracted and find their focus already in *f*. This property of refraction is called the spherical aberration.

If we cause the rays which proceed from a small luminous body to pass through such a lens, we receive at F the image projected by the central rays. It is not sharp, however, but is surrounded by a halo or circle of diffusion, caused by the peripheral rays, which have become divergent again. If we apply a dark disk with a circular opening, a so-called diaphragm or stop DD, the peripheral rays are intercepted, and we receive a distinct though faintly illuminated image at F; also at *f*, when we obstruct the central rays, and thus permit only the peripheral rays to pass through the lens. Such annular diaphragms are extensively employed in practical optics, for the purpose of improving the images.

We here mention another, for the theory of the microscope, important effect of this spherical aberration. When very small cones of rays arrive at a convex lens of considerable diameter, as is the case with the ocular O in fig. 8, those which pass through the periphery of the lens necessarily receive a stronger refraction than the more central ones. The peripheral points of the image must, accordingly, appear nearer each other than the internal portions. A wire network, fig. 10, presents an aërial image, such as is represented by fig. 11. If we observe the quadrangular network through such a loup we receive a diametrically opposite phantom, after the manner of our fig. 12. In both cases arises, therefore, a distortion of the image.

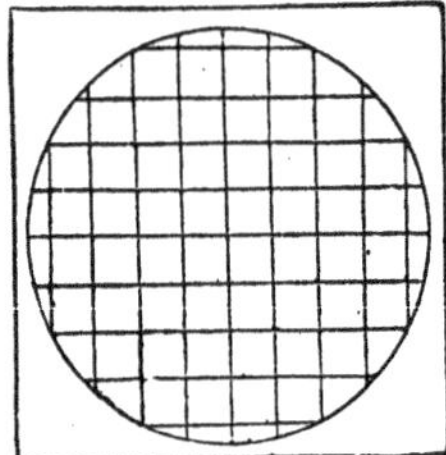

Fig. 10. Quadratic network.

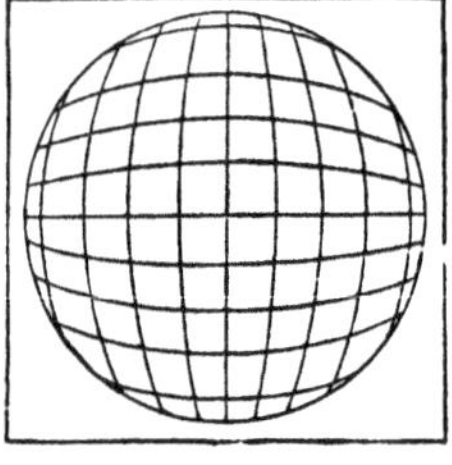

Fig. 11. Image distortion.

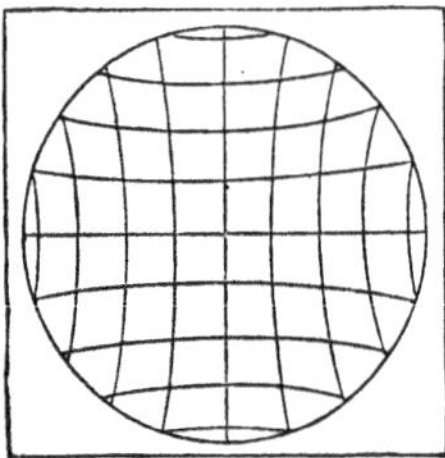

Fig. 12. Image distortion.

A second not less palpable inconvenience in the use of such lenses arises in consequence of their chromatic aberration.

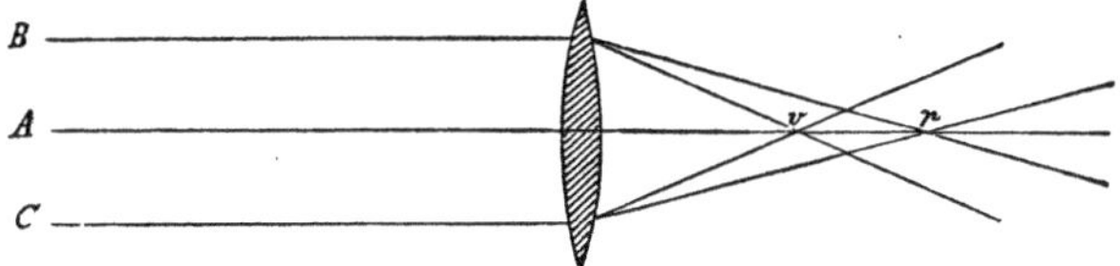

Fig. 13. Chromatic Aberration.

A ray of white light (fig. 13, B or C), in passing through a convex lens would not be refracted as a whole, but would be decomposed into rays of various colors, which suffer deviation in varying degrees in the direction of the plane of refraction, and thus form a spectrum, at one extremity of which the strongest refracted, the violet (*v*), and at the other, the least deviated, the red ray (*r*) appears.

From what has just been said it follows, that with ordinary convex glass lenses we perceive the object indistinctly defined and surrounded with colored borders. Both faults increase rapidly with the increase of curvature of the lenses. The older microscopes, therefore, produced images which were faintly illuminated, insufficiently defined, and surrounded with colored borders. The image projected by an imperfect object lens was still further magnified by the likewise imperfect ocular lens.

Achromatic lenses have now taken the place of the old useless glasses. By this name is indicated such with which the foci of the various colored rays of light are united, in other words, those which show the object free from colored borders.

The powers of refraction and of dispersion are not united in equal proportions in any single refractive medium, as has been known for a long time. One medium gives, in the same power of refraction, a greater deviation to the colored rays than another. There are two different kinds of glass which act in this manner with regard to each other—crown-glass and flint-glass. The latter is partly composed of lead, and has a considerably greater power of dispersing the colored rays than the former.

If we combine a bi-convex crown-glass lens with a plano-concave flint-glass lens (both are generally cemented together with Canada balsam) we obtain a combination in which the refraction of the convex crown-glass lens is lessened, but not destroyed, by the dispersive action of the flint-glass lens. At

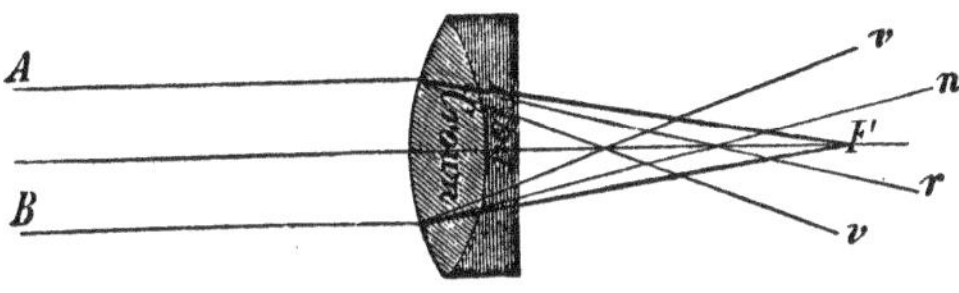

Fig. 14. Achromatic lens.

the same time, however, the color dispersion (*v r*, fig. 14) of the crown-glass lens is neutralized by the contrary action of the flint-glass lens, so that the violet and red rays are accurately united in F, at the central focus of the lens. The image here produced will either be colorless or have its natural color.

Such a combination presents, at the same time, an expedient by which the spherical aberration may also be substantially improved.

A double lens, with which the spherical as well as the chromatic aberration is annulled, is usually called aplanatic. But in reality it is impossible to completely obviate either the spherical aberration (for reasons, to discuss which here would lead us too far) or the chromatic, for even when the violet and red marginal rays are made to unite, the ratio of the dispersion of the various colored rays of the spectrum is never entirely equal.

Should, therefore, even with a double lens, the violet and red rays of light be united, the edges of the image would still exhibit traces of the ununited central rays of the spectrum. The edges appear greenish yellow. It is customary, therefore, in the construction of double lenses for the microscope, to give a slight preponderance to the flint-glass lens, so as to obtain a bluish tinge, which is more agreeable to the eye; the double lens is then said to be over-corrected. A double lens is under-corrected when a reddish border is perceptible.

As, in regard to color dispersion, one speaks of over- and under-correction, the same mode of expression is also used in speaking of the correction of spherical aberration.

While the discovery of achromatism had already, in the middle of the previous century, led to the production of improved telescopes, the smallness of the object lens discouraged the microscope-makers from making the same experiment on the latter.

According to Harting's statement, the Hollander, Hermann Van Deyl, produced the first achromatic microscope, in a very satisfactory manner, in the year 1807. Four years later, Fraunhofer, the renowned optician of Munich, supplied achromatic instruments. In the year 1824 the two Chevaliers of Paris, under the direction of Selligue, combined for the first time several achromatic objective lenses into one system. The Italian Amici, of Modena, then acquired an immortal renown in the sphere of microscopical improvements. Other opticians followed him with worthy emulation, among which, for the end of the first half of the present century, we will particularize only Plössl, of Vienna; Schiek, of Berlin; and Oberhäuser, of Paris. The instrument soon became as useful and complete as that of the eighteenth century was unserviceable and incomplete. The commencement of the great and brilliant era of modern microscopy is cotemporary with these improvements of the instrument. But the recent past has also presented many permanent and important improvements, as we shall hereafter see.

Let us return, however, to the mechanism of our instrument. If we glance at fig. 8, we shall perceive that the image of the arrow produced by the achromatic lens is now free from colored borders and the spherical aberration essentially corrected, but the incurvation and distortion of the same, as well as the diminutiveness of the field of vision, that is, the plane surveyed by the eye-piece, remains as before.

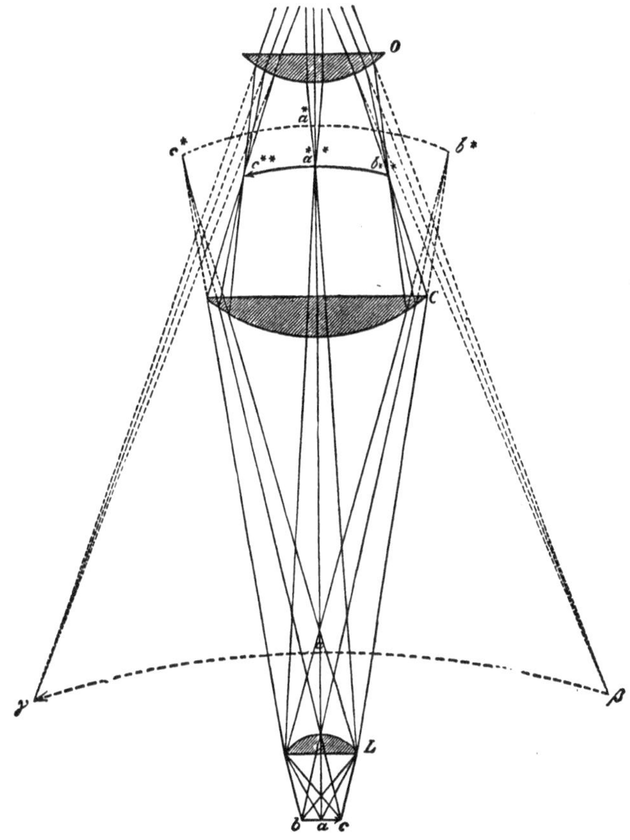

Fig. 15. The compound microscope with a field-glass.

Among the accessories which are employed for further correction is a very old one, namely, the introduction of another convex lens into the tube of the microscope. This, fig. 15, C, is placed between the objective L and the ocular O, so, however, as to occupy a position beneath the point of union $c^* a^* b^*$ of the cones of rays refracted by the objective lens from the object.

The advantage obtained by the introduction of such a convex lens or field-glass is manifested in various ways. Firstly, the rays proceeding from the points *b* and *c* of the arrow are refracted by the convex lens towards its axis, as may be seen in the figure. Without the field-glass the image would be projected at $c^* a^* b^*$, too much extended to be surveyed from the ocular lens. An image is now projected at $c^{**} a^{**} b^{**}$ which, though not so large, still comprises the entire arrow. Secondly, the clearness of the image is increased by the field-glass, as all the rays, which, without this lens, would have produced the image $c^* a^* b^*$, are now united in the smaller space of the image $c^{**} a^{**} b^{**}$. Thirdly, such a field-glass, in connection with the ocular, may serve to improve the correction of the spherical and chromatic aberration. Fourthly—and in this lies a great advantage—the field-glass obviates the distortion of the image and the unequal enlargement of the various portions of the field of vision connected with it. As we have already learned, the rays passing through the marginal portions of the lens are more strongly refracted than those passing through its central portions, in consequence of the spherical aberration, and the axial and peripheral points of the image approach each other proportionately (fig. 11). Now, as the ocular lens, for observing the aërial image $c^{**} a^{**} b^{**}$, exerts exactly the contrary effect (fig. 12) by its considerable diameter, the entire correction may be obtained by the proper use of an ocular and a field-glass (fig. 10).

These various, and, for the most part, highly important advantages secured by the addition of a field-glass render it appreciable that the latter is no longer omitted in any of the compound microscopes of the present time, and that it has become an integral element of all its combinations.

We have remarked above that, since the year 1824, the individual achromatic double lenses are combined with each other into, so-called, systems. Various advantages are thereby obtained. It is very difficult to construct a double lens of crown and flint glass with a short focus, while several weaker ones combined give the same magnifying power as the simple objective, and are much easier to make. Then, as we have also seen, by the combination of a single crown and flint glass lens, whereby, also, a small angle of aperture must always be given to the lens, the spherical and chromatic aberration are essentially lessened, but not entirely obviated. By a suitable combination of several double lenses, where the aberrations of one lens are made to correct the opposite ones of another, a considerable further correction is obtained, and a much larger angle of aperture may be used; in this manner the greatly improved lenses of our microscopes of the present day are produced. In these only two, or, at most, three double lenses are combined with each other (fig. 16).

Fig. 16. An achromatic objective and its angle of aperture.

The earlier opticians generally designated the several double lenses by a series of numbers, 1, 2, 3–6, the weakest bearing the lowest number, and screwed them together into a system (for example, 1, 2, 3, and 4, 5, 6). In this way, with a moderate number of single lenses, a series of systems may be constructed, it is true; but two things which are of greater importance, the accurate centering (that is, the falling together of the axes of the lenses in a single straight line) and the proper distancing of the single lenses from each other, cannot be so accurately accomplished as where these are permanently combined with each other in systems. The preference has, therefore, very properly been given to the latter arrangement, and, though the former is the least expensive, it should be entirely discarded. The permanent systems are variously designated by the opticians; either by numerals increasing with the strength of the combination, or by a series of letters. The manner of expression of the English opticians is peculiar. They speak of $\frac{1}{4}$-, $\frac{1}{8}$-, $\frac{1}{12}$-, $\frac{1}{25}$- of an inch combinations of lenses, making the increase of the strength of their systems the same as that of a simple lens of $\frac{1}{4}$, $\frac{1}{8}$, $\frac{1}{12}$, $\frac{1}{25}$ inch focus.

The combination of the three achromatic lenses with each

other is such that the strongest and smallest lens is turned downwards and the weakest upwards (fig. 16). A somewhat greater focal distance is thus obtained, and such angles of aperture may be given to the lenses, that all the rays of a cone of light *c a b* received by the lower lens may also pass through the entire combination of lenses. Only in this way has it been possible to give the above-mentioned high angle of aperture to the objectives, which must naturally increase the brightness of the image, and besides, as we shall see hereafter, considerably increase the intrinsic power of the combination also.

The ordinary eye-piece of our microscopes (fig. 17, O), also called the Huyghenian or negative eye-piece, consists of a longer or shorter tube carrying at its upper end the plano-convex lens A, whose plane surface is turned towards the eye of the observer, while the plano-convex lens C, with its curved surface also turned downwards, is screwed into the lower end of the tube. The aërial image P* falls, as we have seen, between the field and ocular glass. Every microscope is supplied with several such eye-pieces of various strengths, designated by numbers. The eye-pieces become shorter in proportion to the increase of their magnifying power. Another form is called the Ramsden, or positive eye-piece. It also consists of two plano-convex lenses; but their curved surfaces are turned towards each other, and they lie nearer together. Here the image does not fall between the field and ocular glasses, but lies at a short distance beneath the former. The latter eye-piece is, however, but little used.

Kellner's orthoscopic eye-piece is a modification of that of Huyghens, or the negative; its field-glass is bi-convex. It presents a very large field free from image distortion without, however (and in this I must agree with Harting), appreciably increasing the optical power.

Hartnack has recently constructed a new, strong eye-piece, the *oculaire holostère.* It consists of a single conical-shaped piece of glass, after the manner of the Coddington lens, and magnifies about ten times. It has thus far, however, presented me no considerable advantages.

In order to render the Huyghenian eye-piece as free as pos-

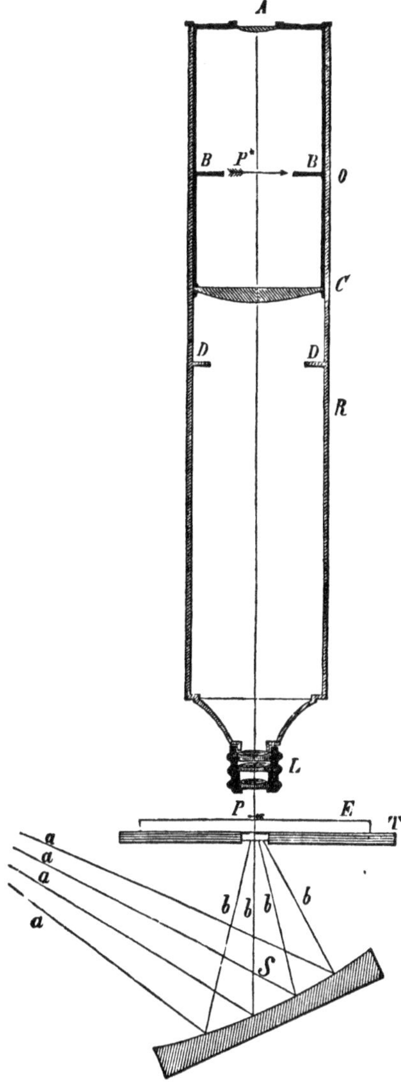

Fig. 17. The compound microscope.

sible from spherical and chromatic aberration, it has been proposed to make it aplanatic and to combine it with an aplanatic objective system. Such aplanatic eye-pieces are to be found with many instruments. Their magnifying power is weak and the field of vision small.

The usual arrangement is different. It consists of the use of eye-pieces which are by no means entirely aplanatic, and the aberration present in the eye-piece is made to correct the opposite aberration of the lens system. Objectives of somewhat over-corrected chromatic, and also spherical aberration are combined with under-corrected eye-pieces. An objective which was rendered as aplanatic as possible would, on the contrary, if combined with one of the ordinary eye-pieces, produce an imperfect image. While therefore aplanatic lenses are necessary for the loup and the simple microscope, the art in the production of a compound dioptrical microscope rests directly in the removal of the aberrations of the objective by means of the contrary aberrations of the eye-piece. It is only thus that an image free from defects may be obtained, in a manner similar to that already mentioned of correcting one double lens of an aplanatic system by means of the other.

In eye-pieces the distance of the field-glass from the ocular lens is of importance. If the former glass is made to approach the latter, the aërial image is greater; if the latter glass be made to approach the former, it is smaller. The opticians, as a rule, fix both the glasses of an eye-piece in an unchangeable position; they select the ones offering the most advantageous action. The length of the tube of the microscope, which in increasing also increases the magnifying power, is likewise of importance to the advantageous combined action of the ocular and objective systems. Less increase in the length of the microscope tube is permissible with a higher degree of over-correction of the lenses than with a weaker one.

Still another element is associated with the optical relations enumerated, for a knowledge of which we have to thank Amici. At the present time it receives due attention, whereas for a long time it was entirely ignored. We refer to the thickness of the scale of glass with which it is customary to cover the object for microscopical examination. The thickness of the glass cover exerts considerable influence on the sharpness of the image, especially when strong lenses are used. An object which, uncovered, or with a very thin glass cover, presents a sharp image, becomes somewhat dim and foggy, and the appreciability of its details diminishes, if a thicker cover be used. Inversely, many lenses only manifest their highest capabilities when a covering glass is used.

Now, on what does this influence of the cover depend, and what are the means for its correction?

Let P, fig. 18, be a thick glass plate, and *a* a point of light from which proceeds a cone of rays. The rays on entering the

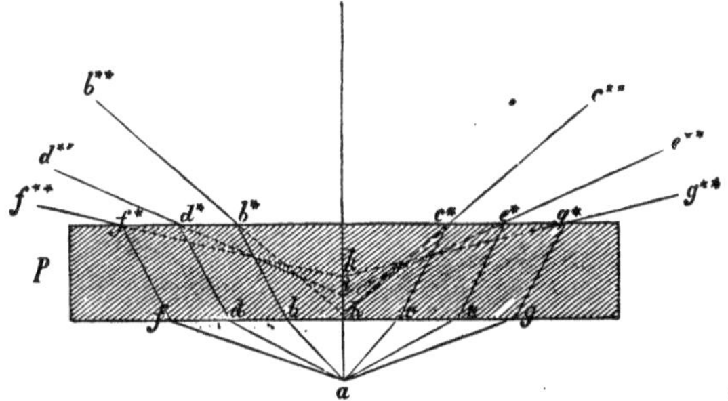

Fig. 18. Effect of the covering-glass.

glass will be refracted in various degrees. The external, most obliquely incident rays *a f* and *a g* will be the most strongly refracted and will assume the directions *f f** and *g g**, the more internal rays *a d* and *a e* will be less influenced, and the more central rays *a b* and *a e* will be still less deflected from their course. On emerging from the glass the most external rays are refracted in the direction *f* f*** and *g* g***, the more internal ones in the direction *d* d*** and *e* e***, and those most internal in the direction *b* b*** and *c* c***. The luminous spot will appear to the eye as though it were seen nearer in the glass, and instead of one luminous point a series of points lying over each other will seem to be present; as *h* for the rays *b* and *c*, *i* for *d* and *e*, *k* for *f* and *g*. If, instead of a point, we have an object, it will make an impression as if it consisted of a series of images lying over each other. We receive, therefore, the same effect as from spherical aberration, and in a degree which increases with the thickness of the glass covers. It will therefore be appreciable how imperfect such a course of the rays of light must render the image of an object with a lens which is arranged for uncovered objects; for the same reason, an objective constructed by the optician for a covered test object will only develop its perfect action when a suitable cover is used. Weaker combinations manifest this influence of the glass covers only in a minor degree, however; stronger ones, on the contrary, in a very appreciable manner.

This influence of the cover may be counteracted by changing the length of the microscope tube, and also by altering the distance between the ocular and field glasses. It is advisable, in a practical point of view, to use a system with its appropriate covers only, and to have special thicknesses of glass for each system.

Another method is now becoming more generally adopted. By changing the position of the individual lenses of a combination this action of the glass cover may be obviated, and thus one and the same system may be employed either for uncovered objects or for such as have covers of various thicknesses. For this purpose the individual double lenses of a system are arranged so that their relative positions may be changed by means of a fine screw; thus the observer is enabled to make the necessary change at any moment. Such combinations are called objectives with correcting apparatus. They are naturally more expensive than ordinary objectives and require in their use a certain amount of practice and some outlay of time, but the arrangement is almost indispensable for very high powers.

The rule is, that with increasing thickness of the cover the individual lenses of the system must be brought nearer to each other, while for very thin covers they must be moved farther apart. In fig. 19, 1, the objective represented with a correcting apparatus has a small metallic slider, *s*, which, moving up or down, indicates the various positions of the lenses. It is represented at 2 *a b c* in its various positions.

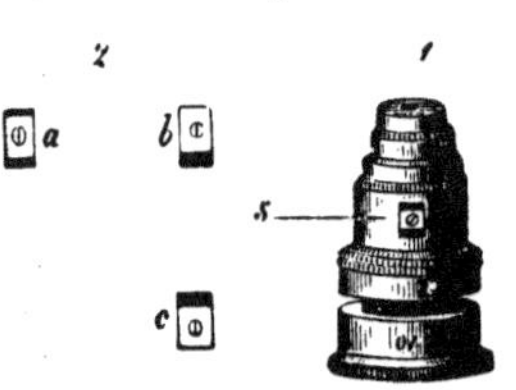

Fig. 19. Achromatic objective with correcting apparatus.

Having familiarized ourselves with the objective and the eye-piece, we are now in condition to examine more closely the construction of a modern compound microscope.

The optical portion is of the greatest importance; the arrangement of the stand is, on the contrary, of much less consequence. Good lenses with suitable eye-pieces, placed on a very imperfect stand, would enable an observer to recognize subtile structural conditions which would be concealed from another, whose instrument combined an imperfect optical apparatus with a very superior mechanism. Nevertheless, disregarding the tediousness of the manipulation, poor, incomplete stands exert an immediately injurious effect on the optical performances of a microscope, by not permitting of the necessary modifications of the illumination.

Every modern instrument requires several objectives; namely, a weak, a medium, and a strong one; the combinations of each should, preferably, be permanent. Large microscopes have a more abundant supply of objectives, five or six, and even more, and among them the most powerful ones, in the production of which great proficiency has been developed of late, as we shall see further on. These very high powers are not required for ordinary investigations, and can, therefore, be more readily dispensed with than combinations of medium strength.

A few eye-pieces, at least two, are necessary; a weak one, magnifying about three or four times, and a stronger one of double that strength.

It might readily be believed that a considerable number of eye-pieces with increasing, and, finally, with very high magnifying power would be of advantage to our instrument. But this is erroneous. Let us remember that an enlarged image would be projected by the objective L, fig. 20, into the tube R, and would not be without defects, as it is impossible to produce mathematically correct lenses. The image would be magnified by the eye-piece, and, naturally, its imperfections in the same proportion also. The eye-piece does not, therefore, like the objective, permit us to penetrate more deeply into the structure of the object; it only enlarges the images of the latter. The use of somewhat stronger eye-pieces has the advantage, however, of enabling us to recognize many things more readily, because they are more enlarged. But in using still stronger eye-pieces, we soon arrive at a limit where the image is impaired. The

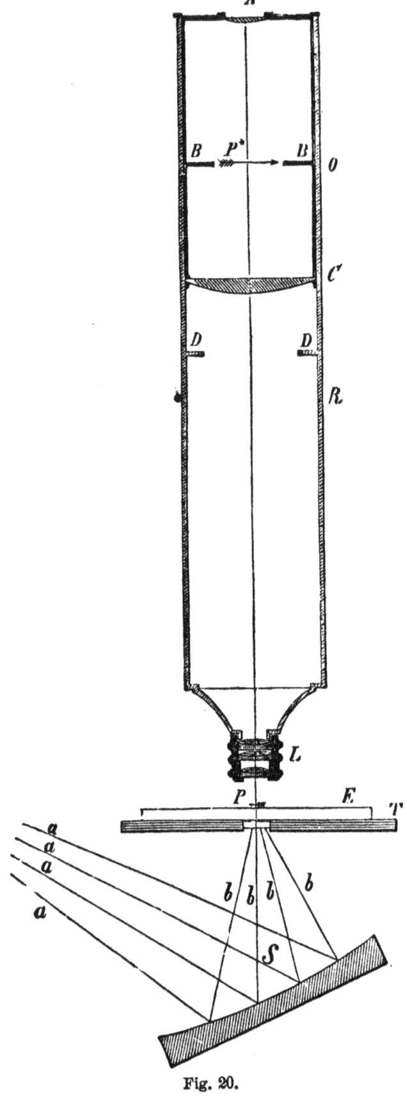

Fig. 20.

most beautiful and elegant images are obtained with very weak eye-pieces. Nevertheless many of the more modern lenses bear considerably stronger eye-pieces than those of a former epoch, which must always be regarded as a proof of superior optical perfection.

No further observations are necessary, therefore, to show that it is impossible to compensate for the poverty in lenses of a microscope by a profuse endowment of eye-pieces. It is also evident that the value of an enlargement, obtained with a stronger objective and a weaker eye-piece, stands higher than that of another, where a strong eye-piece is used with a weaker objective. Older German microscopes frequently have only weak objectives, but are, on the contrary, furnished with several eye-pieces of too great strength. This is to be regarded as a fault. At the commencement of the fifth decade of this century, for example, the instruments of Schiek compared very disadvantageously, in this regard, with those of Oberhäuser.

The tube of the microscope, as well as that of the eye-piece, has its inner surface blackened, and is either in one piece (fig. 20, R) and therefore incapable of extension, or it is composed of two pieces which glide over each other, after the manner of the telescope. That the latter is to be regarded as the better arrangement is proved by several previously mentioned optical principles.

The objective (L) is to be fastened on to the lower end of the tube by means of a screw.

The stage (T) is the metal plate perforated in its centre, already described in speaking of the simple microscope, for the reception of the object (P. E.) to be examined. The stage should not be too small, and especially not too narrow.

Objective and object must be capable of being moved from or towards each other as circumstances may require. All compound microscopes have arrangements for this purpose, that is, for focussing the object. Shoving the microscope tube with the hand through a metal sheath is a very primitive contrivance, and is only practicable with weak powers.

Various expedients are employed for more accurate alteration of the focus. This may be accomplished, to a considerable extent, by means of a single mechanism, when the latter is carefully constructed. In fact, the older instruments frequently have no other. As a rule, the tube of the microscope is made to screw up and down on its support; less frequently employed, and still less to be recommended, is a movable stage, the tube being immovable.

With the more carefully constructed modern stands a double mechanism is provided, one of which serves for the coarse, and the other for the fine adjustment. Such a distribution of the work naturally deserves the preference. The coarser movements are made either by machinery, or, what answers just as well and is more practicable by reason of its greater simplicity, the tube of the microscope may be moved in a sheath which surrounds it, by the hand. The finely constructed micrometer screw, which moves the microscope, is used for more accurate focussing; the practised observer almost never removes his hand from it in making delicate investigations and when using the higher powers.

Ordinary incident light is rarely used for the illumination of the object, and then only when the lowest powers are employed. When strong illumination is required, a convex lens (*a*, fig. 21) with a long focus is employed. It should be movable in various directions, and placed either on a stand *d b c*, or on a ring which is to be shoved over the tube of the microscope.

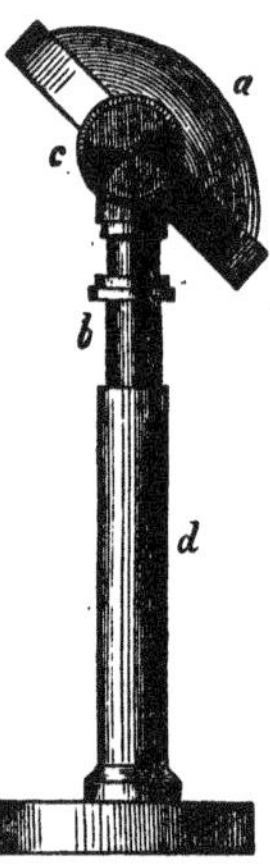

Fig. 21. Illuminating lens.

Objects are most frequently illuminated by means of transmitted light; the light being received on a mirror (fig. 20, S), placed beneath the stage, and reflected through the opening to the object (P).

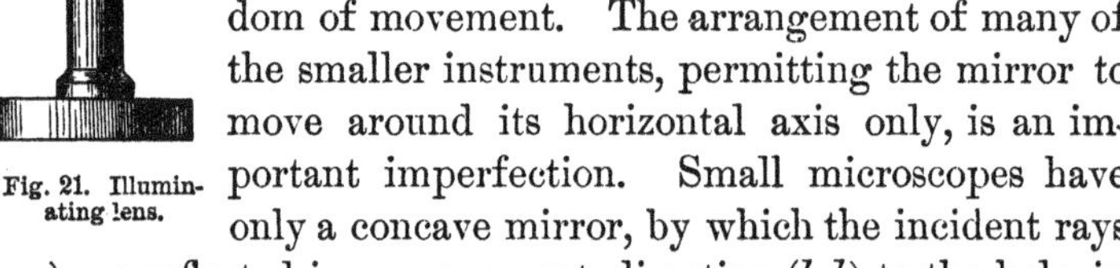

The mirror must be fastened to the stand in such a manner as to permit of the greatest freedom of movement. The arrangement of many of the smaller instruments, permitting the mirror to move around its horizontal axis only, is an important imperfection. Small microscopes have only a concave mirror, by which the incident rays (*a a*) are reflected in a convergent direction (*b b*) to the hole in the stage. Larger instruments have a mirror with one of its surfaces concave and the other plane. The latter surface gives a less intensive illumination than the former and is, therefore, more frequently used with the weaker powers.

Careful illumination is a very important accessory in microscopical examinations, and is not to be obtained with the above-mentioned contrivances alone. Other special apparatuses are consequently necessary. For many examinations, especially of delicate, finely bordered objects, the light reflected through the opening of the stage would give a much too dazzling illumination. A portion of the rays must therefore be cut off. This is accomplished by diminishing the opening of the stage; for this purpose the so-called screens or diaphragms are used.

Two forms are employed; the rotary and the cylindrical diaphragm. The rotary diaphragm (*a*, fig. 22) has a circular form and is fastened under the stage by means of a button. A series of circular openings diminish, with the exception of the largest, the aperture of the stage. The smallest holes are employed with the highest powers.

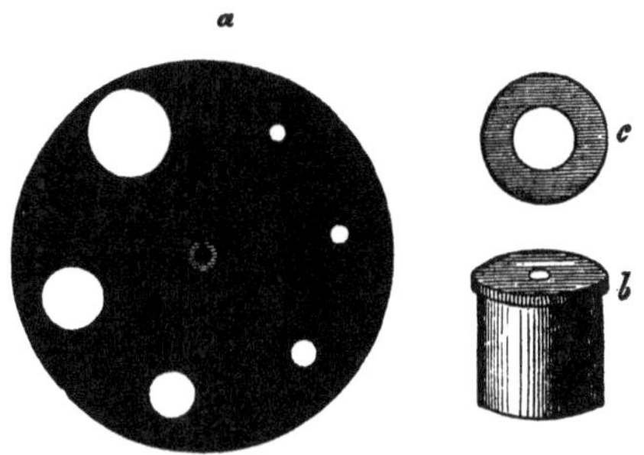

Fig. 22. Diaphragms. *a*, the rotary diaphragm; *b*, *c*, cylindrical diaphragms.

Cylindrical diaphragms are cylindrical tubes which have at their upper ends a circular disk with an opening of varying size (fig. 22, *b*, *c*). They are inserted into the opening of the stage, either immediately or by means of a socket. To develop their complete action, some contrivance is necessary, by means of which they can be raised or lowered.

Both arrangements accomplish their objects; but the cylindrical diaphragm deserves the preference, as it permits of finer gradations of illumination. On many of the older instruments we find both these kinds of diaphragm combined.

For many purposes, instead of the ordinary illumination, generally called central light, it is necessary to reflect the light from beneath in a more or less oblique direction on to the object, called oblique illumination. For this purpose, the mirror should have the utmost freedom of motion, because it is sometimes necessary to give it a very lateral position.

A further modification of the illumination is obtained by inserting a convex lens or a combination of lenses into the aperture of the stage. By elevating or depressing the lens, we may cause the rays of light coming from the plane mirror to unite in a focus at the object, or to arrive at it in a divergent direction, either before or after their union. A concave mirror, combined with such lenses, sometimes affords very good illumination.

Such an illuminating apparatus, consisting of achromatic lenses, was made many years ago by Dujardin. Considerable attention was afterwards directed to these, especially by the English opticians, who called them condensers, resulting in their essential improvement. A condenser of perfected construction is shown in fig. 23. Below it is a rotatory diaphragm which is capable of covering a greater or lesser portion of its margin and, by means of several openings, darkening the central portion of the lens, which causes peculiar effects, resembling many of those of oblique light.

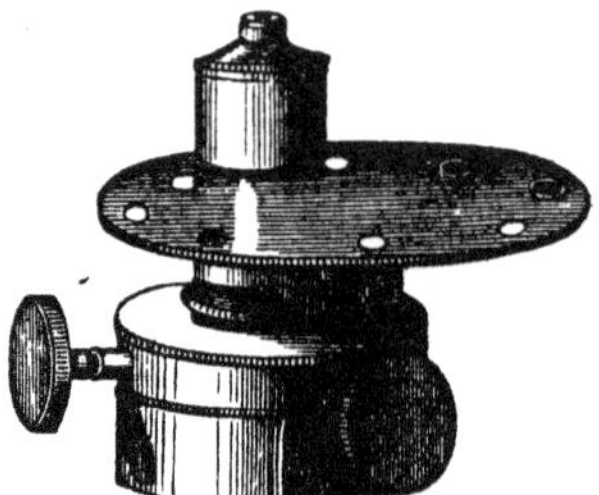

Fig. 23. Achromatic condenser of Smith and Beck.

I have recently received from Hartnack an efficient condenser, quite similar to the illuminating apparatus formerly constructed by Dujardin, consisting of three achromatic lenses. Diaphragms may be screwed on to the upper lens. The apparatus is to be inserted into the stage in the same manner as a cylindrical diaphragm.

As an achromatic condenser is expensive, it may be substituted, to a certain extent, by an ordinary plano-convex lens. Fig. 24, 1, shows such a one inserted into the tube of an ordinary cylindrical diaphragm. In 2, it is covered by a black ring, so that only the central portion remains free

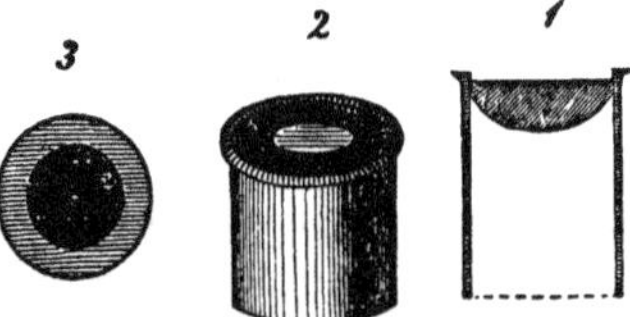

Fig. 24. Ordinary condenser; 1 in section; 2 covered by a black ring; 3 with a central disk.

for the passage of the rays of light; while in 3, a small black disk obscures the central part of the lens, leaving only the peripheral portions free. The latter arrangement is to be recommended to those whose microscope does not permit the mirror to be placed obliquely. The whole arrangement is, besides, the least expensive of any.

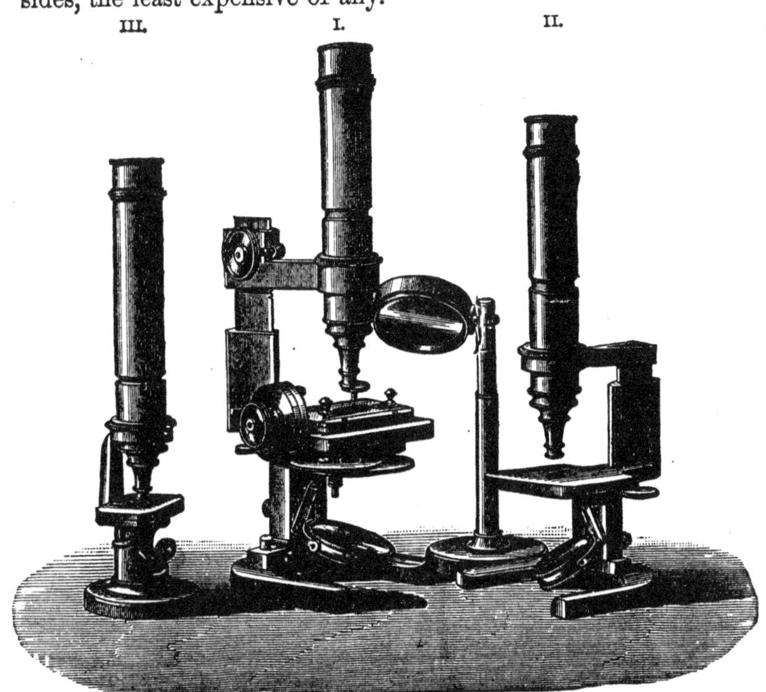

Fig. 25. Microscopes of Merz, of Munich. III. smallest, II. medium, I. large instrument.

As we shall see hereafter, such convex lenses are necessary for investigations with polarized light, as well as when the microscope is used as a micro-photographic apparatus.

At the close of this section it may be expedient to glance at several microscopes, and thus obtain a few examples of the various methods adopted by opticians for fulfilling the various indications.

Fig. 25, III. shows a microscope of the smallest sort made by Merz, of Munich. The coarse adjustment is made by shoving the tube through a spring sheath, the fine movement is (inexpediently) accomplished by the elevation and depression of the stage. The concave mirror permits of central illumination only. Fig. 26 represents a small instrument made by Nachet, of Paris, with a stand which, though simplified, is much more efficient and suffices for most investigations. Here, also, the microscope tube may be shoved up and down in a spring sheath, and thus serves for the coarse adjustment. The fine movement

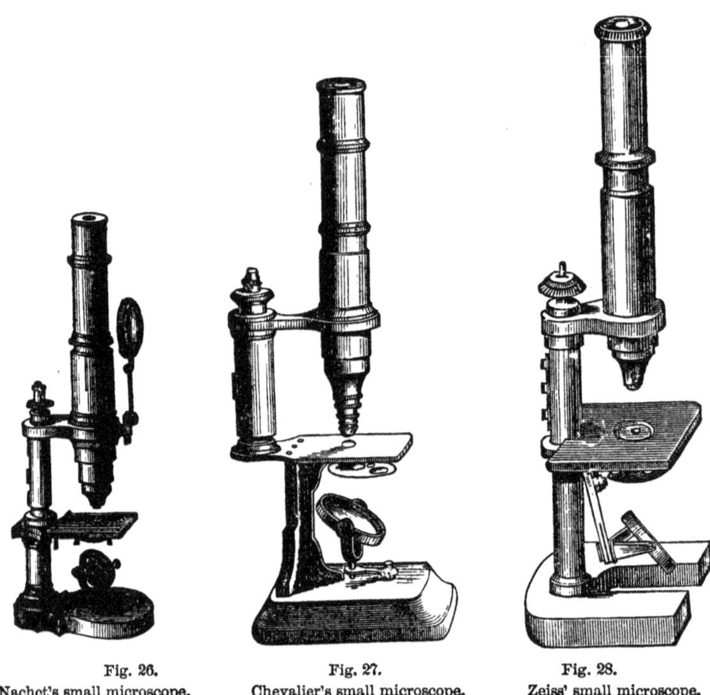

Fig. 26. Nachet's small microscope. Fig. 27. Chevalier's small microscope. Fig. 28. Zeiss' small microscope.

is obtained by means of a screw-head placed at the upper end of the stem. The stage is sufficiently wide, and beneath it is a rotary disk for moderating the illumination. Several clamps on the stage, intended for holding the glass slides, may be removed if necessary. The mirror is fastened to the round foot, and permits of great freedom of movement. It may be moved away from the axis, and thus be employed for oblique illumination. The illuminating lens (placed erect in the figure) serves for illumination with incident light. The smaller microscopes of Chevalier, of Paris, fig. 27, as well as those of Zeiss, of Jena, fig. 28, have a quite similar arrangement.

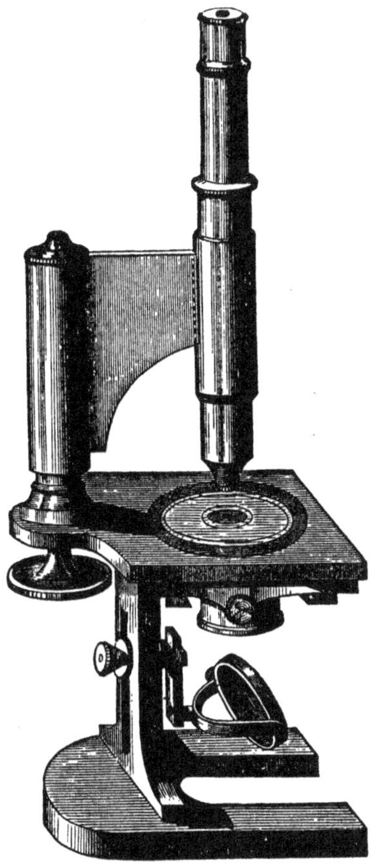

Fig. 29. Older large horse-shoe microscope of Oberhäuser and Hartnack.

In the latter instrument, however, the suspension of the mirror is different, and the rotary disk under the stage is convex on its upper surface, so that the aperture of the diaphragm may come as close as possible to the object. Such instruments, among which the medium-sized microscope of Merz, fig. 25,

II., is also to be reckoned, have very suitable stands, which are frequently imitated by other microscope-makers with slight modifications. Naturally a stand may be still further simplified, but its adaptability to various kinds of examinations is impaired, for example, when the oblique illumination is omitted.

The large horse-shoe microscope (fig. 29), invented by Oberhäuser, of Paris, has one of the most efficient stands. It has been more frequently imitated than any other with which I am acquainted, and combines the advantage of the greatest adaptability with simplicity of construction.

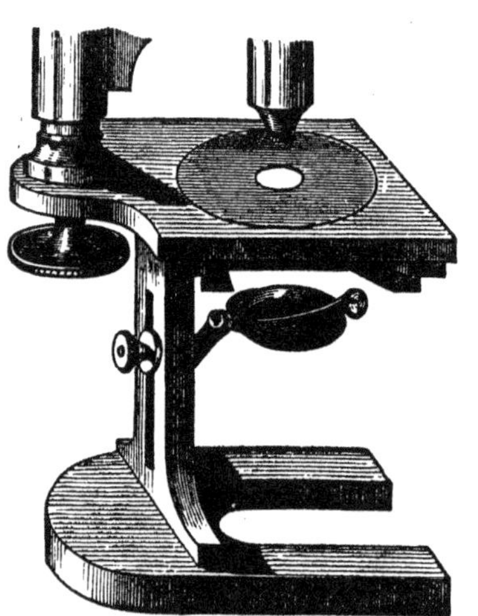

Fig. 30. The same with oblique illumination.

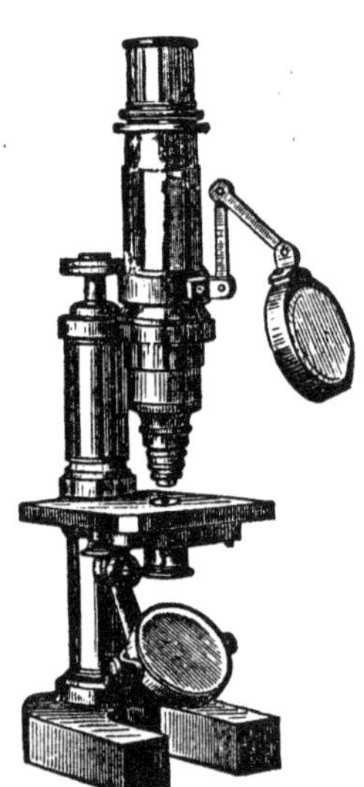

Fig. 31. Hartnack's small horse-shoe microscope.

In the old stand, the coarse adjustment is also made by sliding the tube in the spring sheath, but his latest instruments are furnished with a mechanism for this purpose. The tube itself is capable of being shortened. The fine adjustment is made with a micrometer screw which projects beneath the stage and runs in a hollow tube containing a spiral spring. The screw moves a second tube which surrounds the former and is joined to the sheath of the microscope tube. The diaphragms, surrounded by a cylinder, are carried by a so-called sliding plate, and are adjusted by raising or depressing the cylinder. When one diaphragm is to be replaced by another, the cylinder is to be drawn out, armed with a new diaphragm, and replaced from beneath the stage. If oblique illumination is necessary, the sliding plate with its entire apparatus is to be removed.

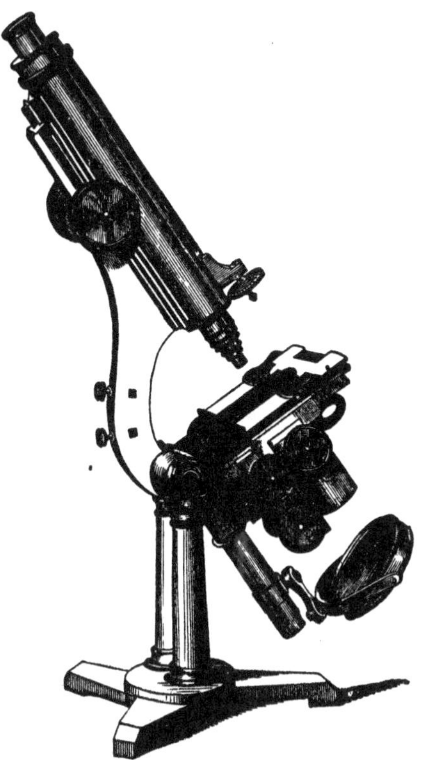

Fig. 32. Large microscope of Smith and Beck.

With the latter illumination the stage may be rotated, so that the obliquely incident rays of light may come from all sides on to the object. The mirror works on a square piece fitting into the embrasure of the double bar which supports the instrument, and permits of the most varied changes of position. The large, heavy horse-shoe foot supports the whole. An illuminating lens on a separate stand (after the manner of fig. 21) may be placed before the instrument.

A smaller form of the same stand (fig. 31) dispenses with the rotary stage, and does not permit the mirror to be moved up and down in a slit, though the oblique position is still possible. This stand of Hartnack's is very good and at the same time exceedingly cheap.

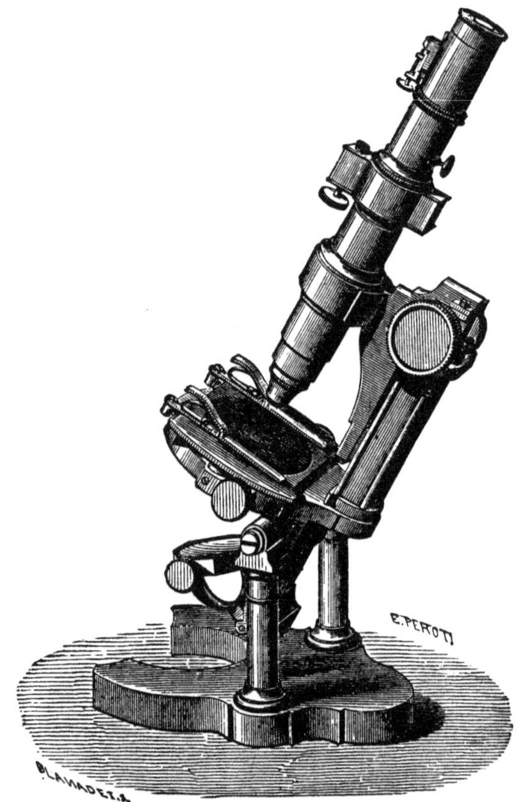

Fig. 33. Nachet's large microscope, most recent pattern.

Both stands may also be obtained furnished with a hinge for inclining.

Merz's large instruments are also quite similar to these, as is shown in fig. 25, 1.

We also notice a large microscope, fig. 32, of Smith and Beck, of London, as an example of an instrument which is much more complicated (too much so, according to our Continental notions) in its structure. Much is here allotted to screws which, in Oberhäuser's stand, is done by the human hand. The instrument hangs between two pillars and can, therefore, be given an oblique or horizontal position. The mirror permits of a pretty free movement. The stage is too profusely covered with appurtenances, but permits (and in this lies a great advantage over Oberhäuser's instrument) of the introduction of a perfected condenser.

Nachet's large microscope of latest construction (fig. 33) is likewise remarkably complicated, but its mechanism is admirable.

Section Second.

APPARATUS FOR MEASURING AND DRAWING.

It is unnecessary to mention how important the measuring of objects seen under the microscope is for scientific work. In fact, various and, in part, ingenious methods were proposed in the infancy of microscopy for determining the size of objects. The reader will find more on this point in the excellent work of Harting.

At the present time, we have measuring apparatus of relatively greater accuracy. Two forms of micrometers are to be particularly discriminated; namely, first, the screw micrometer, and, second, the glass micrometer.

The screw micrometer is a somewhat complicated but, when the workmanship is good, very accurate, and, therefore, also very expensive implement. Its arrangement rests upon the following. If a cobweb thread is drawn across the eye-piece, it is self-evident that a microscopical object may be so guided through the field of vision, by means of a stage moved by screws, that its anterior border first cuts the thread, and then gradually passes beyond it, till at last only the posterior margin of the object exactly touches the thread. Now, the screw micrometer is such a movable stage. It has a double plate, the upper one of which is movable, by means of a very fine micrometer screw, over the lower one, which is fastened to the stage of the microscope. A partial view of this arrangement is afforded by fig. 25, 1. The extent which it is necessary to turn the screw, in order to move the object, in the manner indicated, through the field, may be read from the index of the upper plate and the divisions on the drum of the screw. The unities of these screw micrometers vary. Those of Plössl give $\frac{1}{10000}$ of a Vienna inch, those of Schiek, $\frac{1}{1000}$ and $\frac{1}{10000}$ of a

Parisian line. The ocular screw micrometer is an advantageous modification of the screw micrometer, especially the improved form described several years ago by Mohl.

Nowadays, however, the expensive screw micrometers are rarely used; the much simpler and less expensive glass micrometers are used in their stead.

It is well known that the art of engraving five divisions on a glass plate, by means of the diamond point, has made great strides, and in a later section we shall see the marvellous manifestation of this skill in Nobert's test plates.

The line is now divided, with the greatest elegance, into 100, 500, and 1,000 parts. In some of these micrometers the lines are all of equal length; those are better in which the greater divisions are indicated by longer lines, as is the case with our ordinary rules. A modification, which is convenient for many purposes, consists in the crossing at right angles of one series of lines by a second series, generally so that quadratic spaces result.

Now, such micrometers are capable of the simplest application as object bearers. Let us assume that we have a division where the value of each space is $\frac{1}{500}'''$, it is self-evident that a microscopical object which occupies two of these spaces is $\frac{1}{500}'''$, and another, which covers five, is $\frac{1}{100}'''$ in size.

However efficient these methods may seem, at the first glance, to be, they are nevertheless very inconvenient, and it is therefore no longer customary to use them. First, because the minuteness of many objects requires a very finely divided and therefore very expensive micrometer. Then, in cleaning them they become injured in a comparatively short time, and gradually become seriously impaired. Further—and this is of much greater moment—the objects to be measured very often lie obliquely, and not parallel to the lines, however fortunately they may be placed on the micrometer. Finally, the case often arises where it is necessary to estimate the value of a fraction of a space, whereby the eye may be deceived.

From the above mentioned it will be conceivable that the glass micrometer, in the form of an object bearer, has been discarded except for certain special purposes.

These micrometers are now placed in the eye-piece, as ocular micrometers, in the form of circular glass plates. They lie on its diaphragm, between the field glass and the ocular lens (fig. 20, B).

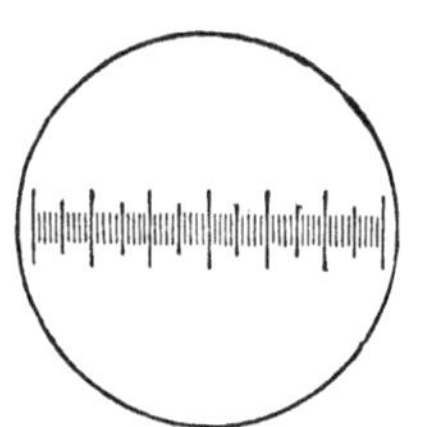

Fig. 34. Eye-piece micrometer.

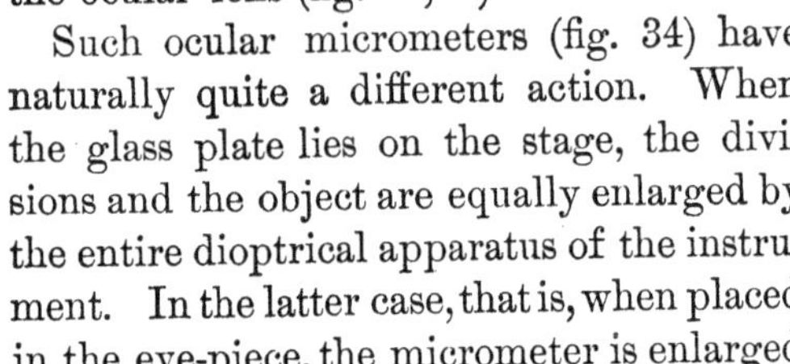

Such ocular micrometers (fig. 34) have naturally quite a different action. When the glass plate lies on the stage, the divisions and the object are equally enlarged by the entire dioptrical apparatus of the instrument. In the latter case, that is, when placed in the eye-piece, the micrometer is enlarged by the weak ocular lens only, and appears to the eye simultaneously with the image of the object to be measured, which is enlarged by the objective and again somewhat diminished by the field glass. This may be accomplished with glass micrometers which are coarser, and therefore more accurate, and also less expensive to construct. They do not wear out, and any object in any position on the slide may be instantaneously measured, so soon as the eye-piece containing the micrometer has been substituted for the ordinary one, and adjusted by turning it in the tube. But with more opaque objects an inconvenience arises in the difficulty of seeing the micrometer divisions over the object to be measured. No microscope should be without such a micrometer, which may be obtained for a few (4–5) thalers. In consequence of the unequal visual distances of different observers, it is necessary to vary the adjustment of the ocular micrometer by means of a screw arrangement, so that it and the object may simultaneously appear alike distinct to any eye.

It should not be forgotten, in using the eye-piece micrometer, that its value is a relative one, depending on the strength of the objective used (therefore variable with immersion lenses), and the length of the microscope tube, which always determines the size of the image. It is most advantageous to have the latter completely drawn out when measuring.

We have a very simple procedure for determining the value of the micrometer in the eye-piece; we avail ourselves of the aid of a glass micrometer on the stage. Assuming that it has a

Paris line divided into 100 portions; with the objective A, perhaps five spaces of the eye-piece micrometer will exactly cover one space on the stage micrometer; the value of one of its spaces for the objective A is therefore $\frac{1}{500}'''$. To obtain greater accuracy, various portions of the stage micrometer should always be used for the measurement, and the mean of 10–15 single measurements drawn. Always keep in the middle of the field to avoid any distortion of the image that may be present. In this manner the value of the eye-piece micrometer for the various objectives of a microscope is reckoned and tabulated.

Besides this most simple eye-piece micrometer, serving completely nearly all the purposes of measurement, various other modifications have been produced, but we cannot at present enter into their consideration. Those who are interested in the subject may read the respective section in Harting's work.

All statements as to the size of microscopical bodies naturally depend upon the unity of measure on which they are based. Microscopists, as a rule, use the measure in general use in their country; those of England use the English inch, which is divided into decimal and duodecimal lines; those of France use the Paris line or the millimetre. In Germany one of the two last mentioned unities of measure is generally used, though the Vienna and Rhine lines are also used. The Paris measure is most convenient, and the millimetre deserves the preference. It is very convenient to use the thousandth part of a millimetre as a unity, under the name of micromillimetre, *mmm*, as proposed by Harting.

A millimetre is 0.4433 of a Paris line.
0.4724 of an English duodecimal line.
0.4587 of a Rhenish line.
0.4555 of a Vienna line.

The Paris line is 2.2558 millimetres.
" English " 2.1166 "
" Rhenish " 2.1802 "
" Vienna " 2.1952 "

For further comparison we give a small table for reducing Paris lines to millimetres and *vice versâ*.

1.				2.		
Millimetre.		Paris lines.		Paris lines.		Millimetres.
1.	=	0.4433		1.	=	2.2558
0.9	=	0.3990		0.9	=	2.0302
0.8	=	0.3546		0.8	=	1.8047
0.7	=	0.3103		0.7	=	1.5791
0.6	=	0.2660		0.6	=	1.3535
0.5	=	0.2216		0.5	=	1.1279
0.4	=	0.1773		0.4	=	0.9023
0.3	=	0.1330		0.3	=	0.6767
0.2	=	0.0887		0.2	=	0.4512
0.1	=	0.0443		0.1	=	0.2256
0.01	=	0.0044		0.01	=	0.0226
0.001	=	0.0004		0.001	=	0.0023

The goniometer is an apparatus used for measuring the angles of crystals. A simple and efficient arrangement (fig. 35), contrived by C. Schmidt for this purpose, consists of the following:—A circular plate, *a b c*, divided into thirds of a degree, is placed around the (fixed) microscope tube, at its upper end. To the outer edge of an eye-piece (*p*), provided with a crossed thread, a vernier (*d*) is fastened. The angle of the crystal to be measured is brought to the centre of the cross, and the threads are made to cover both sides of the angle in succession. The extent which it is necessary to turn the eye-piece in order to effect this is read off at the vernier, above which a plano-convex lens (*e*) is placed.

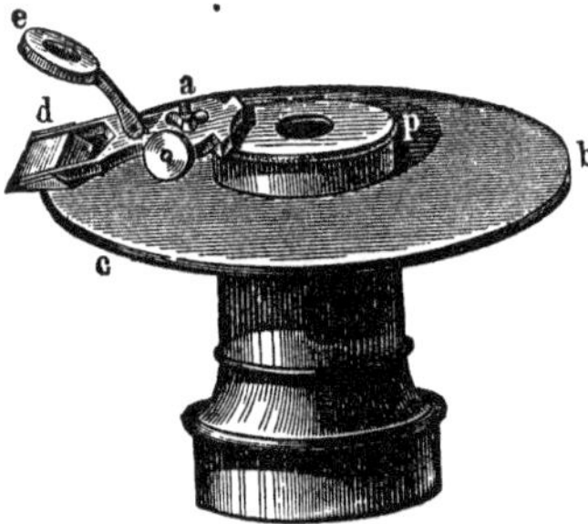

Fig. 35. C. Schmidt's goniometer. *a b c*, graduated disk; *d*, nonius at the border of the eye-piece *p*; *e*, lens for reading off.

Drawing the investigated object is not less important than its measurement with the microscope. It would appear to be superfluous to speak further of its value. It is, indeed, generally acknowledged in the study of all branches of natural history, and a successful illustration is often much more rapidly comprehended than the most detailed description.

Every one occupied with natural science, and especially with medicine, should, therefore, be able to practise this art, at least in a slight degree. This qualification is all the more necessary in consequence of the peculiar nature of microscopical vision. While an object which is perceived by the naked eye may be grasped and portrayed by an artist who is experienced in the guidance of pencil and brush, seeing correctly with the microscope is itself an art which must first be learned before thinking of making a successful drawing. Although the inquirer who understands his specimen, even though no great master of the art of drawing, would be able to produce a tolerable and useful representation of the object, this would not be the case with a much more proficient artist who ventures, for the first time, to represent a microscopical image. Misunderstandings and errors would not be wanting. The latter is deficient in comprehension, while the microscopist is often enough in the fatal position of understanding his specimen thoroughly, and yet not able, with his unpractised hand, to reproduce it faithfully or with artistic conception.

The simpler accessories for drawing, such as the pencil, the rubber, and water-colors, are generally sufficient for the microscopist. Much that one draws to assist the memory during an investigation has only the character of simple sketches; likewise, many things that are only incidentally noticed, are considered worthy of being drawn in a note-book. It is not advisable to draw everything, on account of the great outlay of time which would be necessary. Since we have learned to preserve specimens in fluid, in such a manner as to retain their natural appearances, they will render better service during a prolonged investigation than a volume filled with simple sketches. One should be particular in selecting drawings for publication. Not every preparation, not every view is characteristic. A well-selected representation is of more service than a whole series of less pregnant ones.

More explicit directions as to details would here be out of place. Rough paper may be used for the larger sketches; a very fine English drawing-paper is necessary for the reproduction of very delicate textural relations. A series of various sorts of pencils, of the best manufacture, should be selected. One should become accustomed to trace the first outlines as delicately as possible, then proceed to the darker shades, and only at last bring in the strong shade lines. The greatest care should be devoted to keeping the point of the pencil in order, for which purpose a file is best, if it be desirable to make any approach to the delicacy and fineness of many microscopical preparations. The use of the rubber should be learned from an expert teacher, thereby saving much consumption of time in shading. It should not be forgotten to lay in the shading symmetrically on the right side, as it is only thus that elevations and depressions can be brought forward in the picture. The intensity of the same is to be kept carefully in mind, and rendered as faithfully as possible, as this is the essential foundation of the natural disposition of many microscopical appearances.

In using water-colors, the more transparent ones are, as a rule, employed; more rarely those of a thicker consistence. Their application is soon learned. Too dazzling colors are not to be used; one should accustom one's self to make fine colored lines with the point of a brush, which, for many purposes, are preferable to lines made with a lead pencil.

In the course of time many kinds of apparatuses have been invented for rendering assistance in making drawings from the microscope, and, in fact, it is necessary for the microscopist to have such an appropriately constructed arrangement, especially when laying out a somewhat complicated figure, and for accurately rendering the various relations of form and size of its constituent parts.

All the respective apparatuses aim, by means of special contrivances, to throw the microscopical image on to a sheet of paper, placed near the microscope, where its outlines may be traced with the point of a pencil.

Glass prisms are generally used for this purpose. The simple drawing prism may be placed over the eye-piece by means of a ring which fits over the tube of the microscope. It must be movable over the former so that it can be approached towards, or removed farther from the eye-piece. The paper may be placed on a drawing-desk, similar to a note-desk, behind the microscope.

The camera lucida of Chevalier and Oberhäuser is more convenient than the simple drawing-prism for our vertical instruments, but is also somewhat more expensive, costing from 30 to 50 francs. It consists of a complicated eye-piece, which contains two prisms, and causes a complete inversion of the image. A glance at fig. 36 will readily enable us to comprehend the arrangement of this instrument. A tube A, bent at right angle, has at *d* a prism. In front of it is placed the eye-piece B, with the field-glass *f* and the ocular lens *e*. At a short distance from the latter is the small glass prism C, surrounded by a black metal ring. The course of the rays of light is clear. They pass through the external prism to the eye of the observer. The latter sees not only through the small prism, but also through the opening of the ring, a paper placed beneath, where the microscopical image is projected, and can be easily traced with a pencil.

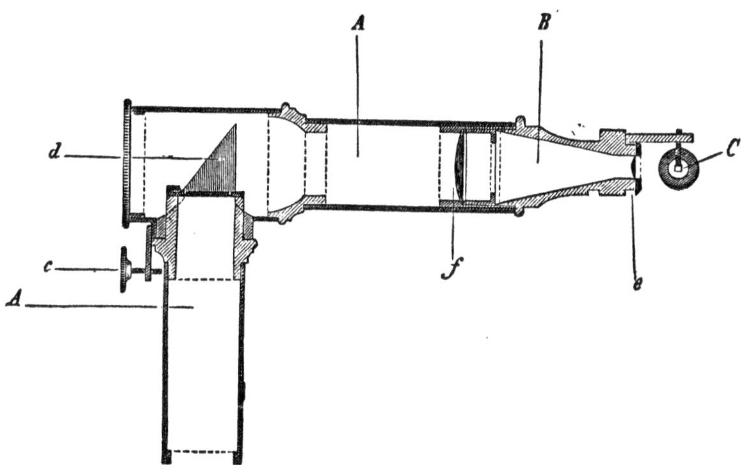

Fig. 36. Camera lucida of Chevalier and Oberhäuser. The piece B is turned 90°.

The camera lucida, when used, takes the place of the eye-piece, and is fastened on to the miscroscope tube with the screw *c*. The illumination must be carefully regulated, so that the point of the pencil may be accurately seen, which is indispensable. A black pasteboard screen, placed in front of the drawing-paper, has a good effect.

The place where the image is received, that is, where the

paper lies, is naturally of importance. The farther from the instrument this takes place, of course, the larger it will be. It should be made a rule to have the drawing-paper placed at as nearly the same elevation with the stage as possible, that is, about 25 centimetres beneath the prism. If the magnifying power of the objective and of the camera lucida be ascertained, by shortening the microscope tube and raising the drawing-desk, round numbers may be obtained, which is certainly convenient. The camera lucida, however, cannot readily be used with advantage for more than tracing the outlines. It is very convenient to use the knee-shaped tube with the prism, after replacing its eye-piece with another, thus converting the microscope into a horizontal one, though there is a loss of light.

The magnifying power used when drawing should always be noted, preferably near the drawing in the familiar way; as $\frac{20}{1}$ (20 diameters), $\frac{300}{1}$, $\frac{650}{1}$, etc. It is only practicable in a very few cases to draw everything with the same enlargement, as has been proposed by many persons. What pictures would often result; dwarfs by the side of giants!

We can readily conceive that photography, that grand discovery of modern times, has not been ignored by the microscopist; its value for giving a true objective representation of a microscopic specimen must be self-evident. Nevertheless, the number of observers who have worked at it, either for themselves alone or, which has generally been the case, in connection with a professional photographist, has not, as yet, been very considerable. The greater part have been deterred by the want of familiarity with the technology of photography, and the generally very much overrated difficulties of micro-photographic manipulation. What may be thus accomplished, what a future photography also has in store for microscopical investigations, is shown by many examples of the present time.

As early as the year 1845, Donné, a French observer, published an *Atlas d'anatomie microscopique*, the illustrations for which were copied from those taken on Daguerre's metal plate by means of the sun-microscope. In more recent times an immense progress has been made in photography by taking the negative on a glass plate covered with iodized collodion. We have received from Paris many beautiful micro-photographs, which were taken, in part, with very high powers. Within a few years Hessling and Kollmann, in connection with Albert, the renowned photographer of Munich, have commenced the publication of a folio of photographs deserving, in every respect, the highest encomiums. Unfortunately, it remains uncompleted. Professor Gerlach, of Erlangen, whom we have to thank for a number of valuable contributions to microscopical technology, has published a small but interesting guide to micro-photographic manipulation. (Die Photographie als Hülfsmittel mikroskopischer Forschung. Leipzig, 1862.) Beale and Moitessier have more recently treated of the same theme in a very thorough manner. B. Benecke published Moitessier's work, enriched by many additions of his own, in the German in 1868. (Die Photographie als Hülfsmittel mikroskopischer Forschung. Braunschweig.) It is the best work that we have on this subject at the present time.

The ordinary compound microscope may be readily and, as Gerlach informs us, with very slight outlay of money, turned into a micro-photographic apparatus working with sunlight (fig. 37).

Concentrated parallel light, which is afforded by the concave mirror (*q*) in connection with a plano-convex condensing lens, is used for the illumination. Cylindrical diaphragms with small apertures are to be used with the stronger powers. The ordinary objectives are used, but they must be scrupulously cleaned before taking the impression, as every particle of dust will cause a spot on the negative. The eye-piece is to be removed and the photographic apparatus placed on the tube of the microscope and held by a ring (*i*). A tube (*g*) supports a wooden case (*d*), in the upper end (*c*) of which the sensitive glass plate may be introduced (at *b*). It is better that the wooden frame (*b*) of the focussing screen should contain paper rendered transparent by oiling instead of the ground-glass plate of the ordinary apparatus. The usual black cloth, thrown over the head, serves to darken the same while focussing; the cone (*a*) on the chest contains a magnifying glass, to permit of the most accurate focussing. In order that the weight of the chest may not depress the microscope tube (*a*) in its sheath (*m*), a ring (*l*) surrounds the latter, and may be tightened by means of the screw (*k*). The brass capsule which covers the objectives of the ordinary apparatus is replaced by a black, horizontal tablet,

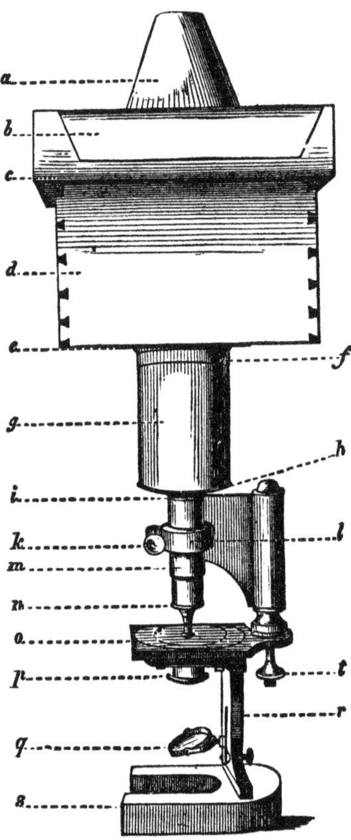

Fig. 37. Gerlach's micro-photographic apparatus: *a* hollow cone, to be placed on *b*, the focussing screen; *c* projection at the upper part of the case *d*; *e* metal ring at its bottom; *f* metal ring at the upper part of the wooden tube *g*; *h* metal plate at the lower end of the same; *i* ring at the upper end of the metallic tube; *k* screw of the metal ring *l*, which serves to clamp the spring sheath *m*; *n* tube of the microscope with the objectives; *o* stage; *p* the metallic cylinder for carrying the diaphragm and illuminating lens; *q* the mirror; *r* the metallic bar which supports the stage; *s* the horse-shoe foot; *t* the micrometer screw.

which may be placed between the mirror (*q*) and the condensing lens (*p*).

That this apparatus, afterwards still further improved by the inventor, suffices for obtaining excellent representations, may be learned from Gerlach's beautiful photographs. However, it is still of a somewhat primitive character, and has many defects. The illumination is not sufficient for strong magnifying powers, and, as the length of the tube is unalterable, the magnifying power of an objective cannot be changed. The tube of the microscope is too heavily loaded, which restricts and endangers the working of the micrometer screw.

A similar, but improved arrangement of Moitessier's appears, therefore, to be more suitable (fig. 38).

A folding camera (B) is supported by a strong three-legged wooden stand (A), which rests on a small table. The camera, like the bellows of an accordion, is capable of being lengthened or shortened, so that the impression may be taken at various distances from the objective. A sheet of white paper, stretched in the frame (D), which may be seen from beneath at the side, when the door is open, takes the place of the usual ground-glass plate, which, as I know from personal experience, renders the accurate adjustment of the focus very difficult. The tube of the microscope projects freely into the camera, through the opening in its bottom. This aperture should fit closely around the tube. The illumination is obtained from a silvered mirror and a condensing lens, both of which play through a sliding arrangement on a horizontal wooden ledge. The light from the mirror is concentrated by the lens on to the mirror of the microscope, into the stage of which an achromatic condenser is to be inserted.

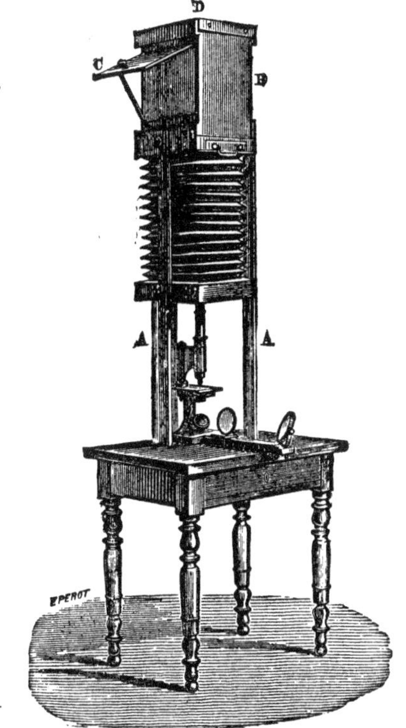

Fig. 38. Moitessier's Apparatus. A Pillars of the folding camera B; C its door; D frame.

Another arrangement (fig. 39) appears to be still more efficient, although it can only be accomplished with a microscope which is capable of assuming a horizontal position. The removal of the mirror from the microscope permits of the employment of direct sunlight. The illumination is obtained from the silvered mirror H, the diaphragm F, the convex lens E, and a plate of very delicate ground glass D, which are placed in a sliding arrangement. The ground-glass plate should have such a position that a small circle of light will be thrown on it.

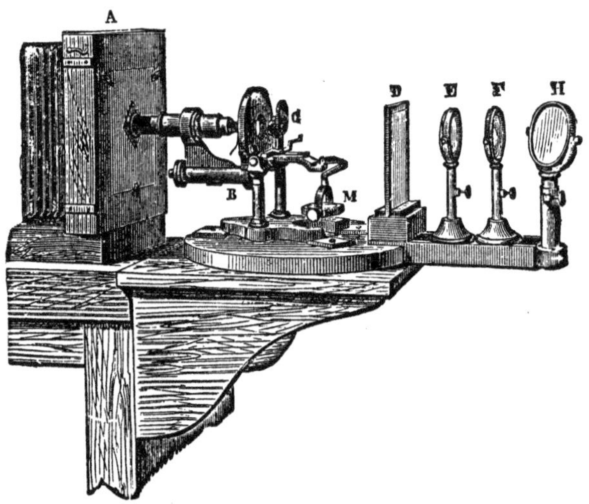

Fig. 39. Horizontal apparatus. A folding camera, B microscope; C achromatic condenser; M the mirror of the microscope turned aside; H the silvered mirror; F the diaphragm; E convex lens; D ground-glass plate.

A temperature of 14–18° R. is best suited for taking the impression. Natural light is used to produce the photographic picture. The duration of the exposure, naturally varying according to the intensity of the light, increases with the strength of the magnifying power employed, and with full sunlight lies, according to Gerlach's observations, between five seconds for a magnifying power of 5–25 diameters, and 40 seconds for one of 250–300 diameters. Among the methods of artificial illumination, that with magnesium light deserves mention above all others. A photogenic lamp, together with some additional contrivances, also affords good illumination (S. T. Stein). The duration of the exposure depends on the manner of treating the sensitive glass plate. The wet method with collodium requires the shortest time; the dry method and that with albumen require a much longer time.

Gerlach, Beale, Moitessier, and Benecke have entered fully into all the remaining details of the process. The limits of our little work prevent us from noticing the subject further, and we must therefore refer to those authorities.

It is self-evident that the trouble of photographing should only be bestowed on preparations which are irreproachable and free from every contamination. It is important to have but a small number of bodies in the field; for example, only a few blood-corpuscles, or a few epithelial cells. Compact tissues require the thinnest sections. Pale-bordered objects require stronger shading. Therefore, Canada balsam preparations are less suitable, as are also objects mounted in glycerine, though some assistance may be rendered in such cases by tinging them with carmine. Preparations injected with carmine or Prussian blue afford admirable pictures, and Gerlach has reproduced them even with a repetition of their colors!

If a micrometer of known value is photographed at the same time and with the same enlargement, the size of the object represented may be ascertained by measurement with a pair of compasses with exceeding facility and accuracy.

Such micro-photographs are less adapted for furnishing large works to be issued in large numbers, as a certain inequality of the positive prints is unavoidable. They are admirable, on the contrary, for purposes of instruction. Judging from the photographs we have seen of microscopic objects, we must doubt whether such photographs as are now made will be useful for deciding subtle questions of texture. Only a few French and American representations of diatomaceæ make an exception.

In recent times, as is well known, such extraordinarily small photographs have been produced, that the picture can only be recognized with a strong magnifying glass or a microscope. Here the silver precipitate is of such fineness that a considerable magnifying power is necessary to render it visible.

These minimal photographs have led Gerlach to make a peculiar application of photography to microscopical purposes; to an increase of the enlargement by photographic means.

The first negative of an object obtained by means of the microscope is hereby subjected to a new enlargement. A second negative is thus obtained, which presents light and shadow in the same manner as the object, and therefore cannot be converted into a useful positive image. This is quite possible, however, when the second negative is exposed to a new enlargement and the tertiary is thus obtained, which corresponds to the light and shadow of the first one. The enlargement may be increased till the silver precipitate becomes visible. By diluting the photographic solutions, as well as by a peculiar treatment of the sensitive glass plate, this visibility may be very much deferred. In Gerlach's work three such photographs of the scale of a butterfly (Papilio Janira) by 265, 670, and 1,460 fold enlargement may be found. Parisian and North American photographs of the pleurosigma angulatum, which I have obtained through Lackerbauer and Woodward, show the hexagonal areolations very beautifully, enlarged to 2,000 and 2,500 diameters. Impressions of the latter with 19,050 fold enlargement I certainly do not comprehend. It remains for the future to show what practical advantages such applications of the micro-photographic apparatus may present, that is, how far structural relations, which by the first impression are not recognizable by the eye, can be made visible by the following ones.

Section Third.

THE BINOCULAR, THE STEREOSCOPIC, AND THE POLARIZING MICROSCOPE.

The idea of producing microscopes through which several persons are able to observe simultaneously one and the same object is sufficiently obvious, and, without doubt, such instruments must be very convenient for a teacher in his demonstrations.

By the application of prisms over the objective, the rays of light which pass through it may be divided into two, three, four bundles. This is accomplished either by dioptrical means, through an achromatic compound prism (fig. 40), or by catoptrical, by total reflection, as, for instance, the prism combination shown in fig. 41. If several microscope tubes, each provided

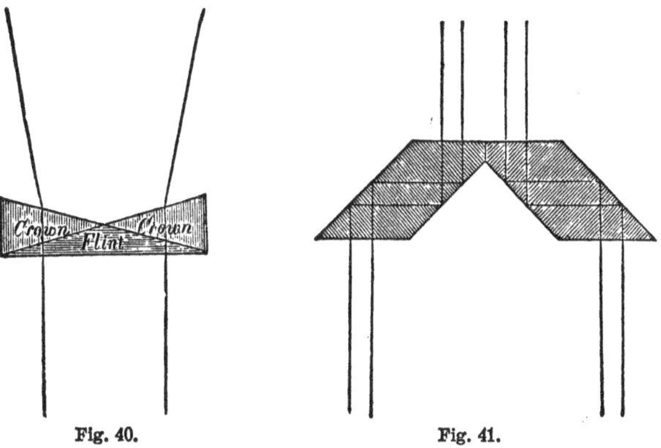

Fig. 40. Fig. 41.

with its own eye-piece, and corresponding to the number of bundles into which the rays are divided, be placed over the prism, it will be possible for a number of persons to observe simultaneously. The eye-piece must be movable in its tube, by means of a screw, to permit of individual focussing.

The division of the rays which have passed the objective into two, three, or four bundles, is naturally combined with a corresponding diminution of the intensity of the light; there is also some loss of light in the prisms. Only the weaker objectives can therefore be employed with such multocular microscopes, as they have been called, and the images leave, as a rule, much to be desired. Such binocular, triocular, and quadrocular microscopes have recently been constructed and brought into commerce, especially by Nachet, of Paris.

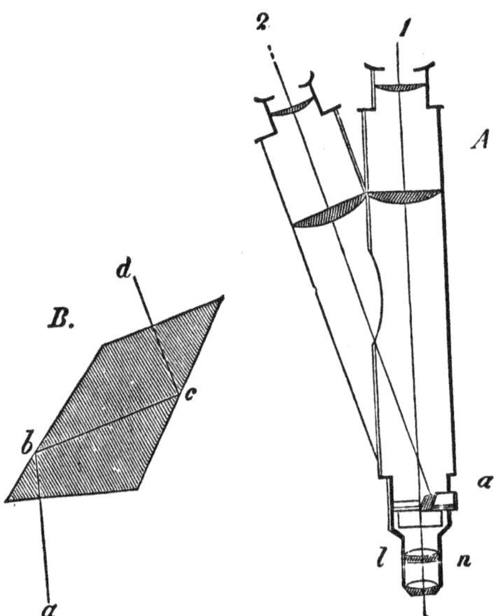

Fig. 42. Wenham's arrangement of the stereoscopic microscope.

The binocular microscope may also be so constructed that its two tubes can be used for both eyes of one and the same observer. When they are so placed as to correspond to the convergence of the optic axes, the two images cover each other, and the consequence must necessarily be that the object no longer seems flattened, but assumes a corporeal appearance. In this manner is constructed the stereoscopic microscope, the only efficient application of the binocular principle. We have to thank Riddell, an American, for the production of the first instrument of this kind. Since that time, English opticians especially, such as the firm of Ross & Co., of London, have, with a certain predilection, constructed these microscopes, and have contrived arrangements by means of which ordinary microscopes may be readily converted into stereoscopic instruments. Wenham's very excellent arrangement, in general use there at present, is represented by our fig. 42. With the main tube A 1 of the instrument is movably connected—that is, capable of being approximated towards, or separated from it—the secondary tube 2. A small prism *a* projects as far as the optical axis of the tube 1; its form may be recognized more accurately in the enlarged drawing B. Each bundle of rays is so divided, after its passage through the objective, that the one passes unbroken through the tube 1, the other through the prism B, in the direction *a b c d*, into the secondary tube 2. Nachet has also, for years, supplied such stereoscopic microscopes; likewise Hartnack, whose stereoscopic eye-piece is represented by our fig. 44. Opinions are divided with regard to the utility of these instruments; they have certainly been over-estimated by many. We must leave it for the future to decide whether science is to derive any benefit from them. As examples, we have represented in our fig. 43 such an instrument by H. and W. Crouch, of London, and in fig. 45 one by Nachet.

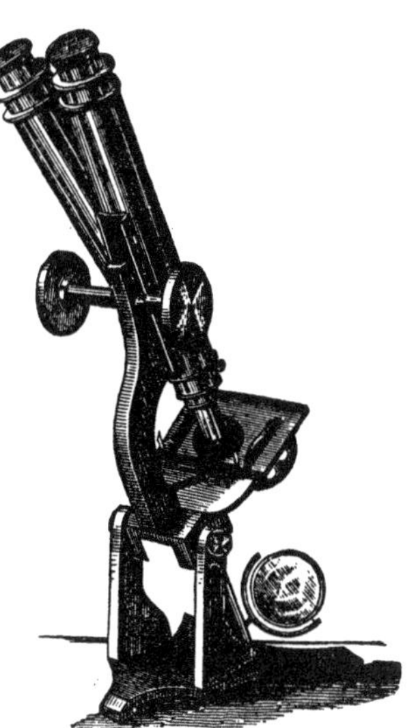

Fig. 43. Crouch's stereoscopic microscope.

The examination of tissues by polarized light has, on the contrary, a high scientific value, as, by this means, molecular relations become evident which, by investigation with ordinary light, remain entirely concealed. The interpretation of what is seen is, in many cases, difficult, and generally lies within the province of optics, with which the medical observer is usually but little familiar.

Every ordinary instrument may be changed to a polarizing microscope, in a very simple manner, by adding to it a polarizer and an analyzer. For this purpose Nicol's prisms, consisting of double refracting calcareous Iceland spar, are used. They are so constructed as to transmit only one of the two rays into which a beam of ordinary light is made to separate on passing through this substance, while the other is lost by reflection.

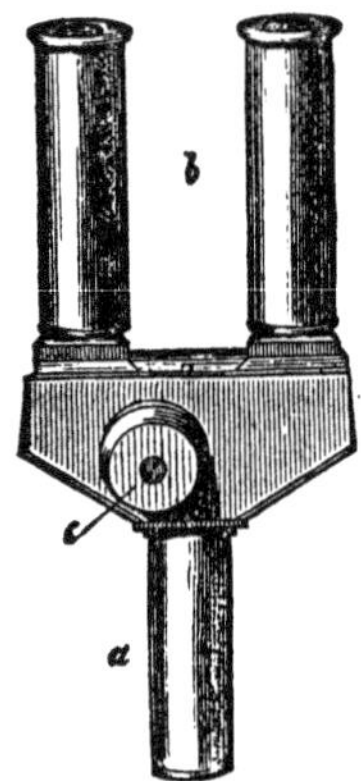

Fig. 44. Hartnack's stereoscopic eye-piece. The two tubes *b* may be adjusted according to necessity by means of the button *c*; *a* for insertion into the tube of the microscope.

The polarizer is placed close beneath the object, preferably in the opening of the stage with the addition of a convex lens (fig. 46). The analyzer, on the contrary, receives various, and by no means equally good positions. As a rule, it is placed by the opticians over the objective in the tube of the microscope; an arrangement, however, which causes too great loss of light, which becomes very unpleasant in investigations where the double refraction is weak. It is much more advantageous to place the analyzer over the eye-piece, enclosed in a metallic tube. Although by this means the field is extraordinarily diminished, especially when the Nicol is small, it presents much more light than the larger field obtained by the first-mentioned arrangement. Hartnack has recently placed a plano-convex flint-glass lens of short focal distance (fig. 46, *a*) over the polarizer. The analyzer (fig. 47) he has placed in the eye-piece (*b*) and the latter is made to rotate within a graduated disk (*a*). By this means he has essentially increased the efficiency of his polarizing apparatus.

The two Nicols are to be, at first, so arranged that their polarizing planes are parallel to each other, which gives an illuminated field. This cannot be made too intensely bright, especially with weak double refraction. A condenser, such as we have mentioned above, placed over the polarizing calcareous spar prism, is very serviceable, as was pointed out years ago by H. von Mohl.

When the polarizing planes are placed at right angles to each other, by turning the analyzer 90°, the field is darkened (it should appear entirely dark with a good apparatus), and doubly refracting bodies appear either illuminated or in colors.

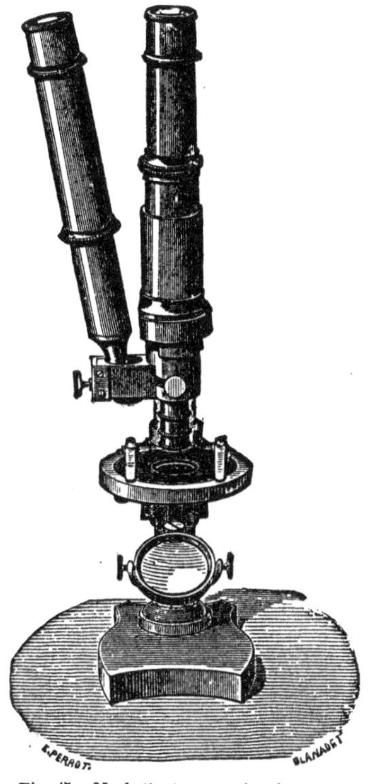

Fig. 45. Nachet's stereoscopic microscope.

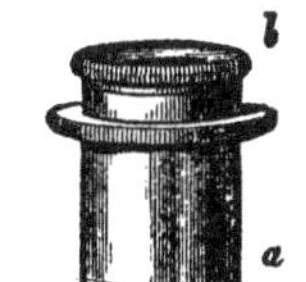

Fig. 46. Polarizer. The tube *a* fitted into the stage; *b* convex lens of flint glass.

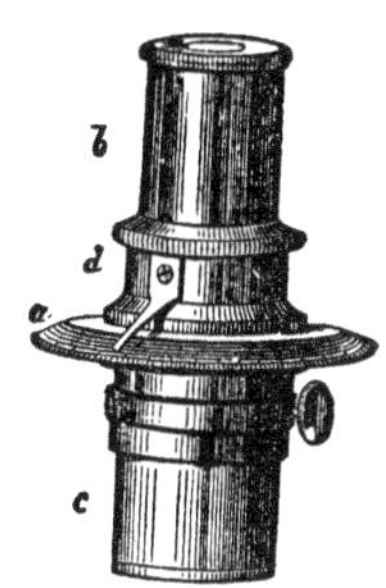

Fig. 47. Hartnack's analyzer, new construction. The eye-piece *b c* turns in a sheath, which is to be fastened to the microscope by means of the screw at the right; it also has a graduated circle *a*; *d* vernier.

The rotation is made in various ways: either the analyzer placed upon or within the eye-piece is rotated, or the stage is rotated, if capable of that motion. When the stage is immovable and the analyzing prism is placed within the tube over the objective, the opticians introduce an especial mechanism, by means of which it may be rotated in its sheath.

The objects to be investigated should be made as transparent as possible, when the recognition of weak double refraction is in question. Mounting with Canada balsam, which would perhaps render the object so transparent as to be totally unserviceable for the ordinary methods of examination, would here render excellent service.

In delicate investigations, the incident rays of light must be carefully excluded, by placing a hood over the stage.

Thin scales of selenite or mica, of various thickness, placed over the polarizer, are the means generally used for developing a lively play of colors with polarized light, and for deciding as to the character of double refracting animal tissues. They are examined at an angle of 45°. A film of selenite produces more lively colors than one of mica. In using these plates, it is preferable to have them of the thickness which gives a red of the first order. At the same time, the sharpness of the microscopic polarizing apparatus may also be increased by the introduction of a film of such thinness as not to cause any coloring of the field.

Section Fourth.

TESTING THE MICROSCOPE.

In testing and critically examining the optical performances of a microscope, with which, naturally, the extent of its magnifying power must also be included, a number of things have to be taken into consideration; and when the appreciation of very fine distinctions, especially with the stronger objectives, is concerned, it becomes a difficult business.

To ascertain the magnifying power of a microscope, the focal length of the objective and that of the lenses composing the eye-piece may be measured, and from this the enlargement reckoned. This subject is further elucidated in the text-books on physics.

It is much more convenient, however, to measure directly the joint magnifying power of the several combinations.

For this purpose, an ordinary glass stage-micrometer with fine divisions is used; a rule is also placed on the stage. By means of the power of double vision, which, however, requires practice, that the head and eyeball may be kept quiet, the image of the micrometer divisions will be seen projected on to the rule which lies on the stage, and the relative size of their spaces may be thus compared. Granted the rule is divided into millimetres, and that the micrometer has one such millimetre divided into 100 parts. Two of the divisions of the rule are covered by one space of the micrometer image. The magnifying power of the microscopic combination measured is, therefore, 200-fold.

The distance of the eye-piece from the stage must also be taken into consideration in order to obtain a precise expression, corresponding to the visual distance accepted as the normal medium, which is, as was already remarked, from 8 to 10 inches, or 25 centimetres. Let us accept the latter as the visual distance. If now, for example, the distance between the image and the eye over the eye-piece is 20 centimetres, the magnifying power, with a visual distance of 25 centimetres, would be 250-fold.

It is necessary to determine in this manner the magnifying power of the various eye-pieces with one and the same objective. It is then only necessary to obtain the magnifying power of each of the remaining objectives with one of the eye-pieces,—for instance, the weakest one,—to find by calculation that of the others.

In making this measurement, only those divisions lying in the middle of the field should be used, to avoid any incidental distortion of the image.

The image of the micrometer projected on to the stage may be readily measured with the points of the compasses, and its size determined with the rule.

It is also convenient to employ the various projecting apparatuses, especially prisms on the eye-piece.

Every serviceable modern instrument should have received a careful correction of the spherical aberration of its lenses. Various means have been used for testing this. They are more fully treated of in the larger works on the microscope by Mohl and Harting. A slide thickly smeared with India ink, in which small circles or other figures are scratched with the point of a fine needle, may be recommended, when it is desirable to make a few rapid tests of the lenses. If the instrument is adjusted with transmitted light for such a circle, it should appear sharply cut on the black ground, and not surrounded by a halo of light. If the circle is then brought out of focus, it gradually enlarges, while its sharp borders disappear, without spreading a strong halo of light either inwards or outwards over the black field.

Secondly, adequate correction of the chromatic aberration should be observed. This cannot be complete, because there is no means by which the secondary spectrum can be removed. Therefore, reference is here made only to a correction which is as complete as practicable. Modern objectives are, for the most part, over-corrected with regard to chromatic aberration, and show a bluish border. Under-corrected lenses present, under the same conditions, a reddish margin, which is less agreeable to the eye, though the sharpness of the image remains the same.

The flatness of the field is of great importance to the advantageous use of the instrument. Here, as we have already found, two things are to be separately considered, namely, the incurvation of the field, and the distortion of the image.

If we strew a very fine powder over a flat plate of glass, we should, if the field is flat, be able to see the molecules at the centre and at the periphery equally distinct and simultaneously. If there is any incurvation present, deeper focussing is requisite to see the molecules at the periphery of the field.

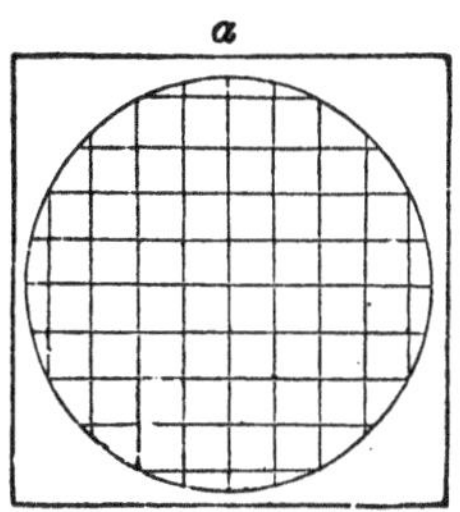

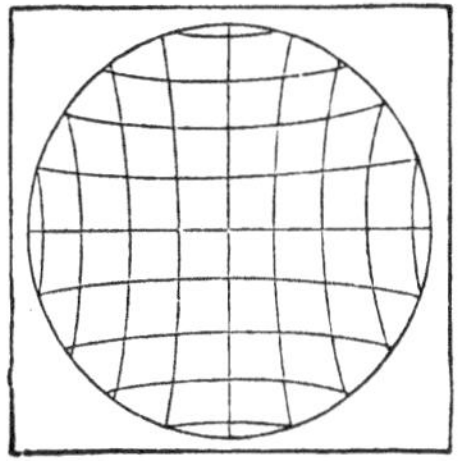

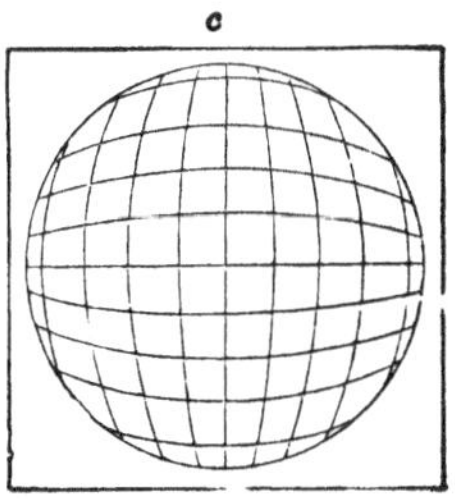

Fig. 48. Quadratic glass micrometer.

A glass micrometer divided into quadratic fields, placed on the stage, should appear as fig. 48 *a*, if the image is not distorted; while, on the contrary, if any distortion is present, the squares assume the appearances represented in our figure at *b* and *c*, according as the enlargement from within outwards increases or diminishes.

If restrained by purely practical considerations in testing a microscope, regard must always be paid, in deciding on the merits of an objective, to the purpose for which the optician has constructed it; whether for incident light or for light reflected from the mirror; and, when the latter is the case, whether for central or oblique illumination. An objective may, for example, perform very well with the latter, and yet be very indifferent with central illumination; inversely, many opticians construct objectives which are very good in the latter regard, but are defective with oblique illumination. It is quite impossible to construct a combination which will be equally serviceable for all the different requirements, resting, in part, on opposite physical conditions. The testing of an objective should, therefore, never be restricted to the use of a single test object.

Two attributes may be distinguished in an object glass; first, its defining, and second, its penetrating or resolving power. Mohl was right in saying that on the first depends the distinct recognition of the outlines and forms of bodies; on the latter, the appreciation of its finer structure.

1. The defining power of an objective depends upon the complete correction of its spherical and chromatic aberration. Such an attribute, and of adequate extent, must be expected from every superior modern objective, for whatsoever purpose it may have been constructed. Good definition may be more easily obtained with lenses of small or moderate than with lenses of larger angles of aperture; and by aiming to extend the aperture, the perfection of the definition is not unfrequently impaired.

A certain amount of practice is necessary to recognize a good defining objective. The outlines of the image obtained by it appear fine and sharp; objects lying near each other, and those which are pushed over each other in the same optical plane, show their individual outlines distinctly and may be readily appreciated; the entire image has something clear and elegant about it, like a good copper plate or a print with sharp letters. To recognize the opposite condition, it is only necessary to furnish the microscope with a pretty strong eye-piece. Thick, confused contours and diminished distinctness of the image would be met by the observer; the whole would appear like a print with dull, disconnected letters. It is just this sharpness and neatness of the image which at first prepossesses one in favor of such an objective, whereas an objective with greater penetrating power usually gives paler, more milky images, and only unfolds its high superiority to the connoisseur.

The best defining objectives are a prime necessity for all microscopes intended for scientific work.

2. The penetrating or resolving power of an objective depends

on its capability of bringing the very fine details of the surface and interior of an object into view. Its perfection has become the aim and the pride of the microscope-makers of the present day, and to it is due, in a great measure, the calling into existence of the superior objectives of modern times.

The resolving power of a combination depends, however, on the extent of the angle of aperture, and, consequently, on the obliquity of the rays of light which the system is capable of receiving from the various points of the surface of an object. In regarding a transparent surface containing lines placed close to each other, whether these appearances be due to elevations or to furrows, we are made to appreciate the value of oblique illumination. It is clear that rays of light which are transmitted axially through the object would yield less information with regard to such inequalities than those which fall on its surface obliquely. Thus, by means of objectives of medium power, but with considerable angles of aperture, one may see with oblique illumination things of which no trace can be recognized with central illumination. An object-glass of very wide aperture, however, will receive, even with ordinary illumination, so many rays of great obliquity that the same kind of effect will be produced as by oblique illumination with an objective of smaller aperture; but when with such an objective oblique illumination is used, a greater resolving power is obtained than any combination of smaller angular aperture *can* possess.

It will be appreciable, from the remarks which have just been made, why it is just this enlargement of the angle of aperture which has been, of late, the chief aim of the optician.

Thus we see that older instruments have only the slight angle of 50° or 70° in their strongest systems. But, even as early as the year 1851, the renowned London house of Andrew Ross had given their strongest systems an aperture of 107° and 135°, a few years later 155°. But the limit was not yet reached; for more recently apertures of 160, 170, and even 176° have been obtained, in which the actually available portion remained at about 130 to 146°.

Such objectives are of the greatest value when penetrating power is required; but the defining power is usually relatively greater with a combination having a smaller angle of aperture.

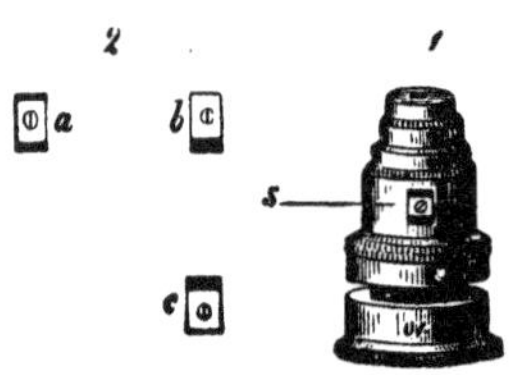

Fig. 49. Objective with correcting apparatus. The metallic slider *s* in 1 shows the changes in the position of the lenses which 2 *a b c* represent.

We have already (p. 19) mentioned the influence which the thickness of the covering glass exerts on the sharpness of the microscopic image. It is customary to combine with all of the stronger objectives the apparatus for correction, fig. 49, which was spoken of in a preceding section, so that the lenses may be brought nearer together or moved farther apart, as necessity may require, according to the thickness of the covers used. Some of these objectives are only to be used dry, that is, with a stratum of air between the upper surface of the glass cover and the lower surface of the lower lens; others, only with a stratum of water in the place of the stratum of air, and are then called immersion lenses. Other modern combinations can, however, be used in both media.

These immersion lenses are properly greeted as a great advancement, and Hartnack, of Paris, has obtained a brilliant reputation within a few years by producing excellent combinations of this kind, of very high power and very low price. The immersion lenses of Hartnack may be divided into those with single and those with double correction. In the first, the two lower lenses, fixed with regard to each other, are shoved up towards the upper one (the one turned towards the eye-piece). In those produced more recently, with apparatus for double adjustment, the middle lens also changes its relative position to the lower lens in a determined ratio during the turning.*

*A few additional remarks on the use of immersion lenses may here be in place. With a glass rod or a camel's-hair pencil a drop of water is placed on the covering glass, and a second one on the under surface of the lens. The lens is then carefully approached towards the object till the two drops flow together and the focus is accurately adjusted. By turning the screw, it will soon be ascertained whether the image assumes sharper or less delicate contours, and thus the best adjustment will soon be found. With Hartnack's arrangement, after each correction of the objective, the focus is naturally to be readjusted; this

In indicating the basis on which the optical advantage of such an immersion system, in contradistinction to the ordinary "dry" combinations, is founded, we will permit one of the greatest authorities to speak. Harting, in an interesting paper, remarks as follows:—

"As the water is a stronger light-refracting medium than air, the reflection of the rays of light is much diminished at the upper surface of the cover and at the under surface of the objective, indeed, it almost entirely ceases. Hence, more rays of light pass into the microscope, and the thin stratum of water has nearly the same effect as an enlargement of the angle of aperture. This favorable modification influences chiefly the peripheral rays, which fall most obliquely. The peripheral rays have most influence on the formation of the image, which takes place in front of the eye-piece; and as, by their passing through a transparent object, they are for the most part deflected from their course, and the slight deviations thus caused become visible in the image, the defining power of the microscope must necessarily be increased by the stratum of water."

As this stratum of water exerts the same effect as an increased thickness of the glass cover, it must produce an entire change in the spherical and chromatic aberration. We also notice that objectives intended for immersion give only inelegant and obscure images when used without the stratum of water. The intercalated stratum of water is, therefore, an integral constituent, a new optical element of the combination, and may exert an advantageous influence in the removal of the residuary secondary aberration.

In a third manner, finally, the optical power of an objective system is increased by the stratum of water. As the latter acts like a covering glass, and, as we have seen above, the lenses must approach each other in proportion to the increase in its

---

is not the case, however, with those of the English opticians, in which the position of the lowermost lens remains unaltered during the correction. The middle position of the correcting apparatus of Hartnack's immersion system corresponds to a thickness of the covers of about 0.1 mm. After being used, the under surface of the objective is to be carefully dried with a fine cloth.

thickness, the magnifying power and the angle of aperture are thereby also increased.

Harting shows what may be obtained by this means. In testing one of Hartnack's objectives, made in the year 1860, he obtained, with the various adjustments of the apparatus for correction, an angle of aperture of 166 to 172°, with an available portion of 135 to 140°, and a focal distance of 1.8 to 1.6 mm. A stronger objective of Powell and Lealand, of London, had an angle of aperture of 175 to 176°, with an aperture of 145°, and a focal distance, with the closest approximation of the lenses, of 1.36 mm. Its performance was the same as Hartnack's objective, and if any difference, however slight, existed, Powell and Lealand's objective was, according to Harting's test, the strongest.

Ten years have passed since that time and meanwhile many changes have taken place. Hartnack's immersion objectives Nos. 9 and 10, with angles of aperture of about 170 and 175°, and the nominal focal distances of $\frac{1}{12}$ and $\frac{1}{16}$ of an inch, have obtained the most universal acceptance. A still stronger system, No. 11, $\frac{1}{18}''$, with a total angle of aperture of 176°, was soon afterwards introduced by this optician. Hartnack has recently constructed an entire series of very powerful objectives. No. 12 corresponds to $\frac{1}{24}''$, No. 16 to $\frac{1}{40}''$, and the highest, No. 18, to $\frac{1}{60}''$, of the English.

It is, self-evidently, of great practical value to find objects which are as homogeneous as possible, and of such delicate and fine texture that, in their resolution, the optical, or, more correctly speaking, the penetrating power of a lens may be accurately estimated. They are called "test objects." Their study is of interest and importance. To the beginner, who is desirous of ascertaining the capacity of the instrument which, perhaps, he has but recently obtained, such test objects are to be recommended as discipline; since their resolution is by no means easy, and with them the accurate adjustment of the focus, and the skilful application of the illumination, may be learned. Some of these test objects, the finer ones, are so difficult as to occupy the beginner for hours in vain, and may occasion much labor even for the practised. By careful practice one may arrive at a certain virtuosoship, and thus, in a few minutes, appease the novice, who possibly begins to despair of his instrument, by showing him an example of what it is capable of performing in skilful hands. Then the endeavor to discover finer and more difficult test objects, thus always holding a higher aim before the optician, has led to the great emulation existing at the present time in the construction of objectives. It is therefore unjustifiable to regard the study of such tests with contempt, as is occasionally to be observed among notable microscopists.*

In the course of time, such test objects have been frequently highly commended and, with the increasing perfection of practical optics, again abandoned. Therefore, all those which were recommended before 1840, all the various hairs and scales of butterflies and wingless insects,† may be regarded as conquered territory. To attempt to test a first-class modern microscope with these expedients of a former epoch, would be an insult to the optician from whose establishment the instrument has proceeded.

In the year 1846, H. von Mohl, one of the first judges of the microscope, called attention to the brighter scales of the anterior wing of the Papilio Janira ♀, a knowledge of which he had obtained through the Italian, Amici, the most renowned constructor of microscopes of that epoch. Together with the familiar longitudinal lines, fine, closely approximated ($\frac{1}{1200}$ mm. apart), sharp, and not granular transverse lines appear, fig. 50. Mohl remarked, at that time, that with a magnifying power not exceeding 200, noth-

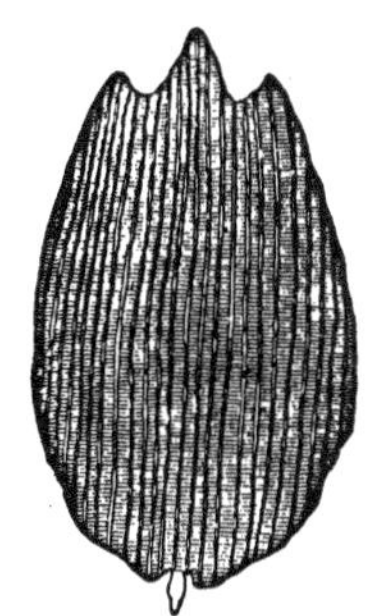

Fig. 50. Scale of Papilio Janira.

* M. Schiff has expressed the same sentiments with regard to the value of test objects. We cannot agree, however, with many of his views regarding the diatoma scales.

† It is well known that the trichina disease has, in our day, led to the production of an innumerable quantity of cheap instruments, intended only for the microscopic examination of meat. The well-known scales of the Lepisma Saccharinum, a wingless insect, are useful for testing them. We shall refer to this subject at the examination of the muscles.

ing was to be seen of these transverse lines, and that it was necessary to have an instrument with very strong and very good lenses to recognize the transverse markings, sharply and distinctly, with 220 to 300 fold linear enlargement. He cited only the microscopes of Amici, Plössl, and a single one of Oberhäuser, as at that time standing the test thoroughly. I still remember very well how I, as a student, with a, for that time, very serviceable Shiek's microscope,—my companion for many years,—was obliged to vex and trouble myself to obtain only a passable view of these transverse markings.

Nowadays an instrument would be called bad which, with a magnifying power of 200, left anything to be desired in resolving a Janira scale. By means of a large Hartnack instrument, made in the year 1861, I see them (in a test object coming from Kellner) without any precautionary measures, even with 120-fold enlargement (objective No. 5, eye-piece No. 2). At the present time, the scales of the Papilio Janira deserve to be regarded as a means of testing objectives of medium strength only.

The silicious envelopes of the Diatomaceæ have taken the place of the butterfly scales; those with the finest and most closely placed markings are employed.

The fineness of the markings may be represented in a table collated by Harting from English sources.

The	Pinnularia nobilis	has	from	4 to 6	striations in the	$\frac{1}{100}$	of a millim.
"	Pleurosigma formosum	"	"	12 to 14	"	"	"
"	" attenuatum	"	"	15 to 16	"	"	"
"	" angulatum	"	"	22 to 23	"	"	"
"	Grammatophora marina	"		25	"	"	"
"	Nitzschia sigmoidea	"	"	30 to 31	"	"	"
"	Navicula rhomboides (affinis, Amicii)	"		30	"	"	"
"	Surirella gemma (longitudinal lines)	"	"	30 to 32	"	"	"
"	Grammatophora subtilissima	"	"	32 to 34	"	"	"
"	Frustulia saxonica	"	"	34 to 35	"	"	"

Of the numerous Diatomaceæ there are several which deserve mention as being of particular importance; namely,—the Pleurosigma angulatum and Nitzschia sigmoidea, already mentioned in the table; then, the Navicula Amicii, Surirella Gemma, and the Grammatophora subtilissima, made known by the deceased Professor Bailey, of North America. The last two objects (we have these always in mind as they are to be obtained from Bourgoyne of Paris) are extremely difficult, and in resolving them the microscope withstands a hard trial. Reinicke (Beiträge zur neueren Mikroskopie, 3. Heft. Dresden, 1863) has called attention to the Frustulia saxonica, mounted in Canada balsam, as a very subtile test object. Its transverse lines do not stand very close to each other, but are very delicate and difficult to perceive. At the last London Industrial Exhibition the Navicula affinis, mounted in Canada balsam, was used as a test object. Their longitudinal striations are resolved without difficulty, while, on the contrary, their transverse lines are very sharp and fine, so that I must pronounce their solution (in Bourgoyne's preparations) more difficult than the Surirella Gemma and Grammatophora. Bailey has also recommended the Hyaloidiscus subtilis.*

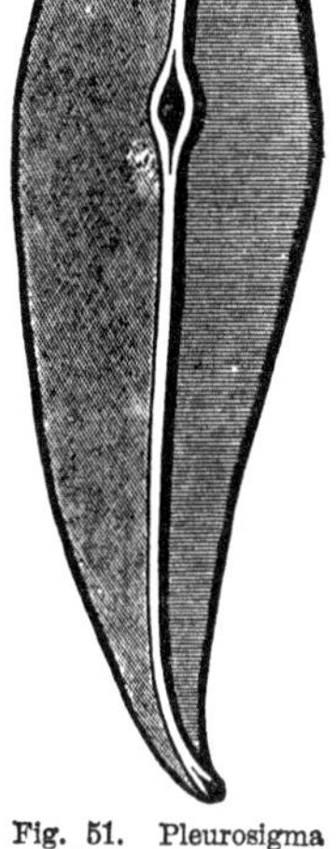
Fig. 51. Pleurosigma angulatum.

The Pleurosigma angulatum, fig. 51, furnishes, with oblique light, an excellent means of testing the resolving power of good objectives of medium and greater power, but should expose all its

* J. D. Möller, of Wedel, Holstein, has recently produced a very excellent but expensive Diatom test-plate. Each one contains 20 Diatoms, arranged according to their value as test objects, as designated by Dr. Grunow, namely: 1. Triceratium Favus; 2. Pinnularia nobilis; 3. Navicula Lyra var.; 4. N. Lyra; 5. Pinnularia interrupta var.; 6. Stauroneïs Phœnicenteron; 7. Grammatophora marina (more coarsely marked than Bourgoyne's variety); 8. Pleurosigma Balticum; 9. P. acuminatum; 10. Nitzschia amphioxys; 11. Pleurosigma angulatum; 12. Grammatophora Oceanica subtilissima (marina); 13. Surirella Gemma (transverse lines); 14. Nitzschia sigmoidea; 15. Pleurosigma Fasciola var.; 16. Surirella Gemma (longitudinal lines); 17. Cymatopleura elliptica; 18. Navicula crassinervis, Frustulia Saxonica; 19. Nitzschia curvula; 20. Amphipleura pellucida. Rodig, of Hamburg, also issues a similar Diatom plate.

delicate markings with a good immersion lens with simple central illumination. With oblique light this test object is entirely too easy for immersion lenses.

If the examination of the Pleurosigma angulatum is commenced with a weak objective, it appears smooth and without markings. Passing, together with the application of oblique illumination, to stronger lenses, a period arrives when systems of lines sparkle forth, which run, in part, diagonally over the scale, in part obliquely and crossing each other. Sometimes the one, sometimes the others of these lines are most distinct, according as the oblique light passes through the scale.

They come forth gradually and quite sharp, and in fortunate cases one may distinguish all three—the two oblique ones cutting each other at angles of nearly 60° (not 53)—at the same time with perfect distinctness, all lying, according to my view, in the same plane. It is still believed that we have here to do with perfectly straight lines.

By using an immersion lens with central illumination, they

Fig. 52. Areolations of the Pleurosigma angulatum; from a photograph.

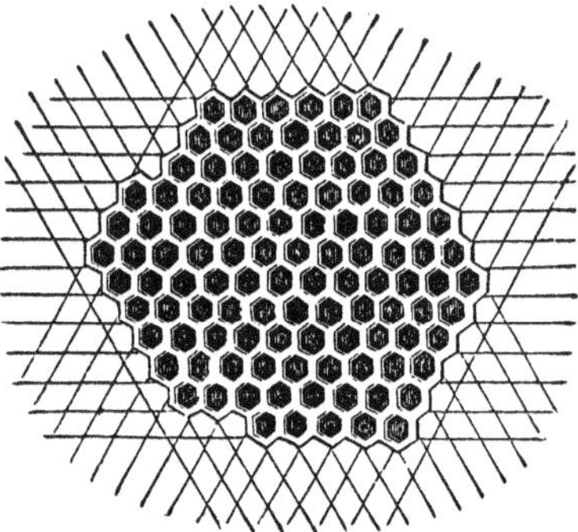

Fig. 53. Areolations of the Pleurosigma angulatum.

appear to surround a series of extremely small and very delicate hexagonal areolations in close approximation, fig. 52. These appear, according as the focus is altered, either dark and surrounded by bright margins, fig. 53, or bright, with dark margins, fig. 52. So much may be stated with entire certainty. Now arises, however, the difficult and by no means definitely settled question:—Are the areolations concave and their margins elevated, or, on the contrary, are the latter furrows between projecting areæ? Both propositions have been sustained by distinguished observers. I formerly regarded the depression as probable, and also that the focus was correctly adjusted when the areolations appear dark. M. Schultze has also expressed the same opinion, in accordance with certain rules (see below) established by Welcker. I afterwards adopted the contrary view. This does not appear to be the place to enter further into this subject.

A good objective, magnifying about 80 or 100 times, should, with the proper oblique illumination, enable one to recognize the systems of lines sharply and distinctly on all the scales; while weaker objectives, magnifying 40 or 50 times, should show something of the lines. When it is found impossible to obtain oblique illumination, this inconvenience may be remedied by means of a condenser, with its central portion obscured. Oblique illumination and a stage capable of rotation are of great assistance. Hartnack's immersion lenses Nos. 9, 10, or 11 show the areæ very distinctly and beautifully, with central illumination, and even with an unfavorable sky. Other opticians, Amici, Nachet, and some English and German artists, have also been able to resolve them with their strongest lenses in the manner last mentioned. Hartnack's newly constructed objective No. 9, not intended for immersion, accomplishes the same, as I have myself witnessed, likewise his latest No. 8; even an excellent No. 7, received several years ago, gives the same result, with similar central illumination and elevated concave mirror.

The other test objects already mentioned, Nitzschia sigmoidea, Surirella Gemma, Grammatophora subtilissima, and Navicula rhomboides are much more difficult, and can only be resolved by means of suitable oblique illumination and very accurate correction of the objective. The first is the easiest; the last three, on the contrary, serve to test the best and most powerful modern immersion lenses.

As was just mentioned, the Nitzschia sigmoidea is the easiest of these objects to resolve. With oblique illumination, the long and narrow valve shows a series of very fine and compact transverse lines. Bourgoyne's preparations of the Nitzschia sigmoidea are mounted dry.

The Surirella Gemma, fig. 54, is a very delicate test object, and only to be mastered with much pains. Seen from its broad surface, the oval disk shows parallel ridges running as far as the central line. Between these appear very readily a series of fine, but distinct, transverse lines. It is these latter lines, cutting the transverse ones at right angles, which give the Surirella Gemma its value as a test object of the first class. Undulating curved lines of extreme fineness should appear, which give to the whole an interwoven appearance like basketwork (fig. 55). With the aid of his best lenses, Hartnack even succeeded in resolving these undulating lines into a series of very narrow hexagonal areolations (fig. 56). Bourgoyne's preparation is also mounted dry.

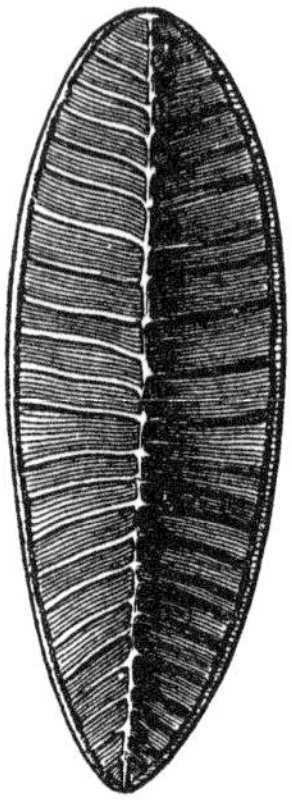

Fig. 54. Surirella Gemma.

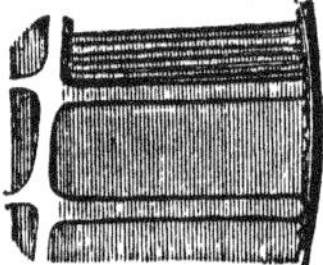

Fig. 55. Longitudinal lines on the silicious envelope of the Surirella Gemma.

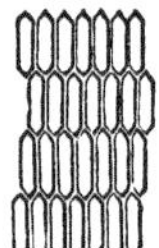

Fig. 56. The same, resolved into areolations.

The Grammatophora subtilissima, mounted by Bourgoyne in Canada balsam, is equally difficult. I do not know whether it is identical with the species first used by the American microscopist, Professor Bailey, of West Point, U. S. Besides, it seems that two kinds of valves of unequal difficulty have there been pronounced to be Grammatophora subtilissima.

Seen from its broad surface, the silicious envelope presents the appearance of an oblong square, with blunted corners (fig. 57, 1). It is divided into three regions by the two peculiarly curved longitudinal furrows. The two lateral regions (*a*) of every valve should exhibit very fine and compact transverse lines (2*a*), with the aid of good oblique illumination. The central portion does not show any markings.

This is, however, only a portion of the markings we are at present able to recognize. Other sharper and more coarsely marked species of the genus Grammatophora show these transverse lines, intermingled with a double series of oblique lines crossing each other at an angle of 60°, so that exactly the same markings result which we have previously described in the Pleurosigma angulatum. These oblique lines of the Grammatophora subtilissima also appear to be quite separated. Hartnack informs me that he has succeeded in resolving them with one of his strongest objectives, and I believe that I have myself caught at least a glimpse of them, with an immersion lens No. 10.

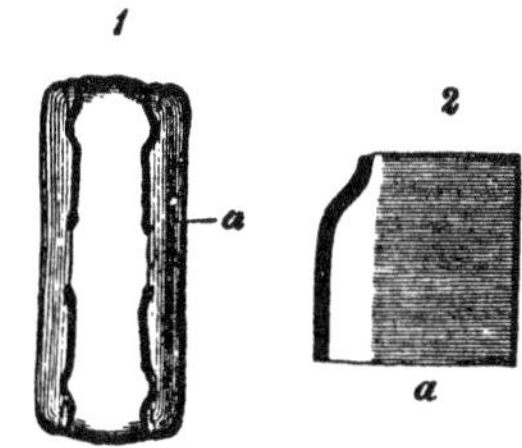

Fig. 57. 1 Grammatophora subtilissima. 2 Transverse lines of the same.

We add, finally, a few remarks on the Navicula rhomboides, sporangial form* (fig. 58). Its somewhat undulating, longitudinal lines (*a*) may be recognized with oblique light and a good immersion lens, without much trouble. They may be from 0.0002 to 0.00018 of a Paris line distant from each other. The elegant transverse lines (*b*) of the specimen in Canada balsam appear much more compact and extremely delicate. To recognize them, very oblique light and the most accurate correction of the immersion objective is necessary.†

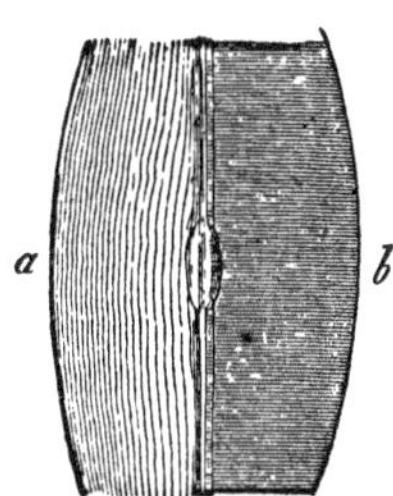

Fig. 58. Navicula rhomboides. *a* longitudinal, *b* transverse lines.

All organic test objects have, as a fault, the peculiarity that they are not exactly alike, but are, in the most fortunate cases, only very similar. It was, therefore, a fortunate idea of No-

* The Navicula in question was used at the last London Industrial Exhibition, as N. affinis, and was given to me in the form of a preparation by Bourgoyne, as N. Amicii. I have to thank Th. Eulenstein for the definition given in the text.

† The recognition of these transverse lines may be accomplished almost instantly by an expert, with a Hartnack No. 11 immersion objective. I have succeeded in doing this, though with some trouble, even with a No. 9 of this optician. Let me remark, incidentally, that the latter combination should also resolve the Surirella Gemma and Grammatophora subtilissima.

bert's to produce glass plates with bands of parallel lines, the distances between the lines constantly decreasing. The oldest of these plates, made about 1845, presented ten bands. In the first band, the distance between the lines was $\frac{1}{1000}'''$, in the last, $\frac{1}{4000}'''$. At the present time, with the progress of practical optics, such plates would no longer afford a means of testing first-class microscopes. Nobert afterwards made plates with 30 bands; but these marvellous productions of art cost 30 thalers. Quite recently he has issued a plate with 19 bands; the lines in its last division are $\frac{1}{10000}$ of a line apart. Thus markings as fine as those of the Diatomaceæ have been made by art. Nevertheless, these wonderful plates of Nobert's also have the fault of not being identical, although, in the most recent ones, the differences are almost imperceptible. Diversity of opinion still prevails regarding the resolution of the last bands, and this is connected with the still unsettled question, as to where the limits of accurate vision with our modern microscopes lies. We here introduce a table of the divisions of the last two test plates:—

Plate with 30 bands.

1.	Band	0.001000	of a Paris line.
5.	"	0.000550	"
10.	"	0.000275	"
15.	"	0.000200	"
20.	"	0.000167	"
25.	"	0.000143	"
30.	"	0.000125	"

Plate with 19 bands.

1.	Band	$\frac{1}{1000}$	of a Paris line.
2.	"	$\frac{1}{1500}$	"
3.	"	$\frac{1}{2000}$	"
4.	"	$\frac{1}{2500}$	"
5.	"	$\frac{1}{3000}$	"
6.	"	$\frac{1}{3500}$	"
7.	"	$\frac{1}{4000}$	"
8.	"	$\frac{1}{4500}$	"
9.	"	$\frac{1}{5000}$	"
10.	"	$\frac{1}{5500}$	"
11.	"	$\frac{1}{6000}$	"
12.	"	$\frac{1}{6500}$	"
13.	"	$\frac{1}{7000}$	"
14.	"	$\frac{1}{7500}$	"
15.	"	$\frac{1}{8000}$	"
16.	"	$\frac{1}{8500}$	"
17.	"	$\frac{1}{9000}$	"
18.	"	$\frac{1}{9500}$	"
19.	"	$\frac{1}{10000}$	"

The resolution of these lines with oblique light has been employed as a means of testing objectives. Harting was able, years ago, with a Hartnack's immersion lens No. 10, to recognize the lines in the 30th band of the older plates, and the resolution of the 25th, 26th, and even the 27th band is not an unusually great achievement. M. Schultze succeeded in resolving the 15th band of the more recent plate, and I afterwards resolved the 17th band with objective No. 11. In the year 1869, an American, Woodward, whom we have to thank for excellent photographs of test objects, also mastered the 19th band of this wonderful test plate.

Several years since, Schultze tested a series of the best modern objectives with central illumination. The highest performance consisted, at that time, in resolving the 9th band with an immersion lens No. 10 of Hartnack, and one of Merz's $\frac{1}{24}''$. I have repeated this experiment. My immersion objective No. 11 resolved the 12th, less distinctly the 13th, No. 10 the 11th, and the combination No. 7 of most recent construction, the 7th band of this test plate.

We have, finally, to discuss the question as to what precepts and advice are to be given to those who desire to procure a microscope; how should the instrument be constructed, and which optical establishment deserves, at present, to be most recommended.

He who desires to possess a first-class instrument will generally select one of the large microscopes with a horse-shoe stand, fig. 59, as constructed by Oberhäuser and imitated by other opticians. Its convenience of manipulation, combined with a certain simplicity, render it a truly model stand. The large stage, its capability of being rotated (which, however, requires very accurate workmanship, and is therefore expensive), the micrometer-screw for the fine adjustment, and the mobility of the mirror, are extraordinary advantages. The illuminating apparatus might, it is true, be improved, still it suffices for most purposes. When the stands which the English opticians select for their larger instruments (see page 31, fig. 32) are compared with it, they appear to be unpleasantly loaded with screws and unessential appurtenances, which inconvenience one who daily works with the instrument, as it is better to do with the human hand much which is there allotted to mechanical contrivances.

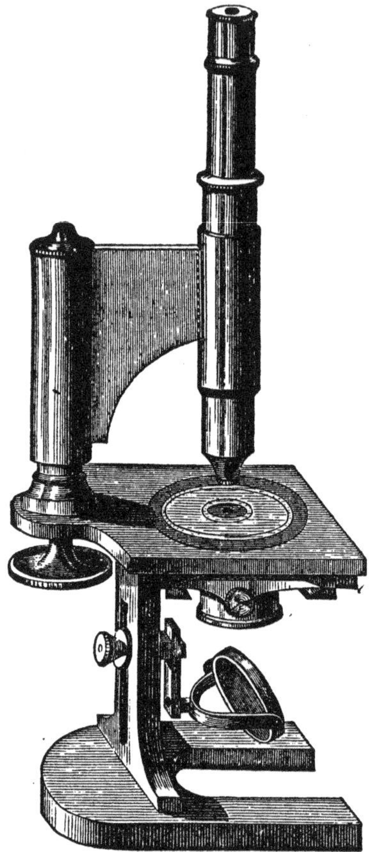

Fig. 59. Hartnack's large horse-shoe microscope.

For medical purposes the rotary stage may be readily dispensed with; less readily the oblique illumination; and this, which may be added with little expense, should, indeed, no longer be omitted from any instrument of medium class. Smaller horse-shoe stands, similar in construction to the larger stand, but without the rotary stage, deserve, therefore, to be especially recommended. Still smaller stands should possess a plane and concave mirror, and at least a rotary diaphragm to regulate the illumination, or, which is better, several cylindrical diaphragms, as well as a stage an inch and a half wide. When there is no oblique illumination, a simple condenser, similar to the one represented in fig. 24, may be used as a substitute. When there is but a simple mirror, no rotary diaphragm, and the stage is very narrow, as is the case in Hartnack's older microscope à l'hospice, the stand is certainly deficient.

Nevertheless, the mechanical portion of a microscope is a secondary consideration and of minor importance; in the optical apparatus is founded the actual value of the instrument.

One or the other form of instrument will be selected, according as a greater or lesser price can be afforded. Beginners, especially, should not have recourse to the largest, most expensive microscopes, as their manipulation is more difficult, and considerable practice is required before very powerful first-class lenses can be used.

The strangest notions not unfrequently prevail with regard to the optical portion. How often is the question still heard: How much does this instrument magnify? How often are microscopes ordered from an optician, with a magnifying power of from 5-600 diameters. Nothing shows a greater misconception of the optical performance of our instrument, as it is only necessary to add a perhaps uselessly strong eye-piece, to change a serviceable magnifying power of 400 diameters into a completely unserviceable one of 800, and, therefore, of no value to the instrument.

The individual objectives with the various eye-pieces form, each for themselves, a particular microscope. For this reason, one should have at least a twofold combination of lenses, if possible, three,—a weak, a medium, and a strong one. A double combination of lenses may be obtained from one system, in the most economical manner, by removing its lowermost lens. Many of the most simply constructed instruments have only one such objective, with two eye-pieces. Good microscopes of this kind may be obtained for 20 thalers. It is better to have several objectives with inseparable lenses.

We would here call to mind what we have previously said with regard to the great value of weaker powers. They should never be wanting. At least one objective of medium strength is also a valuable addition. Finally, a stronger objective, which, with a weak eye-piece, magnifies from 200 to 250 times, and, with a stronger one, affords a good and thoroughly serviceable magnifying power of 300 to 350, should not be wanting on any microscope.

These, usually, completely suffice, especially if another eye-piece with a glass micrometer is added. Such instruments are to be purchased for 30, 40, and 50 thalers, according to the stand, and, when obtained from the best modern establishments, stand higher, as to their capabilities, than the large microscopes constructed fifteen years ago at three or four times their price.

Stronger objectives are very seldom required; their addition naturally increases the cost considerably. For the commencement we would advise the omission of the very strongest objectives, especially those with apparatus for correction which are delicate to manipulate, as well as immersion lenses (fig. 60), and to select in their place a combination which works dry. With this, the magnifying power might be increased to 450 or 600, and rarely, even in extended scientific researches, would a higher magnifying power be missed. Such instruments, of excellent quality, may be bought on the continent for about 70 or 80 thalers.

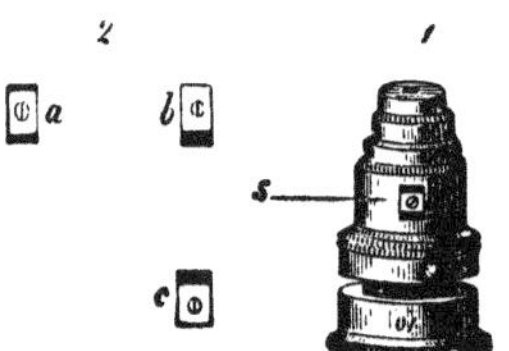

Fig. 60. Hartnack's immersion objective No. 10.

Other more or less expensive accessories, such as drawing and polarizing apparatuses, are, as a rule, added to the larger instruments only.

The value of a microscope being founded on its optical portion, on the excellence of its lenses, the question here arises as to the present productions of the various optical establishments. It is very difficult to render an impartial decision on this point. Disregarding a certain amount of odium in which one would be placed with the opticians not accorded the first rank, one should have just made a long journey, instituted for this purpose, through Germany, France, England, and North America, for in this department our industrial epoch exhibits a steady progress, one maker being surpassed by another.

The problem of the construction of weak, medium, and ordinary stronger objectives has been solved in a perfectly satisfactory manner by a considerable number of modern opticians, so that every year a large number of excellent microscopes, thoroughly adequate to all the requirements of the physician, are put in the market. It is true that certain objectives of one maker are superior to the same lenses of another maker; but these differences do not appear to exert any great influence on their practical requirements, and are only to be discovered by the practised eye. The effort to obtain a large angle of aperture has stamped modern objectives with a peculiar character. We would give the practical advice, not to purchase an instrument of an unknown optician, or, at least, not without having it tested by an expert, and to have the greatest mistrust of all charlatanical recommendations, whether they come from the optician himself or from a writer glorifying him.

The various optical establishments differ greatly in the construction of very powerful, or the most powerful combinations, as to the greatest excellence which may be accomplished in this department. Therefore, he who would procure a first-class instrument should proceed with circumspection.

Within twenty years several large firms in England maintained a higher rank in this department than the Continental opticians had obtained, if we disregard the Italian savant and distinguished microscope-maker, Amici († 1863). No impartial person, who knows how to test a microscope, could deny this, if he were to compare first-class instruments originating in that epoch. Since that time the emulation of the Continental opticians has spurred the most skilful on to even higher productions; the difference has become less and less, and has finally disappeared. Indeed, a few which have been produced among us of late deserve, perhaps, to be placed higher. At the same time there is a very considerable difference in price between the larger kind of English instruments and those of Germany and France. For example, a single objective with a nominal focus of $\frac{1}{16}$ inch, made by Powell and Lealand, of London, costs somewhat more than 16 pounds, while Hartnack, of Paris, furnishes a combination equally strong, No. 10 à immersion, for 200, and a still stronger one, No. 11, for 250 francs. The strongest objective, $\frac{1}{50}$ inch, of the London firm mentioned is charged at 31 pounds 10 shillings in the price-list; in that of the Parisian optician, at 500 francs.

Large modern microscopes of the most renowned English makers have not been accessible to me. I am, therefore, unable to say how far the achievements of former years have been surpassed. Several years ago, Harting, one of the first and most profound judges of the microscope, passed the highest encomiums on the powerful and most powerful objectives of Andrew Ross, as well as of Powell and Lealand. Several years ago, the $\frac{1}{25}$ inch objectives of the latter firm became quite numerous in England, and obtained great appreciation at the Industrial Exhibition in 1862. Another of $\frac{1}{50}$ inch is announced in the new price-current. Beale has praised it very highly. I became acquainted with it in the year 1866; so slightly, however, that I am unable to express an opinion.

Among the Continental opticians, Hartnack, of Paris, the successor to Oberhäuser (Place Dauphine No. 21), stands first, according to my views. Not only that his immersion lenses have not as yet been equalled by any Continental microscope-maker, but the weaker objectives, which are so very important, have also been very much improved, and from the industry and carefulness of this highly accomplished artist, further improvements are to be expected. Thus, objective No. 5 has already an angle of aperture of about 80°. Hartnack's Nos. 7 and 8, especially, are excellent, and, like all of his apparatuses, to be recommended for their slight expense. The former has within a few years been brought to an ever higher stage of consummation, as well in penetrating as in defining power, as I know from numerous comparisons and tests, and with an angle of aperture of about 100°, forms a wonderful combination for histological investigations. No. 8 has 125–130°, No. 9 (dry), 155–160° total aperture.

The smallest microscope à l'hospice, with objective No. 7

and a sufficiently broad stage, may be obtained for the low price of 65 francs; though deficient with regard to the illuminating apparatus, it is nevertheless very serviceable for medical purposes.

A somewhat larger instrument with a rotary diaphragm and a wide stage, with a weak objective and the No. 7 just mentioned, together with several eye-pieces, costs 115 francs, which is increased, when the objective No. 8 is added, to 165 francs. Disregarding the absence of oblique illumination, we should scarcely wish for anything further. It is very convenient for travelling, on account of its small size.

The small horse-shoe microscope, No. viii., is a very convenient stand, permitting of oblique illumination. With three objectives, 4, 7, and 8, together with the necessary eye-pieces, it costs 275 francs. During a series of years a considerable number of instruments of this kind have passed through my hands, and I know of no other modern microscope which I should be more inclined to recommend to physicians and students who are able to afford the moderate price. If, instead of a No. 8, an immersion lens No. 9 is taken, the price is increased to 390 francs. Besides this stand, Hartnack has recently introduced a still more simplified form, with a rotary diaphragm and a bronzed foot. With objectives 4 and 7, and two eye-pieces, it costs 140 francs. If the foot is replaced by a simple slab, the price is reduced to 120 francs.

Hartnack makes his large microscope only in the larger form, and with a rotary stage; together with four ordinary objectives it usually receives a No. 9 immersion lens, costing, with this addition, 750 francs, at present the best Continental instrument.

Nachet, of Paris (Nachet et fils, Rue St. Séverin No. 17), has also obtained a considerable reputation as a microscope constructor. Several large microscopes, constructed several years ago, resembling the English pattern, capable of being inclined, and furnished with a condenser, were very good for that time. What progress Nachet has since made in the construction of the most powerful objectives, I have unfortunately not become sufficiently informed. I had recently in my hands an immersion objective No. 7, a little weaker than Hartnack's No. 10; it was very good. Several small microscopes which I formerly tested were, as well in their mechanical as in their optical portions, excellent and very cheap, costing only 200 francs. Nachet's prices are as follows:—The large stand, fig. 33, fashioned after the English microscopes, and arranged for inclining, with very numerous accessories and seven objectives, costs 1,300 francs; the older large instrument 1,150, and more simply furnished 650 francs. Smaller instruments, with various, in part very convenient stands, may be obtained from Nachet for 500, 380, 200, 150, 125 and 70 francs.

The older firm of Chevalier has recently taken a new start, through the son, Arthur Chevalier (Palais Royal, No. 158). A competent judge, von Heurck, has recently given prominence to Chevalier's optical productions. Unfortunately, I have not as yet seen anything from this establishment.

Among the opticians residing in Germany (who have developed the most commendable and successful emulation), we will first mention a firm of Munich, G. & S. Merz, into whose hands has passed the renowned establishment of Fraunhofer and Utzschneider. We would especially commend one of their smaller instruments. It is a horse-shoe stand, and is provided with three eye-pieces and two objectives, with the nominal foci of $\frac{1}{3}$ and $\frac{1}{12}''$. The latter objective is of excellent construction and magnifies from 240 to 480 and 720 diameters. This microscope may be bought for 77 guldens; it is one of the most praiseworthy of which I know. Merz's stronger objectives, furnished with a correcting apparatus, have received well-merited recognition through Harting and M. Schultze. Another renowned Munich establishment is that of Mr. Bader. His smaller instruments cost 45 guldens.

Zeiss, of Jena, has also supplied compound microscopes for some time. Several years since, Schacht and M. Schultze gave us accurate information concerning these instruments, bestowing upon them the highest praise. Zeiss has, at present, eight different efficient stands, varying in price from 8 to 55 thalers. His objectives are distinguished, according to their strength, by the letters A to F. The first costs 5 thalers, and the following ones from 7 to 18 thalers; No. F. is reckoned at 25 thalers. All of his objectives with which I have been latterly acquainted are very well constructed. No. F. is such a strong and excellent combination that one rarely requires to use one of greater power.

In Wetzlar, about 1840, C. Kellner supplied instruments which were excellent for that time. His immediate successors, Belthle and Rexroth, have in their price-currents microscopes from 35 to 120 thalers. Belthle showed me good instruments years ago. Since Belthle's death, the business has passed into the hands of Leitz. His productions merit entire recognition. Another establishment in the same place is that of Engelbert & Hensoldt. Their instruments are likewise very good.

Möller and Emmerich, of Giessen, have supplied microscopes for several years.

In Hamburg, Schröder (Holländischer Brook No. 31), has made himself a name as a microscope-maker. A strong immersion objective, with apparatus for correction, which I tested several years since, was good, but ranked considerably below Hartnack's. The angle of aperture is large in his strongest objectives. The price of the stands is from 12 to 60 thalers. The objectives cost from 14 to 20 thalers. Immersion lenses cost from 20 to 32 thalers.

Hasert has made his appearance in Eisenach as a microscope-maker. He has produced very strong immersion systems which have been very highly praised by several for oblique illumination.

F. W. Schiek (Halle'sche Strasse No. 14) is the oldest firm in Berlin. Some of his productions, which I have recently seen, were equal to any made at the present time, being very good, at a moderate price. The stands and objectives resemble those of Hartnack in form and designation. Microscopes costing from 38 to 65 thalers may be recommended as being very suitable for students and physicians. Strong objectives, some of them of gigantic strength, and deserving of great praise, have issued from another workshop, that of E. Gundlach, in the same city. I have also seen some beautiful and very recent productions by Bénèche (Belle Alliance Strasse No. 33)

I mention, finally, one other firm, that of Schmidt & Hänsch (Dragoner Strasse No. 19).

Nobert, of Barth, in Pomerania, is to be mentioned as a microscope-maker. Unfortunately I am not acquainted with any of his instruments.

S. Plössl (alte Wieden, Theresianumgasse No. 12) was the first maker in Vienna. Nearly thirty years ago Plössl's microscopes were reckoned among the best which were known. I am unable to give any information regarding his later productions. The present address is Plössl & Co.

The excellent instruments of Amici, of Italy, have obtained great renown. They were the best Continental microscopes from 1840 to 1850, the deceased Amici having at that time acquired the greatest merit in the construction of improved microscopes. I know nothing further of his instruments produced more recently.

The three most renowned London firms are: Powell & Lealand (170 Euston road), Andrew Ross (7 Wigmore street, Cavendish square, W.), continued, since the death of the founder, by the son, Thomas Ross, and Smith, Beck & Beck (6 Coleman street). Among the remainder we will also mention Pillischer (2 New Bond street), W. Highley (70 Dean street, Soho square 10), and Baker (44 High Holborn). It is highly commendable that, for a series of years, the English have endeavored to produce instruments which might be as cheap as possible and, at the same time, good. Thus, for example, a number of establishments furnish very fine instruments even for £5, as Pillischer, Smith, Beck & Beck.

Among the microscope-makers of North America, the most important are Spencer, Tolles, and W. Wales. Zentmayer has recently produced very good stands. The optical performances do not surpass those of our best European instruments; the prices, however, are enormous (H. Hagen).

[The history of the microscope as an American instrument commences at a very recent date.

I am informed by Mr. T. H. McAllister, optician of this city, that in the year 1840, when the United States Exploring Expedition to the South seas, under Commodore Wilkes, was fitting out, it was thought necessary to have a microscope. It was then discovered that none was to be had. The various makers of scientific and philosophical instruments were applied to, but none of them could furnish the expedition with the desired microscope. In this dilemma a private individual was applied to, and an instrument was finally obtained from Dr. Paul Goddard, of Philadelphia. It was a French microscope, which would now be considered very inferior, but was the best instrument then to be had in this country.

Since that time the instrument has come into general use, and in certain departments of the manufacture of microscopes this country has become pre-eminent. Scarcely had the English microscope-makers published those inventions and discoveries which rendered achromatic microscopes really possible, and elevated the instrument from the position of a mere scientific plaything to that of an instrument calculated for the most accurate investigations, before Charles A. Spencer, of this State, succeeded in producing lenses which at once took a front rank among the art productions of the world. Spencer and his pupil Tolles, also Wales, Grunow, Zentmayer, and perhaps a few others, have since that time kept up the reputation of the American lenses, and to-day there is no country in the world in which are produced finer object glasses than those of domestic make.

Previous to Spencer's time, some few microscopes and objectives had been constructed by amateurs, but their authors have never become celebrated in this department.

Spencer was induced, while still a lad, by the perusal of the article on optics in the "Edinburgh Encyclopædia," to construct a compound microscope. His first lens was made when he was about twelve years of age; this first attempt was followed by others, at intervals, during subsequent years. After making several compound microscopes, and a refracting one upon the original plan of Prof. Amici, he constructed several Gregorian and Newtonian telescopes with specula of six and eight inches diameter, some of which were quite successful.

It was not till the publication of the "Penny Magazine" and the "Library of Useful Knowledge," however, that he became aware of the improvements which had been made in Paris and London in the achromatic microscope. The results obtained by Goring and Pritchard in both the achromatic and reflecting microscopes excited his attention especially. The discovery by the former of the effects of angle of aperture was a powerful inducement for Spencer to perfect himself more thoroughly in this branch of optical science. About this time he also learned of the successful researches of Guinaud, Fraunhofer, and Faraday in the manufacture of optical glass. By laborious and protracted experiments, frequently working over the furnace for eighteen consecutive hours, he succeeded in improving the homogeneousness and other qualities of the glass considerably, which enabled him to make an evident advance upon his previous efforts in constructing lenses.

A few instruments were made for personal friends, but it was not till later, about 1847, that Spencer became a professional microscope-maker. About this time he visited New York City, and was introduced by Dr. John Frey to the late Prof. C. R. Gilman. Dr. Gilman had a microscope, constructed by Chevalier, of Paris, which he showed to Spencer and induced him to make one like it. The result was, that Prof. G. sold his Chevalier instrument and replaced it with the one made by Spencer. This instrument was completed in November, 1847. In bringing it to New York, Mr. Spencer stopped at West Point, and showed his microscope to the late Prof. Bailey, then the acknowledged chief of microscopical observers of this country. It was with this instrument that Prof. Bailey resolved the Navicula Spencerii, noticed in the first edition of Quekett on the Microscope. This author, at page 440, in speaking of the N. Spencerii, says: "that an object glass, constructed by a young artist of the name of Spencer, living in the backwoods, had shown three sets of lines on it, when other glasses of equal power, made by the first English opticians, had entirely failed to define them."

The information which Quekett's treatise contained concerning the discoveries of Lister and the labors of Amici and Ross, was extremely useful to our "Yankee backwoodsman." In the account given of Ross's discovery of the effect of thin glass covers upon the correction of an objective, the announcement was made that "on several occasions the enormous angle of 135° had been obtained," and that, "135° is the largest angular pencil that can be passed through a microscopic object glass." This statement, coming from a source so generally considered authoritative, arrested Spencer's attention and led to an immediate theoretical and practical examination of its validity. The supposed theoretical grounds of the assumption not having been found to sustain Mr. Ross's position, conclusive evidence of its incorrectness was speedily obtained by the construction of a $\frac{1}{12}$ in. objective, having an angle of aperture of 146°.

An increase of the angle of aperture of the higher powers had been made from time to time, until the maximum angle of 178°, for the $\frac{1}{12}$, was obtained in June, 1851; and subsequently the medium and lower powers were correspondingly improved. An investigation into the practicability of so far increasing the defining and resolving powers of the objectives of medium focal lengths was made, and results have been obtained which could not, *à priori*, have been expected. The angle of aperture of the $\frac{1}{4}$ has been increased to 175°, and its defining and resolving powers are such that it bears oculars bringing its amplifying powers up to twelve hundred diameters, without any considerable deterioration in the sharpness of its images. With it the 19th band of Nobert's test plate has been resolved by Spencer with ordinary daylight illumination and with artificial light. The residual errors have been the subject of continued investigation since then, and they afford an ample field for the exercise of the highest mental powers and manual skill.

We have now to speak of another Automath, Mr. J. Grunow, who came from Berlin to this country in 1849, and settled in New Haven, Conn. He was induced by Drs. Henry Van Arsdale and C. R. Gilman to study optics and to commence the manufacture of microscopes. Grunow was his own teacher, and had been engaged in an entirely different business previous to his arrival in this country. He constructed his first microscope for Dr. Van Arsdale in 1852, and soon afterwards a second one for Prof. Gilman. About this time Grunow's brother became associated with him and constructed the stands, which were models of good workmanship.

It may be interesting in this connection to remark that Prof. Riddell of this country, the inventor of the binocular microscope, used for his experiments in this direction prisms made by Fitz, of New York. The first binocular microscope, however, was constructed for Prof. R. in 1853, by Grunow.

A very valuable improvement, made by Grunow, consists in letting the rotary diaphragm into the upper surface of the (immovable) stage in such a manner that it is just below the level of the same, and can be rotated without disturbing the slide on which the object is placed. In this way the full optical effect of the diaphragm is obtained exactly at the point where it is needed.

Mr. R. B. Tolles became a pupil of Spencer in 1843. In 1856 he commenced business for himself, and after several years removed to Boston, Mass.

Mr. Tolles is the author of a number of valuable improvements in microscopical accessories; among these his stereoscopic binocular and solid eye-pieces, his method of adjusting for cover by making the front lens stationary and no back lash, as well as his method of making two fronts to an objective—one immersion, and one dry—deserve especial mention. The excellent quality of his objectives has earned him a world-wide reputation.

Mr. J. Zentmayer, of Philadelphia, was introduced to the American microscopic public by Mr. T. H. McAllister. The first instrument which he made was for Dr. Paul Goddard, of that city, in the year 1858. This instrument is now in the possession of Dr. Squibb, the Chemist, of Brooklyn, N. Y., and is almost identical with the present "Grand American Stand."

Zentmayer is the inventor of several valuable improvements in microscopical accessories, which will be mentioned in the price-list at the end of this book. The elegant workmanship of his stands is unsurpassed by those of any other maker.

Mr. Wm. Wales was a pupil of Smith and Beck, in London. He came to this country about 1862. After remaining here for a few months he went back to England, but soon returned to Fort Lee, N. J. Since this time his lenses have constantly improved in quality, and are considered by many competent judges to be equal to, if they do not excel, those of any other maker in the world.

Mr. L. Miller was formerly in the employ of Tolles, but commenced business for himself in 1868. Some of his lenses which I have seen were very good.

In addition to the firms above mentioned, there are three others who also furnish very excellent instruments; namely, Messrs. McAllister, Pike, and Queen.

The microscopes made in this country are generally to supply the home demand, and but few have been exported. Some have found their way to Europe, where they have been critically examined by the French and English makers, and various important improvements, which originated with American makers, have been appropriated by them. For instance, in "Carpenter on the Microscope," London, 1868, will be found, on page 68, a description of a piece of microscopic apparatus, invented by Zentmayer in 1862, but which was copied by a Paris maker, to whom Dr. Carpenter gives the credit of being the inventor.

In speaking of this instrument (Nachet's student's microscope) Dr. Carpenter says:—"The chief peculiarity of this instrument, however, lies in the stage, which the author has no hesitation in pronouncing to be the most perfect of its kind that has yet been devised." The instrument from which Nachet copied the circular stage was made by Zentmayer in 1864 for Dr. W. W. Keen, of Philadelphia, who showed it three different times to M. Nachet, and had it packed by him, in the spring of 1865, for transportation.

The American microscopes are characterized by extreme simplicity, combining all that is necessary for a good working instrument, and rejecting numerous complicated movements and much superfluity of workmanship which some foreign makers seem to consider essential.

The form of stand which has found most favor in this country is the one devised by Mr. G. Jackson, of London. It consists mainly of a stout bar which carries the body, stage, accessory box, and mirror; securing steadiness, equal distribution of tremor, and facilitating the centring of the accessories and achromatic illumination. It will be found, on examination, that modifications of this principle have been applied in nearly all of our American microscopes.

As complete descriptions of the various instruments, objectives, and accessories will be found in the price-lists of the different makers, at the end of this book, it is unnecessary to allude to them in this place.

Microscopes of American manufacture, from their comparative cheapness—to the cost of importation must be added the duty, which is 45 per cent. ad valorem—and the facility with which they can be procured, offer inducements to students and others to procure their instruments at home, and thus save time to themselves, at the same time that they stimulate the manufacturers to make increased efforts to attain even greater excellence.]

Section Fifth.

USE OF THE MICROSCOPE.—MICROSCOPIC EXAMINATION.

Practical directions for learning to work with the microscope may be given pretty rapidly and without difficulty, while it is painfully troublesome for the beginner to acquire this from written instructions. We shall therefore limit ourselves to rendering some of the chief points prominent, and must leave many other things for the microscopist to study out for himself.

Suitable illumination is of great value for microscopic work. As most examinations are made with transmitted light, and the application of natural light is, in this case, to be preferred to any artificial illumination, the selection of a working room is not a matter of indifference. When possible, one should be selected which lies towards the northwest or northeast, and affords an outlook, so that a larger portion of the sky may be used for the reception of the rays of light. In the narrow streets of cities, only the upper stories of the houses can generally be used. It is convenient to have windows on two sides of the room; but those of the side opposite to the windows which are in use should be closed by a dark curtain or shutters.

For ordinary investigations, one may without disadvantage place the instrument on a table standing near the window, and thus make the preparations and examine them on one and the same table. But when the best possible illumination is required, such a position should not be selected for the microscope; the instrument should be placed at a considerable distance, from six to nine feet or more, from the window. A dark shade, which can be placed over the stage by means of a ring fastened to the microscope tube, will shut off all incident light from the object, and essentially improve the image. Such shading of the stage should never be neglected when making observations with polarized light, or resolving very difficult test objects with oblique illumination.

The condition of the sky is of importance for the illumination. A clear blue sky gives a very fine, soft light which does not tire the eye, and is sufficiently bright for all but the very strongest objectives. A dull, white, uniform cloudiness is still more preferable. Bright white clouds, which lie near the sun, should not be selected, on account of their dazzling light. The rapid passing of white clouds over a blue sky, when the atmosphere is strongly agitated, is very unpleasant and troublesome. When the sun shines through the window, a white curtain drawn over it or lowered from a roller is of service.

To illuminate the field, the microscope is turned towards the window, and the mirror is rotated and moved with one hand, while the observer is looking through the instrument. When the best light has thus been found, the object to be examined is placed on the stage of the microscope and the further correction of the field is commenced; for example, lowering the cylindrical diaphragm or slightly altering the position of the mirror, the object being constantly kept in sight. When the mirror is freely movable it is unnecessary to alter the position of the instrument; but the limited movement of the mirror which many of the smallest microscopes permit, often requires a turning and moving of the microscope.

The beginner generally thinks that he can accomplish most with a brightly illuminated field, and thus he works, dazzled by a sea of light, with weeping, rapidly tiring eyes. The experienced observer is accustomed, as a rule, to diminish the intensity of the light considerably. Together with the protection of the organ of vision, it is only in this way that the finest details of the microscopic image can be perceived. The skilful application of the illuminating apparatus and the use of the diaphragm should therefore at once be practised by the beginner as much as possible. When the instrument has a mirror with plane and concave surfaces, the former is used with the weaker objectives and a bright light, the latter with the stronger objectives and a light which is less intense. A very perceptible deficiency is always connected with instruments not having such an arrangement. This may be remedied in a measure, it is true, by turning the microscope, or by holding the hand in certain positions before it.

Considerable practice is requisite with the oblique illumination (fig. 61). The aperture of the stage must be freed from diaphragms, or any other apparatus which may be under the stage, and the various positions of the mirror are to be tried while the eye is looking into the microscope. At the same time, while the mirror is brought up close beneath the stage, the illumination is made as oblique as possible. Truly diabolical illumination is thus sometimes obtained, which, however, shows many fine details in an astonishing manner. When the microscope has a well-centred rotary stage, its rotation is of great importance with this illumination. An observer who is familiar with his instrument and well versed in this department of microscopical technology, will be able to show many things, to the astonishment of the unpractised investigator, which the latter was unable to accomplish after hours of unsuccessful labor. The resolution of the systems of lines of the Pleurosigma angulatum into areolations, with strong objectives, and the exhibition of the markings of the Surirella gemma and Grammatophora subtilissima by means of the strongest immersion lenses, may be designated as specimens of the art of oblique illumination. This is, however, only of minor value for our purposes.

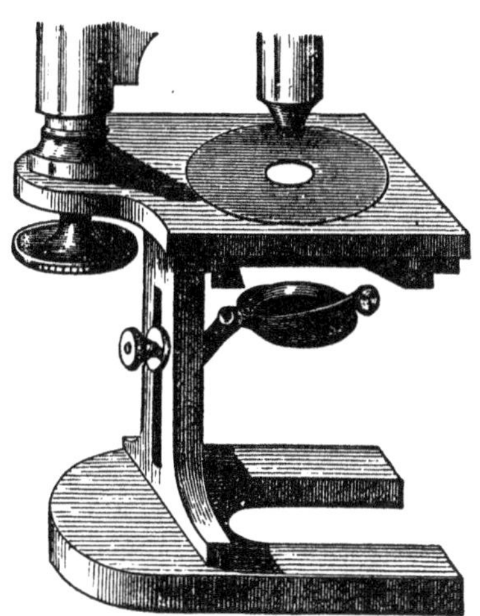

Fig. 61. Oblique position of the mirror on the horse-shoe stand.

No one should make any protracted microscopic investigations by the aid of the artificial light of a lamp or gas flame if he can possibly avoid it, or would spare his eyes. In Northern Europe, during the winter, there are days when the natural light is entirely unserviceable, and, vexed by the miserable light, one finally has recourse to artificial illumination. When it is necessary to resort to artificial light, an ordinary *modérateur*, which should not be too high, an Argand or a petroleum lamp, with a globe of opalescent glass, is worthy of recommendation. A petroleum lamp (fig. 62) provided with a large condensing lens, recently constructed by Hartnack, is very convenient; it should also have a shade. Properly constructed gas lamps may also be employed with advantage. A number of these, with very judicious arrangements, have been invented and recommended by English microscopists.

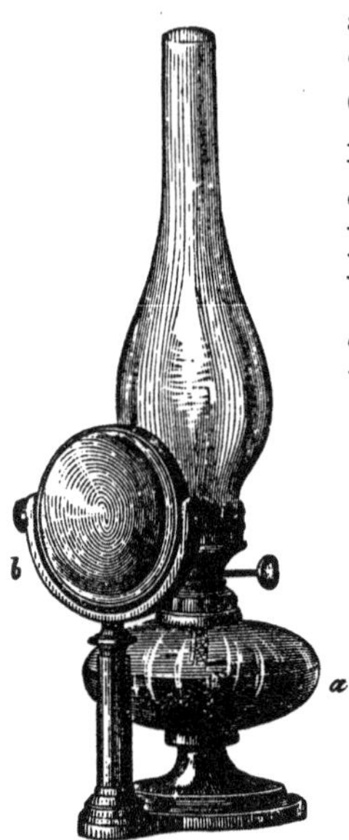

Fig. 62. Hartnack's microscope lamp.

A proper moderation of the light is here urgently necessary. The illumination may be essentially improved by placing a cobalt-blue glass of greater or lesser intensity between the lamp flame and the object. It may be placed on the mirror or, better, on the stage. A black pasteboard screen with apertures of various sizes, which may be placed in front of the microscope parallel with the mirror, and on which the blue glass may be fastened with wax, forms a cheap accompaniment of the large Oberhäuser-Hartnack microscope, and deserves to be highly recommended for its important action. A rotary diaphragm should be placed behind the screen. In place of the above blue glasses of various sorts, which can be shoved into a metallic ring, may be inserted into the stage, as necessity may require. Such an arrangement may be readily applied to all of the larger stands.

[A very ingenious arrangement of the illumination has been contrived by my friend Dr. Edward Curtis, of this city.

A small lamp, similar to the one represented in fig. 62, is placed in a cigar-box, which stands on one of its ends. On one side of the box is cut a small aperture, in which is placed a piece of blue glass, to soften the light as it passes to the microscope mirror. Another larger opening is made in the front of the box and is occupied by three different glasses. The one nearest the lamp is a square piece of ground glass; the next one is also square and flat, but colored blue. Finally, a plano-convex glass lens of long focus is placed at such an inclination as to condense the rays of light, thus softened, on to the work-table for use in dissecting or arranging preparations.]

Although direct sun and lamp light is to be entirely rejected for ordinary investigations, these most intense of all illuminating methods must, on the contrary, be selected for many investigations with polarized light. Opaque objects require illumination by incident light with seclusion from transmitted rays. Ordinary daylight is sufficient for very weak powers; with stronger ones, more intense illumination is necessary. Sunlight may be used in certain cases. Numerous contrivances are in use for concentrating the light on the object. A plano-convex lens of large focus (fig. 21), which is placed before the instrument, is generally sufficient; this may also be accomplished with a glass prism. Lieberkühn's illuminating apparatus also deserves mention as a good and very suitable contrivance, although it is rarely employed in medical investigations.

The object to be examined will have to undergo a preliminary preparation, if it be not already a permanent preparation. This process, which naturally varies considerably according to circumstances, generally rendering the examination possible with transmitted light, however, we shall soon treat more in detail. Let the remark here suffice, that the preparation is to be made with care and the observance of the greatest cleanliness; and then, on the other hand, we would say, not to do too much of a good thing, that is, not to select too large pieces for examination. Beginners fail in this very generally, and place under the microscope masses which, divided, would have furnished a dozen serviceable preparations. One rarely examines with incident light alone, in which case the object can be placed uncovered and dry on the stage of the microscope. As a rule, it is necessary to moisten the preparation (with water, preserving fluids, glycerine, etc., see below). With weak powers the object may still remain uncovered, and, in fact, many things are thus examined, although the preparation is generally placed in a watch-glass, a glass box, or a cell, instead of on a simple slide.

If, however, one has recourse to higher powers, it will be necessary to cover the object with a plate of glass. This should be thin, and as clean as possible. All fluids must be prevented from running over its free surface, as the image becomes somewhat dim and indistinct with ordinary objectives, although, as previously mentioned, with the new immersion systems a drop of water must be placed on the upper surface of the covering glass. In the application of the covering glass, all contact of its surfaces with the fingers is to be avoided; held by its sides, it is to be laid over the object. Some caution is necessary with very delicate objects; for example, a primitive mammalial ovum might be crushed by covering it awkwardly; the elements of the fresh retina might have their connection destroyed, etc. Simple contrivances serve for the protection of such preparations; a piece of hair or bristle, or the fragment of a thin film of glass, may be placed between the slide and the covering glass.

The adjustment is made, while looking through the microscope, by sinking the tube. This is done either by simply shoving it down through its sheath with the hand, or, when there is a coarse screw, by moving it downwards with the latter. In doing this, thrusting the lens against the preparation is to be avoided, because the latter may be spoiled and its covering glass broken, and in certain cases the lens may also be injured. It is well for beginners to make this movement in the contrary direction, that is, elevating instead of depressing the tube, which is to be so adjusted that there is only a small space between the covering glass and the lens, and then moved upwards. Accurate adjustment requires some practice, and is not very easy with strong objectives. That the correct adjustment has been made, is shown when the contours of the object are sharpest and finest. Here the fine adjusting screw comes in play.

The preparation is to be first examined with low powers and transmitted central light, gradually passing to higher powers; very weak eye-pieces being constantly employed. In some cases the tube of the microscope may be shortened with advantage.

[The changing of the objectives is facilitated by the employment of a "nose-piece." This is an apparatus having two or more arms capable of revolving and carrying the objectives at their peripheral extremities. The mechanism is to be screwed on to the lower end of the microscope tube, the same as a simple objective. The various objectives are brought into position successively by simply turning the arms. Further movement and accurate centring is controlled by means of a catch. The original "nose-piece," invented by Brooke, of London, revolved on a horizontal plane, but by a more recent improvement the objectives which are not in use are elevated obliquely, and only become vertical at the moment they are adjusted to the axis of the microscope tube, so that they do not in any manner interfere with the manipulation of the stage. This accessory is almost indispensable for those who work much with the microscope.]

It is a general mistake of beginners, who under-estimate the value of low powers, to use high powers at the very commencement of the examination. As, however, only the weak objectives afford a somewhat extended field of vision, while with stronger lenses it is extremely small, it results that the employment of weak combinations is of great importance for the simultaneous survey of the whole, as well as to give the observer the first ideas as to the relation of its several parts.

One then gradually passes to the employment of stronger objectives; at first, always with very weak eye-pieces. When working with cylindrical diaphragms it is necessary to vary them, exchanging those with large apertures for those with smaller ones, likewise occasionally exchanging the plane mirror for the concave, and in all cases adjusting as accurately as possible with the micrometer screw.

When the observer has found it necessary to make use of his higher powers, he may proceed to employ somewhat stronger eye-pieces; but he should be sparing in their use. One is soon convinced that less is obtained with them, which results from their optical nature, than would be at first believed. The image is larger, whereby some points may at first appear more distinct. An enlargement is soon arrived at, however, which does not show any more, but rather less, than the weaker of the previously employed eye-pieces, the brightness of the field and the sharpness of the image having considerably diminished. Very strong eye-pieces, which are added to the larger instruments as an optical supplement, are articles of luxury, and are scarcely of any use.

Objectives which are well made as to their optical portions bear stronger eye-pieces than those which are less fortunate in their construction. Nevertheless, even in this case, one should be careful of forcing the magnifying power by means of the eye-piece. The latter, it is true, might be still more improved, and it is to be wished that capable opticians might turn their attention to this subject. The orthoscopic eye-pieces which, so far as I am aware, were first constructed and sold by the unfortunately so early deceased Kellner, of Wetzlar, give a very flat image, but have shown me nothing further, even in their stronger numbers.

It follows from what has just been said, that he who can obtain about the same magnifying power in a double manner, that is, by means of a weak objective and a strong eye-piece, or by means of a strong objective and a weak eye-piece, should always have recourse to the latter. The effort of the older opticians to combine weaker objectives with relatively stronger eye-pieces cannot, therefore,—we repeat it,—be approved of, and is at present being more and more abandoned.

The objects of histological and medical investigations will seldom require the application of oblique illumination. If it be desired to learn the effects of the latter, the directions given above are to be followed.

When reagents are to be used, it is customary, as a rule, to add a drop of them to the preparation by means of a pointed glass rod, either by removing and replacing the covering glass, or by placing the drop at its edge, so that it may flow under the cover and unite at this point with the fluid in which the specimen is mounted. A gradual streaming in of the reagent may be obtained by means of a thread of lint which lies half under the glass cover, half free on the slide where it receives the drop.

For the protection of the instrument, it is necessary to observe due caution with reagents, especially when using strong acids, alkalies, and all substances which attack the lead of the flint glass. Concentrated muriatic and nitric acids are to be avoided as much as possible, and care should be used with volatile acids and ammonia. Sulphuretted hydrogen should never be used. All of these reagents require the use of the largest possible covering glasses. If a lens should unfortunately become moistened with the reagent, it must be immediately dipped into distilled water. Chemical processes which develop vapors should in no case be undertaken in the microscopic work-room. The destructive effects of such influences are best shown by the unfortunate condition in which the microscopes of chemical laboratories are usually found.

The repeated packing and unpacking of the microscope is too troublesome for those who use it daily, and not at all beneficial to the mechanism of the stand. It is therefore preferable to place the instrument on a thick piece of cloth, on the work-table, and under a bell-glass or a glass-case, which affords sufficient protection from dust. The eye-pieces, the objectives shut up in their case, and such other things as are daily used may be kept under a second smaller bell-glass. It is advisable to heat the room during the winter, to prevent dampness.

The instrument should be re-examined each time that it is used, especially by the beginner, before being replaced under the glass case. Stains are to be removed from the brass work with a linen rag; dust which has settled on the mirror, eye-piece, etc., by means of a fine camel's-hair brush. Although these procedures consume some time, they are still of great value for the protection of the instrument and the preservation of its original power of performance, especially if the objectives are also examined each time they are used.

The objectives are best cleaned, after removing the dust with a camel's-hair pencil, by means of a piece of fine linen rendered soft by frequent washing. Fine leather or elder pith may also be used. Some stains are to be removed with distilled water; others, as glycerine for example, require a cloth freshly moistened with alcohol. A larger quantity of alcohol is to be avoided, as the fluid might possibly get into the setting of the lens and reach the Canada balsam which cements the crown and flint glasses together.

This wetting of the lenses rarely happens to the more expert. It is obvious, that where reagents are being used it is to be particularly avoided, and the greatest caution is therefore requisite. The objectives used should be as weak as possible with long foci, and, when much of this kind of work is to be done, the stage is to be covered with a glass plate, which latter may be fastened with clamps, when there are any on the stage. Broad slides for the specimens also afford some protection.

Notwithstanding every precaution, the optical portions of the instrument require cleaning, after a time, in consequence of a fatty coating which settles on the objective and eye-piece, and renders the image quite dim. Instruments which have been used for years almost always show this coating. One should not be too anxious about such a cleaning process, as by using a good brush and fine linen the glasses of the microscope do not suffer in the least.

The microscopist's work-table should be large and massive, so that it may stand sufficiently firm. A hard-wood board, at one or both sides of which small slabs of slate may be inserted for objects which require preparatory manipulation to rest upon, is most to be recommended as a table-top.

A series of drawers is a valuable addition to the table. A number of smaller accessory apparatuses which are necessary to the microscopist may be kept in these, and are best preserved in this way from dust and other contaminating influences.

In these are kept the slides, the various sorts of glass covers, glass vessels, drawing arrangements, accessory apparatus for the microscope, the linen rags for cleaning, etc.

A few bell-glasses and glass cases are necessary on the work-table to protect things which have been temporarily set aside from dust.

Reagents are to be removed from the table after being used, and kept in another place.

The question as to what corporeal and psychical qualities the microscopist should possess, is discussed with great profoundness in many works. We think that it may be here omitted. Acute mental organs, calmness, love of truth and talent for combination are qualities which the physician and the naturalist should always possess. He who has not these, whose perceptive faculties are clouded and the impartiality of whose observations are constantly disturbed by a lively, excited imagination, should keep away from the microscope as well as from the profession of medicine.

For microscopical observation and work it is necessary to have visual organs capable of moderate endurance. Somewhat short-sighted, light eyes are generally the best adapted. He who is so fortunate as to possess two equally good eyes, should accustom himself to employ them alternately. Every microscopist who uses one eye for a long time continuously in looking into the microscope while the other eye, though remaining open, is unemployed, knows how much the acuteness of the one has increased while the passive eye has acquired a certain irritability, so that when the latter is used in order to relieve the other, the field of vision appears much brighter and weariness soon makes its appearance. Where one eye is evidently weaker than the other, the microscopical work naturally falls to the latter. One should accustom one's self from the beginning, while looking with one eye into the instrument, to keep the other open also. The attention is concentrated so predominantly in the active organ that the observer is no longer conscious of the impressions made on that which is unemployed.

For the protection of the visual powers, one should not work too continuously, avoiding the earlier morning hours, as well as the time immediately after dinner. Leave off as soon as fatigue commences. For beginners especially, whose eyes often become rapidly tired from the unusual nature of the visual act, this is advisable, until later when long practice has accustomed them to more continuous labor.

Whether to stand or sit while working, is to be determined by one's previous habits. Bending the head down over vertical microscopes usually causes but little inconvenience. English microscopists, however, as a rule lay great weight on the oblique or horizontal position of the tube and of the whole instrument, to prevent the neck from becoming tired, or a flow of blood to the head, so that not only their large, but also quite simple microscopes have such an arrangement. But, according to our

Continental notions, the inconvenience of an oblique or vertical position of the stage is too great, when more than the examination of tests is concerned; this arrangement has not, therefore, become very generally adopted.

Very important for the protection of the eyes is the previously mentioned judicious shading of the field, and the skilful application of the diaphragm (fig. 22, page 25).

The gift of seeing and observing with the microscope is, like all human abilities, unequal, greater with one person, less with another; but with a little perseverance it may be acquired to a sufficient degree by most persons.

The peculiar nature of the microscopic images causes some difficulties for every commencing observer. The compound microscope shows us only that stratum of the object which lies directly in the focus, and everything else, which lies in other planes, either not at all or only indistinctly. At the same time, in the usual manner of examination, the whole specimen is transparent, illuminated from beneath and not from above, as in ordinary vision. Things which lie in other planes, higher or lower, only appear after altering the focus, and this condition is much more appreciable with strong objectives of a high angle of aperture than with weaker objectives of a lower angle of aperture. Hence it follows, that we are able to recognize immediately the outline of an object, and the relation of length and breadth, but not its thickness or its entire form. We are able to obtain these only by a combination of the various microscopic images received by varying the focal adjustment. Here the beginner frequently meets with considerable difficulties, and errors may arise from improperly combining the images. In this kind of vision we are deprived of the aid which enables us, in ordinary vision, to judge rapidly of the shape of the object. The form of a microscopic object, regarded with incident light, is, for this reason, generally more readily comprehended. The appreciation of the form of a blood-cell is not difficult for those who are somewhat practised, but the contrary is the case in ascertaining the polyangular shape of many diatomaceæ, or the form of a cavity in an organic part. The comparison of a number of sections made in horizontal, vertical, and oblique directions, a means resorted to by botanists especially, is here, when practicable, of great value.

The estimation of the form is difficult in still another manner, namely, in consequence of the extraordinary diminutiveness of an object. With a little practice it is not difficult to recognize the relations of relief in a microscopic object, for example, to distinguish a somewhat larger concave surface from a convex one, if only by means of a combination of various images. When such surfaces are extremely small, as is the case, for example, with the delicate areolations of the so frequently employed test object, the Pleurosigma angulatum, the discrimination is very difficult. Thus, as has been previously remarked, these last-mentioned areolæ have been declared by some excellent observers to be convex, by others to be excavated, and the matter has not yet been definitely decided.

Welcker has given us a good means for discriminating between convex and concave bodies. The former act as convex lenses, the latter as concave. When we start with the tube in a medium position, a convex body will appear lighter by raising the microscope tube, a concave by lowering the tube ; a globular structure and a hollow sphere, a ridge and a furrow, may thus be discriminated.

It is much easier to recognize the shape of microscopic objects by means of weaker objectives than by the employment of very strong combinations with large angles of aperture, so that herein also lies a weighty argument in favor of the former. Although the practised microscopist may accomplish his purpose with very strong objectives, still it is frequently desirable to have a well-constructed medium power, with the smaller angle of former days, added to one's instrument. The English opticians have sought relief in this direction by adding a diaphragm to the lenses with large angles of aperture.

One soon learns to appreciate the foreign substances which contaminate the microscopic image, much of which may be avoided by neatness and carefulness in making the preparation. One should make one's self familiar, and as soon as possible, with the appearance of air bubbles, oil globules, starch granules, fibres of linen and cotton, etc.

It is also important to compare the image which an object presents by transmitted light with that which it affords by incident light. Among other things, the appearance of one and the same object in media of various refracting powers is also to be studied.

The optical portion of microscopic work is much more difficult to acquire than the manual, such as the cautious use of the adjusting screws and the mirror, and the steady and not jerking movement of the object through the field. In this place the important principle should be impressed, that movements which can be securely accomplished by the human hand are to be left to it, and are not to be executed by screws and other mechanical contrivances. Every experienced microscopist will regard the massive accessory apparatus of the large English microscopes as being somewhat superfluous and inconvenient.

The inversion of the image by the compound microscope causes some difficulty for the beginner. One soon becomes accustomed to this, however, and finally to such a degree that it is no longer noticed, and one is only reminded of it when using an erector (where the inverted image is again inverted by means of a lens placed within the tube of the microscope, or by a prism on the eye-piece). As this inversion is attended with optical disadvantages, such instruments have not been extensively adopted, and are only convenient for microscopic dissections when furnished with weak lenses.

Finally, one word more is necessary concerning the phenomena of movement visible under the microscope. Not everything which is here seen in motion is, for that reason, to be pronounced vital.

Currents sometimes occur in water, with which one should become familiar in order to guard against error in other cases. For instance, if alcohol is mixed with water, the small bodies suspended in it acquire a rapid motion, which continues until both fluids are equalized, that is, have become thoroughly blended.

Then very small particles of substances which are insoluble in water present an uninterrupted dancing motion, the cause of which is still unexplained, but is, at all events, a purely physical phenomenon. This movement is called the "Brunonian molecular motion."

Finely powdered charcoal, small crystals, the granules of a coloring material exhibit the same peculiar dancing movement as the molecules of fat and melanine taken from the animal body. In certain cases we can observe the same motion in the fluid contents of cells as are taking place in the surrounding fluids.

On the vertebral column of the frog, at the points of exit of the spinal nerves, lie small white collections of columnar-shaped crystals of the carbonate of lime. These, deposited in a drop of water, present one of the finest examples for the study of the molecular movement. Larger crystals, of about $\frac{1}{150}$ to $\frac{1}{200}'''$, lie perfectly quiet, so long as there is no current in the fluid. Those of half this size are seldom found to have the dancing motion. The smaller the columns are, the more generally the movement is to be met with, and the smallest, of $\frac{1}{1000}'''$ and smaller, on which we are no longer able to recognize the columnar form, are engaged in a continual, restless motion.

The examination of the molecular movement is instructive for the beginner in still another regard. One readily forgets how much the excursions of a moving object are enlarged by the optical apparatus of the microscope. The dancing of a small molecule will appear slight to the eye with 200-fold enlargement, but very energetic, on the contrary, with an enlargement of 1000–1500 diameters.

This is repeated in the vital movements which are seen with the instrument. An animalcule which we examine with very strong lenses shoots quickly through the field, while with the lowest powers it does not swim with any considerable degree of rapidity through the water. When the circulation in the web of the frog's foot or in the tail of its larva is examined with high powers, the blood-corpuscles hasten through the capillary passages, while in reality the current in the capillary vessels is quite slow.

There is still another consideration which is not to be disregarded in observing the phenomenon of microscopic motion. When a series of movements follow each other with great rapidity, we may readily recognize the total movement, but not the single motions; these are separately appreciable to the eye only when the entire phenomenon is retarded. We shall become acquainted with an example of this, the ciliary movement, in a later section.

Finally, we have to mention still another series of movement phenomena which has recently attracted the attention of investigators more and more,—we refer to the changes in shape of the living animal cell.

Isolated examples of this marvellous change of shape, especially in the bodies of the lower animals, were known long ago. At present it is known that the young animal cell, so long as the cell body still consists of the original substance, the so-called protoplasma, is endowed in the highest organisms also with a capacity for independent vital contraction. Numerous cells of the normal superstructure, likewise pathological new formations —so long as they possess the character of youthfulness—present the changes mentioned. Such cells have been seen to pass out (after the manner of the amœbæ) through the walls of the capillaries (A. Waller, Cohnheim), to wander through the lining tissue, and to take up into their contractile cell-like bodies small particles, such as molecules of indigo, anilin, cinnabar, and carmine, the finest milk globules, and even extravasated colored blood-corpuscles; so that the view here opens into a new world of minute actions, and it has already furnished extremely important information, to which we shall again refer.

If in any microscopic investigations the most conservative preparation is requisite, it is just here.

In order not to kill the cell prematurely, one must employ a truly indifferent fluid medium. Whoever proceeds to make such investigations with the old idea of possessing such indifferent fluids in solutions of sugar and salt, albumen and water, or humor vitreus, will soon become convinced of the contrary. In general only those fluids which surround the cell in the body can be called truly indifferent. In many cases the indication may be fulfilled with iodine-serum (see below), or a similar composition. The greatest caution is necessary to prevent pressure and evaporation. The very thin covering glass is to be supported by placing beneath it the fragments of one of its predecessors, which are usually not very rare with the microscopist, or —what is for many cases still better—the covering glass is entirely omitted.

Recklinghausen has invented a very efficient little apparatus for preventing the evaporation of the fluids. This, the "moist chamber," the reader will readily comprehend by glancing at fig. 63. The object is placed on the somewhat broad ground slide (*d*) in the usual manner. The glass ring (*a*), with its under surface likewise ground off, rests on the slide at some distance from the object. This ring may, in certain cases, be made higher. A tube (*b*) of thin rubber is fastened as firmly as possible about the ring. The upper end (*c*) of the tube is fastened round the neck or tube of the microscope with a small rubber band. In order to keep the space thus enclosed saturated with moisture, two strips of elder pith or bibulous paper, saturated with fluid, are to be placed at the inner surface of the glass ring, and the external surface of the lower border of the ring is to be surrounded with several little pads of moist blotting paper.

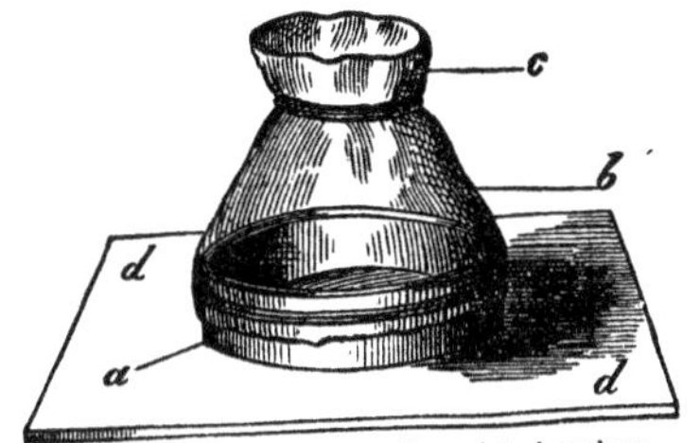

Fig. 63. Recklinghausen's moist chamber.

A moist chamber may be made in still another, more simplified manner (fig. 64). A glass ring (*b*), a few millimetres in

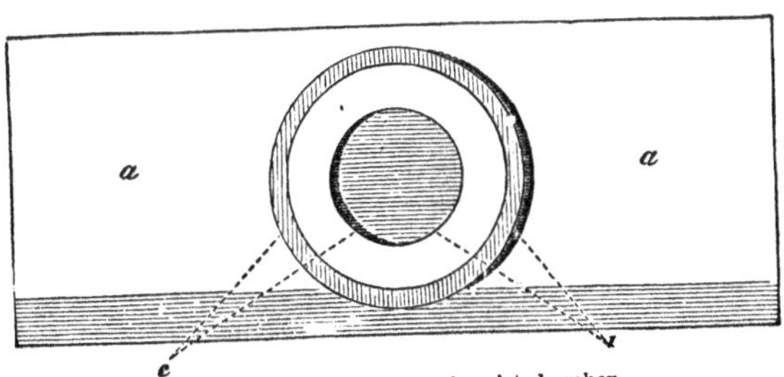

Fig. 64. A more simplified moist chamber.

height, is to be cemented on to an object slide (*a*). A few drops of water are to be cautiously placed at the inner edge of the former with a brush. The object is to be placed on a circular covering glass (*c*) and the latter then turned over and placed on the ring; in this way all pressure is necessarily avoided.

One may thus—with the aid of an immersion objective—follow the movement of these cells for hours, and even days.

The last-mentioned simple apparatus may be readily converted into a gas-chamber (fig. 65). The thick glass plate shows a

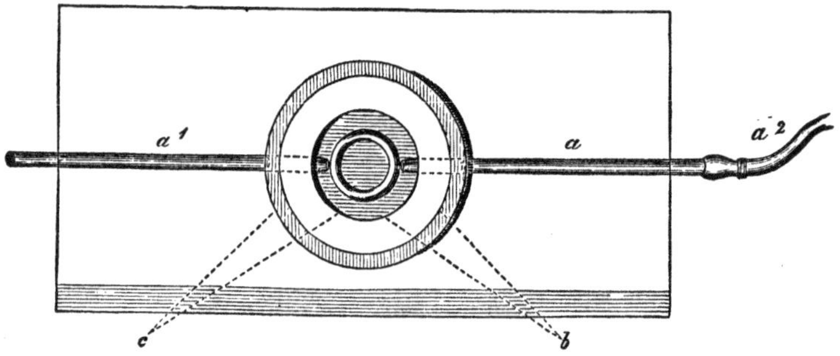

Fig. 65. Stricker's gas-chamber.

ring ground out at the bottom of the chamber. Two glass tubes are cemented into the two half canals. One of these tubes (*a*), connected with the caoutchouc tube a^2 serves for the entrance of the gas, the other (a^1) for its exit. The covering glass may be more securely adapted to the glass ring with a cement.

Although we can in this manner, at the ordinary temperature of the room, study the cell life of a cold-blooded vertebrate animal, for example, in the connective tissue, cornea, blood, and lymph of a frog, we cannot, with the same success, study those from the body of a warm-blooded animal. The movement is too rapidly retarded by the surrounding coldness. A condition as to temperature resembling that of the living organism must be attained for the success of the observation. The older microscopists helped themselves in this dilemma, so far as it was possible to do so, by using warmed slides. Beale afterwards constructed a warmable stage, but of rather crude form. Quite recently a celebrated investigator, M. Schultze, has rendered a great service by producing an apparatus of this kind which fulfils the indications more completely.

Schultze's apparatus* is represented by our fig. 66. A brass plate A, which is notched (*c*) posteriorly, so as to fit on to the support of the microscope, is to be fastened on to the microscope stage with clamps. It is perforated at *a* for the illumination, and at its front part, in the middle, the thermometer (*d*) is placed slantingly; at the corners are the two arms *b*. Two small spirit lamps under the latter supply the heat. The lower extremity of the thermometer is enclosed in the brass case B*a*, which has

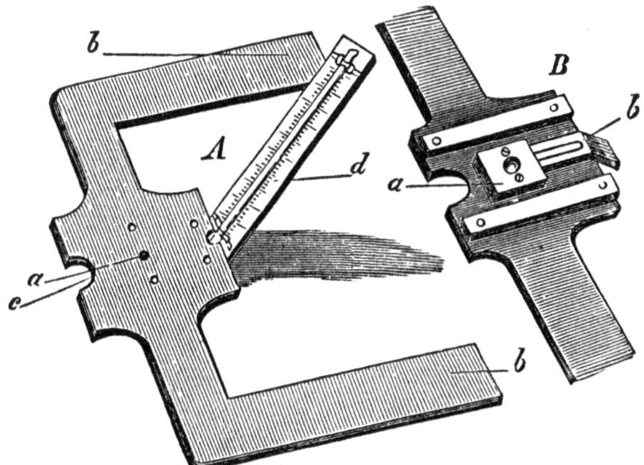

Fig. 66. Schultze's warmable stage.

two somewhat thicker wooden ledges at its sides. The thermometer winds round the aperture in the stage, passes uncovered and horizontally, for a short distance, on its under surface, and then bends to pass through the opening *b* to arrive at the face of the graduated metal plate. It has been ascertained by experiment that the actual temperature of the object is indicated by the thermometer.

It is scarcely necessary to remark, that it is most advantageous to use immersion lenses and the moist chamber with the hot stage.

Unfortunately, this apparatus has an unpleasant defect, as Engelmann has shown. The temperature of the object is occa-

* It may be obtained of the mechanician Giessler, in Bonn, for nine thalers, Prussian.

sionally reduced very considerably by the metallic setting of the lens and the microscope tube, so that, in this case, the focal distance of the objective exerts a marked influence. The insertion of a bad conductor of heat between the lens and the microscope tube has been proposed. An ivory tube, 30 mm. in height, applied in this manner, lessens this defect very materially.

Various arrangements have been contrived for the purpose of conducting electrical currents through an object under the microscope. We introduce, as an example, the simple one of Harting (fig. 67).

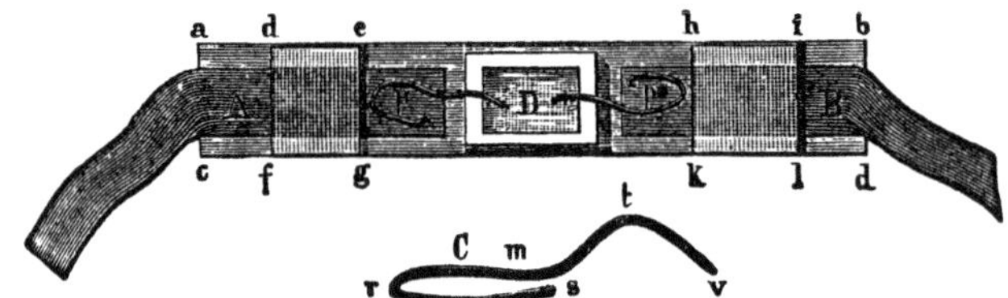

Fig. 67. Harting's electrical apparatus.

Two somewhat narrow strips of tin-foil A B are fastened with starch paste on to a glass slide *a b c d;* a portion of the tin-foil projects beyond the ends of the slide, so as to connect with the conducting wires of the galvanic apparatus. The central portion of the slide remains free. The two glass plates *d e f g* and *h i k l* are to be cemented over these strips of tin-foil with marine glue or a mixture of pitch and rosin, for the stage clamps to rest on. The two polar wires *p* and *p* (for which platinum is the best material) are not fastened; they receive the curve shown in the figure at C. The part *m r s* rests on the tin-foil, the other curved portion *m t v* (to which may be given any curve desired) dips its point into the examining fluid, which in our drawing is surrounded by a cell D. Harting's apparatus may be readily modified.

Section Sixth.

THE PREPARATION OF MICROSCOPIC OBJECTS.

WITH the exception of the finished preparations of a collection, in most cases the objects to be examined require preparation, which, as we have already remarked, should be as careful and cleanly as possible. It is only when the blood, mucus, pathological fluids, etc., are examined, that the mere spreading out of a drop of the same is sufficient.

The object slides, which are simply strips of glass, serve for the reception of the object to be examined. Several dozen of them should be kept on hand in a clean condition, protected from dust in an accurately closing box. Good slides should be made of pure glass, preferably without any color, and the edges should be ground, for the protection of the instrument. Too great thickness of the glass renders the use of the stronger objectives and the cylindrical diaphragms, which then become necessary, inconvenient. Therefore, the thickness of the slides should not exceed $\frac{1}{3}$–$\frac{1}{2}'''$. The most convenient form is that of a long square (3 inches by 1 inch), but when the stage is narrow they should be of a corresponding width. Square slides are less suitable. One should also accustom one's self to place the object to be examined in the centre of the slide. It is rarely examined in the dry condition, but, as a rule, with the addition of some fluid; as, water, glycerine, etc. This is to be added at the commencement of the preparation. One soon learns to judge of the quantity which is necessary.

The object to be examined, if it is large, and especially if it is thick,—as for example, when one desires to examine a small embryo or an injected specimen of considerable size,—is to be placed with some fluid in a watch-glass under the microscope.

Small quadratic glass boxes, about an inch or an inch and a half in size and two or three lines in depth, are more convenient for this purpose. Glass cells, as made by the English (see further on, at the preparation of microscopic objects), may also be employed with advantage. Thick slides with an excavation in the centre are less suitable.

Fig. 68. Glass box.

The preparation is seldom examined uncovered; such a method of examination is confined almost entirely to the cases last mentioned. The covering glasses or covering scales, which have been so frequently alluded to, serve for a covering. Pieces of pretty thick glass were formerly used with low powers; at present these have gone out of use, since thin and even very thin glass may be obtained from England at slight cost.

As we have seen in an earlier section, the thickness of the covering glass exerts considerable influence on the optical performance of the stronger objectives. It is well, therefore, to have these glasses arranged in a series of various thicknesses, which are kept in separately designated boxes. It is necessary to have them from $\frac{1}{5}$–$\frac{1}{6}'''$, to those of $\frac{1}{10}$–$\frac{1}{15}'''$ in thickness, according to the objectives with which they are to be used. Even the pressure of this thin glass is occasionally too great for very delicate objects, if one desires to avoid crushing or splitting them. In such cases it is necessary to insert a harder substance between the slide and the cover, a precautionary measure which has already been alluded to on a preceding page. Thicker covers may be cut from thin plate glass.

A few special instruments are requisite for making preparations. But one should not think that they are indispensable. In practised hands the same and even more may be accomplished, in a shorter time, with simple tools than with complicated ones. A number of microscopic knives, small forceps, and scissors have been invented, it is true, but they are generally used by their inventors only, and are, as a rule, quite worthless trash.

First of all, one should have several fine forceps terminating in thin points for seizing objects. Such should be selected as have light springs, and not the stiff ones which many anatomists are accustomed to use. The points should be either quite smooth or only slightly grooved. A hooked point is unsuitable. Many objects, especially those of a delicate nature, are more conveniently moved with a camel's-hair brush.

The scissors are most frequently used for dissecting. A fine pair of so-called eye scissors is indispensable. A small pair with curved blades is very convenient for many purposes; here and there, a pair of fine elbow scissors also renders good service.

A few small knives, though useful, are of relatively inferior value. Several very fine scalpels with narrow-pointed blades, when possible, of somewhat strongly hardened steel, are more serviceable. The ordinary anatomical scalpels are much too clumsy, and, as a rule, are made from too soft steel to be useful for the microscopist.

When a still finer cutting instrument is necessary, the ordinary cataract needle is to be used. It is also extremely useful for moving small objects.

The tearing of microscopic objects is frequently necessary in histological investigations. This may be accomplished with very finely pointed steel needles, of medium length, let into wooden handles. When this picking process is necessary it should, in consequence of the minuteness of the elements of the human body, always be performed with accuracy; devoting the few minutes required, as one will be rewarded for the little pains, by a good preparation. Beginners very frequently fail in this. They stop picking too soon on a preparation which was too large at the commencement.

Not unfrequently, in such cases, the work is so fine that one must have recourse to magnifying glasses, to the loup or the simple microscope. A very great inconvenience is connected with the latter when used with stronger lenses; the shortness of the focus soon renders needle-work impossible. Zeiss deserves credit, therefore, for having produced the useful microscope represented in fig. 69. It has, on a short tube, an objective consisting of three lenses, and has a concave lens as an eye-piece. It permits the use of needles, even with 150–200 fold enlargement.

Fig. 69. Zeiss' new dissecting microscope.

[A very economical and efficient substitute for the simple microscope has been devised by Dr. Curtis. It consists simply of a binocular ophthalmoscope, the mirror of which is replaced by a biconvex lens of one or two inches focus. The whole rests over a small aperture in a little wooden box, the front and part of the sides of which are removed for convenience of manipulation with the preparation which is placed on a support in the box. As will be readily appreciated, this gives an upright, stereoscopic image, and a considerable magnifying power.]

It is often found necessary to make very thin sections from

fresh, and especially from artificially hardened tissues. Knives with double blades running parallel, and close to each other, have been used for this purpose. The double knife invented by Professor Valentin has become the best known. It is not

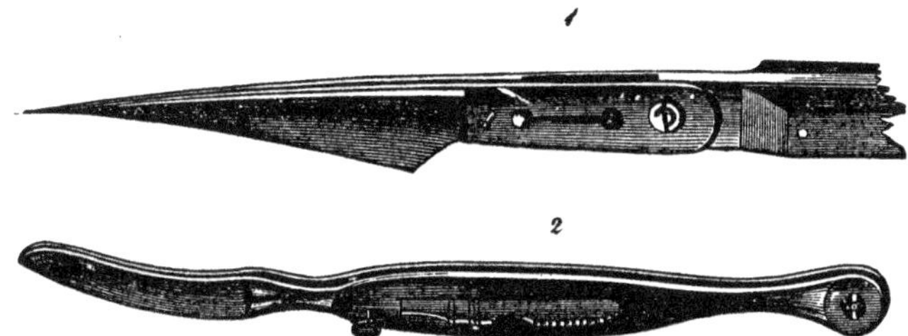

Fig. 70. Double knives. 1 that of Valentin; 2 the improved English instrument.

easy to produce a good instrument, such as is represented by our fig. 70 at 1, and when not well made it is entirely unserviceable. This instrument has received a judicious improvement in the hands of English cutlers. We see such an improved form of the double knife represented in the same cut, at fig. 2.

It is much more desirable to make thin sections, with the free hand, by means of a good razor. Any one who has such an one at his command, and has acquired the necessary dexterity, will discard the double knife. Good English razors, of light construction, with small blades, are the most suitable. For many purposes, it is well to have them ground flat; but it is preferable to have the blade ground hollow for very thin and fine sections. To preserve it in a proper condition, it is necessary that it should be well sharpened, and the strap should be frequently used. The blade, as well as the preparation to be cut, must be well moistened, for when they are dry a good section can never be made. The fine section is best removed from the wet blade by means of a brush; it is then to be carefully and cautiously spread out on the slide. For making very large sections with sufficient delicacy, however, the razor is unserviceable, in consequence of the thickness of the back of its blade. In such cases, according to Thiersch, another instrument may be advantageously employed; it consists of a knife-blade of the thickness of paper (about 1 cm. in breadth by 20 cm. in lengt'), which is to be stretched in a watchmaker's ordinary saw-bow.

[The best razors which I have seen for anatomical work are those made by Le Coultre, in Switzerland. The blades are as thin as paper, and are made of good material. The price varies according to the width and number of the blades. Those with one wide blade cost $1.75, gold; with two wide blades, $2.50. They may be purchased of the agents, Taylor, Olmstead & Taylor, No. 5 Bond St., N. Y.]

Henson has also invented an apparatus, the "section cutter," to be used under the microscope, for making very fine sections. I have no personal experience in its use, and quite as little of an instrument recently made by His, of Basel, and recommended by others. Such apparatuses are certainly convenient in certain cases; as a rule, the same end may be accomplished by a practised hand.

[A very efficient "section cutter" has been contrived by Dr.

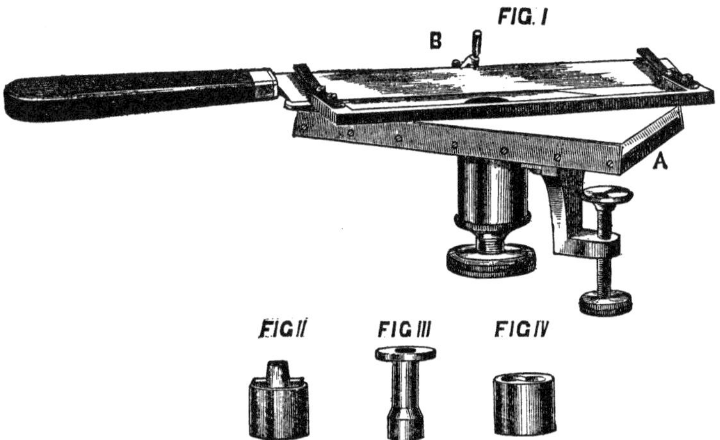

Fig. 70*. Curtis' "section cutter."

Edward Curtis, of this city, to whom I am indebted for the accompanying wood-cut.

The following description of the apparatus, and the manner of using it, is condensed from an article presented by Dr. C. at the annual meeting of the American Ophthalmological Society, held at Newport, July, 1871.

The apparatus is shown in fig. 70*. "It consists of the part A (fig. I.), for holding the object to be cut, modelled after a form of section cutter in common use; and of the cutting part B, after my own device, composed of a long straight knife held in a frame. The holder, A, consists of a heavy brass plate, faced with glass, six inches long by three and three-quarters wide, from the centre of which is sunk a hollow cylindrical barrel, one inch and three-quarters in depth, and one inch and one-quarter in diameter, inside measurement. Through the bottom piece of this barrel (which unscrews for convenience) works a screw shaft, furnished with a large milled head. The threads of the screw are fifty to the inch, and the circumference of the milled head is marked off into eighths, so that, if desired, the thickness of the sections cut can be measured. Attached to the under surface of the face plate, near one end, is a screw clamp for fastening the apparatus to the edge of a table, so as to leave both hands of the operator at liberty to work the cutter.

"The principle of this 'holder,' as it might be called, is very simple. The object to be cut is first embedded in a cast of wax and oil, or paraffine, made to exactly fit the bore of the barrel; this mass is then pressed into the barrel, pushed upward by turning the milled head of the screw-shaft until it projects slightly above the level of the face-plate, when the projecting part is cut off by a sweep of the knife. Another turn, or a fraction of a turn, of the milled head is then made, and again the projecting portion is cut off, this time as a smooth, even section of microscopic thinness. For a cutter I had a straight knife made, with a blade eight inches long, one and a quarter inch wide, and three-twentieths of an inch thick at the back, and with the sides concave like a razor. This I at first used by itself, sweeping it over the surface of the holder in cutting the sections, but I found that, from the length and thinness of the blade, it was apt to bend under pressure, and so fail to cut an even section, and also that the edge soon got dull from friction upon the face-plate. To obviate these difficulties, I conceived the idea of having the knife fixed in a frame which should answer the double purpose of holding the blade stiff, and carrying it with the edge raised slightly above the level of the surface of the holder, so that the under surface of the arms of the frame should be the bearing surface upon the holder, and the edge of the knife be allowed to touch nothing but the tissue to be cut. The design of the frame will be seen at once from the figure; it is made of brass, and the knife is pushed into place from behind, under a couple of springs, which hold the blade down, and, when pressed home, the knife is kept from slipping back by a little fastening, which is pushed against the back of the blade and then fixed by a turn of a quick screw. Besides the advantages of the frame in holding the knife stiff, and keeping its edges from scraping over the surface of the holder, its weight and broad tread make it sweep much more steady and true than that of a light knife used by itself."

This apparatus is so efficient that Dr. C. has often cut with it sections of an entire human eye, thin enough for microscopical examination.

In preparing the tissue and cutting the sections, however, there are one or two points which it is necessary to observe in order to get the best results. The tissues are to be hardened in the usual manner, with bichromate of potash and alcohol. When the consistence is suitable for cutting, the piece is to be transferred to oil of cloves, where it is allowed to stay until thoroughly impregnated with the oil; this takes from half an hour to several hours, according to the size and solidity of the specimen. The piece is now to be embedded for cutting.

"In order to get a mould for the paraffine which shall yield a cast of the right size to go into the barrel of the section-cutter, a solid brass plug (fig. IV. of the woodcut), one inch high, and made to exactly fit the bore of the barrel, forms part of the apparatus. In one end of it an oval excavation is countersunk, to let the cast of paraffine take a firm hold of its surface, and to keep the same from turning round under the pressure of the knife. A strip of letter-paper, about two inches and a half wide, is now wrapped tightly around this plug, the plug being in the middle of the strip, so that the paper projects at both ends. The part projecting beyond the flat end of the plug is folded down over it, and the opposite projecting cylinder of paper then forms a cup, with the excavated face of the plug for a bottom, of the exact calibre of the barrel of the section-cutter. Into this the melted paraffine is to be poured. A capsule containing paraffine, or wax and oil, in considerable excess of the amount required for the embedding, is held over a lamp till the mass is just melted, when it is taken off and the piece of tissue dropped into it and stirred about for a few minutes to rinse off the excess of oil of cloves. Then some of the melted material is poured into the paper mould, the piece of tissue immediately transferred to the same, arranged in proper position, and the whole set aside to cool. When perfectly cold and hard, the paper is unwrapped, care being taken not to loosen the paraffine cast from the brass plug, and cast and plug together are then pushed into the barrel and holder, and each section is cut by a sidelong sweep of the cutter. It is usual in cutting sections to flood the surface of the tissue and the blade of the knife with alcohol, so that the section, in cutting, floats freely over the blade.

"In this apparatus, however, from the great length of the knife, and from the fact that its edge is raised off the surface of the holder, this procedure is impracticable; but I find that I get even more perfect sections by the plan of cutting dry—that is, without flushing the surface of the tissue with any fluid. It will be seen in the figure that the knife is set at a slight angle in the frame; and this obliquity seems to give the section a tendency to curl away from the knife-blade in the cutting, so that in this dry process there is not only no more, but there is actually less danger than by the wet method of the section clinging to the blade, and so getting torn. But here the preliminary impregnation of the tissue with oil of cloves is essential; for, were it in alcohol, the microscopically thin section would instantly become ruinously dry as soon as cut. The oil of cloves, however, from its very slight volatility, keeps the tissue of the section moist until it can be transferred to a fluid. But it will not do so long, and hence the moment a section is cut it should be promptly seized and dropped into fluid; and if, from imperfect impregnation before embedding, the surface of the cut tissue looks dry it should be *moistened* only by a touch of a camel's-hair brush dipped in oil of cloves. If the sections are not to be stained, they are dropped into turpentine as soon as cut; this dissolves the adhering wax or paraffine very promptly if slightly warmed, and the sections are then ready for examination or mounting. If they are to be stained, they are put into alcohol instead of turpentine. This in a few minutes dissolves out the oil of cloves, and the sections are then put at once into the carmine staining fluid.

"The wood-cut represents two accessory pieces of apparatus that have not been alluded to. Fig. III. represents a secondary barrel with a half-inch bore, which can be screwed into the bottom-piece of the main barrel to diminish its size when any small pieces of tissue are to be embedded. It has a plug made to fit it similar to the large plug of the main barrel. Fig. II. is a simple and ingenious contrivance, devised by Mr. Wale, the maker of the instrument, for holding hard substances which will bear squeezing, and which, therefore, it is not necessary or desirable to embed, such as cartilage, horn, wood, etc. It consists simply of a brass plug, made to fit the barrel of the section-cutter, hollow, but with the bore slightly conical, and with a screw-thread cut on its face. A few wedge-shaped pieces of soft wood of different sizes, roughly whittled out, complete the apparatus. Supposing a stick of wood is to be operated on, it is grasped between two of the wedges of the right size, being allowed to project somewhat above their tops, the whole pressed firmly into the conical bore of the plug, and with a turn or two the soft wood of the wedges is tightly grasped by the screw-thread of the plug, and the object to be cut, tightly jammed between the wedges, is immovably fixed. Sections can be cut from the projecting portion in the usual way. It may be remarked, in passing, that the cutter for anatomical tissues must not be used for hard substances. For these a strong, heavy knife or chisel of less brittle temper is to be employed.

"In conclusion, it may not be amiss to state that the section-cutter and accessories can be obtained (to order) of Messrs. Hawkins & Wale, physical instrument makers, Stevens' Institute of Technology, Hoboken, New Jersey. Price $30, or, without the knife-frame, $20. The knife is a simple affair and can be made by any first-class manufacturing cutler from the dimensions given above. Mine was made by Mr. A. Eickhoff, 381 Broome Street, New York, price $3."

This instrument, when constructed and employed in the manner above described, affords better results than any other with which I am acquainted.]

Peculiar difficulties present themselves in making thin sections from very small objects, as they cannot be held in the fingers of the left hand, like more substantial masses. Moist substances are therefore placed in others of greater size; thus, for example, the spinal cord of a small mammalia is placed in that of a larger creature. Small objects may also be advantageously placed in a thick solution of gum-arabic, or embedded in a mixture of wax and oil (Stricker), in paraffine, or in a mixture of glycerine and gelatine (Klebs).

1. *Embedding in gum.* A paper cone or box is to be filled with a very concentrated solution of gum-arabic; in this mass is placed the object, from which the water has been drawn out by means of alcohol. The whole is then to be placed in alcohol for two or three days, and is then in a condition proper for cutting. The sections are to be washed out with water.

2. *Embedding in a mixture of wax and oil.* Equal parts of each are to be warmed in a porcelain dish till they become fluid, and the mixture is then poured into a paper box. The preparation is to be deprived of its water with alcohol, and rendered transparent by means of an ethereal oil; it is then placed in the mixture, and is ready for cutting as soon as the latter has become cold. The sections are to be washed out in oil of turpentine.

3. *Embedding in paraffine.* A cavity, made in a piece of paraffine, is partially filled with melted paraffine. In the latter is placed the object, which has been hardened in chromic acid and alcohol. Paraffine is again poured in, and the whole may afterwards be placed in alcohol. In many cases it is sufficient to drop a little melted paraffine on to a slip of gutta-percha, place the object on it, and cover the latter by dropping on a little more paraffine (His).

4. *Embedding in a mixture of glycerine and gelatine.* Alcohol or chromic acid preparations may be placed in a mixture composed of about one volume of a very concentrated solution of isinglass and half a volme of pure glycerine. The whole, when cooled, is to be replaced in chromic acid or alcohol, where the preparation and the gelatine become sufficiently hard.

For very hard substances, such as bones and teeth, the knife is no longer serviceable for making thin sections. In such cases a small saw with a watch-spring blade is to be used, and the section is to be ground down on a whetstone. This can be best and most rapidly accomplished with a rotary stone.

The ordinary camel's-hair pencil is an indispensable implement for the histologist. In addition to its usefulness for removing dust from the lenses of the microscope, it is also very extensively used in preparing specimens. It is the best thing to use for removing foreign bodies and fragments of tissue from the surface of preparations, and for spreading out thin and delicate sections on the slide. When it is necessary to remove from an object the cellular elements, which often occur in such great profusion as to conceal the entire arrangement of its superstructure, this may be much better accomplished by the pencilling method, originated by His, than with the stream from a wash-bottle. The specimen is to be thoroughly moistened and

Fig. 71. Pencilling microscopic objects.

covered with fluid (generally glycerine and water), and then brushed with a camel's-hair pencil of medium size, the perpendicular strokes rapidly following each other (fig. 71). The fluid gradually thickens and the tissue becomes transparent. After a few minutes the preparation is to be turned over, and the process repeated on its other surface. In this manner, and occasionally removing the old fluid and replacing it with new, the isolated frame-work gradually makes its appearance. It is also very serviceable to pencil an object while it is floating in a larger quantity of fluid—for example, in one of the above-mentioned glass boxes. A considerable amount of patience is necessary to make good preparations in this way, and still more to obtain the proper consistence of the object to be prepared. When it is not sufficiently hardened, it becomes filled with rents, even when the brush is carefully used. Such tissues generally become quite serviceable after hardening for a day or two longer. It is much worse when an over hardened tissue is to be prepared in this manner. In such cases one can obtain only an imperfect preparation or none at all; it is no longer possible to remove the cells. As a rule, the thing is then to be entirely abandoned, for a subsequent softening rarely leads to success. Billroth has also given some particular directions on the pencilling method.

Fig. 72. The pipette.

A strip of bibulous paper may be used for removing superfluous fluids from the slides. It is more convenient to use a small pipette (fig. 72), an instrument which can hardly be dispensed with in making permanent specimens.

Section Seventh.

FLUID MEDIA AND CHEMICAL REAGENTS. TITRITION.

ANIMAL tissues are comparatively seldom examined in a simple dry condition, but, as a rule, a fluid is added. This fluid may act indifferently, although this is more rarely the case than is generally imagined; it may act chemically on the object; it may withdraw fluid from it, or allow fluid to pass into it, so that shrinking or swelling results; finally, it may produce changes in the refractive conditions of the tissue.

Let us first investigate the latter. The greater the contrast between the refractive power of the object and that of the surrounding medium, the sharper will the former appear. Thus many delicate structures may be most distinctly recognized when dry and surrounded by atmospheric air, while the addition of water, by changing the refraction of the light, perhaps entirely prevents the details from appearing, or renders them very indistinct. Many textural relations of animal tissues are exceedingly difficult to recognize, in consequence of the slight difference between their refractive power and that of the surrounding water. We must, therefore, coincide with Harting, who says, the discovery of a fluid medium of less refractive power than water would afford very valuable assistance in many investigations. Other methods of rendering many things darker and more distinct, such as tinging the tissues, the application of reagents which coagulate and therefore darken them, are discussed further on. Certain reagents, acetic acid for example, act very advantageously by rendering a constituent part, as for instance the nucleus of a cell, darker, while the refractive power of the surrounding substance is diminished. The action of acetic acid on connective tissue affords us an instructive example of how little one is justified in assuming, from a single method of investigation, that there is nothing in the field because there is nothing to be seen there. By causing the finest fibres into which the intercellular substance of connective tissue is split up to swell, their refractive power and that of the surrounding fluid is rendered similar, so that one might think that the fibrillæ had been dissolved by the reagent, were it not for other methods which cause the reappearance of the fibres which had been rendered invisible by the acid.

On the other side, the necessity often makes itself felt of rendering objects which are too dark, and therefore no longer recognizable, as transparent as possible by means of a fluid which refracts the light strongly. Hence strong solutions of sugar, gum, or albumen may be used, when it is necessary to clear up tissues which are saturated with water. In modern times we have learned to recognize in glycerine an invaluable accessory of this nature; creasote also deserves recommendation. Tissues which are free from water are more permanently cleared up by means of turpentine oil, Canada balsam, or anis oil. While the index of refraction of water is 1.336, glacial acetic acid has that of 1.38, pure glycerine 1.475, equal parts of glycerine and water 1.40, oil of turpentine 1.476, Canada balsam 1.532–1.549, and that of anis oil is even 1.811.

How much the appearance of a microscopic object is determined by the refractive power of the fluid medium is self-evident. A small glass rod lying in water can be readily recognized with exactness, in consequence of the difference of the exponents of refraction. When it is placed in Canada balsam, whereby they become nearly similar, the glass rod ceases to glisten and can only with great attention be distinguished from a flat band. If anis oil be selected as a medium, an image is received as though there was a cavity in the oil (Welcker).

The necessity for the discovery of an actually indifferent fluid medium, that is, one which does not change the tissue, cannot be impressed with sufficient force on the hearts of microscopists. We have fallen into the beaten track of ascribing, with generous credulity, to water such a rôle, but which, in fact, it does not play. It is conceded that a small fraction, at the most, of the animal tissues make an exception, and the energetic action of water on the colored blood corpuscles and the elements of the retina cannot be denied. That the number of tissues affected by water is much greater, and that very few can remain indifferent to it, is very clear to a few persons, but is by no means generally known. While so many have recently occupied themselves with the endosmatic processes of physical physiology, in the particular domain of microscopy, no investigation of this process has yet been commenced.

Theory requires that each constituent of the body should be examined in a fluid medium which resembles, in respect to quality and quantity, the fluid which saturates the living tissue. Naturally, this requirement cannot be completely fulfilled in practice; our aim should be to approach it as nearly as possible.

Saliva, vitreous humor, amniotic liquor, serum, and diluted albumen are generally recommended as suitable media for the investigation of delicate changeable tissues, and, in certain cases, they accomplish their object in a satisfactory manner. But do not expect them to suffice for every case. Not unfrequently one and the same tissue of different species of animals reacts differently with the same fluid medium, as may be seen with the blood corpuscles. M. Schultze has communicated to us an important and readily proved observation of Landolt's, that animal fluids may be preserved from decomposition for a long time by the addition of a small piece of camphor.

A physical examination made by Graham presents us with a key to the nature of these indifferent fluids.

In an exceedingly interesting work (*Annalen der Chemie und Pharmazie*, Bd. 121, S. 1), this scholar some time ago called attention to the fact, that two groups of substances, which he has designated by the names of Crystalloids and Colloids, are to be distinguished according to their power of diffusion. The former, belonging to the crystalline bodies, diffuse rapidly and remind one, in this regard, of more volatile substances; the latter, characterized by their inability to assume the crystalline condition, show a very slight diffusibility. Among the organic bodies may be numbered, for example, gum, starch, dextrine, mucus, albumen, and gluten.

When a column of water is placed over a solution which contains both these varieties of substances, chloride of sodium and albumen, for instance, the salt will penetrate to the uppermost stratum of the fluid, while the albumen, in consequence of its slight diffusibility, will not pass anything like so far upwards, so that the upper strata remain free from it. Gelatinous matters from the colloid series, such as mucus, for instance, permit a very easy passage to readily diffusible matters, but resist very energetically that of less diffusible ones, and do not let other colloid matters pass through. By means of suitable membranes of this kind, crystallized matters may be separated from colloid substances, and the latter may be thoroughly purified in this way. According to Graham's observations, readily diffusible substances, such as chloride of sodium, even spread themselves through a stiff jelly with almost the same facility as in pure water.

It is self-evident that these investigations are of great significance in connection with the processes of diffusion in the tissues composed of colloid substances.

The above-mentioned indifferent fluids now appear to us in a new light. They always contain colloid and crystalloid substances. Vitreous humor contains 987 parts of water to about 4.6 parts of colloid matter, 7.8 of crystalloid substance (that is, chloride of sodium). In amniotic liquor about the same proportions are met with. In 1000 parts occur about 3.8 of colloid substance (albumen), 5.8 of salts, together with 3.4 of urea. In serum we have about 8.5 per cent. of colloid and 1 of crystalloid substances.

After what has been said it is unnecessary to remark that fluids which contain only crystalloid or only colloid matters can make no claim to the character of truly indifferent media, although they may not perceptibly alter the contours and forms of the tissue elements for a long time.

Accordingly, it has very properly been suggested that the microscopist should always have such indifferent fluids in readiness, the more so as solutions of albumen or liquor amnii may readily be preserved from decomposition for months by placing a piece of camphor in them (M. Schultze). A solution of albumen, of known quantitative composition, purified by means of

Graham's dialyser, and to which a certain quantity of chloride of sodium is to be added, may be preserved with the aid of a piece of camphor, and will be very serviceable if diluted with water each time that it is used. It is useless, however, for the preservation of large pieces of tissue.

It is obvious that the addition of colloid substances to solutions of the salts ordinarily used by microscopists also deserves a trial.

Schultze has more recently recommended, in the warmest manner, an albuminous fluid tempered with iodine—and in fact according to my own experience it is exceedingly serviceable. "Jod-serum," as he calls it, consists of the amniotic fluid of the embryo of the ruminantia, to which a concentrated tincture of iodine or a strong solution of iodine in hydriotic acid is added. About six drops are to be added to the ounce while shaking the mixture. The color of the solution is at first wine yellow, but after a few hours it becomes paler; this paleness afterwards increases, and the subsequent addition of a few drops of the iodine solution becomes necessary. Our mixture forms an excellent fluid for the examination of delicate fresh tissues, and is also a very good and very preservative macerating medium, acting in this way even for hours or days. We must here give a piece of advice which is of great importance in the numerous macerations of this kind which are necessary, namely, to have the piece which is to be placed in them very small, and the quantity of the fluid as large as possible. An artificial mixture, composed of 1 ounce of the white of an egg, 9 ounces of water, 2 scruples of chloride of sodium, with the corresponding quantity of tincture of iodine, appears to form a substitute.

In the use of water, in which case distilled water should be employed, the swelling of the delicate tissues is, possibly, very considerable; not unfrequently they may even be more permanently altered; so that it is advisable for any one who would protect himself from deception to try other fluid media also, in order to decide what has remained unaltered in his microscopic image, and what has been acted upon by the water.

Glycerine has already been mentioned several times in these pages. Together with its property of rendering tissues transparent, which is of inestimable value for such as have been hardened and rendered opaque by reagents, it forms a preservative, though not indifferent medium for many tissues, and even for the prolonged preservation of larger pieces. Its power of rendering tissues transparent may be restrained by the addition of water. A mixture recommended by Schweigger Seidel, composed of 1 part of pure glycerine to 9 parts of distilled water, is useful for the examination of numerous objects. Many delicate structures shrink in glycerine, it is true, but after longer action they again become filled out and clear. A number of really chemical reagents—for example, acetic acid, formic acid, iodine, tannin, and chromate of potash—may be advantageously combined with it, and it also forms an ingredient of cold injection masses (see below). Finally, it presents the best fluid for mounting moist tissues.

Nowadays chemical reagents are very frequently employed in microscopic investigations, and the number of these which are necessary for various histological and medical purposes is by no means small. They are the same as are generally used for zoochemical investigations.

They are chiefly employed in microscopical investigations when we wish to ascertain the nature of amorphous and crystalline deposits, the disposition of elementary granules, or the constitution of tissue elements. The ordinary solutions, naturally from a reliable source, are used for these processes. Their application, however, requires great foresight with regard to the microscope, if one would protect it from injury. We therefore repeat certain precepts which have already been given. The lenses should never be wetted with the fluids. Only the weaker objectives, with long foci, are to be used, and the covering glasses should be as large and broad as possible. In order to prevent the fluids from running on to the stage, the slides should not be too narrow. I generally cover the stage completely with a glass plate of the same size with ground edges, a precautionary measure worthy of recommendation to those who have the protection of their instruments at heart. When the stage consists of a plate of ground black glass, as is the case with some of the older microscopes, it is very convenient for chemical examinations.

The reagent is either simply added to the microscopic preparation by means of a pointed glass rod, the covering glass being previously removed, or the fluid is allowed to flow under the edge of the cover to the object; it may also be allowed to enter gradually, in order to observe the successive changes which occur during its action. A thread of lint, one end of which is placed under the cover, may be used as a conductor, or two very narrow strips of bibulous paper may be placed at opposite borders of the cover, one of which serves to remove the old fluid while the new is being introduced by the other, whereby, however, the entrance of the reagent is more rapid and its effects are more energetic.

More important than this momentary use of chemical reagents is the continuous application of the same for a longer period as hardening, preservative, or macerating fluids. Animal tissues are frequently allowed to remain in the solutions for hours or even days consecutively. This method has been frequently used in modern times, and to it is due most of the knowledge that has been obtained in latter years concerning the tissues, etc., of the human body. Its perfection should, therefore, be of great interest to every investigator. Its application, however, requires an exact procedure. Above all things, one should avoid all such old delusions as when putting a tissue in acetic or sulphuric acid, or in solutions of potash or soda, to disregard the strength of the solutions, or the relative volume of the tissue to that of the fluid. Hence it is the duty of every one who employs any of these chemical methods, or recommends a new one, to describe his process accurately.

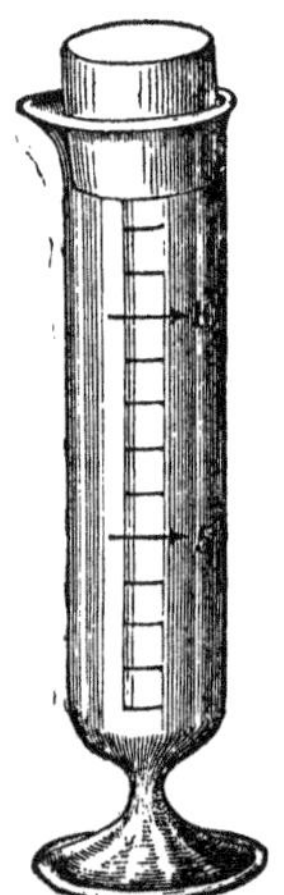

Fig. 73. Graduated cylinder glass.

A watch-glass, or a small shallow glass box may be used when the process is to continue but a few moments. For more continued action, small bottles with wide mouths and ground glass stoppers are to be used, or, still better, small graduated cylinder glasses (fig. 73). The vessels should always be labelled, to avoid confusion, to remember the date, etc.

We will now proceed to the consideration of the reagents which are at present most frequently used.

1.) Among the strong mineral acids, sulphuric, muriatic, and nitric acids, in a concentrated form, act destructively on most histogenetic substances. Nevertheless, they afford an important means of isolating certain tissues, inasmuch as they dissolve their connecting or cementing substance, and also, in part, the connective tissue which occurs in them. In a more diluted condition they form useful hardening solutions for various tissues, while with a still higher degree of dilution we obtain the action of weak acids on various tissue elements, causing them to become transparent, to dissolve, or to swell up. In this way these acids may constitute very important macerating mediums.

Sulphuric Acid.—The purified concentrated English sulphuric acid, non-fuming, with a specific gravity of 1.85–1.83, is to be used.

The concentrated acid is but rarely used. Still it is an excellent auxiliary in the investigation of the horny structures (the cornified epithelium, the nails, and hair), for isolating the cells of these tissues. It also forms alone, or combined with iodine, a good reagent for cholesterin ; the latter combination is also useful for cellulose or amyloid substances. Sugar and sulphuric acid redden many organic substances, such as albuminous and amyloid bodies, oleic acid, etc.

Strongly diluted sulphuric acid hardens albuminous tissues, acting similarly to chromic acid (see this). It has the advantage over the latter of rendering gelatinous and connective tissues transparent, and, at the same time, so consolidated that thin sections of them may be made. Moreover, less depends on the accurate concentration of sulphuric acid than on that of chromic acid.*

When connective tissue is treated for twenty-four hours with highly diluted sulphuric acid, 0.1 grm. to 1000 grammes of

* M. Schultze, who has presented us with these statements, employs an acid of 1.839 specific weight, of which about 18 drops make 1 gramme and 22 a scruple. He recommends, as a mean, 3 to 4 drops to 1 ounce of water (with extremes of 1 to 10), and praises its effects for hardening the supporting substances of the central organs of the nervous system, of the retina, and also the reticulated tissues of the lymph glands and kindred organs.

water, and warmed to a temperature of from 35° to 40° C., it is resolved into gelatine; so that in this way other elements are spared as much as possible, and may be isolated from connective tissue. Kühne has employed this method with good success for muscular fibres.

Sulphurous Acid.—It has been recommended by Klebs to add a small quantity of sulphurous acid to a solution of cane-sugar of 5 per cent. (one drop of a pretty concentrated solution of the former to one ccm. of the latter fluid) for loosening epithelium, and for rendering connective tissue transparent without causing infiltration.

Nitric Acid.—The pure concentrated nitric acid of the chemical laboratories of 1.5 specific weight may be used, or acids containing more water, with a specific weight of 1.4 to 1.2 (the latter is the officinal nitric acid).

The former (1.5), mixed with chlorate of potash, destroys connective tissue in a short time, and is therefore a good medium for isolating muscular fibres (Kühne). This may also be accomplished, though more slowly, with weaker acids. This reagent, recommended by Schultze, is frequently used by botanists; it deserves further trial for animal tissues. Caution in its application is always advisable.

The property of nitric acid of coloring albuminous matters yellow is, in general, rarely made use of in microscopical investigations.

Strong nitric acid serves to isolate connective tissue corpuscles, bone corpuscles and their processes, as also the dentinal canals.

Nitric acid of 20 per cent. was recommended many years ago by Reichert and Paulsen as a medium for the isolation and recognition of the elements of the smooth muscles.

Diluted nitric acid (5 to 10 per cent.) is also used for the extraction of the so-called bone earths (a mixture of lime and magnesia salts) from calcified cartilages and bones; although hydrochloric acid and, still better, chromic acid (see this) may be employed for this purpose.

In a condition of extreme dilution (0.1 per cent.), nitric acid has recently been tried by Kölliker for rendering muscles transparent. It does not present any special advantages.

Muriatic Acid.—Pure muriatic acid, thoroughly saturated with chlorine, of 1.19 specific weight, is not at all, or only rarely, used undiluted for histological investigations. Strong muriatic acid is frequently used for dissolving the intercellular substance of connective tissue organs, and for isolating the connective tissue corpuscles and their radiating tubular systems, as in the cornea, the teeth, and bones. This action generally requires some time, occasionally a number of days. By this means the intercellular substance of the muscles (Aeby) and of the urinary tubes (Henle) has been dissolved. An acid which has been diluted with water till it ceases to smoke, is frequently used for this purpose. The time required is generally 12–14 hours; weaker acids act more slowly. The object is then to be washed out and macerated, for a day at least, in distilled water. When this process succeeds, the whole may be rapidly and beautifully isolated by the careful use of the needles. An important modification of the above-mentioned process consists in boiling pieces of the kidney for 6–8 hours in alcohol of 90 per cent., to which $\frac{1}{2}$ to $\frac{3}{4}$ per cent. by volume of purified muriatic acid of the greatest possible strength has been added. The operation is to be conducted on the water-bath, in an alembic provided with a cooling apparatus (Ludwig and Zawarykin). This process is also useful for other glands. Tomsa has recommended boiling for one or two days, and subsequent washing in water for isolating the cutaneous nerves. Size injections with Prussian blue retain their color and the consistence of the vessels with both methods. Muriatic acid, of the same dilution as the nitric acid, may be used for extracting the bone earths. In a high degree of dilution (0.1 per cent.) it forms a medium for macerating connective tissue and rendering it transparent, the cells and elastic tissue of which then appear very beautifully. Our acid also dissolves the fleshy substance of muscular fibres, and may thus be advantageously employed in the examination of muscular tissues.

Chromic Acid.—Since Hannover, in the year 1840, recommended chromic acid to microscopists as a medium for hardening animal tissues, its reputation has been constantly increasing, especially since the inaccurate method of estimating the strength of its solutions from their color has been abandoned for that by means of the scales.

It is extremely useful for hardening the brain and medulla oblongata, as well as the peripheral nervous apparatus. Not unfrequently it is more serviceable than alcohol, which exerts too great a change on these tissues, while the latter is either equal or preferable to the former for other organs, such as most glandular structures, the intestinal canal, etc.

Well-crystallized chromic acid, as free as possible from sulphuric acid, should always be used. It should be kept in a well-closed vessel, in a dry place, and the portion to be used should be dried over sulphuric acid previous to its employment. To economize time, a considerable quantity of a strong solution may be kept on hand, which may be rapidly diluted to any degree desired by means of a graduated measure. I dissolve 2 grammes in 98 grammes (or cubic centimetres) of distilled water, so that a 2 per cent. solution stands ready.

For hardening purposes a chromic acid solution of from 0.5 to 1, or, at most 2 per cent. is necessary. In no case should a higher degree of concentration be used, and generally the weaker ones work better. Very fresh tissues usually require weaker, older pieces somewhat stronger solutions. Very fine results may be obtained, especially when the pieces are not very voluminous, by commencing with a weak solution (about 0.2 per cent.), and then, after a few days, changing the fluid for one of stronger concentration (0.5 to 1 per cent.), in which the object is to remain for days and weeks, until it has obtained the desired consistence. Then—in consequence of the readiness with which fungi are formed in chromic acid solutions—the hardened preparation should be kept in diluted alcohol.

When a voluminous organ is to be hardened it is advisable, before placing it in the chromic acid, to drive the same solution through the blood-vessels.

However, with all chromic acid operations, very much depends on the proper degree of concentration, and this is not always hit upon even by the most experienced; this is all the more so, as the contamination with sulphuric acid is quite variable. Very voluminous organs may present a hardened peri phery, while the interior is decomposed. Portions which are over-hardened have their tissue elements very much shrunken, and are often found to be so hard and brittle that it is no longer possible to make thin sections from them. An improvement may sometimes be obtained by laying the piece of organ in glycerine for several days. It is preferable to add some of this to the chromic acid at the very beginning.

So much for these concentrated hardening solutions of chromic acid. This reagent has, however, when highly diluted, still another and more important property, namely, of preserving the finest textural relations while exerting a somewhat macerative action on them. So that in this way very delicate organizations, especially in nervous tissues, may be made visible which remained completely hidden in the examination of the fresh tissue. For this very reason it has exerted a very enduring influence in the histology of the higher nerves of sense, to which fact the works of M. Schultze especially testify. It has since been used with success for the investigation of the central organ of the nervous system, the ganglia, and also for glandular structures.

In general, according to present experience, only a concentration of from $\frac{1}{8}$ to $\frac{1}{4}$ of a grain to the ounce of water, that is, a solution of from 0.025 to 0.05 per cent., is applicable for this purpose, and in fortunate cases the desired effect is accomplished in from one to three days. Others (Deiters, J. Arnold, Kühne) have even descended to solutions of from 0.02 to 0.01 per cent. and less—and even to these an effect cannot be denied.

The volume of the fluid and that of the portion of tissue placed in it are here of greater importance than for simple hardening. Generally, when the latter is small and the fluid plentiful, the action is naturally more energetic and rapid, so that in this case the limits may easily be exceeded. It is proper, therefore, not to select too small a piece, and not to add too large a quantity of fluid. When the object is too small, the same effect takes place as with stronger solutions, they receive a lively yellow color and become opaque; when of the proper proportions they become paler and semi-transparent.

We have already mentioned above the interesting and, in their consequences for practical microscopy, very important observations of Graham on the colloid and crystalloid substances. Schultze (who was the first among German histologists to comprehend the full significance of Graham's observations) has properly called attention to the fact that the action of chromic acid is not alone concerned in this case, but to this is also added, when the piece is large and the quantity of fluid moderate, the preponderating effect of the colloid substances of the tissue; such as blood, mucus, and albumen. Hence a fluid results which consists conjointly of crystalloid and colloid substances; while a small piece of tissue placed in a larger quantity of chromic acid solution is subjected, almost entirely, to the action of this crystalloid substance alone.

At present microscopic technology is only in its youth, not to say childhood. In a riper period such combinations will certainly play an important rôle. Schultze informs us that he is making investigations relative to this subject, and that a watery solution of gum-arabic appears to be suitable as a colloid substance. May he soon give us further information on this subject!

Similar, but much weaker and more slowly commencing effects are also produced by bichromate of potash, which will be spoken of further below.

Finally, still another very advantageous use has been made of chromic acid; namely, for extracting the earthy salts from so-called ossified cartilage, and also from bones. It is especially commendable for fœtal tissues. Generally it is necessary to have a higher degree of concentration (about 2 per cent., Thiersch), and, during an exposure of several weeks, the fluid should be frequently changed. It is well to add a little glycerine. The effect may be increased by a little hydrochloric acid, without injury to delicate textures. The decalcified specimen, after being washed, is to be placed in absolute or strong alcohol for further hardening. A similar decalcifying effect may also be obtained with pyroligneous acid (see below).

Oxalic Acid.—Oxalic acid was formerly but little or not at all used by histologists. Some time since M. Schultze instituted a series of experiments with it which assign it a not unimportant rank among the microscopist's reagents. A cold saturated solution of oxalic acid (one part of pure crystalline hydrated acid requires for its solution 15 parts of water) causes connective tissue structures to swell and become transparent, while the tissue elements which are formed of albuminous substances retain their sharp contours, become somewhat hardened and permit of convenient isolation. Extremely delicate elements of the body, such as the rods of the retina and the olfactory cells, are preserved in it excellently. The length of time is of relatively slight importance, so that the examination may be commenced after a few hours or several days.

According to Schultze's experience, an alcoholic solution of oxalic acid acts more strongly than the watery, and appears to present certain advantages for many purposes.

Finally, oxalic acid is used in carmine tingeing in the same manner as acetic acid, although more circumscribed in its application, of which mention will be made later.

Acetic Acid.—The hydrated acetic acid, thoroughly pure acetic acid, acidum aceticum glaciale, should always be used when an accurate estimation is necessary (since the so popular specification of the specific weight affords no definite conclusion as to the amount of water present), and combined drop by drop, or in greater quantity, with water.

Acetic acid, which is so rapid in its action, is one of the oldest and most frequently employed reagents in animal histology. Its property of rendering nuclei within the cells visible, or of causing them to appear isolated after the destruction of the envelope and cell body; and further, of giving to connective tissue a crystalline transparency, and disclosing its admixture in the cells, elastic fibres, vessels, nerves, etc., has especially led to its general employment.

Only at a later period were quantitatively defined solutions of acetic acid, as well as combinations of the same with other fluids, especially alcohol, employed for more prolonged action on animal tissues. Even a few drops of the acid to the ounce of water is sufficient to induce considerable transparency of the connective tissue in a few days. In this way, for example, the intestinal ganglia lying in the submucosa; furthermore, the marvellous ganglionic plexuses, discovered years ago by Auerbach, between the muscular layers; also muscle cells in the mucous membranes, on vessels, etc., are made to appear distinctly. For the recognition of smooth muscles, Moleschott used a 1 or 1½ per cent. solution of acetic acid for a few minutes. One part by measure of strong acid, of 1.070 specific weight, is to be mixed with 99 of water, that is, 1½ to 98½.

More recently, Kölliker has used very dilute acetic acid for rendering the muscles of the frog transparent, in order to discern the nerve terminations, and the reagent accomplishes this exceedingly well. He recommends the addition of 8, 12, or 16 drops of the acidum aceticum concentratum of the Bavarian pharmacopœia, of 1.045 specific weight, to 100 cubic centimetres of water. I have substituted 1 to 2 drops of hydrated acetic acid to 50 cubic centimetres of water. Acetic acid of from 0.3 to 0.2 per cent. has been employed by others for many purposes.

Acetic acid, diluted to an extreme degree, is also to be recommended for softening thin sections from parts which have been dried in the air; also for washing out specimens after they have been tinged in carmine, in order to fix the red in the nucleus. This will be again alluded to further below.

Maceration in acetic acid presents a certain difficulty in the recognition of delicate structural relations, in so far as that the part should be examined at the right time. Before this period, the swelling and transparency are still too little developed; later, however, the changes induced in the tissue by the acid have become too considerable.

Beale has recommended the combination of acetic acid with glycerine.

Vinegar.—The employment of ordinary cooking vinegar offers no kind of advantage. In it connective tissue becomes transparent like glass, after 6, 8, or 12 hours. If the tissue becomes too soft to permit of sections being made, this may often be remedied by placing it, supplementarily, in a solution of chromic acid. It will frequently be found useful to boil in vinegar animal tissues which are to be dried.

Pyroligneous Acid.—Pyroligneous acid (none but the purified, or acidum pyrolignosum rectificatum, should ever be employed) has frequently been used for rendering connective tissue structures transparent, and especially with a certain predilection for pathological tissues. It exerts a similar, though not entirely the same effect as diluted acetic acid, inasmuch as it possesses, together with the macerating action, a hardening effect (from admixture of products of the dry distillation of the wood). Diluted pyroligneous acid should always be used for macerating, if it be desired to avoid marked textural changes of the elements which then become visible through the connective tissue. Pyroligneous acid, diluted according to circumstances with an equal, double, or quadruple volume of water, is a good accessory for many structural conditions; for example, in the recognition of the corneal cells and their contents, the course of the nerves in the sub-mucous connective tissue, etc., and especially structures which are embedded in connective tissue, such as glandular elements, vessels, pathological new formations, etc. The desired effects generally take place after one or more days, often enough again disappearing, as a result of continued maceration. Consequently, without regarding the smell or its injurious effects on the blades of the knives, there is an inconvenience in the employment of our reagent. Besides, pyroligneous acid preparations do not usually keep well when put up in glycerine. We have, therefore, discontinued the use of this fluid for many investigations. Still, it is useful for extracting the bone earths from calcified cartilage and from normal and pathological bone-tissues.

Osmic Acid (hyperosmic acid).—Within a few years this has come into frequent use through M. Schultze and others; it is readily reduced by several tissues and substances. It shares this property with several similarly applicable salts of the nobler metals, which will also be mentioned hereafter.

Picric Acid.—Recommended in part as a means of tingeing, by Schwarz (see below), in part for hardening tissues. According to Ranvier's experience, a concentrated solution produces an excellent consistence, even in 24 hours. Neither shrinking nor coagulation of albumen occurs, and lime salts are extracted at the same time. I can only coincide in these statements.

Iodine.—A solution of iodine, about 1 part iodine (it is well to combine with it 3 parts of iodide of potassium) to 500 of water, may be used for tingeing animal cells. Nevertheless, we have better and newer methods of tingeing. A solution of iodine serves the microscopist for the recognition of amylum, and, in combination with sulphuric acid, of amyloid and cellulose. For this purpose a watery solution, which should not be too strong, is allowed to act energetically, and then a drop of concentrated sulphuric acid is added.

It has already been remarked above (page 124) that iodine forms a constituent of an important mixture, the so-called "jod-serum," recently invented by Schultze.

2.) Among the alkalies, solutions of potash, soda, and ammonia are frequently used. They are of quite inestimable value for the investigation of animal tissues; this is especially true of the first two substances. There is one disadvantage, however, connected with the use of alkalies, namely, that objects which have been macerated in them can hardly be preserved permanently.

Caustic Potash (hydrate of potassa).—The melted form, the kali causticum in baculis is used. As this attracts water and carbonic acid from the air with great eagerness, it and its solutions must be kept in well-stoppered bottles.

The kali causticum in baculis of commerce contains, in addition to carbonic acid, a variable and not inconsiderable quantity of water, which constitutes an inconvenience in its application.

The strong solution of potash softens the substance of many elements, and induces in them a condition which is very favorable to the imbibition of water, which afterwards penetrates very rapidly, so that the cells swell up, burst, etc.

Manifold use has been made, in the investigation of tissues, of the resolving and destructive properties of potash solutions. The manner in which potash solutions act varies entirely according to their strength; a subject to which Donders first drew attention many years ago. A saturated, or at least very strong solution softens many elements, without dissolving or attacking them very strongly. Although diluted solutions produce this effect more or less rapidly, they also frequently dissolve the intermediate connecting substance, the tissue cement, and thus become an extremely important, in many cases invaluable accessory. Credit is due to Moleschott for having more recently recommended 30 to 35 per cent. solutions of potash as excellent reagents. To make a 32.5 per cent. solution of potash he uses 32.5 parts by weight of kali causticum in baculis, which is to be dissolved in 67.5 parts by weight of distilled water. An exposure of $\frac{1}{4}$ to $\frac{1}{2}$ an hour or more is an extremely useful means of isolating muscular and nerve elements, glandular passages, and even ordinary ciliated and olfactory cells. Schultze, who, together with other histologists, has likewise made use of potash solutions, employed for the last-mentioned delicate cell formations solutions of the strength of 28, 30, 32, 35, and 40 per cent. For other purposes, weaker solutions of 5 to 10 per cent. are necessary, as will be indicated in speaking of the individual tissues. Naturally, in the histological examination, the same solutions should be employed as a fluid medium, and the use of water avoided, as otherwise the rapidly dissolving effect of diluted solutions would take place.

Caustic Soda (hydrate of soda).—The white melted mass is used for making the solution. Although solutions of soda have been used experimentally, they present no superiority, when concentrated, to those of potash. Weaker solutions are generally necessary, about two-thirds of the potash quantity (corresponding to the atomic weight) being used.

Liquor Ammonia.—The action of ammonia on animal tissues is similar to that of potash and soda. Ammonia is useful for neutralizing acids which have been applied to a tissue; also as a means of dissolving carmine.

Lime-water.—Rollett has recently made us familiar with lime-water, which was previously but little noticed, as an important accessory for the investigation of connective-tissue structures, and especially tendons. After remaining in it for six or eight days, a piece of connective tissue may be readily divided into its fibrillæ by the use of the needle. It is therefore one of the animal cement substances which is here dissolved.

Baryta-water.—The same result may be obtained with connective tissue by means of the much more energetically acting baryta-water, in from four to six hours, as is afforded by lime-water after several days' action. At the same time, the swelling is rather greater, and the transparency somewhat more considerable. In both cases, before the application the tissue is to be washed with distilled water, or, still better, with distilled water to which a minimum of acetic acid (just enough to neutralize it) has been added.

3.) Salts.

Chloride of Sodium.—Formerly, weak solutions of common salt were commonly regarded as indifferent media. According to Graham's observations, a colloid substance (albumen or gum-arabic) should always be added. A special application of the chloride of sodium is also made in the impregnation of tissues with nitrate of silver, which will be alluded to hereafter. It is likewise an ingredient of various preservative fluids.

Chloride of Calcium.—Chloride of calcium in solutions of medium strength (one part of dry chloride of calcium to two or three parts of water) has been recommended as a fluid medium for microscopic preparations, in consequence of its well-known property of attracting water. It has also been recommended for rendering sections of the spinal cord, etc., transparent, for which purpose it is not very serviceable. It has a peculiar effect on muscles.

Acetate of Potash, in a nearly concentrated watery solution, has recently been recommended by M. Schultze as an excellent preserving medium.

Chlorate of Potash.—This is only used in combination with nitric acid (see this), as Schultze's reagent. Extremely varying degrees of concentration of this mixture have been made use of in animal histology, and the desired effect has naturally been obtained in very unequal spaces of time.

Phosphate of Soda.—Solutions of phosphate of soda of 5 to 10 per cent. have been considerably used by microscopists. According to my experience thus far, they do not present any advantages.

Bichromate of Potash (Red Chromate of Potash).—The purest possible crystallized material should be used. The action of this salt, which may be very suitably combined with glycerine, is similar to that of chromic acid, but weaker, and not so rapid in its appearance. For hardening many tissues it is extremely serviceable, and perhaps it is better than the free adulterated acid; it also exerts a much less coagulating effect on albumen. Besides this, the solutions of this salt have the advantage of not readily developing mould, which is a great fault of chromic acid solutions. It has also been recommended to commence the hardening process with our salt, and then to continue it with the free acid (Deiters).

Where one part of chromic acid would be sufficient, it is requisite to have several parts of chromate of potash. Thus, where a given effect might be obtained with a fluid containing from $\frac{1}{8}$ to $\frac{1}{4}$ of a grain of free chromic acid, to accomplish the same without the acid, the fluid would require from 1 to 4 grains of the salt. However, for delicate investigations, much less depends on the accurate concentration of the solutions of chromate of potash than of the chromic acid.

A mixture of the salt in question with sulphate of soda has been recommended by H. Müller for hardening the retina. The tissue should be exposed to its action for at least two weeks.

Bichromate of potash,	2—2$\frac{1}{2}$	grammes.
Sulphate of soda,	1	"
Distilled water,	100	"

This mixture, the "Müller's eye-fluid," is also very useful for preserving many other tissues, such as mucous membranes, glands, and even ciliated cells. It preserves delicate embryos exquisitely, and may naturally be modified according to necessity.

Bichromate of Ammonia.—This has been recommended by Gerlach in the place of the previous salt, in solutions of 1 or 2 per cent., for hardening the central organs of the nervous system.

Molybdate of Ammonia.—This was recently recommended by Krause as an indifferent medium for tingeing.

Chloride of Iron.—This iron salt was formerly used by Führer and Billroth for hardening the spleen, which becomes sufficiently hardened in from 1 to 2 hours in a solution of the color of Madeira or Malaga wine. Chloride of iron is at present supplanted by superior hardening media.

Chloride of Mercury.—The chemical effects of the sublimate are well known. Macerating for several days in a solution of this salt may be advantageously used for hardening and isolating the axis cylinders. Although this reagent has found but little application, still it forms an element of several very serviceable preservative fluids.

Nitrate of Silver.—This has recently come into use for a peculiar tingeing process of the tissues, especially through His and Recklinghausen (see below).

Chloride of Gold.—This was advantageously employed for a similar purpose by Cohnheim, Kölliker, Eberth, Gerlach, and many others.

Chloride of Gold and Calamine.—This has been employed by Gerlach.

Chloride of Palladium was first used by F. E. Schulze.

Chloride of Platinum.—As Merkel informs us, this salt hardens tissues and gives them, at the same time, a diffuse yellow tinge, especially those of flattened organs. Equal portions of solutions of chromic acid and chloride of platinum (each 1 : 400) are recommended for the connective-tissue frame-work of the retina.

4.) *Alcohol.*—Alcohol, the most common of the preservative fluids for animal tissues, is of inestimable value for histological investigations. The use of alcohol has come more into the foreground chiefly within a few years, since we have learned to recognize in glycerine an incomparable means of rendering transparent animal tissues which have been hardened and hence become cloudy. It is only for certain purposes that chromic acid deserves the preference. Either small pieces of the entirely fresh organ are placed in a relatively considerable quantity of alcohol free from water, or several sorts of alcohol are employed. Weak alcohol is used for the first few days; this is then replaced by stronger, and perhaps, later, a still stronger one is employed. I know of no better reagent for hardening glandular organs, the digestive canal, or injected preparations, and for rendering them fit for sections and brushing. Latterly, entire series of investigations have thus been made almost exclusively with alcoholic preparations. The circumstance that the specimens do not spoil in well-closed vessels constitutes an advantage over chromic acid, which so readily develops the formation of fungi. The latter is, on the contrary, preferable to alcohol for the recognition of many of the finest structural conditions, for the central organs of the nervous system and for the organs of sense.

Alcohol is also frequently applicable in other ways. First of all, for microscopic objects which are to be deprived of their water with the utmost possible sparing of the texture, for the purpose of being afterwards mounted in Canada balsam or similar resinous masses. In such cases the thin sections are to be placed for one or two days in absolute alcohol. From this they go into oil of turpentine.

Furthermore, alcohol forms an ingredient of Beale's cold injecting fluid, which will also be alluded to further on.

Finally, alcohol is also an ingredient of several recently recommended mixtures, the description of which follows:—

L. Clarke and Beale's Mixtures.

These serve to make delicate parts hard and at the same time clear. The fundamental idea consists in employing two sorts of substances, one of which hardens the albuminous elements of the tissues, while the other renders them transparent. Beale, who has occupied himself considerably with the action of these solutions, remarks that they must be varied according to necessity, also that by the addition of glycerine to the mixture its refractive power may be increased according to circumstances. He recommends in general alcohol, glycerine, acetic acid, hydrochloric acid, potash, and soda. The last two acids as well as alcohol coagulate albuminous matters; acetic acid, potash, and soda render them transparent; alcohol dissolves fat. When several of these materials are combined in a solution, the above mentioned effects are obtained.

(*a.*) *Alcohol and Acetic Acid.*—L. Clarke used in his investigations a mixture of acetic acid and alcohol, which, as I have also convinced myself, renders sections of the spinal cord marvellously clear, even in a few hours, and permits many things to be better recognized than by other methods customary for this purpose. Lenhossek also appears to have made use of this process in his investigations on the spinal cord.

Clarke's recipe, naturally to be modified according to necessity, is to combine three parts of alcohol with one part of acetic acid.

(*b.*) *Moleschott's mixture of Acetic Acid and Alcohol.*—Moleschott recommends the following modification of Clarke's method :—

Strong acetic acid (1.070 sp. wt.)......	1 volume.
Alcohol (0.815 sp. wt.)...............	1 "
Distilled water......................	2 "

He calls this his strong acetic acid mixture. This fluid is very serviceable for hardening many organs, causes the connective-tissue portions to become transparent, and renders those formed of albuminous matters distinctly prominent. Delicate textures do not, as a rule, tolerate it so well. Another weaker acetic acid mixture was afterwards recommended, consisting of

Acetic acid (same as above).........	1 volume.
Alcohol........................	25 "
Distilled water..................	50 "

(*c.*) *Alcohol, Acetic Acid, and Nitric Acid.*—Beale recommends the addition of a little nitric acid to the mixture of alcohol and acetic acid for the examination of epithelial structures. This is also to be varied as necessity may require. A recipe given by the author runs as follows :—

Water...........................	1 ounce.
Glycerine........................	1 "
Spirit...........................	2 "
Acetic acid.......................	2 drachms.
Hydrochloric acid..................	½ drachm.

Alcohol and Soda.—Beale obtained excellent results, in many investigations, from the use of a fluid composed of alcohol and a solution of caustic soda, in the proportion of from eight to ten drops to each ounce of alcohol. Many tissues are, at the same time, rendered very hard and transparent in such a mixture, and it is particularly adapted, according to his experience, for investigations upon the character of calcareous matter deposited in tissues in various morbid processes, also in tracing the stages of ossification in the early embryo. It renders all the soft tissues perfectly transparent, but exerts no action on the earthy matter of bone. The most minute ossific points can therefore be very readily discovered. A fœtus, for example, prepared by being soaked for a few days in this fluid, and preserved in weak spirit, forms a very beautiful preparation. This fluid will also be found very useful in investigations upon soft granular organs. Beale found it of special service when working at the anatomy of the liver.

Methyl Alcohol.—In England, where the high spirit duty imposes an obstacle to the employment of the ordinary (ethyl) alcohol, methyl alcohol (pyro-acetic spirit) is frequently used as a substitute; this is however unnecessary on the Continent. Methyl alcohol has found special application as an addition to Beale's cold injection fluid (see below), and also in mounting microscopic specimens in Canada balsam.

Sections which have been deprived of their water by means of absolute alcohol are placed for a short time in strong methyl alcohol, then taken out of this and, just as they are commencing to dry, thrown into oil of turpentine. The latter penetrates sections which have been in methyl alcohol, as I have learned by experience, somewhat more readily than those which are brought from the absolute alcohol directly into this oil. Nevertheless, methyl alcohol may be readily dispensed with for this purpose.

Chloroform.—Thus far it has been very little used for histological investigations, but forms the best medium for dissolving and thinning the Canada balsam, which is so important for practical microscopy.

Æther.—This serves to dissolve fat in microscopic work. It also dissolves Canada balsam.

Collodium.—So far, this has only been used for recognizing the axis cylinder of nerve fibres. According to the statements of Pflüger and my own observations, it acts instantaneously.

Oil of Turpentine.—This comes next to chloroform as a

medium for thinning Canada balsam. It also forms the most important medium for rendering transparant dried sections, or those which have been deprived of their water by absolute alcohol. We will return to this subject more in detail further below.

Creosote. — Creosote forms an element of preservative mounting fluids (Harting).

It has recently been recommended by Stieda, after the example of Kutschin, as a very rapidly acting medium for rendering microscopic sections transparent. The property which creosote has of rapidly making preparations which still contain water transparent, is of great importance. By this means, objects which have lain in ordinary spirits, and even chromic acid, can be used after a few minutes. But when a preparation is to be arranged for mounting in Canada balsam, good oil of turpentine deserves the preference, decidedly, according to our experience.

Oil of Cloves.—This was first made known by Rindfleisch, as a medium for rendering tissues transparent, in the place of the oil of turpentine; it has also been warmly recommended by others. It renders tissues which contain water transparent in the same manner as creosote, but more slowly. A series of other ethereal oils, such as cinnamon, anise, bergamot, and rosmarine oils, act similarly to it, while others, like oil of turpentine, render only objects which have been deprived of their water transparent; as orange, juniper, mentha crispa, citron, and cajeput oils (Stieda).

Benzine.—This has been proposed for dissolving and thinning Canada balsam in the place of chloroform and oil of turpentine (Bastian). Toldt recently recommended pure benzine as an excellent medium for making fat tissue transparent, after the previous momentary action of alcohol.

In what has been mentioned above, we have been obliged to adhere to the methods which are still generally used by microscopists for determining the proportions of their reagents. Titrition is a far more certain and a much more convenient method for ascertaining the strength of a solution, and for producing them of definite proportions.

In order to estimate the acid and alkaline contents of such fluids the following is necessary:—

The apparatus (fig. 74), which is quite indispensable for the

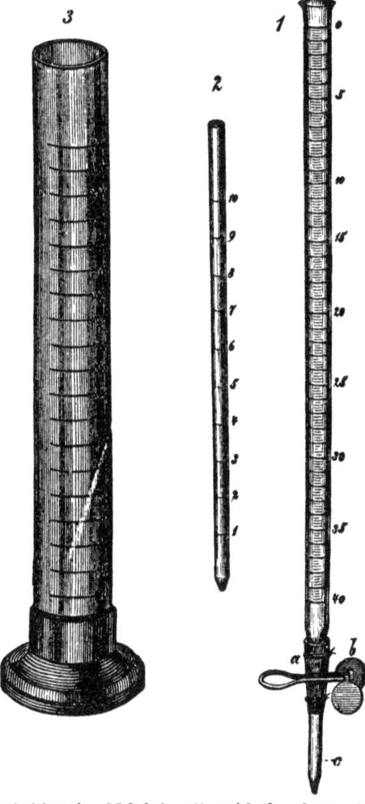

Fig. 74. Apparatus for titrition. 1, a Mohr's burette, with the clamp at *a*, which is opened by pressing on the two metallic buttons at *b*, allowing the fluid to escape from the tube *c*; 2, a pipette; 3, a cylindrical measure.

investigation, consists of:—*a*) two Mohr's burettes (1) of about 60 ccm. capacity divided into $\frac{1}{5}$ of a ccm.; *b*) a pipette (2) which will allow from 10 to 15 ccm. to run out, and is divided into $\frac{1}{10}$ of a ccm., and finally *c*) a cylindrical measure (3) of 100 or several hundred ccm. capacity. The divisions of the latter are from 5–5, or 10–10 ccm., and must retain the designated quantity of fluid, and not allow it to flow out; while the burette and pipette are so divided as only to designate the number of ccm. which they allow to flow or drop out. (Such burettes, pipettes, and cylindrical measures may now be purchased everywhere).

The use of the pipette is self-evident. As to the burette, it is filled to the zero division at its upper part with the reagent (the test acid or the test alkali), and, by making slight pressure on the clamp, the fluid is allowed to flow out either in a stream or by single drops, as may be necessary.

The normal acid and normal alkali solutions are used for the construction of the test fluids, so far as determining the ordinary reagents (acids and alkalies) is concerned. Under these are understood solutions which contain an equivalent weight of the active substance of the reagent, expressed in grammes, dissolved in 1000 ccm. (1 litre) of fluid.

1. *The Normal Oxalic Acid Solution.*—In its formation 6.4 grammes of pure crystallized, uneffloresced oxalic acid is to be dissolved in water, and this solution diluted sufficiently to make 100 ccm. of fluid. (The volume is always to be measured at the same temperature at which the solution is to be used, therefore at 14 to 16 R.) This normal oxalic acid solution is, however, only indirectly employed, that is, in the preparation of other normal acid and normal alkali solutions. For this reason the greatest accuracy and care should be employed in the preparation of this first and most important solution.

One ccm. of this oxalic acid solution contains, as we are already aware, 0.064 grm. of oxalic acid. For its saturation the corresponding equivalent of bases is necessary, therefore of—

a. Soda	0.031	grm.	NaO.
b. Potash	0.0472	"	KO.
c. Ammonia	0.017	"	NH_3.
d. Lime	0.028	"	CaO.
e. Baryta	0.0765	"	BaO.

2. *Normal Potash Solution.*—From a freshly prepared solution of potash, free from carbonic acid, take, with a pipette, 5 ccm., color it to a weak blue with a few drops of tincture of lacmus, and, while stirring, allow the normal oxalic acid solution to flow into it from the burette till the color begins to turn

red. Given, we had used 8 ccm. of normal acid solution; we then add to each 5 ccm. of our potash solution, 3 ccm. of water. In this case we have a normal potash solution; 1 ccm. of the same is just sufficient to saturate 1 ccm. of the oxalic acid solution; it therefore contains the above-mentioned quantity of potash, or 0.0472 grm.

It is clear that with the aid of this potash solution the quantity of acid present in any fluid may be determined at pleasure. By the neutralization of 1 ccm. of our normal potash solution is indicated the presence of—

a. Sulphuric acid = 0.04 grm. SO_3.
b. Nitric " = 0.054 " NO_5.
c. Muriatic " = 0.0365 " HCl.
d. Acetic " = 0.06 " $C_4H_4O_4$.

We limit ourselves in this citation to the investigation of the most important acids.

3. As actually pure oxalic acid belongs to the more expensive reagents, it is unnecessary to use just this acid in the estimation of alkalies. Sulphuric acid is generally used. Nothing is easier than the preparation of the normal sulphuric acid solution. Sulphuric acid, diluted at pleasure, is allowed to flow from a burette into 5 ccm. of normal potash solution, to which several drops of tincture of litmus has been added, till it becomes red. The acid and alkali solutions are then to be diluted, corresponding to what was mentioned above concerning potash, until an equal number of ccm. of each exactly neutralize each other. Accordingly, 1 ccm. of this normal sulphuric acid contains 0.04 grm. of SO_3, and for its neutralization exactly the same quantities of the bases are necessary as were previously given for oxalic acid.

Finally, we add two other test fluids, namely:—1. The normal silver solution for the quantitative estimation of common salt. One ccm. of the $\frac{1}{10}$ normal solution contains 0.0108 Ag., or 0.0170 $AgONO_5$. It corresponds to 0.00585 NaCl. 2. The normal chloride of sodium solution, used for the quantitative estimation of nitrate of silver. One ccm. of the $\frac{1}{10}$ normal solution contains 0.00585 NaCl. and corresponds to 0.0170 $AgONO_5$. In both cases a precipitate of chloride of silver takes place, which collects in lumps by strong shaking; the operation is completed when a drop of the test fluid no longer induces precipitation. To make the recognition more certain in the first of the two processes, several drops of simple chromate of potash solution may be added to the common salt solution, in which case the complete precipitation of the chloride of silver will be indicated by the red color of the chromate of silver which is formed.

A few examples may serve to make the method of employment clear.

1. We have 10 ccm. of a solution of soda, which required 22.2 ccm. of normal sulphuric acid for its neutralization. Now 1 ccm. of the normal sulphuric acid corresponds to 0.031 grm. NaO. By multiplication with 22.2 the quantity of soda contained in 10 ccm. of the titrated fluid will be found to be 0.6882, consequently 6.882 per cent. (disregarding the specific weight).

2. A solution of ammonia requires for 10 ccm. 12.6 ccm. normal sulphuric acid. One ccm. normal sulphuric acid corresponds to 0.017 HN_3. The quantity of ammonia is, therefore, 2.142 per cent.

3. 5 ccm. of acetic acid solution requires 41.7 ccm. of the normal potash solution, 10 would therefore require double this quantity, or 83.4. But the cubic centimetre of normal potash solution corresponds to 0.06 of acetic acid, and hence the proportion of acetic acid is 50.04 per cent.

4. 10 ccm. of a solution of common salt requires, for example, 12 ccm. of the $\frac{1}{10}$ normal silver solution. Now as 1 ccm. of the $\frac{1}{10}$ normal silver solution corresponds to 0.00585 NaCl., the common salt solution contains 0.702 per cent. of NaCl.

5. 10 ccm. of a solution of nitrate of silver requires 15.5 ccm. of the $\frac{1}{10}$ normal chloride of sodium solution. But 1 ccm. of the $\frac{1}{10}$ common salt solution corresponds to 0.017 $AgONO_5$, and the silver solution contains 2.635 per cent. of $AgONO_5$.

6. Supposing we desire to construct a 40 per cent. solution of acetic acid from the diluted acetic acid mentioned in No. 3. We learn from the proportion $40 : 100 = 50.04 : \times$, that we

have to dilute 100 ccm. of the acetic acid solution, which has been estimated by titrition, to 125.1 ccm.

7. Let us suppose the case, that we wish to prepare a soda solution of 20 per cent., and that a solution which we have titrated showed a proportion of 37.5 per cent. of NaO. We find by calculation that 100 ccm. of the latter solution is to be diluted to 187.5 ccm.

8. We wish to make a 1 per cent. solution of nitrate of silver. For this purpose we use the 2.635 per cent. solution of nitrate of silver of No. 5. It requires diluting with water to 263.5 ccm.

Section Eighth.

METHODS OF STAINING—IMPREGNATION WITH METALS—THE DRYING AND FREEZING PROCESSES.

I. Methods of Staining.

Delicate animal tissues frequently gain an extraordinary distinctness when impregnated with indifferent coloring materials, and complicated structures are frequently essentially cleared up in the same way. The non-reception of the color by other tissue elements is also of great value for assisting our discrimination in certain cases. These staining processes, therefore, form a very important accessory for histological investigations, and science is greatly indebted to Professor Gerlach, the inventor of carmine tingeing.

1. *Gerlach's method of tingeing with carmine.*

Gerlach first gave us this process in a small work ("Mikroskopische Studien aus dem Gebiete der menschlichen Morphologie." Erlangen), which appeared in the year 1858. In his carmine injections he had already noticed the eagerness with which the nuclear structures of the blood-vessels absorbed the carminate of ammonia, and how differently they behaved in this respect to the cells and intercellular substance. The cells also absorb coloring matter, but much more slowly and with greater difficulty, and always in lesser quantity than the nuclear formations. Intercellular substances act nearly indifferently.

Gerlach's first experiments were made on the brain and spinal cord. Fine sections of organs previously hardened in chromate of potash were placed in a moderately concentrated solution of the carminate of ammonia and left in it for 10 or 15 minutes. He then soaked them for several hours in water, which was frequently renewed, then treated them with acetic acid and placed them in absolute alcohol to remove the water. The carmine solution tinges even when it is still more diluted. Gerlach observed this at the very first, having one night left a section of a convolution of the cerebellum lying in some water slightly contaminated with carmine. In this case things showed themselves which were not to be recognized by the first method of staining. Thereupon Gerlach employed 2 or 3 drops of a concentrated solution of ammonia carmine to the ounce of water, leaving his sections in it from 2 to 4 days. This is the purport of the first statement of the inventor.

Since that time carmine tingeing has come into the most frequent use. Several years ago one observer went so far as to assume the presence in the central organs of several kinds of nerve cells, which he assigned different functions according to their capacity for absorbing carmine.

The directions for its use have been more or less fortunate. From what my own experience has taught, two evils are to be especially avoided in carmine staining; one is an excessive tingeing, ultimately inducing a very deep and diffuse red, which does not permit of further recognition of the preparation; the other is an infiltration of the elements of the tissue in consequence of the action of the ammonia.

Solutions as free as possible from ammonia should therefore be used. For this purpose take several grains of carmine, combine it with an ounce of distilled water and a few drops of ammonia. The fluid, in which a portion of the carmine has become dissolved, is to be filtered. Another portion of the carmine, which remains on the filter, may be kept for future use. When the filtrate has any appreciable smell of ammonia, the latter should be allowed to escape by leaving it in an open vessel under a bell-glass for a half or a whole day. If, after a time, granules of carmine are deposited, a drop of liquor ammonia serves to redissolve them. However, all solutions of carmine are very decomposable and their coloring power, unfortunately, very unequal—an inconvenience which cannot be entirely obviated.

The fluid thus obtained is, when used for tingeing, to be added, drop by drop, to water, in order to obtain at pleasure a lighter or more intensive red. For very delicate objects a combination of the coloring water with an equal quantity of glycerine is advantageous.

I recommend for this purpose three to six grains of carmine dissolved in exactly the necessary quantity of ammonia and mixed with one ounce of distilled water. To the filtered fluid, one ounce of good glycerine and two or three drachms of strong alcohol are to be added. The solution is to be used either as it is or with a further addition of glycerine.

The length of time which it is necessary for a portion of tissue to remain in the fluid is determined by the intensity of the color of the latter. Strong solutions tinge sufficiently in a few minutes, weaker ones require several hours. The preparations may be left without disadvantage for twenty-four hours in very weak solutions.

The stained pieces, when taken out, are first washed with pure water. They are then exposed for several minutes to a solution of acetic acid. I employ, as a rule, an ounce of distilled water with two or three drops of glacial acid; although a much stronger acid may be used for a much longer time without harm. The process may be readily modified where it is desirable to avoid the further infiltration of the tissue with water. The stained object may be deprived of its water with absolute alcohol, and then exposed to the glacial acid from one to twenty-four hours (Thiersch), or an acidulated alcohol may be directly applied. This may also be accomplished, as Beale correctly remarks, with glycerine containing glacial acetic acid (five drops to the ounce). Only with tissues of very unequal capacity for swelling, a saturated watery solution of oxalic acid deserves the preference to acetic acid (Thiersch). Nevertheless, it finally extracts the red from the nucleus, and the color is less intensive. Fresh tissues, or those hardened in alcohol, color best; not so good, and somewhat more slowly, those which have been hardened in chromic acid or chromate of potash. Good carmine preparations show the nuclei intensively reddened, likewise the axis cylinder of the nerve fibres. The coloring of the protoplasma is usually less lively; the connective tissue interstitial substance appears colorless, etc.

One soon learns to properly estimate the intensity of the color of the tissue. In general, the preparations destined for wet mounting (in weakly acidulated glycerine) are to be less deeply tinged than those for resinous mounting. The latter (best mounted cold in Canada balsam dissolved in chloroform) often furnish charming preparations for review.

Injected specimens are very susceptible of tingeing with many colors, such as chrome yellow and sulphate of baryta. The best kinds of soluble Prussian blue are also useful for staining; although a somewhat more strongly acidulated washing-water is necessary to retain the lively blue. It is more appropriate for objects injected with carmine to stain them blue or violet; nevertheless very handsome appearances may also be obtained with a very light red carmine.

The employment of ordinary red ink, of which use has here and there been made, is to be little recommended.

2. *Thiersch's Carmine Fluid.*

Professor Thiersch employs several methods of staining.

a. Red Fluid.

Carmine..................1 part.
Caustic ammonia..........1 "
Distilled water............3 "

This solution is to be filtered.

A second solution is to be prepared of—

Oxalic acid.............. 1 part.
Distilled water...........22 "

One part of the carmine solution is to be mixed with 8 parts of the oxalic acid solution, and 12 parts of absolute alcohol are to be added, and filtered.

If the filtrate is orange-colored instead of dark-red, more ammonia is added to compensate for the preponderance of oxalic acid, and the orange becomes red. The orange color may also be used for staining. If crystals of oxalate of ammonia are afterwards formed in the solution, which may take place from the addition of liquor ammonia or alcohol, it must be filtered a second time.

According to Thiersch's experience, this solution stains tissues very uniformly in the short space of from 1 to 3 minutes, without causing them to swell and without loosening shreds of epithelium. After the staining, the coloring material adhering to the surface of the preparation is to be washed off with alcohol of about 80 per cent. When the color has become too dark or diffuse, the preparation is to be washed out with an alcoholic solution of oxalic acid.

b. Lilac Carmine Fluid.

Borax	4 parts.
Distilled water	56 "
Dissolve and add, of carmine	1 "

The red solution is to be mixed with twice its volume of absolute alcohol, and filtered. Carmine and borax remain on the filter; this precipitate, dissolved in water, may be again used.

Thiersch found that this solution colored more slowly than the simple red one, and that it was especially attracted by cartilage and by bones which have been decalcified by chromic acid. Alcoholic solutions of oxalic and boracic acids serve for washing. Preparations may be very beautifully stained by tingeing them in the lilac solution and then placing them for an instant in the red fluid.

3. *Beale's Carmine Fluid.*

This meritorious investigator has recommended the following mixture:—

Carmine	10 grains.
Strong liquor ammonia	½ drachm.
Good glycerine	2 ounces.
Distilled water	2 "
Alcohol	½ ounce.

The pulverized carmine is to be placed in a test tube and the ammonia added to it. By agitation, and with the aid of heat, the carmine is soon dissolved. The ammoniacal solution is to be boiled for a few seconds and then allowed to cool. After the lapse of an hour, much of the excess of ammonia will have escaped. The glycerine, water, and alcohol may then be added and the whole passed through a filter or allowed to stand for some time, and the perfectly clear supernatant fluid poured off and kept for use. Various tissues require very unequal time for staining.

Heidenhain's process (first used for the mucous membrane of the stomach) is a modification of Beale's method. The solution is to be prepared without the alcohol, and the superfluous ammonia almost entirely removed by warming on the water-bath, or the addition of acetic acid. (The proportion of ammonia is proper when, after 24 hours, all the carmine of a solution remaining in a small open dish has become deposited in granules.) The specimen is to be placed in a watch-glass with this nearly free from ammonia solution. This watch-glass, and another containing water and a trace of ammonia, are to be placed in a flat glass vessel which can be accurately closed and left for 24 hours. Afterwards washed in glycerine, and then placed in pure glycerine, the preparations are to be exposed in the same manner to the vapor of a small quantity of acetic acid. Such a protective method of staining certainly presents manifold advantages. Modifications of the same may also be readily made.

4. *Acid Carmine Fluid.*

Schweigger Seidel recommends the following method for tingeing objects previously treated with acids:—An ordinary ammoniacal solution of carmine is to be mixed with acetic acid in excess and filtered. The red solution thus obtained stains diffusely; but after the addition of glycerine, tempered with a little muriatic acid (1 : 200), to the microscopic preparation, the cell body is seen to gradually lose its color, and the carmine is only retained by the nucleus. For mounting in glycerine, the preparation is to be washed with water containing acetic acid.

5. *Staining with Anilin red* (*Fuchsin*).

The idea of employing the anilin colors, so much used at present, for tingeing animal tissues was very natural. A series of experiments which I had undertaken for this purpose showed the pre-eminent usefulness of this coloring material.

Fuchsin (crystallized)....... 1 centigramme.
Absolute alcohol...........20–25 drops.
Distilled water............. 15 cubic centimetres.

A beautiful moderately intense red color results. It colors many of the animal tissues almost instantly, and without altering them. It is especially adapted for the study of epithelium, hyaloid membranes, the lens and corpus vitreum. Diluted with a little water, this solution tinges the vibrating cilia of the ciliated epithelium of the frog, without causing the motion to cease. The colored blood cells also become stained, although slowly. The same solution of fuchsin is also useful for coloring the ganglion cells and the cellular elements of lymphatic glands; but it appears to me to be not so well adapted for cartilage and bones. The nerve tubes, after an immersion of several hours, become slightly red, and their axis cylinders become sensibly darker.

The above examples show that the solution produces effects which are superior in many respects to those obtained with carmine. The promptitude and uniformity of the coloration are qualities which render the fuchsin solution especially valuable as a staining medium for instantaneous demonstrations, and for coloring pale delicate cells, which thus become more distinct without suffering alteration. It is very unfortunate that alcohol soon extracts the color, so that it is impossible to preserve the preparations in Canada balsam.*

6. *Blue Colors for Staining.*

It is desirable in many cases to make use of a blue fluid, especially for staining specimens injected with carmine. Other preparations also appear very beautiful when stained with this fluid, so that for many purposes I am inclined to give it the preference to carmine. Several of these methods are now practised, as with the sulph-indigate of potash (so-called indigo carmine), with anilin blue, and soluble parme.

* The nitrate of roseanilin (Magenta red) in watery solution has been recommended by Roberts and Abbey.

a. Blue Staining with Indigo Carmine.

The following solution has been recommended by Professor Thiersch:—

Oxalic acid....................	1 part.
Distilled water.................	22—30 parts.

Indigo carmine, as much as the solution will take up.

The soda salt also affords an excellent blue fluid. If the blue color is in excess it may be removed with a solution of oxalic acid in alcohol.

This blue fluid (which may also be diluted at pleasure with alcohol), when concentrated, tinges very rapidly and uniformly. According to the observation of its inventor, it is very suitable for coloring the axis cylinders and nerve cells of the brain and spinal cord, previously hardened in chromic acid.

b. Anilin Blue Fluid.

Ordinary anilin blue is insoluble in water. By treating it with sulphuric acid, the soluble blue may be obtained. This may be simply dissolved in water until it assumes a deep cobalt color, or the following solution may be prepared:—

Soluble anilin blue....	2 centigrammes.
Distilled water.......	25 cubic centimetres.
Alcohol..............	20—25 drops.

This fluid stains tissues preserved in alcohol of a lively blue, and in a few minutes, but those preserved in chromic acid are not colored so rapidly. This color may be preserved in water, alcohol, and glycerine, and is not altered by the addition of acids. The lymphatic glands, spleen, walls of the intestines, and more particularly sections of the brain and spinal cord assume, under its influence, a fine appearance. I have made very extended use of it for years and can recommend it thoroughly.

This method of staining has recently been improved by Heidenhain.

He employs the neutral reacting aqueous solution in still higher dilution, so that, when poured into a watch-glass, it shows on a light ground a forget-me-not blue color. The sections (from alcoholic preparations) remain for a day in 4 ccm. of this fluid in a moist place, and are then to be immediately mounted in glycerine and cemented. The color and coloring power of the solution are considerably increased by the addition of a little acetic acid, or even its vapor; the vapor of ammonia, on the contrary, deprives it of its color entirely.

c. *Soluble Parme Fluid.*

This substance, which is obtained by treating diphenylate of roseanilin with sulphuric acid, when dissolved in water in about the proportion of 1 : 1000, gives a gorgeous blue, running into violet, and colors the various tissues in a few minutes. They are then to be washed in water, and either examined in glycerine or, after being deprived of their water by absolute alcohol, are to be mounted in Canada balsam.

7. *Violet Tingeing with Hæmatoxylin.*

Bœhmer has brought to our knowledge a valuable coloring medium in hæmatoxylin. He dissolves 20 grains of the same in half an ounce of absolute alcohol, and prepares a solution of alum of 2 grains to the ounce of water. A watch-glass is filled with the latter, and by the addition of two or three drops of the solution of hæmatoxylin a violet color is obtained, which should act on the preparation for a half or a whole day. I have also obtained the same result in a more simple way. An aqueous solution of the extract of logwood is to be mixed with a solution of alum (1 part of the salt to 8 parts of water) till the deep impure red color has become violet, and then filtered.

Hardened preparations (alcoholic or chromic acid) are to be placed in it for about half an hour, fresh sections requiring a longer time, and then washed in water or alcohol. Most tissues are colored very beautifully and evenly, and objects injected with carmine or Prussian blue afford very good preparations with it, and may be permanently mounted in Canada balsam.

8. *Blue Tingeing with Molybdate of Ammonia.*

Krause has recommended this salt in a neutral solution of 5 per cent. as an indifferent medium for giving a marine blue color to various tissues, as those of the nervous apparatus, lymphatic glands, and ciliated epithelial cells. The staining is completed in 24 hours, at an ordinary temperature and under the action of light. The preparations become brown, and assume a consistence suitable for making sections, by supplementary exposure to the action of tannic acid (1 : 1.5) or pyrogallic acid (20 per cent).

9. *Double Staining with Carmine and Picric Acid.*

To the methods of staining previously known, a new one has been added, by E. Schwarz, for double tingeing. He combines staining with carmine with that by picric acid.

That author places the tissues in a mixture consisting of 1 part creosote, 10 parts acetic acid, and 20 parts water. The preparations are to be immersed in this mixture, while it is boiling, for about a minute, and are then to be dried (for two or three days). Thin sections are to be made and immersed for an hour in water slightly acidulated with acetic acid, and then washed out in distilled water. Next they are to be put in an extremely dilute watery solution of ammoniacal carmine, and, after being again washed in water, are exposed for two hours to a solution of picric acid (0.066 grm. to 400 ccm. water). The sections are then placed on a slide, the superfluous acid is allowed to flow away, and a mixture of 4 parts of creosote to 1 part of turpentine, which has become resinous from age, is dropped on to it. In about half an hour the specimen, which has become transparent, is to be mounted in Canada balsam.

If it is undesirable to use the creasote mixture, the sections are to be removed from the watery picric acid solution to an alcoholic mixture of corresponding strength, in order to deprive them of their water.

A peculiar effect is thus obtained. Epithelial and glandular cells, muscles, and the walls of vessels show a yellowish color with reddened nuclei, while the connective tissue is not colored by the picric acid, and only presents the carmine color.

These preparations are very handsome, and the method promises to be of importance, especially for the demonstration of muscular elements.

10. *Picro-Carmine Staining.*

Ranvier, a meritorious investigator, is the inventor of this method of staining. A saturated solution of picric acid is to be prepared and filtered, and to this a strong ammoniacal solution of carmine is to be added, drop by drop, till neutralization takes place. Slight precipitations, caused by commencing over-acidulation, may be removed by filtration (Flemming).

II. Metallic Impregnations.

Within a few years histological investigation has gained important accessories in several readily reducible combinations of the noble metals. Aqueous solutions of nitrate of silver, osmic acid, chloride of gold and of palladium, have thus far come into use. Their effects are essentially different, so that we shall have to discuss each solution by itself.

a. Nitrate of Silver.

Lapis infernalis, in solution or in substance, has been used for a number of years for obtaining precipitates of silver in the cornea. This method was first extensively used by Recklinghausen ("Die Lymphgefässe," Berlin, 1862) on animal tissues. His then endeavored to ascertain the nature and conditions of the precipitates. A great number of observers have lately made good and bad use of the process.

Only fresh (or but slightly altered) tissues, which are still saturated with the albuminous organic fluids, are suitable for impregnation with silver. As the action of nitrate of silver is, for the most part, confined to the surface, they should be chiefly thin, membranous structures. Artificial surfaces, made by the knife, generally produce only very unsatisfactory results.

It is preferable to employ only very weak solutions, those of 0.5, 0.25, and 0.2 per cent., or still weaker, according to circumstances. Frequently the immersion lasts for only the fraction of a minute, till the fragment of tissue is seen to assume a whitish color. It is then to be washed in water and exposed to the light until a brownish color is noticed. The examination may be made in acidulated water or glycerine. Tingeing with carmine may also be advantageously combined with this process.

To increase its durability, Legros recommends the immersion of the colored object in a solution of hyposulphite of soda for an instant; it is then to be washed in distilled water. Silver preparations which have become too dark may be brightened up again by a prolonged action of this salt.

Although the silver method produces excellent specimens in many cases, there are a number of defects connected with it besides those which have been mentioned. One of these is that the nuclear structures very soon become indistinct, and later disappear entirely. Then, with every precaution, the desired result is not always obtained, and the appearances which the strongly acting nitrate of silver produces are frequently very dissimilar, and are often so heterogeneous that the observer remains completely confused by them. The greatest foresight in the interpretation of such artificial productions is therefore necessary —and this, unfortunately, has been frequently neglected.

The silver treatment gives decidedly the best results with epithelium, especially with unstratified pavement cells, and the membranes and tubes covered by them. Here we have produced a mosaic, consisting of sometimes finer, sometimes broader, dark lines of demarcation, which enables us to recognize the contours of the cells most distinctly. This results either from blackening of the cement substance, or from the formation of the dark precipitate of silver in the narrow furrows between the cells. Such appearances are in no wise liable to misinterpretation so soon as the nuclei can be recognized or the scales isolated. We shall afterwards see to what beautiful discoveries in the structure of the finest blood and lymph passages this method has led.

However, even in this case, instead of the bright field surrounded by dark lines, we sometimes have a diffuse, brownish darkening of the scales, without the black lines of demarcation.

The outlines of smooth muscle cells are also made visible in a beautiful manner by this reagent. How far it may serve for demonstrating the finer structural relations of nerve tissue, future investigations will have to decide.

Opinions have thus far agreed quite as little with regard to the importance of the nitrate of silver solution for connective tissue and relative structures; on the contrary, the views of observers are very widely separated on this point.

According to Recklinghausen's statements, a diffuse coloration of the basis substance takes place, from which the cavities and cells glimmer forth in the form of bright spaces. Exactly the reverse may also take place, a dark, granular precipitate of silver being formed in them, while the interstitial substance remains bright. It has been recommended to use a solution of common salt to obtain the latter appearance (His).

We are inclined not to advise the silver method for connective tissue.

b. Osmic Acid (Hyperosmic Acid).

"The treatment of animal tissues with solutions of osmic acid (OsO_4), introduced by M. Schultze,* permits of a very manifold application. Organic substances reduce the acids from their solutions, and hence a combination of the former with a lower grade of oxidation, or perhaps with metallic osmium results, which combination, sooner or later, and independent of light, assumes a dark bluish black color, and resists decomposition. The reduction does not follow in the tissue as a granular precipitate; on the contrary, the elementary structures, if fresh when immersed in it, retain the same transparency and texture which they have in life, and are only altered so far as color is concerned. These changes of color take place in different tissues with very different degrees of rapidity, and on this is based an important advantage of the method. Evidently this condition is due to a difference in the reducing power, or to the relation of the organic substances to the reduced lower grade of oxidation. Through this peculiarity, osmic acid may render structural relations visible, which could in no other way be so summarily demonstrated. This is the case with the terminal tracheal cells in the luminous organ of the Lampyris. All kinds of fat cells and fat drops, and the medullary matter of

* I am indebted for this notice to the kindness of my highly respected colleague in Bonn.

central and peripheral nerves become rapidly colored black; more slowly, the substance of the ganglion cells and of the axis cylinders, the muscular fibres, and all highly albuminous elements, such as cells rich in protoplasma, red blood corpuscles, and the fibres of the lens. The intercellular substance of connective tissues which afford gelatine and mucus, cellulose, amylum, the watery intercellular fluid of many vegetable cells, which only contain traces of dissolved organic substances, color most slowly of all (M. Schultze and Rudneff). The chief value of the method depends on the property of osmic acid of preserving the most delicate, perishable tissues, which are most sensitive to reagents, in a condition apparently the same as in life; as, for example, embryonic tissues, the cells of connective tissue, the central organs and peripheral portions of the nervous system, the retina, etc. In this respect osmic acid excels all previously known reagents, as, properly applied, it prevents all granular coagulation, and does not allow even the structural changes which result from spontaneous post-mortem coagulation to take place. It is preferable to make use of pretty strong, that is, from one to two per cent., watery solutions. It is best to allow the tissues to remain in this solution but a short time, from a quarter of an hour to twenty-four hours. If allowed to act for several hours or days, the piece becomes quite hard and of a very dark color. As the solution does not penetrate very deeply, very small pieces should be selected for immersion. Weaker, $\frac{1}{10}$ per cent. solutions, which do not harden the tissues so much, may also be advantageously used. The exhalations from osmic acid are injurious to the respiratory organs and the conjunctiva, and are therefore to be carefully avoided."

c. Osmiamide.

Owsjannikow recommends Frémy's osmiamide (1 : 1000), that is, the amide combination of osmious acid OsO_3, which is OsO_2H_2N, as a substitute for the disagreeable volatile acid.

d. Chloride of Gold.

The action of the chloride of gold, which Cohnheim introduced into histology, is much slower and less energetic than that of a solution of nitrate of silver, so that a prolonged immersion of the (freshest possible) tissue is necessary, whereby the quantity of the fluid appears to make comparatively little difference. A 0.5 per cent. solution of the gold salt is to be used; it is preferable to temper it with a minimal quantity of acetic acid. Leave the specimen in the solution from 15 or 20 minutes to an hour or more, till the former assumes a distinct straw color, which here, in contradistinction to nitrate of silver, penetrates deeply. After washing the preparation in ordinary or distilled water, place it for 24 to 48 hours or more in acidulated water, leaving the vessel exposed to the light. If the reduction has taken place, the color varies; in the best cases it is a beautiful intense red, sometimes violet blue or a deep gray. They afterwards grow darker, assuming even a black tone of color.

According to Cohnheim's experience, the chloride of gold does not act on cornified cells and those having no protoplasma, like simple flattened epithelium and the scales of epidermis (nor on their so-called cementing substance); furthermore, it does not act on the interstitial substance of connective tissue and cartilage. Finally, it exerts no effect on the cell nucleus; still this is well preserved, the action of the gold impregnation being in general much milder and much less alterative than that of the silver impregnation. On the contrary, the chloride of gold is energetically and with relative rapidity reduced by the protoplasma of the cell, such as that of the lymphoid and glandular cells and the cellular elements of the connective tissue and cartilage; further, by the capillary vessels and the muscles. The chloride of gold reduces most energetically—and herein the chief value of this new method appears to lie—the elements of the nervous system, the ganglia, the medullary sheaths of the nerves, which assume a dark, almost blue red color, and the axis cylinder, which assumes a brighter, more lively red color.

According to what has just been mentioned, the impregnation with gold promised to be of the greatest importance for the finer anatomy of the nervous system, the originator of the method having also succeeded in making a beautiful discovery in the corneal nerves. Unfortunately, the method soon proved to be almost capriciously uncertain, in many cases leaving one entirely in the lurch, while other observers again obtained favorable results. Many have made use of solutions as weak as 0.005 per cent. Immersion in a solution of sulphate of oxide of iron is said to induce rapid reduction (Nathusius).

[Dr. Francis Delafield recommends that the tissue be dipped for a moment in water containing a minimal quantity of acetic acid, before exposing it to the gold solution—a procedure which I find to improve the results materially.]

e. Chloride of Gold and Potassium.

This was first used by Gerlach, several years ago, in solutions of 0.01 per cent., for sections of the spinal cord which had been hardened in the bichromate of ammonia. It was afterwards used by Arnold, likewise extremely diluted, for the fresh sympathetic of the frog. We shall again refer in a more extended manner to this method.

f. Protochloride of Palladium (Chloride of Palladium).

A few years ago, F. E. Schulze made us acquainted with the action of this salt. In order to dissolve the dry salt in distilled water, it is necessary to add a slight quantity of muriatic acid. He employs a solution of 0.1 per cent. (1:800—1:1500). From a half to a whole ounce of this fluid (of a wine-yellow appearance) hardens a piece of tissue of the size of a bean, in the course of two or three days, to a consistence proper for cutting, and at the same time colors it. According to the experience of the discoverers, chloride of palladium is especially adapted for the recognition of striated and smooth muscles, which thereby become brownish and straw-colored. Cells rich in protoplasma (epidermis and glands) likewise assume a yellow color. Cornified, fat, and connective tissues do not become colored. Furthermore, the medullary matter of the nerves assumes a black color, by direct action. Schulze also praises this reagent for the retina and crystalline lenses; like the chloride of gold, it causes the nuclear formations to appear sharply, and by subsequently tingeing the connective tissue with carmine, very instructive preparations are afforded. It is an unpleasant circumstance that with many tissues, as the brain and epidermis, the action remains quite superficial.

The preparations are to be carefully washed and mounted in glycerine.

g. Prussian Blue.

Leber recommends a peculiar method of impregnation, more especially, however, for the cornea of the frog. The fresh organ is to be immersed for several minutes in a 0.5–1 per cent. solution of a protoxide salt of iron. It is then to be taken out for the removal of its epithelium, after which it is to be replaced in the fluid for a short time, so that the total action amounts to about five minutes. After washing the cornea with water, it is to be seized with the forceps and moved to and fro for a few moments in a 1 per cent. solution of ferrocyanide of potassium till the preparation assumes an intense and uniform blue color. Finally, after again washing the specimen in water, it shows a colored basis substance, while the corneal cells and canals have remained transparent. The color penetrates very deeply, and subsequent tingeing with iodine, carmine, and fuchsine readily succeeds. This method promises to be useful.

III. The Drying Method.

We also add to the methods above mentioned, that of drying animal tissues. The object of this process is to give the parts such a degree of hardness and firmness, by depriving them of their water, that the thinnest sections may be obtained by the aid of a sharp knife, which again swell up by the addition of water, and thus resume their natural appearance. We have already, in a previous section, become acquainted with a series of chemical reagents, such as chromic acid, chromate of potash, and alcohol, which are used for the same purpose.

The drying process is decidedly better adapted for many tissues and parts of the body, as it does not render them opaque. Especially compact structures, organs rich in connective tissue, as the skin, the tendons, and the walls of vessels, also the lungs (even injected preparations of the same), muscles, epidermis, crystalline lens, and the umbilical cord may be treated in this manner with the greatest advantage. The drying process is less suitable for glands, lymphatic glands, and delicate mucous membranes. It is unserviceable for the brain, spinal cord, the nerves, and their terminal expansions in the higher organs of sense, in consequence of their extreme softness and instability.

The management of the parts is very simple. They may be dried on a board or a piece of cork, to which, in certain cases, a convex surface may be given. To avoid wrinkling, many textures may be advantageously stretched out and fastened to the board or cork with pins. The temperature should not be too low, as decomposition might take place; but considerable elevation of temperature is to be avoided on account of the coagulation of the albumen. A temperature of 30 or 40° C. is most suitable. The sun of a warm day may also be very well employed for this purpose. If it be desired to avoid warmth, the sulphuric acid or the chloride of calcium apparatus of the chemical laboratories may be used.

The pieces selected for drying should not be too large and the drying should not be overdone, as they may become so brittle as to prevent one from obtaining fine sections, in consequence of the cracks and flaws which occur. Now and then it will be found most suitable to have a piece not entirely dried, having the consistency of wax. Naturally the blade of the knife should be dry. If the preparation is on cork, this may be used as a support in cutting, hard wood being injurious to the knife.

The thin sections which are made are to be softened in pure water, or in water to which a little acetic acid has been added. If they are to be stained, they may be placed directly into the ammoniacal solution of carmine. It is less convenient to soften the sections on the slide than in a watch-glass or a glass box. This process requires a few minutes' time, in order to allow the air vesicles to escape from the spaces of the tissue.

Dried pieces kept in a box, with the addition of a piece of camphor, constitute excellent material for many histological demonstrations.

IV. The Freezing Method.

This process, which has been made use of more recently, also affords good results, and in a much more conservative manner than that of drying. The preparation is allowed to freeze at a temperature of 6, 8, 10, or 15° C., according to necessity, until it assumes a consistency which will permit fine sections to be made with a cooled razor. The object is more convenient to handle if it is allowed to freeze on a strip of cork; it is also judicious to make use of an artificial freezing mixture. Nerves and muscles have been treated in this manner with good results (Chrzonszczewsky, Cohnheim). Glands, such as salivary glands, livers, kidneys, spleens, the lungs, the skin, and the bodies of embryos also afford excellent appearances (Kölliker); likewise ganglia (Arnold). Indifferent media, such as iodine serum, are to be used in examining the sections.

Section Ninth.

METHOD OF INJECTING.

Of the greatest value for histological studies is the artificial filling of the vascular systems of the part to be examined with colored masses; a procedure which, unfortunately, is still too much neglected by many, inasmuch as without having obtained the necessary practice, it may here and there appear as though such a procedure were somewhat superfluous—a luxurious adjunct. And yet this important accessory should never be neglected in any investigation which is at all accurate of normal or pathological textural relations; for much in the construction of an organ at once assumes the greatest clearness and intelligibility after its capillary system has been filled, and the desired disclosures with regard to the vascular abundance or poverty of a part are at once obtained. The art of injecting can only be learned, and its execution is by no means easy. Much, indeed the greater part, depends on apparently unimportant expedients, on little artifices, as well as on a skill only to be obtained by practice. Nevertheless, with the necessary perseverance, and by not being discouraged by the almost unexceptionably unsuccessful first attempts, one soon attains the desired skill, especially if one renounces the idea of obtaining perfectly beautiful injections at the beginning. Success is gradually more and more readily obtained, and the satisfaction derived from the little work of art which is finally produced, has already been for many the incentive to further investigations.

In the following pages we shall attempt to bring before the reader the most important of the technicalities of injecting, and to render especially prominent that which we have learned from our own experience in regard to this subject. At the same time we are quite willing to admit that others might have replaced many things that are here noticed with much that would perhaps have been better. Although all such directions are not capable of thoroughly supplying that which may be obtained much more rapidly from the practical instruction of an experienced teacher, they will, nevertheless, present serviceable hints to many autodidacts.

It will not be without interest, however, to previously give a cursory glance at the history of the origin of this art.

The art of injecting, filling the canal systems of the body with colored or other readily recognizable masses, is, in its first crude commencement, relatively an old one. Hyrtl, in his important "Handbuch der praktischen Zergliederungskunst," Wien, 1860, has given us an accurate and interesting history of this process. As early as the seventeenth century, wax and also mercury were used for this purpose. Gelatine was first employed for injections in the beginning of the eighteenth century.

Among the older anatomists, the Hollander, Ruysch (1638–1731), through his methods of injecting, obtained great renown —undeservedly, as we are at present, after accurate historical investigation, obliged to say, like so many of the celebrities of older and newer times. Tallow (tempered, in part, with wax) colored with cinnabar formed the mass used by him. N. Lieberkühn (1711–1746), on the contrary, accomplished considerable for his time, as early as the first half of the eighteenth century. Even at the present time his preparations deserve to be called excellent, as we are assured by Hyrtl, the most competent investigator in this department. He made use of a mixture of wax, colophony, and turpentine, and, as a coloring medium, cinnabar. At a more recent period, Sömmering, Döllinger, and Berres accomplished considerable in this department. Among those more recently engaged in this direction, the name of Hyrtl shines above all others. Others may be honorably associated with him, as, for example, Quekett, Gerlach, Thiersch, Beale, etc.

Naturally, the method of injecting interests us here only in so far as it is adapted for microscopico-histological studies, so that we shall pass over with entire silence the technicology of coarser injections.

Among the numerous methods, two kinds may be distinguished.

1. Mixtures which are fluid when warmed, and again become solid when cooled.

2. Mixtures which flow when cold.

Among the materials of the first kind, resinous and gelatinous substances have, as was above remarked, come into use. Hyrtl, who, among the living, has had the greatest experience in this department, informs us that the former renders excellent service in the injection of glandular organs and all capillary vessels of larger diameter; but, on the contrary, in other parts of the body—for example, in filling the subserous blood-vessels or those of the mucous membranes of the air-passages, the œsophagus, the stomach, the perichondrium, the medulla of bones and of the testicle—they are unserviceable. It is altogether an error to believe that a certain injection mixture is equally useful for all organs.

The Vienna anatomist prepares a resinous mixture in the following manner:—He evaporates the purest copal or mastic varnish to the consistence of syrup, and then mixes with it about one-eighth as much cinnabar, which is to be carefully rubbed with the varnish on a grinding-slab. A very slight addition of virgin wax is also made, to give the mixture more consistence.

Some time ago I made use of such a mixture, by way of experiment, and saw how, with a little practice, very handsome objects might be obtained, if the preparations were not required for finer histological studies, but rather for use with weaker magnifying powers.

Those who desire to investigate the finer structure of the organ to be injected should, therefore, have recourse to gelatine. The low degree of temperature at which a gelatine injection is possible, although not sufficient for the resinous injection, is an advantage which cannot be too highly prized. Very properly, therefore, histologists have preferred the use of gelatinous mixtures for their injections, Sömmering and Döllinger having even in olden times accomplished excellent results with the same. The subsequent drying which ensues with the ordinary older methods of preservation is accompanied by a certain shrinking of the tubes which have been filled, an evil which is induced by the loss of water; so that such objects frequently do not exhibit the full, firm appearance presented by the resinous preparations. Nevertheless, the much greater readiness with which the watery solution of gelatine passes through the vessels whose walls are moistened with water is an advantage which can be obtained with no other mixture which solidifies, especially for organs with narrow capillaries. Moreover, this shrinking may be considerably limited by careful mounting.

Disregarding the coloring materials at present, in order to prepare such a solution of gelatine and afterwards make use of it, several precautionary measures are necessary.

Isinglass, as a relatively pure gelatine, has been frequently used. It is in no wise necessary, and its high price and the slowness with which it hardens in cooling are to be designated as disadvantages. More recently I have frequently used the thin, transparent gelatine tablets which are met with in commerce as "Gélatine de Paris," and which constitute, it is true, a mixture which is also not to be numbered among the cheapest. The latter may be formed, however, from the better sorts of ordinary Cologne gelatine.

For dissolving gelatine the process most to be recommended is the following:—

The gelatine is to be broken to pieces and then soaked in distilled or rain water for several hours. The water is to be poured off and renewed, the gelatine is then to be dissolved in a water bath, *never immediately over the fire*, and the solution filtered through flannel into a porcelain dish. The coloring material, the necessary directions for which follow further below, is to be added to the solution while it is still warm. The consistence which is to be given to the gelatine mixture should be dependent on the individual circumstances. A thin gelatine fluid is sufficient, if a pulverized granular coloring matter is added in the form of a thick pulp. If the coloring material is directly precipitated in the injection fluid by pouring together solutions of two kinds of substances, a saturated gelatine fluid should be employed With a little practice one soon learns to hit upon the proper proportions.

In using such an injection fluid it is to be warmed in the same manner, for which purpose the same dish may be used several times in rapid succession. Such a fluid cannot be kept for a long time, however (even in a camphorated atmosphere), without moulding, or even without losing its former homogeneous constitution, which is so important; so that one is often put to the unpleasant, time-consuming necessity of preparing the gelatine fluid anew.*

For the further treatment of specimens injected with gelatine, see the end of this section.

Quite a variety of coloring materials may be advantageously added to the gelatine fluid; they will be mentioned further below.

The use of injection mixtures which solidify on cooling always consumes time, as was remarked, and requires a variety of arrangements. The discovery of a material which is fluid when cold, and which may be used at any time, must therefore appear to be of great value. A number of such mixtures have been invented and recommended in the course of time.

We will first mention a process practised by Bowman, the English histologist, which, although it may be momentarily employed, is less serviceable for producing a good injection than for coloring the blood-vessels of an organ and rendering them visible for microscopic examination. This method consists in forcing two solutions of salts after each other through the same vascular system, in which a lively colored precipitate is thus produced. Bowman used for this purpose the acetate of lead and the chromate of potash. A few experiments which I once made with this method gave a satisfactory view of the course of the vessels. But such a preparation is by no means beautiful.

For his cold injections, as he informs us in his "Zergliederungskunst," Hyrtl also employs the previously mentioned resinous mixture, to which he gives the consistence of an ordinary coarse injection fluid by the addition of a litte wax and red lead. He rubs a portion of the same in a dish, with the addition of

* One of Thiersch's most recent recipes follows further below (p. 182).

ether, to the consistence of syrup. He then adds the coloring material, in about the proportion of 1:8, and again rubs the whole with ether until the mixture becomes completely fluid. Hyrtl commends the facility and convenience of manipulation of this method. In consequence of the evaporation of the ether, the injected organ is ready for examination in a quarter of an hour.

I have of late made the most extended use of a mixture recommended by Beale ("The Microscope in its Application to Practical Medicine." London, 1858, p. 67), which is composed of glycerine, water, and alcohol, for filling the smaller vascular systems. This mixture excels all others with which I am acquainted for its facility of penetration, besides which it affects the tissues less than any of the mixtures in use. As the mixture does not become decomposed or altered in any way, it may be preserved for any desirable length of time. It is *par excellence* the histologist's injecting fluid, and when the preparations are mounted moist they present a most exquisite appearance. It has also afforded me very passable results when mounted dry by means of Canada balsam prepared in a special manner. Still the gelatine fluids are preferable for the latter preparations, and, in consequence of their consistence, they are indispensable for injecting the larger organs.

Having discussed the injecting fluids which are at present generally made use of, let us now pass to the consideration of the coloring materials which may be employed with the same. These coloring substances may be divided into *granular*, permitting of examination only with incident light, and *transparent*, suitable for ordinary histological investigations. Those of the first series are very numerous and were alone employed for the older injections. The number of the latter is, on the contrary, much smaller, consisting at the present time of only a few coloring materials.

If resinous mixtures are used, it is most convenient to employ the finest oil colors for artists, which are to be purchased in thin leaden tubes—a procedure made use of by Hyrtl. Among these "colors in tubes," the Vienna naturalist recommends for red, Chinese vermilion; for yellow, orange chrome yellow; for green, emerald green and verdigris; for white, Nottingham white and Cremnitz white; for blue, a mixture which he prepares from the last white color and Prussian blue.

These colors are expensive, it is true, but are of a fineness such as no one can prepare for one's self. They are therefore to be designated as of the first rank for opaque injections.

If gelatine is used as the solidifying substance, it is customary to employ red, yellow, or white fluids.

a. *Red Mixture* (Cinnabar).

Cinnabar is most commonly employed for this purpose. Commencing with a small quantity of a fine quality, it is to be rubbed up as carefully as possible in a mortar, with the gradual addition of water, and the process continued in this manner. A little carmine may be rubbed up with it, to heighten the color. The coloring matter is then to be gradually added to the warm gelatine solution, which is, at the same time, to be carefully stirred. It is a common fault of beginners that they use too little cinnabar, and hence they obtain an injection fluid with which separated scattered granules of coloring matter afterwards appear in the vessels. A good cinnabar injection should, on the contrary, yield a coherent coralline red color. In consequence of its considerable weight, cinnabar has the unpleasant peculiarity of collecting at the bottom of the gelatine solution, so that it is necessary to stir the mass before its use.

None of the other opaque coloring materials should be used in the form of the commercial preparations, unless they can be given to a professional color-rubber for pulverization, as it is otherwise impossible to reduce the granules to the necessary fineness. It is much preferable to procure them by careful precipitation from diluted solutions.

b. *Yellow Color* (Chrome yellow).

I regard this as the best and most readily manageable of all the opaque coloring materials. In order to obtain a good chrome yellow, 36 parts by weight of sugar of lead may be dissolved in 2 ounces of water, and likewise, in the same quantity of water, 15 parts of red chromate of potash. By carefully mixing these fluids, preferably in a high glass cylinder, a very finely granulated chromate of lead is produced, which is gradually deposited at the bottom of the vessel. This is to be washed with distilled water and then added, in the form of a thick slime, to the gelatine solution.

Harting (in his work on the Microscope, Vol. II., p. 123) gives the following recipe (which I have also found to be serviceable):—

4 ounces $1\frac{1}{3}$ drachm of acetate of lead, or sugar of lead, is to be dissolved in a quantity of water sufficient to make the whole volume 16 ounces.

2 ounces 1 drachm 28 grains of bichromate of potash are to be dissolved in a quantity of water sufficient to make the whole volume 32 ounces. In preparing the injecting fluid, take one part by measure of the solution of sugar of lead, 2 parts by measure of the solution of chromate of potash, and likewise 2 parts by volume of a saturated solution of gelatine. The chrome-yellow is to be first precipitated in a special vessel, and afterwards added to the gelatine. The precipitated chrome-yellow should not be allowed to stand too long, as it assumes a coarse granular form, in consequence of the conglomeration of the colored molecules.

c. *White Color. Carbonate of Lead. Zinc White. Sulphate of Baryta.*

A serviceable white fluid can only be obtained with difficulty, as in most of them the granules are usually too coarse. Harting, who instituted a series of experiments on this subject, gives the following recipe for producing a useful carbonate of lead:—

4 ounces $1\frac{1}{3}$ drachm of the acetate of lead are to be dissolved in water, so that the whole corresponds to a volume of 16 ounces.

3 ounces $1\frac{1}{3}$ drachm of carbonate of soda are to be dissolved in water, and the whole likewise made up to 16 ounces.

For the injection fluid, take one part by measure of the first solution, the same quantity of the second, and combine them with two parts of gelatine solution. Harting remarks concerning this fluid, that it passes through the vessels better than white lead combined with gelatine.

I formerly obtained tolerable injections with finely-pulverized zinc white. I have not used this coloring material, however, for years.

As a very fine white, although the color does not assume such complete uniformity, I recommend the sulphate of baryta. I made the most extended use of this material years ago, and am inclined to prefer it to the chrome-yellow, when it is necessary to have a finely-granulated and therefore readily penetrating fluid.

I employ the following process:—The salt in question is to be precipitated from about 4 to 6 ounces of a cold saturated solution of chloride of barium, in a glass cylinder, by the careful addition of sulphuric acid. After standing for some time, nearly the whole of the fluid, which has again become clear, is to be poured off, and the remainder, with the sulphate of baryta, which is deposited at the bottom of the vessel, is to be added, in the form of a thick slime, to about an equal volume of a concentrated solution of gelatine.

d. Chloride of Silver.

Teichmann, in his excellent work, mentions a new, very efficient, although expensive injecting mixture of chloride of silver. He commends the same as rendering excellent service in certain cases, and says that its molecules possess a very considerable fineness, occasionally similar to those of chyle. It is an unpleasant circumstance that the mixture becomes black from the action of the light and sulphuretted hydrogen. But, like sulphate of baryta, the combination is so fixed that decomposition does not take place with the employment of reagents, and the specimens may be preserved in chromic acid, etc.

Three parts of nitrate of silver in solution are to be combined with the solution of gelatine, and then one part of common salt is to be added.

Considerably superior to these granular substances are the *transparent* coloring materials, that is, those whose particles are so fine that even with high magnifying powers the injected vessels still show a homogeneous color. Such injections are particularly to be recommended for histological investigations, as it is by their use only that it is possible to recognize the remaining structural conditions, while a complete injection with

12

opaque mixtures conceals, more or less, the finer structure of the organ. They will be substituted for the granular coloring materials by any one who has used them a few times when well prepared. The reproach which has here and there been made concerning these colors that they transude, refers only to those which are badly made, but is not applicable to good transparent materials. Unfortunately, the number of these is, as yet, but very small. Until recently, besides the soluble Prussian blue, there was only a red coloring material, carmine, known. Professor Thiersch, who has earned so much credit by his methods of injecting, has recently enriched us with a soluble yellow and green, and was so friendly as to communicate to me their composition.

We shall first speak of such of these coloring materials as are adapted for gelatine injections.

A number of different mixtures are at present known under the name of transparent Prussian blue. Of these, the second receipt deserves but little recommendation, as the blue color gradually fades, especially when the preparation is preserved in glycerine. The first coloring material is, on the contrary, excellent, and the last is also very much extolled.

1. *Thiersch's Prussian Blue with Oxalic Acid.*

The best receipt with which I have become acquainted runs as follows:—

Prepare a cold saturated solution of the sulphate of the protoxide of iron (A), a similar one of ferrocyanide of potassium, that is, prussiate of potash (B), and thirdly, a saturated solution of oxalic acid (C). Finally, a warm concentrated solution (2 : 1) of fine gelatine is necessary. About half an ounce of the gelatine solution is to be mixed in a porcelain dish with 6 ccm. of the solution A. In a second larger dish, one ounce of the gelatine solution is to be combined with 12 ccm. of the solution B, to which 12 ccm. of the oxalic acid solution C is afterwards added.

When the mixtures in both dishes have cooled to about 25 or 32° C., the contents of the first dish are to be added dropwise, and with constant stirring, to the mixture in the latter. After complete precipitation, the deep blue mixture which is formed is to be heated to 70 or 100° C. for a time and constantly stirred; finally, it is to be filtered through flannel.

The injecting fluid thus obtained keeps excellently in Canada balsam. The depth of its color may be readily modified to any desired degree by adding a larger quantity of the gelatine solution.

2. *Prussian Blue dissolved in Oxalic Acid.*

A pure Prussian blue, preferably one that has been obtained by precipitation, is to be dissolved with the necessary quantity of oxalic acid. The color is certainly very intense, so that a moderate quantity is sufficient to give a lively blue color to a dish of gelatine solution. This mixture, like all transparent coloring matters, in consequence of the infinite fineness of its granules, readily passes through the fine capillaries.

Harting recommends the following method (in which the quantity of oxalic acid appears too great):—

Take 1 part of Prussian blue, 1 part oxalic acid, 12 parts water, and 12 parts concentrated solution of gelatine. First, rub the oxalic acid in a mortar, and then add the Prussian blue. Thereupon the water is to be gradually added while constantly rubbing, and finally this blue fluid is to be added to the gelatine.

3. *Prussian Blue from Sulphate of Peroxide of Iron and Ferrocyanide of Iron.*

This color, which was first employed by Schröder van der Kolk and afterwards recommended by Harting, is a good one, although it requires somewhat more time for its preparation. Its granules are extremely fine, and hence it flows very readily. Nevertheless, the older preparations of my collection have lately become considerably faded, so that I rather prefer Thiersch's blue.

I have used the blue made exactly according to Harting's receipt, so that I can only recommend that.

$3\frac{1}{8}$ ounces of sulphate of iron is to be dissolved in from 20 to 25 ounces of water and slightly warmed. Then, by the addition of $4\frac{3}{4}$ drachms of sulphuric acid of 1.85 sp. wt., and the necessary quantity of nitric acid, the iron is changed to an oxy-salt. A sufficient quantity of water is then added to make the whole volume of fluid 40 ounces.

3 ounces 6¾ drachms of ferrocyanide of potassium (yellow prussiate of potash) is to be dissolved in water, and the whole volume of fluid increased to 80 ounees.

1 part by measure of the solution of oxide of iron, 2 parts by measure of the solution of yellow prussiate of potash, and likewise 2 parts of the gelatine solution are to be employed.

In order to prevent the gelatine from collecting in lumps and becoming ropy, I recommend the following method. The solution of ferrocyanide of potassium should be warmed and combined with the heated solution of gelatine. The solution of sulphate of protoxide of iron is then to be added by drops, and while constantly stirring the mixture, which is finally to be filtered through flannel.

4. *Soluble Prussian Blue.*

This is obtained by adding to a solution of ferro-cyanide of potassium in excess, a solution of perchloride of iron, or of another oxy-salt. The precipitate is to be collected on a filter, and after the fluid has filtered away, again washed with distilled water till (after the removal of the salts which were in the solution) a blue color begins to appear in the filtrate. The blue mass which is thus obtained becomes so finely divided in water that the impression of a solution arises.

Several years ago, Brücke recommended the following receipt for preparing such a soluble Prussian blue:—

Ferrocyanide of potassium 217 grammes, dissolved in 1 litre of distilled water.

Perchloride of iron 10 grammes, in 1 litre of distilled water.

Sulphate of soda, a cold saturated solution.

One volume of each of the two first solutions is to be mixed with one volume of the soda solution. The iron and soda solution is then to be gradually mixed with the ferro-cyanide and soda solution, with constant stirring.

An extremely fine blue is obtained, but I am inclined to place it after that of Thiersch, so far as its durability is concerned.

Sections of organs injected in this manner often appear colorless, but subsequently assume the blue color in oil of turpentine.

A good carmine mixture remains unexcelled as a transparent red. This substance requires careful preparation, it is true, and when not properly prepared it is completely useless, as it transudes in all directions.

5. *Gerlach's Carmine Fluid.*

Professor Gerlach, the inventor of the method of injecting with carmine, has had the kindness to communicate to me the composition of the fluids used by him, and to permit me to make them public.

Dissolve 5 grammes of the finest possible carmine in 4 grammes of water and ½ grm. of liquor ammonia. The mixture should be allowed to stand for several days in a vessel not closed air-tight, and then mixed with a solution containing 6 grammes of fine white French gelatine to 8 grm. of water. A few drops of acetic acid are then to be added, and the mixture injected at a temperature of 40–45° C.

I have made the most extended use of carmine fluids for a long time, and recommend, after many trials, the following method:—

Have ready a solution of ammonia and one of acetic acid, of which the number of drops necessary to neutralize each other has been previously determined.

Take 30–40 grains of the finest carmine, a determined number of drops of the solution of ammonia (the quantity may be greater or smaller as may be desired), and about half an ounce of distilled water; all of these are to be put in a mortar, and the carmine dissolved by rubbing. The solution is then to be filtered, which requires several hours, and a considerable loss of ammonia ensues in consequence of evaporation.

The ammoniacal solution of carmine is to be mixed with a filtered, moderately-heated concentrated solution of fine gelatine while stirring. The whole is then to be slightly heated on the water-bath, and the number of drops of the acetic acid solution necessary to neutralize the original solution of ammonia is to be slowly added, still constantly stirring the mixture. By this procedure a precipitation of the carmine in an acid solution of gelatine is obtained. If it be intended to inject organs of strongly alkaline reaction (for instance, those of human bodies which have been dead for some time), the acidity of the fluid may be increased by the further addition of several drops of acetic acid. The proportion of gelatine is to be increased or diminished according as a deeper or brighter red is desired.

This simple procedure never fails, if the carmine used be of a good sort (which is of great importance), and an increase of temperature beyond about 45° C. be avoided during the injection.

For the preservation of such a fluid—and, we might add, of other injection fluids containing gelatine—Thiersch recommends the following precautionary measures:—

Sulphate of quinine is to be added to the water used for dissolving the gelatine, in the proportion of two grains to the ounce of dry gelatine; pieces of camphor are also to be boiled in it; when the preparation of the injection fluid is completed, more camphor is to be added, and, finally, several pieces of camphor are to be placed on the mixture when solidified. Thus, even in the heat of summer, decomposition and the formation of mould may be prevented. I have made use of a camphorated atmosphere for a number of years, and can only recommend the recipe of this distinguished technist.

6. *Thiersch's Transparent Yellow.*

This beautiful yellow, which requires some care, however, to prepare it well, is to be obtained in the following manner:—

Prepare a watery solution of chromate of potash, in the proportion of 1 : 11 (A), and a second solution, equally strong, of the nitrate of lead (B).

Combine one part of the solution A with four parts of a concentrated solution of gelatine (about 20 ccm. to 80) in a dish. In a second dish, two parts of the solution of lead (B) is to be mixed with four parts of gelatine (about 40 ccm. to 80).

The contents of both dishes are then to be slowly and carefully mixed with each other at a temperature of about 25–32°

C., and with constant stirring. This mixture is to be heated to about 70 or 100° C., on the water-bath, for a considerable time (half an hour or more), and finally filtered through flannel.

When a dish of this yellow mixture has stood for some time, it is generally necessary to heat it for a considerable time and filter it again, in order to render it serviceable. I have used a double quantity of the solutions A and B for many purposes with advantage.

7. *Hoyer's Transparent Yellow.*

Hoyer recommends the following mixture as a yellow of the finest division, which also appears transparent in the smaller vessels and has a lively color:—

Equal parts of a solution of gelatine, a concentrated solution of bichromate of potash, and the same of sugar of lead (the neutral acetate of lead), are to be combined with each other in such a manner that the solution of gelatine and that of the bichromate of potash are united, and then heated nearly to the boiling point. To this is carefully added the sugar of lead solution, which should also be previously warmed.

8. *Thiersch's Transparent Green.*

Equal parts of the blue gelatine solution as used by Thiersch, and the yellow one mentioned under 6, when carefully mixed, heated for some time, and then filtered, make a good and handsome green.

Many transparent coloring materials are capable of a more advantageous application, however, than being combined with gelatine. They are combined with a peculiar mixture which flows when cold, and in this manner are obtained the best injection fluids yet known for histological investigations.

As we have made frequent use of them, the compositions used follow.

1. *Beale's Ordinary Blue.*

Dissolve 15 grains of ferrocyanide of potassium with 1 ounce of distilled water in a flask. Dilute from ½ a drachm to 2 scruples of the English muriated tincture of iron with another ounce of water. It is well to have this tincture of sesquichloride of iron accurately prepared according to the British pharmacopœia, by a good apothecary, in sufficient quantity to last for some time. The latter fluid is added by drops to the former, at the same time shaking it smartly. A mixture is then prepared of water 2 ounces, glycerine 1 ounce, ordinary (ethyl) alcohol 1 ounce, and methyl alcohol 1½ drachm.* This mixture is to be carefully added to the blue colored fluid, the flask being smartly shaken during the process, and the charming blue injection fluid is ready for use.

2. *Beale's Finest Blue.*

Beale has recently ("How to work with the Microscope." Third edition, p. 200) given a modified formula for the preparation of a cold flowing blue injection fluid, which, when well prepared, surpasses all others that I know of in fineness, so that after standing quietly for weeks the appearance of a solution remains unchanged, and there is not the least sediment formed. I prepare it, somewhat modified, in the following manner :—

Combine 10 drops of the muriated tincture of iron mentioned with half an ounce of good glycerine in a flask; in another flask 3 grains of ferrocyanide of potassium dissolved in a little water, to which is to be added another half ounce of glycerine. Both solutions are then to be very carefully mixed together, shaking them smartly, and finally half an ounce of water with 3 drops of strong muriatic acid is to be added.

3. *Richardson's Blue.*

B. Wills Richardson (*Quart. Journ. of Micr. Science,* Vol. 8, p. 271) recommends another composition.

10 grains of pure sulphate of iron is to be dissolved in 1 ounce of distilled water, and 32 grains of ferridcyanide of potassium in another ounce of water. As with Beale's blue, these two solutions are then gradually mixed together in a bottle, the iron solution being added to that of the ferridcyanide, and mixture insured by frequent agitation. This makes a beautiful

* The methyl alcohol in this and the third formula is a superfluous addition and in consequence to be omitted.

greenish blue, in which there is as little appearance of granules to be recognized with the naked eye as in Beale's mixture. Then the mixture mentioned under No. 1, consisting of water, glycerine, and the two alcohols, is to be carefully added and considerably shaken.

4. *Müller's Blue.*

W. Müller prepares in a simple way a cold flowing blue mixture by the precipitation of soluble Prussian blue from a concentrated solution by means of 90 per cent. alcohol. The coloring matter is thus precipitated in a state of most extreme fineness, and a completely neutral fluid is obtained.

5. *Beale's Carmine.*

Mix 5 grains of carmine with a few drops of water, and when well incorporated add from five to six drops of liquor ammonia. To this solution about half an ounce of glycerine is to be added, and the whole well shaken. Another half-ounce of glycerine, containing eight or ten drops of concentrated acetic or hydrochloric acid, is to be slowly and gradually added to the carmine solution, frequently shaking during the mixture. The carmine thus becomes very finely granular, and the whole assumes a bright arterial red color. For its dilution a mixture is used consisting of half an ounce of glycerine, 2 drachms of ordinary alcohol, and 6 drachms of water.

6. *White Fluids.*

As I have not, as yet, been able to find a third transparent coloring material for cold flowing injections, I have used an opaque mass, the sulphate of baryta. The mass is, as was remarked, very finely granular, and is capable of being combined with a blue, if it be desired to inject the arteries and veins separately. I employ the following process:—

The salt is reprecipitated from a cold saturated solution of 4 ounces of chloride of barium by adding, drop-wise, sulphuric acid. After standing for some time (12 to 24 hours) in a tall cylindrical glass vessel, it is deposited at the bottom. About half the fluid, which has again become clear, is now to be poured off, and the remainder, well shaken up, is to be combined with a mixture of one ounce each of alcohol and glycerine.

These masses *—we repeat it—are distinguished by their great permeability, so that we prefer them to all gelatinous substances for the injection of lymph passages and glandular canals. They also have the extraordinary advantage of being capable of preservation for months without alteration, so that they are instantly at hand. They are kept in small bottles with well-fitted glass stoppers. The requisite quantity for an injection is poured into a porcelain dish, and it is then ready for use.†

7. *Solution of Nitrate of Silver.*

Within a few years a solution of nitrate of silver has been used for injections, in order to render the cells of vessels visible. The animal is bled to death, a solution of nitrate of silver (0.25, 0.5–1 per cent.) is then injected, which is to be followed after a few minutes by a stream of water; or, a mixture of gelatine and a solution of nitrate of silver is used, in

* W. Müller, in his excellent monograph on the spleen, also mentions a cold flowing, brownish red mass, which is obtained by precipitation from a solution of the chromate of copper with ferrocyanide of potassium. Chromate of copper is obtained by digesting equivalent portions of sulphate of copper with chromate of potash, and washing out the brown precipitate. The latter readily dissolves in chromic acid in excess, and may be precipitated from the diluted solution, by means of ferrocyanide of potassium, in the form of an extremely fine brownish-red sediment of ferrocyanide of copper. It may be at once injected, without further addition than the solution of bichromate of potash which has been formed, and thus, at the same time, serve as a medium for hardening the tissue.

† I have recently employed with advantage soluble Prussian blue simply dissolved in water for the injection of the ducts of glands, urinary canals, and biliary plexuses, also for lymphatic canals. 10 grains of sulphate of iron dissolved in 1 ounce of water, 32 grains of red prussiate of potash in another ounce of water, and both carefully combined (see above), make a good fluid. If the canals to be filled are very fine, double the quantity of each salt is to be added to each ounce of water. Glycerine may be added if desirable. The red mixture recommended by Kollmann is serviceable. 1 gramme of carmine is to be dissolved in a little water with 15–20 drops of concentrated ammonia, and diluted with 20 ccm. of glycerine. An additional 20 ccm. of glycerine is to be tempered with 18–20 drops of strong muriatic acid and carefully added to the carmine solution, at the same time shaking the latter strongly. It may be subsequently diluted by the addition of about 40 ccm. of water.

order to gain a more permanent distention. Sections of the organ are made and exposed to the light, and then hardened in alcohol. By this simple method the entire vascular arrangement may also be recognized with the same distinctness as by the ordinary injections with colored masses.

After having become acquainted with the fluids used for injecting, and the manner of preparing them, we now pass to the consideration of the apparatus and the act itself of injecting. All who have frequently practised this operation will agree with us that a very simple apparatus is all that is required.

Before discussing the method of injecting which is most important and most frequently used, namely, that by means of the syringe, it is necessary to mention several other modern processes which, according to our own experience and that of others, may be practised with facility and success, and will, without doubt, lead to the extension of our knowledge in many directions—we mean the self-injection of the living animal and the injection by means of constant pressure.

The idea was sufficiently obvious of permitting the escape of a definite quantity of blood from the body of the living animal by opening a vein, and replacing what was lost by an innocuous colored fluid, so that the contractions of the heart would fill certain vascular districts with it in a much less injurious manner than can be accomplished by the human hand.

Chrzonszczewsky made us acquainted with these methods some time since. They consist in the introduction of the watery solution of the carminate of ammonia.

10 ccm. of blood may be removed from the jugular of a medium-sized rabbit, and replaced by the same quantity of a solution of carmine, by means of the syringe to be mentioned below, without injury to the blood with which it becomes mixed. An adult animal bears 15 ccm., a dog of medium size, 25. Even during the injection, the reddening may be recognized on certain portions of the surface. If the larger vessels are then rapidly ligated, first the vein and then the artery, a physiological injection of the blood passages is obtained. The kidneys, spleen, etc., may be advantageously treated in this manner. At the same time, this carmine injec-

tion may be accomplished not only from the vascular system, but also from the stomach, rectum, and abdominal cavity, and in amphibia from the lymph cavities.

The inventor recommends the solution of 2 drachms of carmine in 1 drachm of liquor ammonia, the same to be diluted with one ounce of water. Naturally, this solution is to be filtered before using. The organ is to be placed in acidulated alcohol to cause the granular fixation of the carmine.

Such injections obtain a high value in still another manner. Not only this carmine solution, but also a cold concentrated solution of sulph-indilate of soda is rapidly excreted by the kidneys, and the latter substance also into the biliary passages, after large quantities have been injected. If the ureter be ligated immediately after injecting the rabbit, and the animal killed after three-quarters of an hour to one hour, the entire system of urinary canals will be found filled with carmine. In injecting the biliary passages with the blue fluid, it is unnecessary to apply the ligature. In both cases, however, it is necessary to wash the blood-vessels subsequently, and to replace the original coloring-matter which remains in them with another. The organs injected with blue are next placed in a concentrated solution of chloride of calcium, and then in absolute alcohol, where the coloring matter becomes fixed in fine granules.

Injecting by means of constant pressure also has great advantages for many purposes. First, we may learn by this means to estimate the pressure which is necessary to fill certain portions of the blood and lymph passages, or of the glandular canals; then, besides a very high pressure we can also use a very low one, and finally it permits of making the injection with extreme slowness, which the human hand refuses to do, in consequence of fatigue.

Beautiful results have been obtained by this method for the lymphatic passages and also for the glandular canals (kidney and liver).

A simple way of making such injections is by means of a graduated glass tube (fig. 75 *b*), which should not be too small, held by a support (*a*). To the lower end of this is firmly secured an india-rubber tube (*c*), the lower extremity of which terminates in a metallic tube which can be closed by means of a cock (fig. 75 *d*. 76); this tube should fit into the aperture of the canules of the ordinary injecting apparatus. The canule should be tied into the vessel of the organ to be injected, in the manner hereafter indicated, and placed in a convenient position near the glass tube, which is secured in a perpendicular position, having been previously filled to a certain height and the stop-cock turned off. The end of the tube is to be cautiously but securely inserted into the aperture of the canule and the cock opened. The original pressure may be maintained or increased, as necessity may require, by pouring in more fluid. Such an arrangement may be left to itself for a number of hours or even days.

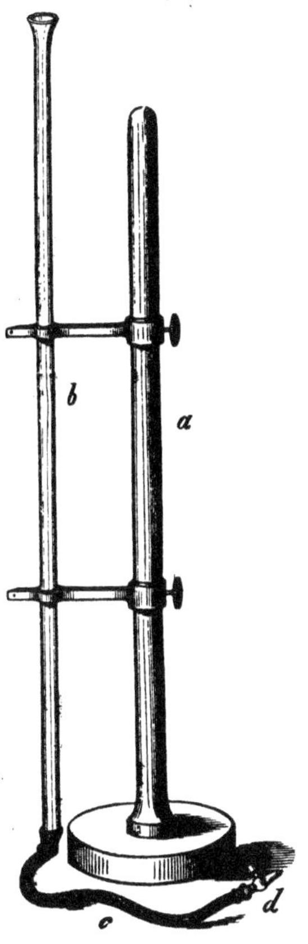

Fig. 75. Simple injecting apparatus, with a glass tube.

If it be desired to use the pressure of a column of mercury, the readily constructed apparatus, represented in less than half size by fig. 77, is to be recommended. A glass bottle (*a*) is to be closed by an accurately fitting cork (preferably of gutta-percha) perforated by two holes. Through these holes pass two glass tubes perpendicularly; one of them (*e*), which is graduated, and slightly funnel-shaped at the top, extends nearly to the bottom of the bottle. A second one (*f*), which is bent on itself, terminates just beneath the cork. The continuation of the external portion of the latter tube is formed by a caoutchouc tube (*g*) firmly secured to it, at the termination of which the above-mentioned metallic tube with a stop-cock (*h*) is inserted and receives the canule (*i*). At the upper funnel-shaped opening (*e*) of the first tube is placed a small glass funnel (*l*), supported by a stand (*k*); the funnel is prolonged by means of a caoutchouc tube (*m*), into the lower end of which is secured a finely pointed glass tube (*n*). It is used for pouring in the mercury, and has on the caoutchouc tube a clamp (*o*), or preferably a screw-clamp.

In preparing the apparatus for use, the lower part of the glass vessel is filled with mercury (*d*), the cock of the delivery-tube

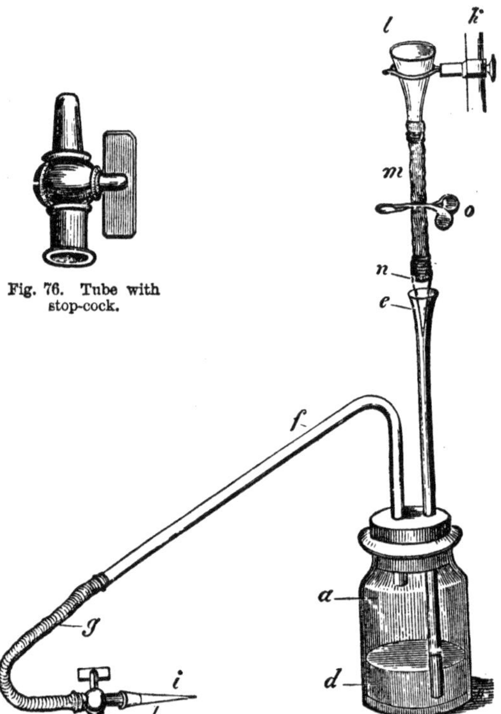

Fig. 76. Tube with stop-cock.

Fig. 77. Injecting apparatus with a column of mercury.

is then opened and the vessel filled to the upper edge with the injecting fluid. The cork with the two tubes is now to be firmly pressed into position, the funnel-shaped opening of the perpendicular tube being at the same time securely closed by the pressure of the thumb, care being also taken that the lower end of the tube dips beneath the mirrored surface of the mercury. If mercury be now poured into the funnel-shaped opening, the knee-shaped tube will become filled with the injection fluid, which soon issues without air-bubbles from the aperture of the metallic tube. The stop-cock is then to be closed and the end of the metallic tube carefully but firmly inserted into the mouth of the canule. The stop-cock should then be opened a second time, when the colored fluid will flow into the organ, and the column of mercury in the vertical glass tube will rapidly sink. This may be raised, by a subsequent addition of the metal, to an elevation of 20, 30, or 40 mm. (in many organs to double this or more), as may be desired. The flowing of the mercury may easily be so regulated, by means of the clamp, that the amount of pressure is constantly maintained.* If the column finally

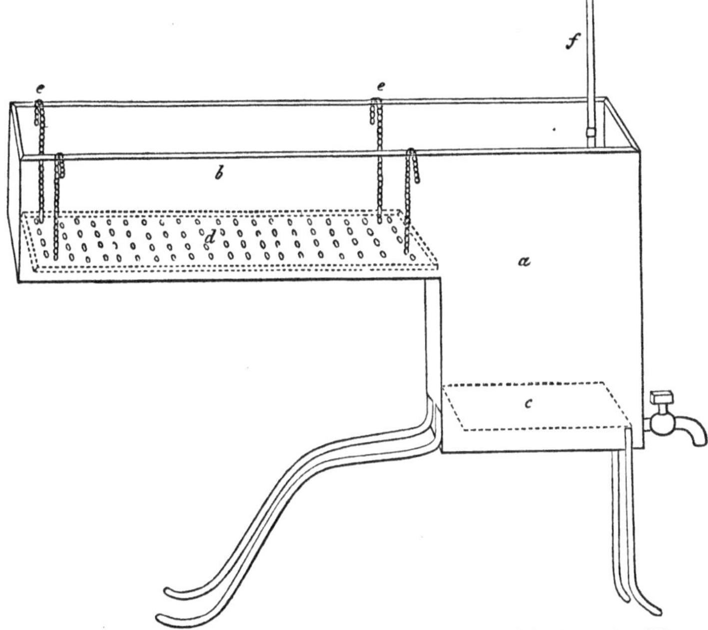

Fig. 78. Harting's injection chest. *a* chest; *c* false bottom for the injection bottle; *f* thermo meter; *b* compartment for the reception of the preparation; *d* perforated plate, the position of which may be altered by means of the chains *e*.

* When it is necessary to employ a very slight pressure of known degree, it is advantageous to bend the tube which is connected with the funnel four

ceases to sink, the injection may be discontinued or the pressure cautiously increased, according to circumstances.

It is unnecessary to remark that cold fluids are here in place. It is preferable to use a watery solution of Prussian blue or the Richardson mixture. The apparatus just described may also be readily used for injecting with gelatine (Ludwig). The bottle is to be placed in a tin chest of considerable size, which rests on feet and contains a table for the support of the organ which is to be injected. This chest is to be filled with warm water and the temperature maintained by means of an alcohol or gas flame.

A well-adapted arrangement for this purpose, designed by Harting, will be at once made intelligible to the reader by our fig. 78. We are indebted to Professor Hering, of Vienna, for an excellent apparatus for such injections. By it the pressure on the fluid may be accurately measured and uniformly maintained. The arrangement is by no means simple, so that we must refer to a description given of it by Toldt.

We now pass to the consideration of the method most extensively used, that of the syringe.

The small German-silver injection syringes (fig. 79, 1), which may be bought of Charrière or Luer, in Paris, for a few thalers, with a half-dozen or a dozen different canules (2, 3), are sufficient for all purposes, and will render equally good service for a number of years if used with some care. It is only necessary to carefully smear the piston from time to time with tallow, in order to preserve the smooth, easy movement which is so extremely necessary. It is also necessary to clean the syringe after being used for resinous injections with turpentine, and after gelatine with hot or cold water; it is then hung up by the ring of the piston-rod to permit the water to drain away. If, after a long interval of time, the caoutchouc of the piston no longer fits closely, the syringe is to be unscrewed and the piston placed for a half or a whole day in cold, or several minutes in hot water. It has then become sufficiently swollen again, and, when rubbed

---

times at right angles, so that it runs downwards and again upwards, outside the bottle and beneath the prolongation of the level of the mercury, somewhat in the form of a manometer (Mac-Gillavry).

with tallow, the piston performs its duty anew. Resinous masses always have the inconvenience of requiring a time-consuming cleansing of the syringe. The canules should also be cleaned with water after having been used, and should be stood upon a warm plate to dry. Nothing further is necessary to keep the larger tubes open. A thin silver wire should always be introduced into the finer and finest ones after cleaning them, as, without this precautionary measure, the narrow passage is found to be closed, that is, rusted, and frequently it causes all subsequent experiments to remain without results.

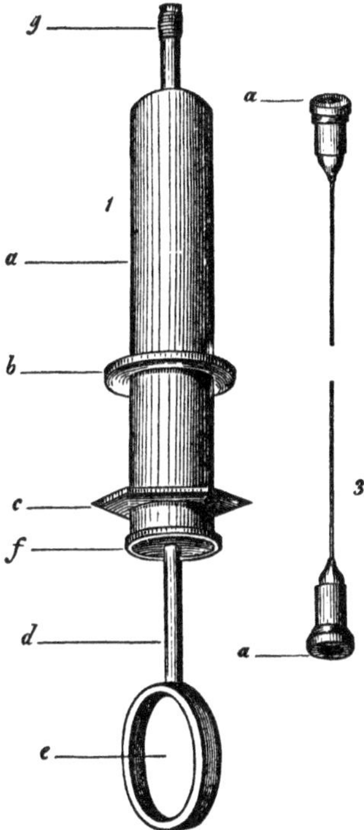

Fig. 79. Syringe for Injections. 1. *a* the tube, with the projecting edges *b* and *c* (for convenience in holding), and the cap *f*, which is to be screwed on; *d* piston-rod with the ring *e*; *g* the aperture (mouth-piece) of the syringe, with a silk thread wound round it. 2 and 3. Canules of the finest variety.

Those who inject much require several of these syringes. It is also very convenient to have a large syringe of about double the capacity of the small instruments, for filling extended systems of vessels, as the repeated removal for filling the syringe is always an unpleasant procedure; and it is just in removing and in replacing the syringe that the beginner so readily meets with a misfortune.

The canules themselves do not require any ring-shaped projection, but rather a notched edge, for convenience in holding them. For frequent work it is necessary to have at least a dozen on hand; it is still better to have even a greater stock of the most varying calibre, from about two mm. aperture to that of capillary fineness. I use those of German silver for large vessels; the finest are of sheet-iron, and therefore unfortunately of a perishable nature.

The remaining contrivances consist of strong, well-waxed silk thread of several sorts, several curved and straight needles, a pair of fine scissors, small ordinary and curved forceps, also several slide forceps (or other clamping apparatus, fig. 80) for possible emergencies. These, together with cold water, are sufficient for cold masses. For gelatine injections it is also necessary to have a kettle with hot water and a double water-bath. The latter are ordinary deep copper basins, which are filled with warm water and kept at an elevated temperature by means of a spirit-lamp burning under them. They serve to receive the dishes of gelatine. Gelatine masses should never be warmed directly over a fire! Together with plates or porcelain dishes, it is also convenient to have an oblong lead chest for the reception of the organ or the body of the animal to be injected warm. The chest should have divergent walls and a drain-tube near the bottom, with a stop-cock.

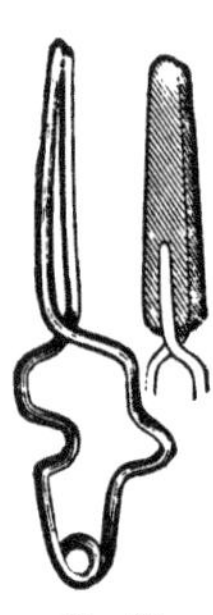

Fig. 80.
Serres fines.

Objects for injecting are generally selected from parts as fresh as possible—that is, from animals which have just been killed. I have frequently used small animals while they were still warm, directly after death, which is preferably induced by bleeding. In this way I have obtained the best results, except where muscular parts were concerned; in which case, especially when injecting warm masses, the rigor mortis, which frequently occurs suddenly, renders the injection impossible. Very soft parts may be previously immersed for a day in alcohol, in order to render them harder. By means of this procedure I have frequently succeeded in injecting the spleen after having been unsuccessful with fresh organs, notwithstanding every precaution. In injecting the blood-vessels of bodies not so fresh, the coagulation of the blood is a great disadvantage, which often ruins the whole process. It has been frequently recommended in such cases to precede the injection mass with a stream of water, and in certain cases this procedure is serviceable. But generally we soon meet with numerous extravasations, and we are compelled to discontinue at an early period, long before a complete injection has been accomplished.

The blood-vessels of pathological new formations are generally difficult to inject. The walls of the vessels are readily ruptured in consequence of their extreme delicacy. It is also frequently necessary to ligate numerous lateral branches. Cold transparent masses should only be used here, if anywhere. But with skill and perseverance much may be accomplished. Unfortunately, this department has been altogether too much neglected by pathological anatomists, with the exception of Thiersch.

In order to inject the lymphatic vessels, for which purpose all bodies are not equally suitable, I have frequently placed the dead body in water for a series of hours, so that these vessels might in this way become more thoroughly distended. One may also frequently have the pleasure of seeing the lymphatic vessels become well filled, after having forced a stream of water through the arteries of the organ for some time. Another method is likewise useful. I kill the animal by a blow on the head or by strangulation, then open the thorax, avoiding the blood-vessels, and ligate the ductus thoracicus high up. The body then remains undisturbed for 2–6 hours. The lymphatic vessels are now sought out, and are, for the most part, found to be filled and distended in a very satisfactory manner. The experiment may be made of ligating the efferent canals or the veins of the larger glands in the living animal, and thus causing the lymphatic vessels to become firm and distended.

The freshest possible material should be selected for injections of the glandular canals. The canule may be inserted directly, or the passage may be made to appear more distended by previously forcing water through the artery, at the same time compressing the vein slightly, also first endeavoring to cause the secretion to flow out. In this case great caution is always necessary.

In searching for a blood-vessel, an artery, or a vein, avoid all unnecessary cutting, as small branches may thus be readily injured, which afterwards makes it necessary to stop the rent with ligatures or the application of sliding forceps, and thus cause an unpleasant interruption to the progress of the work.

In opening the vessel, which is best done under water, avoid making the slit too large, and above all do not make a transverse cut on a small artery, as in the introduction of the canule the vessel might readily be torn in two. By opening the vessel under water the entrance of air, which is always to be carefully avoided in injecting, is, for the most part, prevented. But there is still some air in the canule which is to be introduced. In order to remove this, the tube should be filled with water and the posterior opening closed with a cork before the introduction of the canule, a little precautionary measure which, like so many others, apparently unimportant, renders very great service in injecting. The mouthpiece of the syringe should also be passed beneath the surface of the water, and then introduced into the opening of the canule.

The canule having been successfully introduced into the vessel, it next becomes necessary to secure it by means of a carefully waxed silk thread. The necessary skill is soon acquired, the thread being either seized with the forceps and passed beneath the vessel, or brought round it threaded in a needle. Large vessels should be tied as firmly as possible; with smaller ones more circumspection is required, and with very fine ones, especially embryonic branches, the greatest care is necessary. If the canule has a ring-shaped groove, which should always be the case with large ones, the ligature is to be placed at that part. If there is no groove, the greatest attention is to be paid to the application of the ligature, to prevent the tube from slipping out. In such cases the practised hand of an assistant, who places a finger over the opening of the canule without pressing the tube deeper into the vessel, renders an important service.

The same process is to be followed in tying the canules into the ducts of glands. Lymphatic vessels require greater attention. That it is necessary to inject in the direction in which the valves open is self-evident; although, in a few cases, the resistance which they present may also be successfully overcome. Still it is only rarely and for special purposes that this procedure can be made use of; as, for instance, I succeeded in this way, several years ago, in injecting the lymphatic glands from the vas efferens.

It frequently happens, however, that a well-distended lymphatic vessel, which appears very inviting for injection, can nevertheless not be taken advantage of, especially when the vessel is very fine. The colorless fluid escapes on opening the vessel, and frequently the whole vessel becomes almost unrecognizable. One is sometimes tormented for a long time in endeavoring to introduce the canule within the collapsed walls; attempt after attempt may fail, until after a time, in successful cases, the desired object is attained. Coolness and patience is to be recommended to those who desire to accomplish anything in this direction.

When the fine lymphatic vessels in the interior of an organ are to be injected, Hyrtl and Teichmann's puncturing method is the process chiefly employed. Hyrtl sometimes makes a puncture from the cavity of a blood-vessel into the surrounding tissue in order to open some of the lymphatic vessels which may be present, and thus, in fortunate cases, inject the lymphatic canals from and with the blood-vessel. Another way is to make a small opening directly into the tissue, in order to inject immediately from this into any of the lymphatic vessels which may have been opened, and from these into larger systems.

This may be accomplished in two ways. With larger canules, a needle may be passed through the tube; after having inserted the canule into a small opening in the tissue, the needle may be pressed forward and the tube made to follow till the desired point is reached, when the needle is to be withdrawn.

Where the walls were very thin, I have had better success by another method. A small puncture is to be made with a fine cataract-needle or the point of a fine scissors which has been dipped into the injection fluid. The tube is now to be passed into the small opening, recognizable by the little point of color, and very slowly and carefully pushed forward with an easy rotating movement. When one has the requisite practice and patience for this process, lymphatics may be injected even where the method of preceding the canule with the pricking instrument would fail. Nevertheless, it always remains a difficult piece of work—as, for instance, in the small intestines of the Guinea-pig—to guide the tube along in the submucous tissue, as the slightest awkwardness in the movement causes a perforation of the mucous membrane. Many attempts will fail, till at length, by a lucky hit, the injection succeeds. All who desire to accomplish anything in this direction should first practise on organs which are easy to inject, of which, fortunately, there are many. Try, for instance, the vermiform process of the rabbit, in which the injection is very easy; then the small intestine of the sheep, the testicle of the calf, and the Peyer's glands of the last-named animal, and proceed gradually to more difficult organs. In many cases it is unnecessary to tie the canule, as compression may often be better made with the fingers or fine sliding forceps. If a ligature be used, a very fine needle should be employed, and the loop should be tightened with the greatest precaution, as very frequently the point of the canule is finally thrust through the walls of the vessel. Punctures which are too large permit the escape of the injection fluid. Teichmann, who has obtained great experience in this direction, very properly remarks that a puncture made at random is not sufficient, but that it is to be made in a direction where fine lymphatic vessels are supposed to be. If the extravasation which forms at the commencement remains small, the injection frequently succeeds. If it is large at the very beginning, and increases rapidly, stop, for the procedure has failed. If a rapidly-increasing extravasation subsequently takes place, it is likewise necessary to discontinue at once. Very cautious management of the syringe is for the most part necessary, especially at the commencement of the injection.

But we have wandered from our subject. When the tube has been firmly secured in position, the syringe is to be thoroughly filled from beneath the surface of the injection fluid; and while the canule, which is now opened, is seized and somewhat raised by the index and middle fingers of the left hand, the mouthpiece of the syringe is to be inserted as deeply as possible. The syringe is to be held by the middle phalanges of the index and middle fingers of the right hand, and the thumb is placed in the ring of the instrument. It is important that the forearm should, at the same time, rest quietly and conveniently on the table.

In this manner, therefore, the injection of the mass commences: two fingers of the left hand holding the canule, three of the right the syringe, the piston being pressed forwards as slowly as possible and with the utmost steadiness. Every awkward, spasmodic impulse is to be avoided, especially towards the end of an injection. If the work succeeds, the colored mass is seen to move forward in the vascular system, and it is noticed how in some places the capillary systems become filled, how these latter places constantly become more numerous, and at the same time increase at the periphery until they flow together. During this, the finger feels a gradually increasing pressure, and one soon learns to accommodate the motion of the piston to it. If two or three additional syringefuls of the mass be necessary, the syringe is to be removed, preferably before it is entirely emptied, and the opening of the canule closed with the thumb of the left hand. The syringe is to be refilled either by the operator, with his right hand, or by an assistant. If one possesses several syringes furnished with similar mouthpieces, it is well, when injecting large organs with cold masses, to have several of them lying filled near him at the very commencement, so as to instantly exchange the emptied syringe for one that is filled.

When the injection is completed, whereby it is often well to ligate the opposite vessel in order to prevent an escape of the fluid, the canule is to be closed by means of a stopper of cork, or better of metal, fitted into its opening, or by means of the above mentioned (p. 189) short tube with a stop-cock. The injected vessel is now tied further below, and the other ligature which holds the canule is finally removed, so that the tube may be taken out.

Although the above-mentioned manipulations are soon learned with a little aptness, it is difficult to properly estimate the moment when the injection must be discontinued. Here the beginner very readily errs, and even the most practised now and then has his unlucky day. Too little may be done; the injection is then insufficient, only small places are filled, or fine capillary systems even not at all. Inversely, an injection pushed too far leads to extravasations, and finally to an unserviceable preparation. If it be noticed that numerous though at first small extravasations form, desist, or they will be seen to increase in a frightful measure. That a considerable escape of the injection fluid requires an instant cessation in order to rescue what is possible, is self-evident. If Beale's cold mixture be employed, towards the end of the injection the colorless fluid is seen to be pressed through the walls of the urinary canals and the envelope of the organ, appearing on the surface as a fatty, glistening moisture. Then is the time to leave off; it would be too soon in most cases before this exudation takes place.

The double injection is naturally much more difficult than the single; firstly, on account of the entire procedure, and then while too much should not be sent through one system, that of the vein, for instance, in order that the possibility may remain for the one injection to meet that from the second system in the capillaries. For injecting arteries and veins, such masses should always be used, if possible, which give a pleasant blending of colors when they meet; for example, Prussian blue and carmine, Prussian blue and white, while yellow and green appear less handsome to the eye. Masses which flow when warm and harden when cooled generally deserve the preference for these cases, and with gelatine masses I usually allow some time to elapse between the first and second injection, so that the former may at least acquire some firmness. For most cases, the vein may be first injected and then tied in the usual way. Afterwards, if there be considerable resistance, the artery with its ramifications are to be injected.

For many organs, as for instance the eye, or the spleen, it is well to drive the injection mixture intended for the venous system through the artery first, and then, through the same vessel, the second mass which is to serve for the arterial system. Not unfrequently the injection may be essentially regulated by keeping open or closing the terminal vein.

If, together with the blood-vessels, it be also intended to inject the lymphatics, or, in a glandular organ, its system of canals, the blood-vessel may either be injected first and then the latter, or inversely. If the lymphatics are to be injected by the puncturing method, avoid injuring the injected blood-vessels as far as possible.

For all injections of the glandular passages and the lymphatics, transparent cold mixtures deserve the preference, as was already remarked, on account of their ready permeability, as well as in consequence of the less degree of injury which their employment exerts on the tissues.

Although the directions given are in no wise to be regarded as complete, and require special modifications for special organs, which are only to be obtained by experiment, they will, nevertheless, considerably facilitate the labors of the beginner.

A successful injection having been made, the further question now arises: What is to be done with the specimen in order to prepare it for examination?

As was above remarked, warm injections require, beyond all things, the necessary time for the mass to harden. Resinous substances require a longer time than gelatine. With Beale's cold mixture, the objects may be used at once; with Hyrtl's ether injection, the injected organ may be used after a quarter of an hour.

When a specimen has been injected with a gelatine mass, it should without delay, or at most only sufficient to wash off its surface, be placed in ice-water (in winter, snow), and left till the mass has become hardened. This may be readily recognized when the contents of the larger vessels no longer yield when felt with the points of the fingers. The injected organ is placed, for further hardening and preservation, in weak, and then in stronger alcohol. It is well to let it lie quietly in this for several days before proceeding further. It is better to place very sensitive objects, directly after the injection, in alcohol which has been previously placed in ice, or which has been cooled by putting pieces of ice in it (Thiersch). A few drops of acetic acid are to be added to the alcohol for injections with Prussian blue.

Naturally, even here, numerous modifications are necessary in certain cases. Thus smaller organs may be left in the alcohol without cutting them, as also groups of organs and the entire bodies of the smaller mammalia, which may be prepared after a few days. It is preferable to open an intestinal canal, which has been injected with gelatine, after the injection fluid has hardened. This should be done in water, and the canal washed out carefully. Portions of intestine with the lymphatics injected I cut open, and run a stream of water through the canal to wash out its contents, and then place the preparation for a day or more in alcohol. Large organs after being immersed in alcohol, for example, the kidney of one of our ruminata, should be cut open on the following day at furthest, lest the cortex should harden while the internal portion decomposes. Immersion in chromic acid solutions is also applicable to such purposes, Prussian blue being well preserved in them; but it is rare that alcohol can be altogether dispensed with. I also place organs injected with Beale's mixture, almost without exception, in alcohol, in order to obtain the necessary hardening of the tissue.

After a few days, when the preparation has acquired the necessary firmness, it may be examined by the ordinary methods already given. Thin horizontal and vertical sections, for example, are to be freed from particles of coloring matter which have escaped by washing, or, still better, by means of a camel's-hair pencil. They are then to be reviewed with the microscope, and, if it be desired to preserve them permanently, they are to receive such further treatment as may be necessary.

The old method of mounting dry is to be recommended when the preparation is to be used as an opaque object. It is still better to mount it carefully in Canada balsam, which will be treated of in the following section.

Glycerine is being more and more employed for mounting histological preparations, and, as may be readily conceived, it reproduces the natural relations, although connected with the very great disadvantage of being much less durable.

For the preservation of injected organs for a considerable length of time, alcohol, weak or strong, according to circumstances, is used.

Section Tenth.

THE MOUNTING AND ARRANGEMENT OF MICROSCOPIC OBJECTS.

The reader will have perceived from the preceding sections that it is by no means one of the simplest and easiest things to obtain useful microscopic specimens, even if, at the same time, we also disregard the rareness of many, as, for instance, those of embryonic and pathological occurrences. The desire to preserve for the longest possible time such objects as are only obtained with trouble or the concurrence of fortunate circumstances is also sufficiently obvious; and, in fact, the effort to obtain such preparations is as old as microscopy itself. The value of such collections is quite as great for the study of this branch of natural science as for that of others.

Commencing with crude attempts in the preservation of hard structures, dried preparations of injections, etc., the industry of the investigators has gradually brought better and better methods to light, so that this now constitutes an important section of microscopic technology. At the same time, although much has been accomplished in this department, still more remains to be attained and explored; most of those branches relating to preserving being at the present day still in an incipient condition.

Many portions of the body may be sufficiently well preserved in ordinary alcohol for the purpose of having at hand material from which, in case of necessity, a serviceable preparation may be made with rapidity and little trouble. Hardened glands, intestines, the central portions of the nervous system, tumors, injections with gelatine and cold masses, such as have been described in the previous section, and embryos, may be preserved in the most convenient manner in well-closing glass bottles, and constitute, especially for a teacher, invaluable material for instruction.

But, in most cases, the matter is not so simple when a definite microscopic preparation is to be preserved. For this purpose certain methods are necessary.

Hard structures of many kinds, especially those of a transparent nature, scales of diatomes, thin sections of bone and teeth, and crystals, may be permanently preserved in a very simple way if they are placed on a slide and covered with a thin covering glass, and the latter fastened to the former. Various substances may be used for this purpose; as, thick gum-arabic (a solution of gum with powdered starch is good), wax, resinous substances of thick consistence, and Canada balsam. For the protection of the fragile covering glass, the whole may afterwards be covered with colored paper, through which an aperture has been made with a punch. It is well for those who work much with such objects to have lithographed covers prepared, the posterior surfaces of which are gummed for the sake of economizing time. On one surface of the slide the paper should project beyond its edges in such a manner that they may be covered by it, while the other surface of the glass plate requires a smaller covering. One soon acquires the little dexterity necessary to apply these covers. The gummed surface should only be slightly moistened, so that when pressed on to the slide the gum will not exude and flow over the visible portion of the preparation. Very many such preparations, which are in circulation and to be purchased, may be recommended as models; as, for instance, those of Bourgogne, in Paris; Möller, in Wedel (Holstein); and Rodig, in Hamburg.

But, as we have already remarked, only a small number of objects, which are transparent *per se*, permit of this most simple method of treatment. The greater portion of those which are to be preserved dry require, in order to render them transparent, to be mounted in a substance which refracts the light strongly, in a gradually hardening resinous material.

For this purpose there is none more important or more generally used than the Canada balsam, and, indeed, it suffices for all cases. Other resinous substances, such as copal lack, damar varnish, and mastic are really superfluous, and are, at most, only to be used here and there by way of experiment.

Several sorts of Canada balsam occur in commerce. To be good it should be of thick consistence, nearly colorless, and thoroughly transparent. It is to be kept in wide-mouthed vessels closed with glass stoppers, in order to limit as much as possible its tendency to harden in the air. If, in consequence of the prolonged action of the air, it has become much hardened it may be thinned, after being moderately warmed, with oil of turpentine, or, which is less preferable, with a little chloroform.

The preparation to be mounted must be thoroughly dry. Hence, in many cases a preparatory drying process will be necessary. For this purpose the preparation may be placed over a water-bath, or over sulphuric acid or chloride of calcium. Many preparations may be advantageously placed in oil of turpentine, in which they are to be left for at least a few minutes. If the specimen to be mounted contains air, a longer immersion in turpentine, occasionally in that which is warmed, will be necessary.

The preparation should be mounted in the following manner. The dry, cleanly-washed slide is to be moderately warmed over the spirit-lamp, but never to an extreme degree. A drop of the balsam is then to be taken from the bottle by means of a pointed glass rod and placed on the slide. It will then spread out into a layer which, in fortunate cases, will be quite homogeneous and contain no air-bubbles. But if any of the latter remain in the stratum of balsam (if the slide be too warm they are developed by the boiling of the balsam), they are made to burst by touching them with the point of a heated needle, or are brought to the edge of the layer of balsam with the point of a cold needle. The object to be mounted is now placed in position, and a second drop of balsam is placed over it by means of the glass rod. The two layers of balsam will soon flow together if the procedure be rapid or the slide be again slightly warmed. The clean and moderately warmed covering glass is now to be seized with the forceps and placed in an inclined position, with the side opposite the forceps lowest, over the layer of balsam, and then allowed to gradually assume a horizontal position till it completely covers the object. If there be any air-bubbles still remaining, they may be driven to the margin of the covering glass by careful pressure on its other edge, provided the mounted object be of a nature which permits of the necessary pressure. The preparation is now to be reviewed with the aid of a low power. If several air-bubbles are still to be discovered, it is preferable to place the slide on a warm body (in winter it is best to place it on the cover of an earthenware stove) covered with a bell-glass, and left for several hours, whereby, at the same time, the balsam hardens more rapidly, and on this account it is an advantageous procedure, even where there are no air-bubbles.

If too much Canada balsam has been used, a quantity of it usually spreads beyond the edge of the covering glass, or even on to its surface. In such cases it is necessary to wait till the balsam hardens, after which it may be scratched off with a knife-blade, and the surface of the glass cleaned with a rag freshly moistened with oil of turpentine or benzine.

The hardening of the balsam at the interior of the preparation proceeds very slowly, so that it still remains fluid for days, and even weeks, while the edges have become hard. By an awkward manipulation the covering glass may be displaced and the preparation ruined.

Occasionally a Canada balsam is met with which is at first of a somewhat more fluid consistence. In this case the mounting may be done on a cold slide, which always economizes a certain amount of time. Such preparations should always be placed for a time on a slightly warmed support, so as to dry more rapidly. Although it is quite necessary to accomplish the expulsion of the air-bubbles from most specimens mounted in Canada balsam, there are other objects in which the air contained in the finest canals is of importance for the recognition of certain structural peculiarities, and the air must therefore be retained. If, for example, we place a section of bone directly, or after having been in turpentine, into Canada balsam which is very fluid, the canaliculi and the cavities of the bone become filled with this medium, which gradually penetrates in all directions and forces out the air. But the processes of the bone corpuscles and the canaliculi appear distinct only when they contain air, and it is only in this way that the bone presents a characteristically elegant appearance.

Such preparations must be mounted warm with the thickest possible Canada balsam. For this purpose the balsam may be placed in an open vessel in a warm place, and covered with a bell-glass until it becomes quite hard and firm. It is unnecessary to remark that previous immersion of the object in oil of turpentine is here to be avoided, and that in mounting it is necessary to expose the Canada balsam, slide, and cover to a considerably elevated temperature.

Frequently—especially with histological work—when it is desired to mount a very thin and delicate specimen, as the object is warmed it will be seen, to one's great vexation, to shrink, become curved, and finally break. Here a solution of Canada balsam in ether, or, still better, in chloroform, filtered through ordinary filtering paper, is in place; it may be diluted according to circumstances. By means of a brush or a glass rod it is placed cold on the slide, the object is placed in this, then more fluid is added, and finally the covering glass is laid on. As the dissolving medium evaporates, the air generally enters on one side between the plates of glass. In such cases the preparation is to be held in a slanting position, and a few drops more of the solution added until the process of mounting is finally completed. The whole procedure, which may also be employed for more substantial objects, is very convenient and cleanly.

But how should one proceed when one of the soft watery tissues, such as the greater portion of our body presents, is to be placed in Canada balsam? How are injected preparations to be treated?

That this can only be accomplished by intermediate processes is self-evident. That is, the water must be expelled by a fluid which mixes with it; this is to be replaced by another, etc., until at last the Canada balsam may be used for the final saturation.

Suppose we have a thin section of the spinal cord, kidney, or spleen, which has been tinged with carmine or some other coloring material, or the section of an intestine, brain, or lymphatic gland, with the blood-vessels and lymphatics injected, and we desire to mount the same as a dry preparation, but at the same time to avoid the shrinking occasioned by simple drying, which would change the preparation, in fortunate cases, to a caricature, or, in less fortunate ones, to hieroglyphics. The object is to be placed for a day in very strong, or better, in absolute alcohol; from this it is transferred to strong methylated alcohol for half an hour, although this intermediate step may also be dispensed with. By this means the water has been removed and the alcohol has taken its place. The preparation is now to be removed from the alcohol, preferably by means of a filter, and just as it begins to dry it is placed in oil of turpentine. The previously mentioned small, flat glass boxes are very suitable for this purpose. Sometimes the process of becoming transparent can be very conveniently followed under the microscope. Then, by placing a thick plate of glass which just covers the preparation over it, it will be pressed against the flat bottom of the vessel, and all curving of the object will be prevented, and the shrinking will be limited to a considerable degree, even though the specimen remains in the oil of turpentine for several days. After several hours, all the alcohol is expelled by the turpentine, and the object is ready for mounting in the chloroform solution of Canada balsam. Excellent preparations may thus be obtained when one has once mastered this method. All injected specimens (those with nitrate of silver also) should be mounted dry in this way only. In the same manner many histological details, even cylindrical epithelium and other delicate cells, may be preserved so as to be visible, and if carefully tinged with carmine or blue, they may be rendered still more distinct. All the transparent colors which were mentioned as suitable for combination with gelatine are well preserved in this way. At the same time we would also add, as a precautionary rule, to add a drop of glacial acetic acid to the alcohol used for drawing the water out of preparations injected with Prussian blue.

We would here add still another little precautionary measure. It is best to allow very thin and delicate sections to become sufficiently dry on the filter, then to cut out the portion of paper on which the object rests and immerse it in oil of turpentine. By a slight movement the preparation may then be floated off from the paper.

We have placed this procedure, with all its minutiæ, before the reader, because of its great importance.

Here, as everywhere, the greatest cleanliness, the use of filtered fluids, etc., is necessary.

Thiersch has quite recently made use of colophony for mounting such preparations. He prepares it in the following manner: —It is best for the microscopist to prepare the colophony himself, and a solution of it in absolute alcohol of syrupy consistence should be used. The advantage which this material presents is, that the preparation may be placed in it directly from the absolute alcohol, without becoming cloudy and without prejudice to the durability of the specimen. Venetian turpentine is to be dissolved in an equal volume of sulphuric ether, and the solution filtered through paper. The ether and oil of turpentine are then to be expelled by the heat of a moderate fire, till the residuum shows a shell-like fracture when cold.

But the natural condition of the tissues is completely represented only when mounted in a moist condition. This method permits of the most accurate recognition of delicate textural relations, pale cells and fibres, etc., and should not be omitted with any tissue in the production of histological collections, as, even in cases where good dry preparations can be obtained, it affords an instructive comparison.

Among all the preservative fluids for animal soft parts there is none which stands higher, at the present time, than glycerine. Its strong refractive power, its property of combining with water and of attracting the same from the atmosphere, render it an invaluable medium for mounting animal tissues containing water. It may be truly said, that what Canada balsam is to dry tissues glycerine is to moist ones.

The ordinary impure glycerine may be used in the preparation of a temporary specimen, for brushing, etc., but not, however, for permanent preparations. Here the purified glycerine, containing no lead and as little water as possible, is always to be used. Undiluted, it renders the preparation very transparent; occasionally, after a time, too much so. For many objects it must, therefore, be diluted with distilled or camphor water in about equal proportions, more or less, according to circumstances. It is very useful, indeed almost indispensable, to wash the preparations which are to be permanently mounted for several days in pure glycerine, or a mixture of glycerine and water, in a small vessel, whereby the degree of transparency which they will assume may be ascertained at the same time.

The preparation is then to be mounted in the ordinary manner by means of one of the cements hereafter mentioned. The superfluous glycerine, which spreads beyond the covering glass, may be removed with a fine pipette and dried with a cloth moistened in alcohol. The nature of glycerine is such as to render it unnecessary to be in haste in the application of the cement, so that a number of specimens may be allowed to accumulate before it is laid on to the borders.

For many purposes I have found it well to add two drops of strong muriatic acid to the ounce of glycerine. Objects injected with carmine or Prussian blue always require this addition, otherwise the color will fade and disappear after a time. Acetic acid accomplishes the same purpose, and possibly better. Ranvier has recently proposed the combination with formic acid (1:100).

As glycerine is a constituent of many mixtures, so also may many other materials be added to it, and thus produce more complicated mounting fluids. Thus, for example, gelatine, gum-arabic, etc., may be combined with the glycerine.

Deane recommends a mixture of glycerine 4 ounces, distilled water 2 ounces, and gelatine 1 ounce. The latter is to be first dissolved in the water and the glycerine then added. I have had no experience with tannin and glycerine.

Beale also recommends one of these combinations of glycerine with gelatine. A certain quantity of pure gelatine is allowed to soak in water until it swells up and becomes soft. It is then placed in a glass vessel and melted by the heat of boiling water (that is, on a water-bath). To this fluid an equal quantity of strong glycerine is added and filtered through flannel. The mixture may be kept for any length of time, and only requires to be slightly warmed before being used. Klebs employs 2 parts of a concentrated solution of isinglass and 1 part of pure glycerine slightly warmed.

Bastian recommends a mixture of 15 parts of glycerine and 1 part of carbolic acid for mounting tissues not tinged.

Farrants employs a still more complicated mixture, consisting of equal parts of gum-arabic, glycerine, and a saturated solution of arsenious acid. The mixture is to be used in the same way as the Canada balsam.

Although glycerine is the most important preservative fluid now known, answering all the requirements for many animal tissues, nevertheless one should not believe that everything can be preserved in it with success. Delicate, fresh watery tissues, for example, blood-corpuscles or ganglion cells, soon lose a portion of their water and become distorted. The strong refractive power of glycerine is, therefore, a disadvantage for transparent tissues, however excellent it may appear to be for those which are hardened. Besides glycerine, a whole series of preservative fluids have been tried and recommended, of which one is sometimes here, another sometimes there to be used with success. It is always well in mounting objects not to place implicit trust in such recommendations, but rather to make a series of experiments with various preservative fluids, of which only the best are to be retained after a subsequent examination.

M. Schultze has recently recommended, as a medium for mounting, a substance used by botanists, a nearly saturated solution of the acetate of potash in water, especially for osmic acid preparations which do not bear glycerine. Without removing the covering glass, a drop of this strong solution of potash is added to the microscopic preparation as it lies in water or an indifferent solution. A day later, the water having been in the mean time removed by evaporation, the cement is to be applied; although one may wait still longer. This method has been used for more than two years.

Goadby's solution, the conserving liquor of the English, has obtained a certain renown. It consists of:—

Bay salt...................... 4 ounces.
Alum........................ 2 ounces.
Corrosive sublimate............ 4 grains.
Boiling water.................. 4 pints.

This composition, which brought the discoverer a considerable sum, does not prove suitable for mounting transparent preparations, as they gradually become opaque and are finally rendered unserviceable. On the contrary, I have seen opaque preparations of injections mounted in this fluid, which were made in England, and which left nothing to be desired. Valentin afterwards remarked, that the tissues of sea animalculæ are very well preserved in this fluid. The beautiful preparations of vitreous-like medusæ, salpidæ, etc., in the naturalist's collections also harmonize with this observation.

Pacini has recommended certain preservative fluids, as modifications of this mixture, which contain sublimate, common salt or acetic acid, but no alum, including glycerine, however, as a useful addition, and intended for preserving various tissues. They are incomparably more serviceable and deserve accurate consideration. They are represented by the following two formulæ:—

Corrosive sublimate............ 1 part.
Pure chloride of sodium.......... 2 parts.
Glycerine (25° Beaumé) .... 13 parts.
Distilled water.................. 113 parts.

This mixture is allowed to stand for at least two months. After that time it is prepared for use by mixing one part of it with three parts of distilled water and filtering it through filtering-paper.

Blood-corpuscles are preserved in it exceedingly well, as my own observations have proved. According to Pacini, it is equally well adapted for nerves and ganglia, the retina, cancer cells, and especially delicate proteinous tissues.

A second mixture consists of:—

Corrosive sublimate............ 1 part.
Acetic acid .... 2 parts.
Glycerine (25° Beaumé)......... 43 parts.
Distilled water.................. 215 parts.

This mixture is prepared for use in the same manner as the preceding. It is said to destroy the colored blood-corpuscles, but preserves the lymph-corpuscles of the blood intact.

Further modifications of these mixtures, as they are employed in the Pathological Institute of Berlin, are represented, according to Cornil, by the following :—

1.		2.	
Corrosive sublimate	1	Sublimate	1
Chloride of sodium	2	Chloride of sodium	2
Water	100	Water	200

3.		4.	
Sublimate	1	Sublimate	1
Chloride of sodium	2	Water	300
Water	300		

5.		6.	
Sublimate	1	Sublimate	1
Acetic acid	1	Acetic acid	3
Water	300	Water	300

7.		8.	
Sublimate	1	Sublimate	1
Acetic acid	5	Phosphoric acid	1
Water	300	Water	30

No. 1 is used for preserving the vascular tissues of the warm-blooded animals. No. 2 for those of cold-blooded creatures. No. 3 for pus-corpuscles and related structures. No. 4 for blood-globules. No. 5 is intended for epithelial cells, connective tissues, and pus-cells, when the nuclei are to appear at the same time. No. 6 is applied to the preservation of connective-tissue structures, the muscles and nerves. No. 7 serves for glands, and No. 8, finally, for cartilaginous tissues.

Very dilute solutions of corrosive sublimate are in fact very useful as preservative fluids, but the degree of concentration should be determined every time they are used, for which reason it is judicious to mount several examples of a preparation in solutions of different strengths. Harting recommends solutions of 1 part of corrosive sublimate to 200–500 of distilled water. He remarks that it is only in such solutions that he is able to preserve the blood corpuscles. Those of man and the mammalia require $\frac{1}{200}$ of the sublimate, those of birds $\frac{1}{300}$, those of frogs $\frac{1}{400}$. Some of these solutions which I have tested appear to be useful. His recommendation of these solutions for the brain, spinal cord, and retina appears less justifiable, but, on the contrary, they are useful for cartilage, muscle, and crystalline lens. All solutions of corrosive sublimate readily cause the preparations to become dark and less transparent.

Chromic Acid and Chromate of Potash.—Dilute solutions of chromic acid and of the bichromate of potash, combined according to circumstances with glycerine, may be advantageously employed as preservative fluids. A mixture of equal parts of glycerine and Müller's preservative fluid (p. 139) appears to be very useful. The latter, undiluted, also occasionally forms a very serviceable medium for mounting very delicate textures.

A solution of *chloride of calcium* is a fluid popular with the botanists for mounting. It appears to be less serviceable for animal specimens. Harting praises the saturated solution of the pure salt, or one diluted with the 4–8 fold volume of water. Preparations of teeth and bones and sections of hairs are said to keep well in it. I acknowledge that thus far my experiments, not very numerous, it is true, with the solution of chloride of calcium have afforded me only very moderate results.

Harting recommends solutions of the *carbonate of potash* in 200–500 parts of water as the best medium for mounting nerve fibres. I have had no experience with this solution. According to the same authority, arsenate of potash in solution with 160 parts of water also exerts the same effect on nerve fibres.

Watery Solution of Creosote.—According to Harting's experience, a solution obtained by the distillation of creosote with water, or the saturated and filtered solution of creosote in a mixture of 1 part alcohol of 32° and 20 parts of water is a good preservative medium for many tissues, such as muscle, connective tissue, tendon, decalcified bones and teeth, and likewise the crystalline lens.

Arsenious Acid.—This is to be boiled with water in excess, and, after cooling, filtered and diluted with three times its volume of water. It is used for the same purposes as the solution of creosote, and is also suitable for the preservation of fat cells (Harting).

Methyl alcohol, considerably diluted with water, in the proportion of one to ten, has been recommended by Quekett. If, after several days, the fluid becomes cloudy, it should be filtered. Like the acetic acid solution, it causes most preparations to assume a granulated condition after a time.

Methyl alcohol and creosote are also elements of a complicated fluid mentioned by Beale.

Creosote	3 drachms.
Wood naphtha..............	6 ounces.
Distilled water..............	64 "
Chalk, as much as may be necessary.	

It is prepared in the following manner:—Mix first the naphtha and creosote, then add as much prepared chalk as may be sufficient to form a thick, smooth paste; afterwards add, very gradually, a small quantity of the water, which must be well mixed with the other ingredients in a mortar. Add two or three small lumps of camphor and allow the mixture to stand in a lightly covered vessel for a fortnight or three weeks, with occasional stirring. The almost clear supernatant fluid may then be poured off and filtered if necessary. It should be kept in well-corked or stoppered bottles. This mixture forms a modification of Thwaite's fluid for preserving desmidiæ.

Topping's Fluids.—He recommends the employment of one part of absolute alcohol to five parts of water; and where the preservation of delicate colors is necessary, he dissolves one part of acetate of alumina in four parts of distilled water. The latter mixture, diluted with an equal volume of glycerine, has preserved carmine injections for me very well over three years.

Deane's Fluid.—He praises a mixture of six ounces of pure glycerine, nine ounces of honey, a little alcohol, and a few drops of creosote, for the preservation of animal and vegetable structures. The mixture is to be filtered while warm.

Plain slides and covering glasses may be used for mounting very thin objects. A larger or smaller drop, as may be necessary, of the preserving fluid may be placed immediately on to the former by means of a brush or a glass rod, and the specimen, seized with delicate forceps or a cataract needle, is placed in this, care being taken that it is covered by the fluid. The covering glass, the under surface of which has been breathed on, is then placed over the specimen in the manner indicated for mounting in Canada balsam. One should avoid employing too large a quantity of the preservative fluid, as it is then liable to escape at the sides, or to flow over the edges of the covering glass. In such cases the excess should be removed by means of a very finely pointed pipette, or by the application of narrow strips of bibulous paper. In both cases the edges are also to be dried with a linen rag, care being at the same time taken not to displace the cover. Sometimes air-bubbles remain and can only be removed by slight pressure. It is well to cut a piece of fine writing-paper about an inch long in such a manner as that it will form a long, narrow triangle, measuring about 2''' at its base. The point of this is now to be passed between the slide and covering glass; frequently the air-bubble can be conveniently shoved out with it.

Although with Canada balsam as soon as the covering glass is successfully placed in position the whole process is essentially terminated, a further enclosure of the edges not being in reality necessary, although even here the specimen receives greater protection and a more attractive appearance by means of a finishing touch; it is otherwise with objects mounted moist; they must be cemented;—a procedure which will receive more especial mention further below.

If, however—and this will generally be the case—a somewhat thicker object is to be mounted, or if it be feared that the cement as it hardens will subsequently press the covering glass too much against the preparation and thus injure it, a firm substance should be interposed between the two glasses. Silver wires or narrow strips of paper are to be recommended as simple contrivances; various thicknesses of them may be prepared and placed under two opposite borders of the covering glass, or a narrow paper frame may be substituted. But then the insinuation of an air-bubble is quite possible, and the first layer of cement should not consist of a substance which is very fluid or which hardens very slowly, as it would then either penetrate into the preserving fluid immediately, or the external coating would afterwards in contracting press in the internal layers.

This procedure, in its further development, leads to the con struction of a framework which is sometimes shallow, sometimes deep, and which is fastened to the slide. The flat case thus obtained is called a cell.

Very manifold directions have been given concerning the construction of such cells. The more simple ones will be preferred unless greater cheapness renders another procedure desirable.

Cells may be made of gutta-percha, caoutchouc, and glass. The latter are the best, but also the most expensive.

Gutta-percha Cells.—Gutta-percha occurs in commerce in sheets of varying thickness. A good sheet should be even, homogeneous, and flexible. If crooked or cracked, it may be made to assume the former condition by dipping it in boiling water. With a ruler and knife quadratic or oblong square pieces may be cut out in the same way as from pasteboard; they should, however, be narrower than the slide. In these a

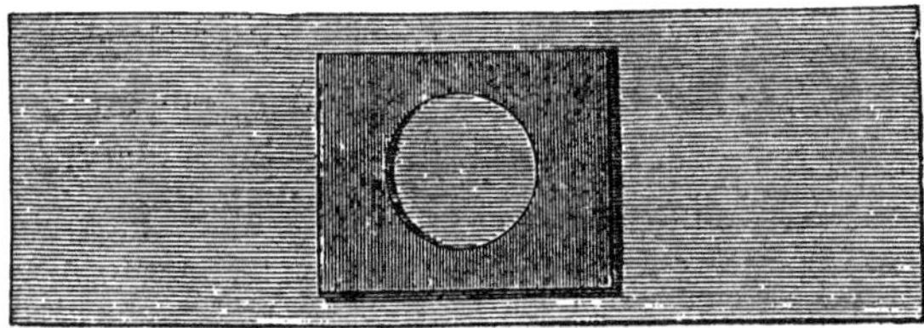

Fig. 81. Gutta-percha cell.

round, oval, or oblate square aperture is to be cut with a punch and hammer, to contain the preparation and the conserving medium (fig. 81).

Caoutchouc Cells.—These are also made from the commercial sheets, which may be placed one over the other and readily made to stick together by heating them, when it is necessary to construct cells with high walls.

Glass Cells.—These deserve the preference, but if bought ready made from the glass-cutter they are rather more expensive. Glass rings of different diameters and varying depths are used, but they have the inconvenience of requiring circular covering glasses. Quadratic or oblong square plates, resembling those of gutta-percha, and provided with circular openings, are useful. Those of half a line in depth and with a round aperture of about four lines in diameter will be found to suffice for most histological purposes.

Excellent glass cells, made in England, have lately been brought to my knowledge through Thiersch. They are made of glass slides several lines in thickness, perforated by a circular aperture of considerable size, with covering glasses cemented to both surfaces. The perfectly injected eyes of white rabbits, divided in halves and thus mounted with their natural curvature in Canada balsam, constitute one of the handsomest preparations which Thiersch has produced.

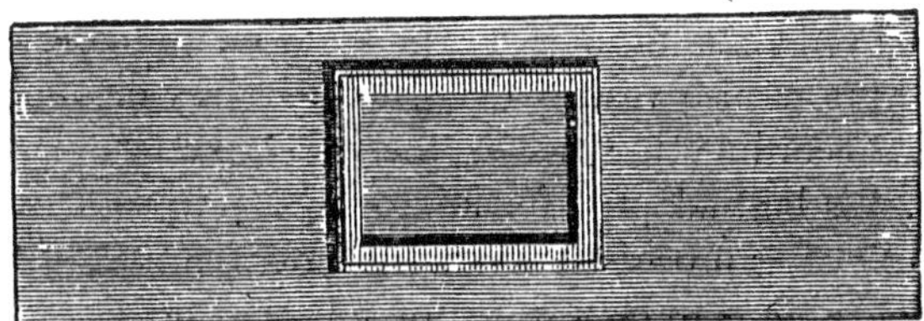

Fig. 82. Glass cell with cover.

Any one to whom economy of time is of less inportance, can himself prepare glass cells in still another way (fig. 82). He should have strips a line or so in breadth, cut from plate-glass (or, if able to use the diamond, he may cut them himself); these should be of two sorts, one of 6–7‴ in length, another form only 3–4‴ in length. With these the walls of the cell are to be constructed.

Beale, who, as is customary with Englishmen, explains the matter accurately, gives several additional practical directions.

A thin glass cover may be used for the construction of very shallow cells. The thin glass may be cemented while warm to a glass ring, or over a hole in a plate of glass, by means of the marine glue which is soon to be described. A hole is then to be forced through it with the point of a three-cornered file, and this is to be enlarged to the margin; the cracks do not extend across that part of the glass which is cemented. The perforated glass may be readily removed when again heated.

A blunt-cornered, square cell may be made by bending a single strip of glass in the blow-pipe flame and melting the ends together. Beale recommends flint-glass for this purpose. In practised hands, this is certainly a very good way of making deep and large cells.

All these cells must be cemented to the slide. But gutta-percha may be warmed in hot water, and its under surface then carefully dried and secured to the heated slide. This method has not proved to be durable.

Marine glue may be used, after the manner of the English, for cementing the glass cells.

This substance is prepared by dissolving, separately, equal parts of shellac and india-rubber in naphtha, and afterwards mixing the solutions thoroughly, with the application of heat. It may be rendered thinner by the addition of more naphtha. Marine glue is also readily dissolved by ether or a solution of potash. According to Quekett, the variety known in commerce as G. K. 4 is best adapted for microscopic purposes.

The following process is necessary for cementing with marine glue: The slide is to be warmed on a heated plate of metal (the English use a table of sheet-iron supported by four feet, with a spirit-lamp burning under it). A narrow strip of the cement is then melted on the heated slide and made to extend over all the places on which the walls of the cell are to rest. Firm pressure is then to be made over the cell, and the whole placed aside to cool. The excess of glue may afterwards be removed with the blade of a knife. A weak solution of potash may be used for cleaning the cell.

According to Harting, the following mixture serves for cementing the india-rubber cell: One part of very finely divided gutta-percha is to be mixed with fifteen parts of oil of turpentine, and dissolved at a gentle heat with constant stirring. It is then to be filtered through cloth, and one part of shellac added to the filtrate; this is also to be dissolved at a moderate temperature and with constant stirring. The heating is to be continued until a drop of the mixture, when allowed to fall on a cool surface, becomes tolerably hard. In this condition, the cement is ready for use. When afterwards used, a little oil of turpentine is to be added.

In order to fasten a caoutchouc cell, it is first to be held under the centre of the slide; and exactly over it, on the upper surface of the slide, a thin layer of the warm cement is to be laid on with a brush. The cell is now to be pressed into position on the upper surface of the heated slide, which is then turned over and allowed to lie on a flat surface till the cement has cooled.

Harting's gutta-percha cement may also be used in the same manner for fastening glass cells, and for building them out of four strips of glass.

The latter may also be accomplished with still another cement. 1 part of india-rubber is to be dissolved in 64 parts of chloroform, and then 16 parts of dried powdered mastic added. A thin layer of the cold mixture is to be placed on the slide with a brush; the cell is then to be warmed and pressed into position.

Whichever method is employed, it is well to fasten the cell as carefully as possible, in order that no leak or entrance of air into the cell may afterwards occur. A glass cell should always have a roughened surface for the cement. The surfaces may readily be roughened by rubbing on a flat stone with emery powder.

Cells of tin-foil have also been recommended, but I have had no personal experience with them.

Fig. 83. Placing the covering-glass in position.

Cells may also be made with certain cements; these are entirely sufficient for many thin objects. Asphalt may be used for this purpose, although I do not esteem it very highly. A white cement, made by the artist Ziegler, in Frankfort-on-the-Main (Friedberger Gasse 23), is better, and serves very well for this purpose. An oblong square, a quadrate, or a circle may be made with it on a slide, and left to harden.

The cell having been filled with the preservative fluid, and the object immersed in it, and care having been taken that there are no air bubbles present, the covering-glass is breathed on and placed in position in the usual manner (fig. 83). The latter should always be somewhat smaller than the cell, so as not to completely cover its outer margin. The superfluous fluid is to be removed with caution, however, as otherwise air-bubbles may again enter.

Now commences the cementing of the covering-glass. This must be done at once, unless the preservative fluids be glycerine or a solution of chloride of calcium, in which case the process may be delayed.

A considerable number of cements have come into use, and preparations may be mounted with several of them with equal perfection and security.

At present a solution of asphalt (Brunswick black) is most frequently used. Various sorts occur in commerce; it consists of a solution of asphalt in linseed oil and turpentine.

Good Brunswick black should have a transparent, homogeneous black appearance. As in the application of other cements, a camel's-hair brush is used, the stroke passing along the edge of the cover, whereby the latter as well as the slide receives a stripe of cement (fig. 84). With a little practice, one soon learns to judge of the proper quantity to use, and to draw a handsome border. If, in the course of time, the Brunswick black becomes too thick, it may be diluted with turpentine. Besides being dirty to handle, it also has the disadvantage of having a tendency to become cracked and fissured, and, after weeks and months, in consequence of its further contraction, to press the fluid out from the cell. It has therefore been recommended to strengthen the margins about every six months with a new layer of cement. This cement may also be considerably

Fig. 84. Making a border with Brunswick black.

improved by the addition of a small quantity of a solution of caoutchouc in benzine.

In consequence of its great tendency to the above-mentioned defects, I have, of late, either entirely ceased to use the ordinary Brunswick black, or only use it for the first coating, especially when strips of paper are placed between the slide and cover, and then, after several days, an external layer of cement is to be placed over this.

I have recently become acquainted with the Brunswick black used by Bourgogne, of Paris, and can only speak of it in the highest terms. Unfortunately, its composition is unknown to me. It dries with comparative rapidity and a single coating is quite sufficient.

[A very good and durable substitute for Brunswick black may be found in the "Liquid Stove Polish," prepared in England and sold in this country.]

A fluid mixture, coming from England under the name of gold-size, is excellent for cementing preparations of thin objects mounted in glycerine, and is also to be recommended as being clean to handle. It is a complicated mixture. Beale gives the following directions for its composition:—25 parts of linseed oil are to be boiled with one part of red lead, and a third part as much umber, for three hours. The clear fluid is to be poured off and mixed with equal parts of white lead and yellow ochre, which have been previously well pounded. This is to be added in small successive portions, and well mixed; the whole is then again to be well boiled, and the clear fluid poured off and kept in a bottle for use.

It is applied with a brush; a second layer may be added after half a day. It is better to leave specimens thus prepared for a considerable time before making the final application of cement.

I use Ziegler's white cement as a final coating; it may also be used alone with the most perfect security. It has been recently improved by Herr Meyer, the proprietor of the Hirsch Apothecary, in Frankfort. It is a somewhat thick mass and may be diluted at pleasure with oil of turpentine, at a moderate temperature. A thin layer applied with a brush suffices for glycerine preparations. A somewhat thicker layer, surrounding the cover like a wall, is generally applied, and is very useful as a protection to the latter, and in no wise detracts from the good appearance of the preparation.

This white cement generally dries very slowly, so that an impression may be made in it even after a long time. One should therefore be careful not to lay such preparations on each other, and especially to avoid all opportunities for anything to stick to them. But when once hardened there is no danger of the occurrence of cracks or flaws, and scarcely any of a leakage. Irregularities of the external margin may be remedied after a few days with the blade of a knife; but portions of the cement which have run over the surface of the cover should be left untouched for months. Oil of turpentine or benzine may be used to clean the brushes or glass.

As the composition of Ziegler's cement remains unknown, we have to thank Stieda for a communication giving the directions for composing a similar cement. Oxide of zinc is to be rubbed up with a corresponding quantity of oil of turpentine, and while rubbing, for each drachm of the oxide of zinc an ounce of a solution of the consistence of syrup of gum damar in oil of turpentine is to be added. If another color than white be desired, cinnabar may be used in the place of the oxide of zinc, using two drachms to the ounce.

Schacht recommended the so-called mask lac, which dries very rapidly, as a cement for wet preparations, and also as a coating for specimens mounted in Canada balsam or copal. The variety of lac used by him is designated as No. 3, at Beseler's lac manufactory in Berlin (Schützen Strasse, No. 66). I made considerable use of this lac several years ago, and do not hesitate to recommend it as being the next best to Bourgogne's cement.

We also mention a cement which is to be used in the manner already indicated as a final coating for Canada balsam preparations. We are indebted for a knowledge of this to a friendly communication from Thiersch.

When the specimens have been mounted for several days, weeks, or even months in pure Canada balsam, or a solution of the same in chloroform, they are surrounded with a border of

Canada balsam dissolved in chloroform, in the manner indicated above (fig. 84) for asphalt. Later—but never before the second or third day, still better after weeks or months—a final coating is to be applied. This consists of a colored and thick varnish of shellac. It is found ready prepared with alcohol at the wholesale druggists. It is to be carefully evaporated to the consistence of thin mucilage and colored with a filtered, concentrated solution of anilin blue or gamboge in absolute alcohol. Finally, about a scruple of castor-oil is to be added to each ounce of the mixture, and after some further evaporation it is to be preserved in a well-closed vessel. If, after a time, it should become too concentrated, this may be remedied by the addition of a few drops of absolute alcohol.

This varnish is to be applied with a brush to the borders of the Canada balsam. It becomes hard in a few hours, and then forms an elegant and hermetic covering for objects mounted in resinous substances.

This blue shellac varnish also forms a good finishing coat for objects which have been mounted wet and cemented with gold-size.

Finally, the form and size of the slides are not unimportant for the elegant appearance of a collection of preparations. Similarity of form, so far as it is possible, is desirable for greater convenience of keeping or of occasional transportation.

The art of grinding the edges of the slides may soon be learned by employing a very thick plate of glass and a fine sort of emery powder, which is to be made into a paste with water.

The slide should not be too small, so that sufficient space may be left at the ends of the preparation to affix two labels, one of which is to contain the general designation, while on the other especial remarks, the number of the collection, etc., may be placed. In certain cases there should also be room left for the indicator.* Such slides will frequently afford room for lar-

* Various indicators or object-finders have been proposed for enabling one to find any particular place in a preparation. Fine divisions, like those of a rule, may be photographed on narrow strips of paper, and one of these strips pasted at one of the broad, and one at one of the narrow margins of the cover; for example, at the right and the lower side of fig. 85. A rectangular plate of

ger objects and thus render it unnecessary to select a different form for special objects; as, for example, when an extensive section of bone or a voluminous injected preparation is to be mounted.

I prefer a glass slide like those of the English collections, three British inches long by one inch wide (72 mm. by 24 mm.), above all others (fig. 85). The preparations of Bourgogne of

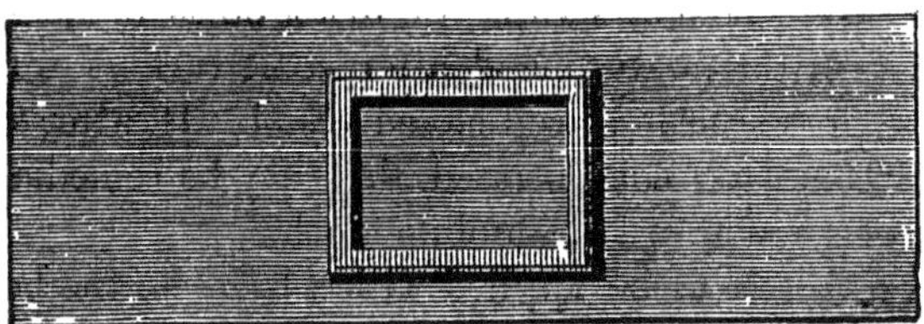

Fig. 85. English object slide.

Paris also have this convenient and handsome form. Slides of a larger size are unnecessary and appear too unwieldy. But those of a smaller size should also never be employed. A pattern, proposed in Giessen, 48 mm. in length by 28 mm. in width, is inelegant, and much less convenient than the English.

If it be desired to place microscopic preparations in layers, for the sake of greater economy of room, either in keeping or in transporting them, the employment of protection-ledges (Schutzleisten) is to be recommended. These are narrow strips of glass which are to be cemented across the slide at either side of the object. They should naturally be higher than the cover and the cell. Although this arrangement is, of itself, quite

---

metal, or, better still, a small square, consisting of two narrow strips of brass meeting each other under an angle of 90°, is used for ascertaining the particular portion of the object; this is to be noted on the preparation, and may be again readily found with the little plate or the square. The best—because the most simple—arrangement has been indicated by Hoffmann. Two crosses are to be scratched at either side of the aperture of the object stage of the microscope, the one standing (+), the other reclining (×). If, now, the portion of the preparation to be marked is situated in the centre of the field, both crosses are to be drawn with ink on the slide exactly over those of the stage. It is only necessary to place these marks over each other again in order to at once find the object.

practical, it nevertheless diminishes the room necessary for the labels to an unpleasant extent.

For preserving and arranging specimens, boxes of wood or pasteboard, with grooved wooden racks at the sides, which hold the slides securely, are occasionally used. As, with these, the slides stand vertically and the preparations may in consequence readily sink, when mounted in resinous substances which have not thoroughly hardened, or when other fluid media have been used, the upright position of such boxes deserves the preference. On the other hand, trays of wood or pasteboard with very low borders, or plain drawers may be used; they may either be arranged to slide out, like a chest of drawers, or they may simply rest on each other and be removed from the chest by means of loops. Slides of the most varying sizes may be conveniently placed in them at the same time, and the sinking of the preparation is also avoided. This arrangement, however, is not serviceable for transportation.

Here, as with all collections (increasing in degree with its growth), regularity and occasional revision are imperatively necessary.

As every assiduous microscopist of the present time possesses his own collection of preparations, so likewise do the various microscopical associations of Germany. For example, those of Frankfort-on-the-Main and of Giessen, as also the Microscopical Society of London.

Among the private collections we enumerate the celebrated one (preparations of injections) of Hyrtl, in Vienna; that of Kölliker, in Würzburg; Gerlach, in Erlangen; those of Thiersch and Leuckart, in Leipzig; Welcker, in Halle; and Schultze, in Bonn. In Holland is the collection of Harting; in London are those of Carpenter, L. Beale, L. Clarke, and others, likewise that of the College of Surgeons; in Manchester, that of Williamson. Among the Swiss collections may be mentioned those of His, in Basel; in Zurich, those of Goll and myself.

Preparations may be purchased of Hyrtl and G. A. Lenoir, in Vienna; J. D. Möller, in Wedel, Holstein; C. Rodig, in Hamburg; Schäffer and Budenberg, in Magdeburg; Bourgogne (9 Rue de Rennes), Paris; of Smith and Beck, also of Topping

(4 New Winchester Street, Pentonville), and of Pillischer (88 New Bond Street), London. Injections and other preparations of the author are to be obtained from the Magdeburg establishment already mentioned, from Lenoir, in Vienna, and from the optician Th. Ernst, in Zurich.

[The histological and pathological preparations of Dr. Edward Curtis, which are models of elegance and neatness, may be purchased of Queen, who also has for sale those of Dr. J. W. S. Arnold.

Dr. Francis Delafield also has a very large assortment of pathological and histological preparations at the "Pathological Laboratory of Bellevue Hospital."]

Section Eleventh.

BLOOD, LYMPH, CHYLE, MUCUS, AND PUS.

The investigation of these cell-containing fluids belongs to the more easy and simple labors of the microscopist, inasmuch as a drop of the same, having been placed on the slide by means of a glass rod, and spread out into a thin layer by the covering glass, suffices for the first examination. But care is to be used in the selection of actually indifferent media, especially in the examination of living cells.

1. Among the animal fluids mentioned the blood is the most delicate substance, so that circumspection is necessary for the recognition of its normal condition.

In order to examine human blood, it is only necessary to prick the point of the finger and allow a drop to exude, which is then received on the slide. For more continued and prolonged investigations a quantity of blood is to be obtained by means of venesection, and beaten to separate the fibrin. The blood of the smaller animals is to be obtained by opening one of the larger vessels or the heart; the blood may be received in a test tube. If allowed to remain in this, or in a cylindrical vessel, the cells gradually sink and the serum which remains above them becomes colorless. This fluid is the best medium for use in the investigation.

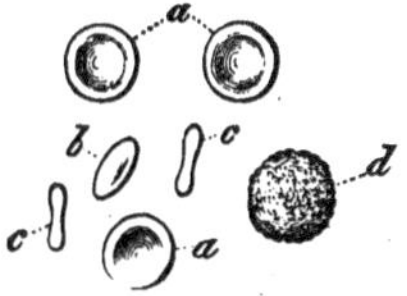

Fig. 86. Human blood-cells. *a a* seen from above; *b* half, *c c* entirely from the side; *d* a lymph corpuscle.

In consequence of the extraordinarily large number in which the colored cells (fig. 86, *a*, *b*, *c*) occur in the blood, it should be spread out in a very thin layer, so as to bring these elements distinctly into view. Slight compression made on the covering glass with the point of the needle will facilitate the examination considerably. In human blood (fig.

86) when these cells have their broad surface turned towards the observer, they present the well-known form of circular disks (*a a*), but when standing on their sides they appear biscuit-shaped (*c c*).

Dilution of the blood requires some little attention. The serum of the blood, if disposable, is best for this purpose. Solutions of salt, sugar, or crystalloid matter may also be used with advantage for a momentary examination, provided they are of the proper degree of concentration. The Pacinian fluid (sublimate, salt, and glycerine with water), which has already been mentioned (p. 213), is very suitable if it happens to be at hand, and I know of no other fluid which is capable of preserving our cells in so excellent a manner for years. The iodine serum is also very useful, and likewise, according to Rollett, a mixture resembling Müller's eye fluid. The latter consists of one part of a cold saturated solution of the bichromate of potash, 5 parts of a similar solution of the sulphate of soda, and 10 parts of water.

Such dilutions will also be necessary when it is desired to cause the colored blood-cells to roll, in order to recognize their form. The pressure of the point of a needle on the edge of the covering glass will then induce the desired current in the fluid.

Fig. 87. Contractile cells from human blood.

Upon carefully focussing, the human colored blood cells will appear to present a yellowish circumference and a colorless centre. If the tube of the microscope be depressed a little, the central portion becomes somewhat darker.

The colorless cells of the blood originate in the lymphatic glands, the spleen, and the marrow of bones. Dilution with an indifferent fluid is also necessary for their recognition, and, in consequence of the small number of these elements, they require some little search (fig. 87).

Even in human blood taken directly from the vein, one may with a 4–600 fold enlargement, and without any further precautionary measures, observe the remarkable changes of form of the living colorless cells, which may slowly pass through the series of alterations which we have sketched (fig. 87). But if the warm stage (p. 105) and a temperature of 38–40° C. be used, and iodine serum added, the play of movements mentioned becomes extraordinarily lively. A portion of the colorless cells now creep about between the colored blood-corpuscles and present, with constantly changing form, the strangest variety of shapes. Granules of carmine, molecules of cinnabar and indigo, which have been added to the fluid, are now readily taken up into the cell-body (Schultze). An extremely fine granulated anilin blue, which has been precipitated from the alcoholic solution by means of water, is very excellent. —If this apparatus is wanting, one may readily perceive the same condition in the lymph corpuscles of frog's blood, with the aid of the moist chamber. Very beautiful appearances may be obtained with the latter animal, if a drop of its fresh blood be allowed to coagulate on the under surface of the covering glass, in a moist chamber (after the manner of our fig. 64). One soon notices, after a zone of serum has formed at the borders of the coagulum, that, in consequence of their lively wandering from the clot, many of these amœboid cells have penetrated the ring of fluid, and that the surface of the coagulum is also thickly covered by them (Rollett).

We will here mention still another method which has recently led to scientific results of the highest interest (Cohnheim).

A small quantity (at most, a few ccm.) of one of the above-mentioned finely granulated coloring materials, suspended in water, is to be injected, for several days after each other, into one of the large lymph spaces which lie under the skin of the frog. A Pravaz syringe, such as is used in practical medicine, may be used for the injection. On examining a drop of the blood, a considerable number of the colorless cells will now be seen to be stuffed ("gefüttert"). We shall afterwards return to this matter.

Some preparation is necessary in order to count the numbers of both kinds of cells. The test blood must naturally be spread out into the thinnest layer, and the space to be surveyed divided. An eye-piece micrometer with a small number of quadratic fields fulfils this object. In consequence of the sparseness of lymph-cells in normal human blood (0.5 to 2–3 per thousand), and likewise in mammalia, it is necessary to count a large number of blood corpuscles in order to obtain even a tolerably accurate result. One should not stop under 10–15,000.

The fluid of the blood, the so-called plasma, appears, as a rule, entirely clear like water, and free from all elementary forms, and therefore is not an object of microscopic examination. As a result of an exuberant reception of fat in the blood, the unsaponified fat of the chyle (see below, at this fluid) may appear in it in the condition of the finest division, in the form of dust-like molecules.

It was formerly hoped that the microscopist might discover changes in the form of the blood-cells in disease, and in this way be able to promote pathological physiology as well as diagnosis. These beautiful dreams have not, in general, been fulfilled. However various its composition may be, the blood presents the same microscopic appearance. It is also to a certain degree the case, that even with regard to the normal life of the blood a considerable obscurity still prevails,—that we have but an extremely incomplete conception of the new formation and disappearance of the cells.

However, although nothing of importance is to be perceived in the endosmatic changes in form of the colored blood-cells, which have been here and there described in processes of disease, and just as little in the shreds of the loosened epithelium of the vessels, the microscope has nevertheless afforded us an interesting glimpse of two pathological processes of our fluid; we mean the so-called leucæmia and melanæmia.

The former, coinciding with an increase in volume of the spleen and often, at the same time, of the lymphatic glands, although but seldom caused by enlargement of the latter organs alone, leads to a constant increase in the number of colorless cells in the blood, so that finally the alteration of the blood no longer remains concealed from the naked eye. A drop of such blood (obtained by pricking the point of the finger with a needle) shows, together with the colored, a considerable number of colorless blood-corpuscles. This may proceed to such an extent that to three colored blood-corpuscles there will be one or even two colorless ones, and in certain cases the number of the latter variety of cells will be greater than that of those containing hæmatin. Transition forms of both kinds of cells may also be met with (Klebs, Eberth).

In malignant forms of intermittent fever the enlarged spleen has been seen to have a blackish appearance. The microscope shows, as a cause of this change of color, granulated lymphoid cells, often of considerable extent, and which contain within them granules of the black pigment. Passing out through the splenic vein, they become mixed with the blood and are seen in this fluid when it is subjected to microscopic examination. In consequence of their size they produce obstructions in certain capillary districts, especially in the brain and liver.

Embryonic blood is to be examined in the same manner. The warmable stage is to be used when it is desired to follow the very rapid progress of the division of the nucleated colored cells. The instability of these cells is, moreover, very great, so that one may often be misled by artificial products.

Recklinghausen communicated a singular discovery to us a few years ago. After a series of days one may see the lymphoid cells become transformed into red blood-corpuscles, in blood taken from the frog, if one understands preserving its vitality.

For this purpose the blood is to be received in a glazed porcelain dish, which is to be placed in a large glass vessel, the air in which is to be daily renewed, and kept constantly moist. After twenty-four hours the coagulation gives place to a process of liquefaction; a few days later, island-like collections of contractile lymphoid cells have become formed; after 11–21 days one may recognize the first of the new-formed blood-corpuscles. Frog's blood may be preserved in this way for thirty-five days without decomposition taking place.

By means of an electrical discharge the colored blood-corpuscles are rendered corrugated, at first with coarse, then with fine indentations. These processes afterwards disappear, and the blood-corpuscle becomes transformed into a smooth bordered globule, which finally undergoes discoloration (Rollett).

The living blood-cells of man and the mammalia undergo a very singular alteration (fig. 88) when exposed to a temperature of 52° C. on the hot stage, represented in fig. 66. A number of deep indentations rapidly take place, which very soon become constrictions, causing the formation of globules on the surface of the cell. These are either separated at once or continue for a time to be connected, by means of a long, slender pedicle, to the remainder of the cell body (*a*). The strangest appearances are thus caused, such as bead-like rods, globules with projections like handles, etc. When these fragments become separated they commence the most lively molecular movement (Beale, M. Schultze).

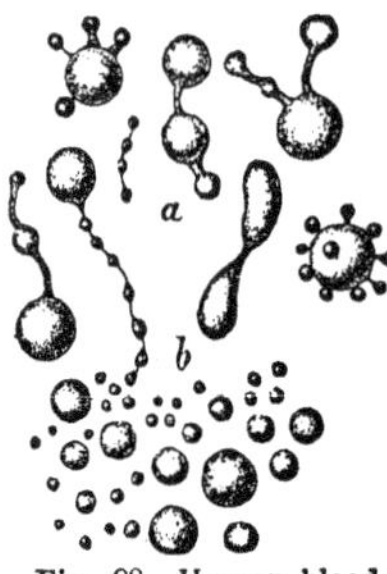

Fig. 88. Human blood-cells heated to 52° C.

The treatment of the blood-corpuscles with chemical reagents is indispensable for the accurate investigation of their structure, and is a very good exercise for the beginner, especially if the large nucleated cells of the naked amphibia are used in the place of the small non-nucleated corpuscles of human and mammalial blood.

Distilled water is used to cause them to swell (fig. 89, *a*). The bright central portion disappears at once and a uniformly yellowish structure is seen which rapidly loses its color, and which, in rolling, permits its globular form to be recognized. In this way the granulated nucleus may be rendered distinct in the blood-cells of fish, amphibia, and birds. Many watery solutions in a condition of extreme dilution exert a similar effect. For comparison, it is useful to treat in a similar manner the large nucleated blood-cells of the first three classes of vertebrates. In order to obtain them in a shrivelled condition (fig. 89, *b*), it is only necessary to leave a drop of blood uncovered on a slide for a few minutes, in which case the familiar corrugated and indented forms make their appearance. A very small drop of blood, taken from the living body by pricking with a fine needle, not unfrequently presents this indented appearance of the cells on the glass slide at once. We can obtain a similar effect by means of numerous concentrated solutions, such as those of salt, sugar, and gum.

If, on the contrary, we rapidly dry the blood-corpuscles on a glass slide, we have the appearance represented by fig. 89, *c*, a form in which the blood-corpuscles may be very well preserved as a permanent preparation.

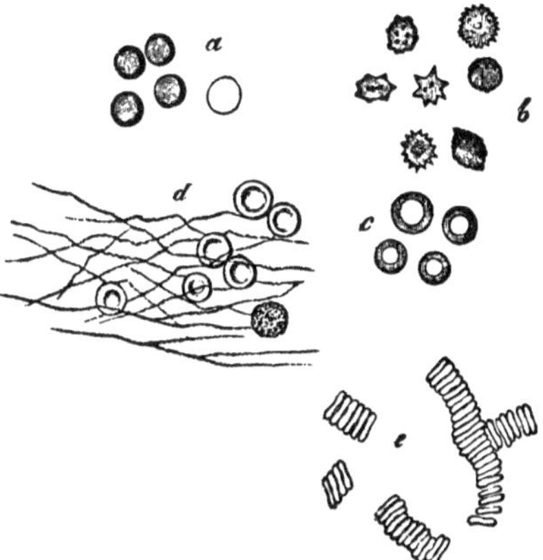

Fig. 89. *a*, Human blood-cells under the action of water; *b*, in evaporating blood; *c*, in a dried condition; *d*, in coagulated blood; *e*, rouleaux-like arrangement.

Other reagents dissolve its substance, and in this way destroy the cell. Diluted acids produce this effect, likewise weak solutions of the alkalies. Although concentrated solutions of the latter cause the blood-corpuscles to swell, they do not destroy them, even after acting on them for hours. A saturated solution of potash is, as Donders found, an excellent medium for rendering the cells of dried blood again visible.

Many materials have a coagulating effect on the cell substance of the blood-corpuscles. Among these are to be enumerated alcohol, concentrated chromic acid, sublimate, and other metallic salts.

In defibrinated, but also very generally in a drop of freshly removed blood, one may observe the familiar joining together of the colored cells with their broad surfaces, the so-called rouleau formation (fig. 89, *e*). This arrangement is missed only in the more distended and globular cells of the blood in the splenic and hepatic veins.

In order to ascertain the appearance of coagulated blood, one may either allow a drop of blood to coagulate on the slide, or the finest possible section may be taken from a clot. The cells will then be seen to be embedded in a homogeneous layer of fibrin which has the appearance of folds or filaments (fig. 89, *d*).

Indifferent fluids are necessary in the examination of blood extravasations, for the proper estimation of the condition of the cells.

The origin of fresh clots of blood, treated in the same manner, may be ascertained by microscopic analysis. One will be able, for example, to distinguish, by their form and size, the cells of the blood of birds from those of human blood, etc., and in this way to detect impostors. It is difficult, and in many cases impossible, to render a decision in old masses of dried blood. The character of a spot which is suspected to be blood may, on the contrary, be determined in the most certain manner by means of Teichmann's hæmine test, a subject to which we shall again refer.

The accessories mentioned under lymph and chyle are to be used when the further investigation of the colorless blood-cells is necessary.

It is only under certain circumstances that the colored blood-cells can be tinged with carmine, but this may be readily accomplished with anilin red; still, nothing is to be gained thereby.

The above-mentioned process of rapid drying may be very advantageously employed for preserving blood-cells permanently as preparations for a collection. I have in my possession preparations of the blood of different animals which are more than 20 years old, and which leave nothing to be desired.

The first-mentioned Pacinian fluid is adapted for mounting the cells of human blood moist; Pacini's second mixture (see p. 213) serves for the colorless cells of the blood.

Solutions of sublimate, as formerly mentioned (p. 215), have also been recommended. Harting employs for the blood-cells of man and the mammalia, 1 part of bichloride of mercury to 200 of water, for those of birds 1 to 300, for those of the frog 1 to 400. Remak employed, for embryonic blood-cells, very weak solutions of the bichromate of potash, of chromic acid (0.03 per cent.), and of sublimate (0.03 per cent.).

We should make ourselves responsible for a considerable deficiency were we to omit mentioning the various crystallizations which are to be obtained from the colored blood-corpuscles. This subject has in our day been zealously and persistently investigated; but, regarded from a scientific point, this matter leaves, even yet, much to be desired.

From the blood of man and the various vertebrated animals, including birds, one may obtain the coloring substance of the cells in a crystalline condition; the so-called blood crystals being formed. This substance has been called hæmoglobin or hæmato-crystalline. Many investigations have been instituted concerning these remarkable structures by Funke, Lehmann, Kunde, Teichmann, Rollett, Bojanowski, and others; Reichert having previously discovered in them a crystallized, colorless, albuminous body.

According to the general acceptation, the blood-crystals present various forms, such as prisms, tetrahedes, hexagonal tables and rhomboids. The prismatic form is regarded as the most common, and appears in man and most of the mammalia (fig. 90, *a*, *c*), together with which, rhomboidal tables (*b*) may also be met with. Tetrahedal (but not regular) crystals are formed by the hæmoglobin of the Guinea-pig (*d*) and, as is generally alleged, of the mouse; rhomboidal crystals are met with in the hamster (*e*), hexagonal tables (*f*) in the squirrel (and mouse ?).*

With regard to the manner of producing blood-crystals, we limit ourselves to the following examples :—

They are to be prepared for microscopic examination according to Funke's directions. A drop of blood is to be placed on the glass slide, where it is left in contact with the air for several minutes. A drop of water is then to be added, and the whole breathed on a few times. A covering-glass is now placed over it, and evaporation allowed to take place slowly, whereby the crystallization is promoted by the action of the light.

---

* In reality, nearly all blood crystals belong to the rhomboidal system, only those of the squirrel to the hexagonal.

Bojanowski recommends the following procedure: Blood, as it escapes from the vein, or still better, such as is taken from the vessels of a dead animal, is to be kept in a vessel for 2–4 days in a cool place, whereby the coagulum begins to dissolve into a thick, fluid, dark red to blackish mass. A drop of this fluid is to be placed on the slide, covered, and exposed to the light for a few hours. The crystals may then be seen. If the blood which is to be used for this purpose is too thick, the drop may be very suitably diluted with distilled water.

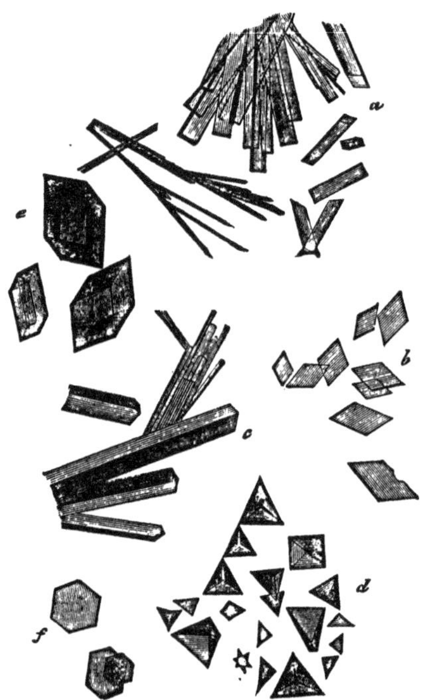

Fig. 90. Blood-crystals of man and several of the mammalia. *a*, blood crystals from human venous blood; *b*, from the splenic vein; *c*, crystals from the blood of the heart of the cat, *d*, from the jugular vein of the Guinea pig; *e*, from the hamster; and *f*, from the jugular of the squirrel.

Rollett, who has also produced a very valuable work on blood crystals, makes use of a blood, the cells of which have been destroyed by freezing and remelting. The formation of crystals also readily takes place in electrified blood, and in that of the Guinea-pig (which of all kinds of blood crystallizes the most readily) this is often so rapid as to appear "as though the crystals had been struck out with the spark." Blood from which the gases have been pumped out is also well adapted for obtaining hæmato-crystalline.

Chloroform with the access of air also causes the formation of our crystals (Böttcher).

Lehmann has taught us how to produce crystals of the hydrochlorate of hæmatin (figs. 91, 92).

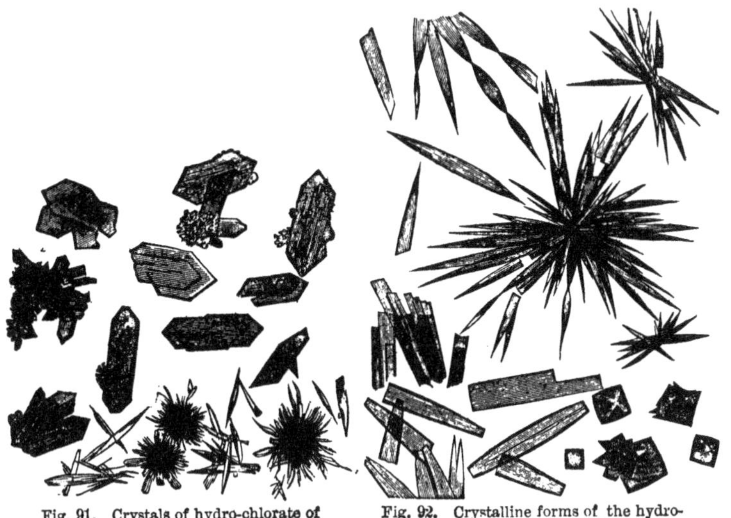

Fig. 91. Crystals of hydro-chlorate of hæmatin.

Fig. 92. Crystalline forms of the hydrochlorate of hæmatin.

They are to be obtained by treating fresh blood or large spots of blood which are two days old with alcohol containing oxalic acid and ether (1 part alcohol, 4 parts ether, and $\frac{1}{16}$ of a part of oxalic acid). Preserved in well-closed bottles, the crystals are gradually precipitated from the fluid; the process is hastened by the addition of chloride of calcium which has become liquefied by exposure to the air. Where the separation takes place more rapidly the crystals are more of the acicular form, as represented at the lower part of fig. 91; if more slowly, either the hexagonal tables of fig. 91 or the crystals which are represented in fig. 92. They appear to have a long and narrow laminated shape and twisted one or two times on their long axis. They are very thin, of a brownish and brownish-green translucency, as represented at the upper half of fig. 92. If allowed to remain for some time in the mixture of alcohol and ether in which they were precipitated, we have produced, as another modification, the crystals given in the lower half (to the right) of the figure, quadratic and also rhomboidal black tables which, by more accurate examination, prove to be flat rhomboidal octahedrons.

Teichmann has produced crystals of the same modification of hæmatin and called them hæmin. The coloring matter of the blood, in its different conditions, is to be dissolved by means of hot concentrated acetic acid, so as to become separated in a crystalline form as it cools. A condition of the precipitation is the presence of alkaline chlorates. The hæmin crystals obtained present the appearances represented in fig. 93, as rhomboidal tables of a blackish-brown, sometimes blacker, more rarely light-brown color.

Fig. 93. Crystals of hæmin.

By proper treatment the crystals with which we are at present occupied may be obtained from blood which is either fresh or decomposed by putridity, from that which is dried, and even from the oldest blood-stains. Hæmin is therefore of great importance in a forensic point of view, and forms the best means of recognizing the origin from blood of a suspected stain.*

If it be desired to produce a somewhat larger number of crystals, a quantity of blood is to be boiled for about a minute or two in the 10–20 fold volume of glacial acetic acid and filtered. As the fluid cools it becomes somewhat cloudy, and a blackish sediment is deposited, consisting of crystals of hæmin. For the momentary demonstration the following process is to be employed:—A drop of blood is to be rapidly dried on the slide,

* With regard to the value of the hæmin crystals in a forensic point of view, as well as the possibility of mistaking other substances for them, see the article of Büchner and Simon (Virchow's Arch., Bd. 17, S. 50).

over the spirit lamp, and then scraped to a powder with the point of a knife. About 10–20 drops of anhydrous acetic acid is to be added and allowed to boil a few times, the slide is then to be set aside for a few moments. A drop of blood, diluted with 15–20 drops of glacial acetic acid and placed in a watch-glass on the stove, also forms the crystals in question, as the fluid evaporates. They are likewise deposited when blood is mixed with an excess of concentrated acetic acid. After a few days a film, consisting of these crystals, is formed on the surface ; after the removal of this a second is formed, and so on.

In order to obtain the hæmin crystals from an old blood-stain, the stained substance is isolated and placed in a test-tube, glacial acetic acid is then poured over it and boiled for a few minutes, it is then filtered into a watch-glass. This fluid, to which more acid is to be added, is then exposed to evaporation in a warm place. I am indebted to the kindness of Dr. A. Schmidt, of Frankfort, for a preparation of hæmin which was obtained from a pocket-handkerchief saturated with blood at Sand's execution.

Hæmin crystals, in consequence of their durability, may be very readily preserved as microscopic preparations. They may be mounted dry or in glycerin.

Fig 94. Crystals of hæmatoidin. (Ordinary form.)

In old blood extravasations, for example, those of the brain, in hemorrhagic infarctions of the spleen, in obliterated veins, in the corpus luteum crystals of hæmatoidin, discovered by Virchow, are formed (fig. 94) ; they differ from the bilirubin which occurs in the bile. They generally occur in small rhomboidal prisms of a lively orange or ruby-red color, with dark carmine-red borders and edges. Together with these, amorphous precipitates of hæmatoidin, in granular and globular masses, will be frequently met with.

Staedeler succeeded, by treating the ovaries of the cow with chloroform, or with sulphuret of carbon, in obtaining uncommonly large crystals (fig. 95) of our coloring material, some of them measuring even 0.2'''. These make their first appearance under the microscope as acute-angled, three-sided tables, with one convex side (*a*), although this convex side may also be replaced by two direct lines, so that deltoid tables (*b*) result. Two such tables usually become united like twins, their convex sides coming in contact with or overlapping each other, and melting together (*b*, *c*). In this manner are formed the rhomboidal tables, usually designated as hæmatoidin (fig. 94). As a rule, there are at first indentations in the place of the obtuse angle of the rhombus, which gradually become filled out (*d*, *d*). Not unfrequently two other crystals also become united with the first two individual crystals, so that four rayed stars now appear (*e*). By the filling out of their re-entering angles four-sided tables (*f*, *g*) are formed, which, by increasing their thickness, finally assume the appearance presented by dice when seen somewhat from the side (Staedeler).

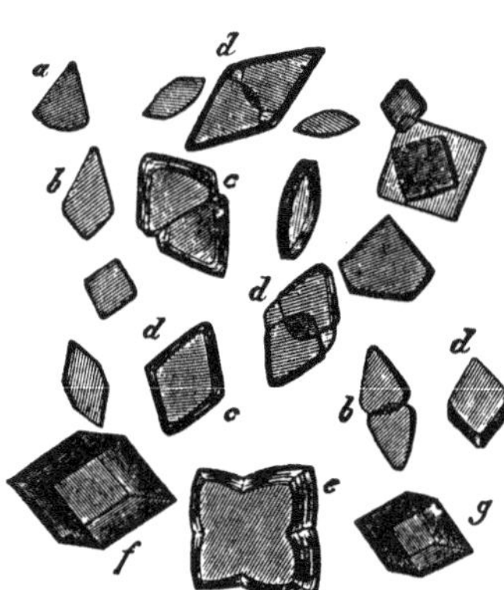

Fig. 95. Very large crystals of hæmatoidin obtained from the ovarium of the cow by treating with chloroform.

Preparations of hæmatoidin may be readily and well preserved dry or lying in glycerine.

We have reserved the most interesting portion of this section for the end; the movement of the blood in the living animal body is still to be discussed.

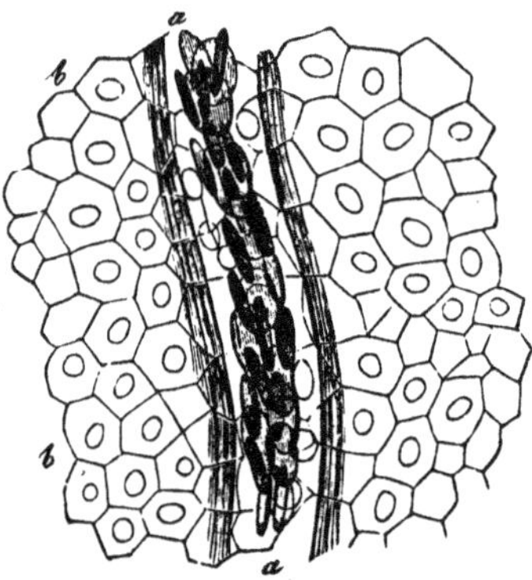

Fig. 96. The blood-current in the web of the frog. *a*, the vessel with the colored blood-corpuscles in the axial port on, and the colorless cells in that portion of the current near the walls; *b*, the epithelial cells of the tissue.

In order to see the blood flowing through the vessels of the living animal (fig. 96) it is necessary to select transparent localities. The web of the hindfoot of the frog, the transparent tail of the larvæ of the frog and salamander, the embryos of fish and small recently-hatched fish are exceedingly well adapted for the first observations.

If the larvæ of frogs are used the anterior portion of the body is to be enveloped with a strip of moistened blotting-paper, and the tail, after being moistened with water, is to be covered with a thin covering-glass. If the frog itself is to be used, a strip of wood or cork containing a glass window, 5 or 6 lines in size, which goes over the hole in the stage, is to be employed. The frog is to be wrapped in a moistened rag, or enclosed in a small linen bag, and secured to the wooden strip. The web is to be expanded by means of pins (but without too great tension), and, after being moistened with water, a thin covering-glass is to be placed over it; it is best to have the latter three-cornered or rhomboidal in shape. Instead of the simple strip of wood, a small table may be very suitably arranged for supporting the frog. Frog-holders have been invented for this purpose; they are quite superfluous. If one has the remarkable muscle-poison, woorara, at one's disposal, it is only necessary to inject a minimal quantity of the same under the skin to render our animal, after a few hours, immovable for a considerable time—one or two days—so that the simplest arrangement is then all that is necessary. An object slide of considerable size, on which is cemented a small, thick, right-angled plate of glass with a cork ring surrounding it, for fastening the toes to, is then sufficient. For tadpoles, the anterior portion of the body may be simply enveloped in a strip of blotting-paper, and the tail, with the addition of water, covered with a thin glass. Object-slides with long, four-sided excavations, as proposed by F. E. Schulze, and represented in section by our fig. 97, are very well adapted

Fig. 97. Object-slide for the larvæ of frogs and salamanders.

for the examination of the larvæ of naked amphibiæ and young fish. The head goes under the edge *a*, and the tail of the animal lies on the inclined surface *b*. The whole is to be covered with a thin glass. The apparatus may be very readily constructed by cementing four pieces of glass on to a slide.

In examining the circulation only very weak lenses should be employed at first, in order to obtain a more considerable view of the relations of the currents. Then proceed to the use of higher powers, with which the details, especially in the capillary vessels, are to be investigated. It is unnecessary to remark that the apparent rapidity of the current is in this way very much increased. In reality, this is by no means considerable in the capillary districts. A corpuscle of frog's blood passes through the fifth or fourth part of a line in a second.

The colored, in contradistinction to the colorless elements of the blood of the adult animal, are without any vital contractibility, as may be best learned by just these observations of the circulation of the frog; it is only possible to recognize here certain passive changes of the so flexible and elastic cells.

A discovery recently made concerning an anomalous condition of the blood cells of the mammalia is of interest. So long as they remain in the circulation they but rarely appear in the above-mentioned form of the passive condition, but rather present the greatest variety of shapes, so that that which in the frog formed an exception has here become the rule. Even this only concerns a passive condition, for as soon as they become quiet they again resume the familiar cup shape (Rollett).

The mesentery of the frog is to be recommended for studying the condition of the circulation during the process of inflammation (Cohnheim). Take a sufficiently large plate of glass, cement to this a small glass disk nearly a line in thickness and of about 12 mm. diameter, and around the latter a narrow (about 1 mm. broad) cork ring. An opening is to be made through the abdominal walls on the left side of a frog which has been paralyzed with woorara, the mesentery is to be drawn out, and the loop of intestine fastened with a few fine needles to the cork ring. The simple irritation of the air produces the inflammation, and if the mesentery be protected from drying the process may be studied for many hours.

Small mammalia, maintained in a condition of narcosis by means of ether or chloral, may also be used for such investigations, although with the assistance of manifold complicated apparatuses (Stricker and Sanderson).

But let us return to the mesentery of our frog! Dilatation of the vessels (at least of the capillaries) gradually takes place, sluggishness of the current follows, and numerous lymphoid cells collect at the colorless peripheral portion of the veins. The lymphoid cells begin to migrate through the uninjured walls of the latter and of the capillaries; colored corpuscles also pass through the capillaries into the neighboring tissues. After a half or a whole day, when the surface of the mesentery is covered by a dull grayish layer of pus cells, these remarkable phenomena have commenced in full force. The pus cells have therefore come from the blood-vessels (Cohnheim). If a finely granular coloring material has been previously injected into one of the lymph sacs of the animal, a part of these cells will contain coloring matter.

If, on the contrary, the circulation be arrested by ligating the crural vein, the blood corpuscles will be seen to be pressed closely together in the vessels of the web of our animal. Here, also, there is a passive escape of colored blood-cells. It is almost impossible, on the contrary, for the lymphoid corpuscles to manifest their vital contracting power, in consequence of the compression exerted by the overloaded condition of the vessel. Here their active emigration generally fails, or is but very slight.

A similar emigration of the lymphoid cells also takes place in normal life. The movable cells, which wander through the spaces in the connective tissue, are to be reckoned among these.

Cohnheim's beautiful observations, which confirm the older views of A. Waller, possess an extraordinary range and have rapidly called forth an entire literature. Opinions are still undefined with regard to this subject. Do all these migratory cells and pus corpuscles originate in the blood, and, having migrated, are they incapable of further increase by division? May not pus cells proceed from the cellular elements of the connective tissue? Finally, are these emigrants capable of being transformed into other tissue elements? The latter is not to be doubted; and very many observers have affirmed the division of the lymphoid cells, as well as their origin from the cells of the connective tissue. We shall again refer to some of these points.

2 and 3. The examination of lymph or chyle is also very easy; only some little preparation is necessary for obtaining the material. In order to obtain lymph, a mammal is to be killed by a blow on the head, and, after carefully opening the thorax, a ligature is to be applied to the ductus thoracicus. The lymphatic vessels will be found to be swollen even after a quarter of an hour, and the distention is increased by waiting for a longer time. If the animal has been killed several hours after a plentiful meal containing fat, the lacteals become filled with a milk-white fluid and stand out in the most beautiful manner. With small herbivorous animals, for instance, rabbits, an elastic catheter may be introduced into the œsophagus, and a considerable quantity of milk injected through this into the stomach. After an interval of several hours the lacteals will be found magnificently filled.

The lymphatic or lacteal vessels are then to be ligated in pieces about one inch in length, a ligature being applied at each end, and the vessel carefully separated from the connective tissue. The separated vessels are to be cleaned by washing in water and again dried, after which they are to be opened over a watch-glass or a slide.

If the lymph-corpuscles are desired for rapid demonstration only, the necessary material may be obtained by pricking any lymphatic gland.

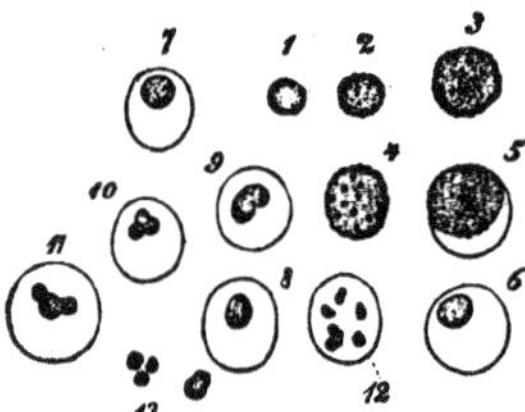

Fig. 98 Lymph-cells.

In lymph and chyle we find, by 2–400 fold enlargement, the characteristic cells (fig. 98), the same with which we have already become familiar in the blood as colorless blood-corpuscles. When examined in their natural fluid, these structures do not, as a rule, show anything further than a granulated globule of varying size (fig. 98 1–4). If this examination is made with the necessary precautionary measures, the same change of form of the cell may here be found, as an evidence of a vital contractibility, which we have mentioned above (p. 231) concerning the colorless elements of the blood.

In order to demonstrate further the manner of its organization (nucleus and cell-body), water or extremely diluted acetic acid may be used. (Stronger acids soon dissolve the envelope of the cell and its contents.) These changes are represented in the figure from 5 to 13. The ammoniacal solution of carmine, fuchsine, and aniline-blue may be used for staining.

Innumerable fat molecules, in the condition of the finest division, occur in the milk-white chyle, as the cause of its color. These atoms require a strong (4–600 fold) enlargement.

For their preservation, the mixture mentioned at p. 214 (No. 3), consisting of sublimate (1), salt (2) and water (300), is to be recommended; Pacini's second fluid may also be employed.

4. Mucus does not require any preparation. It may be either scraped with the blade of a knife from the surface of the mucous membrane, or it may be obtained from the nose or respiratory organs, etc. A moderate quantity is to be placed on the slide. Uncommonly tough masses of mucus are to be divided with the scissors.

The microscopic examination (with a 2–400 fold enlargement) shows us a somewhat irregular constitution. We meet with an extremely variable quantity of the same colorless granulated cells which have just been referred to as colorless blood-corpuscles, and also as elements of the lymph and chyle, the "lymphoid cells." The name of mucous corpuscles has been given them, only in the cavity of the mouth these structures are called salivary corpuscles. In the latter place, coinciding with the thinner and more watery fluid, granular movements are observed in the interior of the cells. Conformably to this, a cessation of these movements may be produced by the addition of concentrated solutions; they may also be incited in all other lymphoid cells by removing them to a very watery neighborhood (C. Richardson). Together with these, is also associated an extremely variable quantity of the desquamated cells of the existing epithelial formation, and likewise the separated cells of the various mucous glands. In consequence of its viscid nature, there are very generally air-bubbles imprisoned within the mucus. Furthermore, there are also very many extraneous admixtures to be seen; as the remains of food, for example, fibres of meat, grains of starch, particles of dust, threads of fungus, and many others. The recognition of the latter ingredients requires a certain amount of practice.

I have tried a number of conserving fluids for preserving mucus, but thus far without any notable results.

5. The same granulated cell formation—we know it already—occurs finally as an element of a pathological fluid, the pus; it receives the name of pus cell, or pus corpuscle.

A fluid may be recognized as pus, not by its constitution, but rather by the number of its cells.

Pus cells are the extravasated colorless blood corpuscles which have collected at the point of irritation. The formation of these structures in the interior of epithelial cells (fig. 99), where, by the destruction of the mother cell, the contained cell was set free, was also formerly assumed (Remak, Buhl, Rindfleisch). These observations are certainly correct. If the thin, watery secretion of the mucous membrane in the first days of a catarrh be examined, one may perceive, according to the variety of the epithelium, together with free pus corpuscles, ordinary desquamated epithelial cells, and others with contents such as are represented in the figure mentioned. But the explanation must be very different. These structures which lie before us are those vagabonds of the body, the wandering cells, which have penetrated from the tissue of the mucous membrane into the epithelial cells.

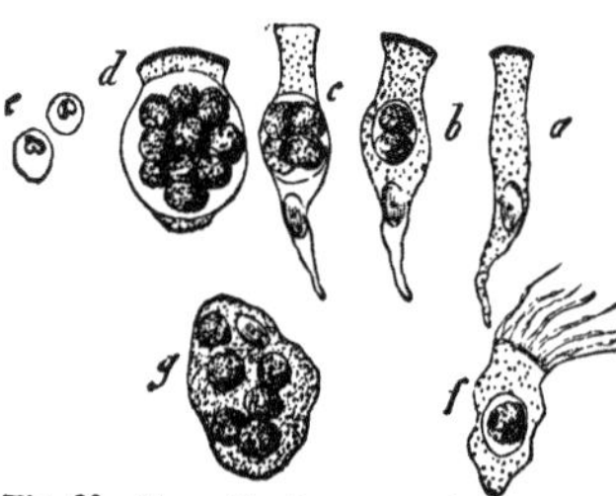

Fig. 99. Pus cells of man and mammalia lying in the interior of epithelial cells.

We shall afterwards speak of a possible and probable formation of pus cells from the connective-tissue cells. Naturally a collection of the pus cells never takes place on the surface of a mucous membrane. They may afterwards be found scattered through the tissue of the interior of an organ (for instance, in the inflamed cornea) also; they may then collect in great numbers beneath the epithelium, finally force off the epithelial covering and thus cause an erosion and an ulcer, or, when in internal parts, they may cause the formation of an abscess in consequence of the melting down of the neighboring tissues.

Pus corpuscles are naturally to be examined in the same manner as the elements of lymph and chyle.

The vital changes of form of pus cells have become known to us within a few years. If after about two days, as the result of the application of an irritant to the cornea of a frog, its humor aqueus becomes cloudy, the latter will show a number of energetically contracting proteus-like cells (fig. 100). Thin, thread-

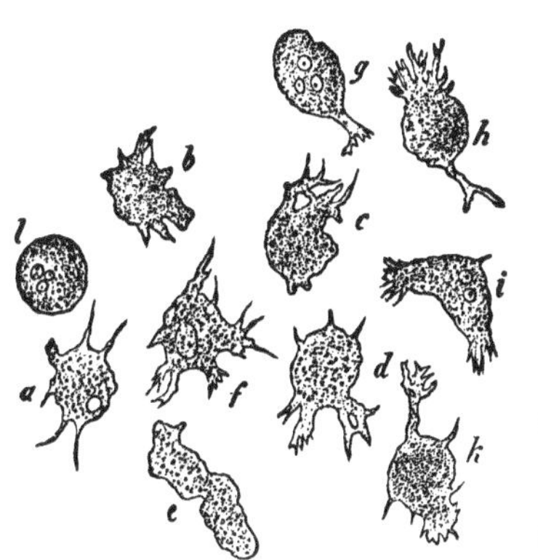

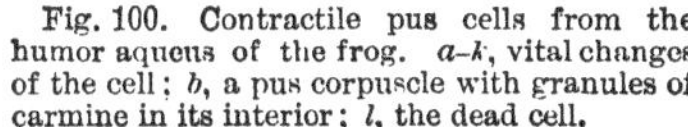
Fig. 100. Contractile pus cells from the humor aqueus of the frog. *a–k*, vital changes of the cell; *b*, a pus corpuscle with granules of carmine in its interior; *l*, the dead cell.

Fig. 101. Acid pus from an old abscess of the upper part of the thigh.

like processes may give the pus corpuscle a radiated appearance (*a*), which may afterwards change to an irregular indentated form (*b*). Not unfrequently the processes become further ramified, and, by the meeting and blending of neighboring branches (*c*), reticular processes result (*c d*). Long, extended forms sometimes show themselves temporarily (*e i*). A reception of neighboring small molecules, such as carmine, within the interior of the cell may also be observed (*b*).

The pus cells of man and the mammalia also possess a similar vital changeability of form.

A similar alteration of shape may be recognized when these

cells are in the spaces of a more compact tissue; as for instance in the cornea. Here, it is true, the cells generally appear stretched out and rendered narrow, being constrained by the limited space.

Such a locality also presents the best opportunity to follow the above-mentioned progression, or wandering of these contractile structures through the passages alluded to. This is not unfrequently quite energetic. However, it is not even necessary to have an inflamed organ, for the same lymphoid cell, with the same variation and the same progression, also occurs in the normal cornea.

The most conservative treatment, the avoidance of positive fluid media, of pressure, and of evaporation are absolutely necessary, if one desires to witness the remarkable phenomena mentioned.

The intermingling of other cells, such as epithelium and blood-corpuscles, is recognized without trouble.

Many transformations take place in pus, which we cannot at present discuss further. We will only mention one of these, the acid fermentation of pus. It precedes the alkaline decomposition, and brings with it anatomical and chemical alterations.

When the reaction is somewhat acid the nuclei of the pus cells become visible. The neutral fats are decomposed, and free fatty acids make their appearance in a crystalline form. Such are shown in our fig. 101, partly in the shape of needles, and partly in pointed, lamellated masses; together with these are the rhomboidal plates of cholesterin.

A mixture consisting of 1 part sublimate, 1 part salt, and 300 water, is used for preserving them. Another preservative fluid has been recommended for causing the nuclei to appear:—1 part sublimate, 1 part acetic acid, and 300 water (p. 214).

Section Twelfth.

EPITHELIUM, NAILS, HAIR.

THE various so-called corneous tissues of the human body require similar methods of examination, in consequence of the resemblance in their chemical constitution, and may therefore be very appropriately discussed together.

1. Under the name of epithelium are understood the coverings of crowded cells which are presented by the various surfaces of the body, partly as simple layers, partly as stratified layers. According to the form of the cells, the pavement or flattened epithelium, consisting of flattened elements, is distinguished from the cylindrical, in which the cells are long and narrow. Furthermore, we have as modifications the ciliated epithelium, in which the surface of the cell is covered with very fine cilia which vibrate during life, and the pigmented epithelium, containing within the cell granules of black pigment, the so-called melanine.

Layers are in general found only in flattened epithelium. The cylindrical forms a single layer, which, it is true, is also the case with many coverings of pavement-shaped cells.

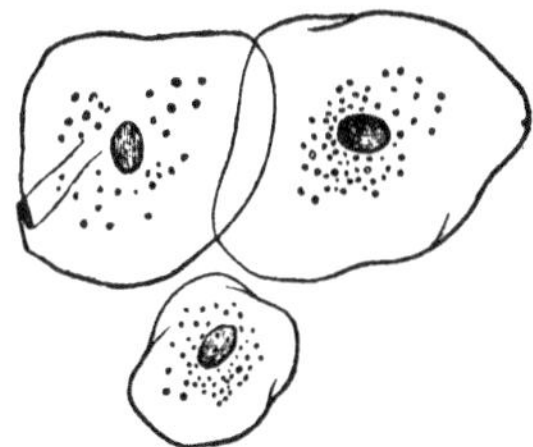

Fig. 102. Flattened epithelium of the mucous membrane of the mouth of man.

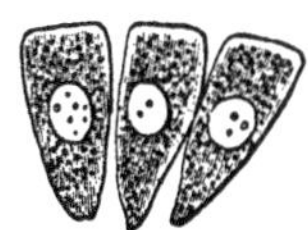

Fig. 103. Cylindrical epithelium of the large intestine of the rabbit.

The adjacent wood-cuts may serve to represent the various forms of epithelium. Fig. 102 presents the flattened epithelium of the cavity of the mouth; fig. 103, the cylindrical variety from the intestinal canal, while fig. 104 shows the ciliated, and fig. 105 the pigmented form.

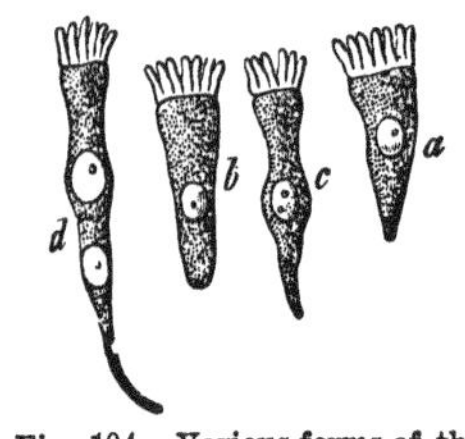

Fig. 104. Various forms of the ciliated cells of mammalia.

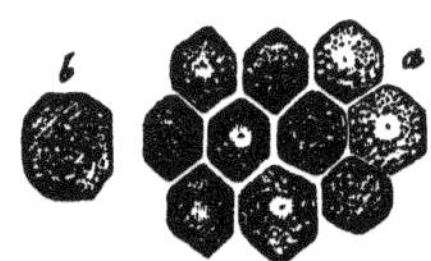

Fig. 105. Pigmented pavement epithelium (so-called polyhedral pigment cells) of the sheep.

It is scarcely necessary to remark, that only stratified epithelium is visible to the naked human eye, while such cellular coverings, when they consist of but few layers or only a single one, are only rendered apparent by the aid of the microscope. Thus, the densest epithelial covering, that of the external skin, was known in olden times, while, for instance, a knowledge of the simple cell coverings of the surfaces of serous membranes and of the vessels is an acquisition of a more recent period.

In order, therefore, to obtain the first view of the epithelial cells of a surface, it is sufficient to separate the cells from their natural connections with the clean blade of a scalpel, and to transfer them with a little fluid to the slide. One will then meet in part with isolated structures, in part with whole shreds of connected cells.

That which is here artificially accomplished is in many cases performed by nature. The pressure and friction which many surfaces of the body undergo, disconnects the epithelium from its basis. In this manner are separated the cells of the epidermis, and those of the various mucous membranes. Old cells fall off spontaneously, as it is said. The mucous coverings of the various mucous membranes exhibit in varying abundance the desquamated epithelium of the membrane in question, as, by way of example, we find in the mucus of the mouth, the oldest and largest cells of the flattened epithelium; in that of the nose and air-passages, the ciliated cells; in that of the intestinal canal, the cylindrical epithelium.

However, much of the epithelium of the body appears to be of a more enduring nature; it is less rapidly renewed, and we do not notice that spontaneous desquamation. As examples, we have the flattened cells which cover the posterior surface of the cornea, the pigmented pavement epithelium of the eye, etc. Moreover, the cells of which these coverings are composed prove to be very delicate when acted upon by reagents, and do not resist decomposition for any length of time.

The cells of all unstratified epithelium are seen to be composed of soft and readily alterable albuminous matters. Hence the greatest freshness of the tissue is necessary for their examination. It would be a piece of folly to look for them in a cadaver which is several days old. In this case they would be either entirely destroyed, or only fragments of them would be found.

Simple pavement epithelium, as it occurs on the posterior surface of the cornea, on the serous membranes, and the inner surfaces of the vessels, is to be obtained by scraping and examined with strong (400-fold) magnifying power. The isolated cells are frequently so pale that, even with considerable shading of the field, it is desirable to have them tinged. Solutions of carmine, hæmatoxylin, aniline blue or aniline red may be used for this purpose. We would especially recommend the latter as coloring instantly, and not exerting any alterative effect. Very dilute acetic acid may be used to render the nuclei more distinct, although there is rarely any occasion for its use. It is somewhat difficult to obtain a simultaneous view of these simple tesselated cells and their basement membrane. Thin vertical sections of the previously dried tissue will rarely lead to the desired results, as the cells, as a rule, become separated by the reapplication of moisture. A characteristic view may be more readily obtained from parts which have been hardened in chromic acid or alcohol. In the vascular system the free border of a valve is a favorable locality for the recognition of its epithelial covering. A view of the epithelium of the serous membranes may be obtained by carefully separating a shred of the membrane from its bed and forming a fold of its free surface. If the posterior surface of the cornea be energetically scraped, inverted pieces of the membrane of Descemet with its epithelial covering may sometimes be seen in the preparation.

Recklinghausen's method of impregnation with silver (see p. 160) is also an excellent means of demonstrating the contours of pale epithelial cells. The boundary-lines become extremely distinct, even after a slight action of the silver solution, as the precipitation takes place in the intercellular or cement substance first, and the cell cavities remain free. If it be desired, the nuclei may be afterwards stained with carmine. The epithelium appears with such distinctness in the small blood and lymphatic vessels that the course of these vessels may be recognized the same as in an injected preparation. Even in the cavernous passages or the sinuses of the lymphatic system, an epithelial covering may in this way be made visible throughout (fig. 106, *a*, *b*).

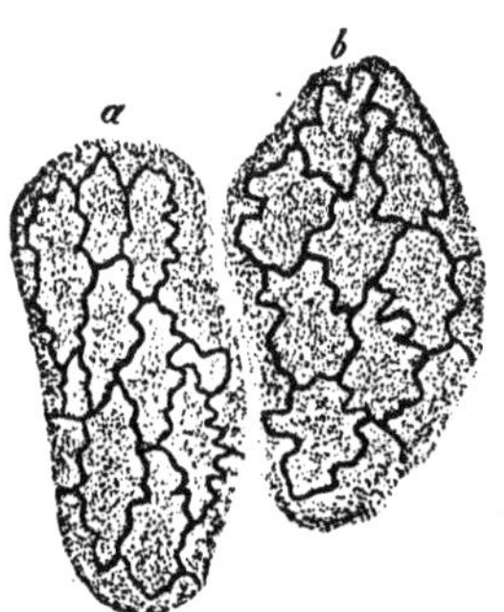

Fig. 106. Cylindrical epithelium from the surface of lymphatic canals after impregnation with silver. *a*, elongated; *b*, broader mosaic.

We shall afterwards learn what remarkable information the silver treatment has recently furnished concerning the structure of the capillaries.

Indifferent fluids are to be recommended rather than water as media for the examination of epithelium. I have not, as yet, been able to preserve these structures very well when mounted moist, but silver preparations keep very well, especially in Canada balsam.

The pigmented pavement epithelium (the polyhedral pigment cells of a former time), which occurs in the eye, and covers the choroid with a simple, the ciliary processes and the posterior surface of the iris with a stratified layer, is to be examined in the same manner. Shreds of the same may be readily removed with a scalpel or a brush, and, when carefully spread out with the brush, they present the beautiful mosaic (fig. 105) to view. Such shreds, when folded, present a side view of the cells. Chromic acid preparations and dried eyes may also be used for demonstrating the epithelial formations of which we are at present speaking.

If one desire to observe the molecular movement of the black pigment granules, it is only necessary to make pressure with the covering-glass in order to liberate the cell contents, which then begin their movements in the water. In this case it is advantageous to use the strongest objectives, so as to magnify the movements of the granules as much as possible.

The highest modern magnifying powers appear to indicate a crystalline constitution of these molecules.

They may be successfully mounted in glycerine, Müller's fluid, and Canada balsam.

The methods of investigating cylindrical and ciliated epithelium are also the same.

Scraping with the blade of a knife furnishes ample views of these cells, isolated and hanging together in shreds (fig. 107). Cylindrical epithelium may be most advantageously examined several hours after death, as they are then more readily separated from their parent tissues. Not unfrequently, groups of cells have their free surfaces turned towards us, and thus present, seen in bird's-eye perspective, the familiar, elegant mosaic (*b*).

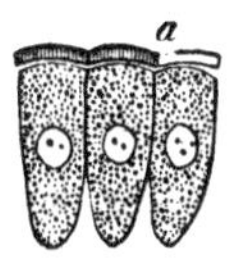

Fig. 107. Cylindrical epithelium from the small intestine of the rabbit. *a*, Side view of the cells, with the thickened and somewhat elevated seam, which is permeated by porous canals; *b*, view of the cells from above, whereby the apertures of the porous canals appear as small points.

The dried mucous membrane may be used in order to obtain a view of the cylindrical epithelium with its attachments. Wet preparations, that is, such as have been hardened by means of chromic acid, bichromate of potash, and alcohol, are much more suitable. If the part was sufficiently fresh when immersed, the epithelial covering will be beautifully seen on thin sections, made with a sharp razor. The appearances are rendered much more distinct by tingeing with carmine.

Indifferent fluids, tingeing, and the application of dilute acetic acid are generally employed for the further investigation of this structure.

An animal killed during the digestion of fat may be used for demonstrating the passage of molecules of chyle through the cylindrical epithelial cells of the small intestine (with this point the previous section, Chyle, is to be compared).

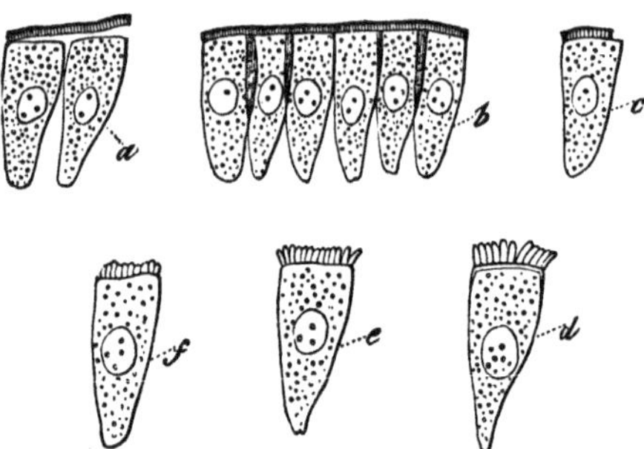

Fig. 108. The same cells. At *a* the seam is lifted up by the action of water and slight pressure; at *b*, appearance in the natural condition; at *c* a portion of the thickened border destroyed; at *d e f* it has become resolved into isolated rods or prismatic pieces by the prolonged action of water.

The thickened seam which occurs on the free surface of the cylindrical cells of the small intestines, etc., has more recently been subjected to an accurate investigation, and found to be permeated by fine vertical lines (compare fig. 108, also 107). Most investigators of the present time accept these lines for the optical expression of fine passages which pass through the seam, so-called porous canals, an opinion which is also entertained by the writer of these lines. Strong magnifying powers, especially immersion systems, are requisite for the recognition of the subtile textural relations. The animal may be used immediately after being killed, but it is better to expose the intestine to the air for several hours, which will facilitate the separation of the cells. Intestinal mucus, blood serum, weak solutions of chromic acid, solutions of salt (2 per cent.), and of phosphate of soda (5 per cent.), may be used as media. The addition of water has a destructive effect on the seam. The individual pieces separate from each other in the direction of the vertical lines; not unfrequently, the appearance is presented as if the cell had a border of cilia (*d e f*), which was also the opinion of the first observers (Gruby and Delafond). A similar effect is produced by maceration for about six hours in phosphate of soda (5 per cent.), or the so-called strong acetic acid mixture of Moleschott (Coloman Balogh).

All that was said concerning the cylindrical cells also holds good for the examination of ciliated epithelium; only, one should commence as soon as possible after death, and make use of indifferent media, as water usually attacks the fine cilia and soon causes them to fall off. On the contrary, a strong solution of potash, from 28 to 40 per cent., preserves them very well, as Schultze ascertained.

Tingeing with aniline red is very useful; it may be rapidly done, and—in the frog at least—does not cause the ciliary movements to cease.

Mounting in strongly diluted glycerine is to be recommended for preserving cylindrical epithelium, especially that which has been hardened by means of alcohol. I have not as yet succeeded in keeping the ciliated cells, with the preservation of their cilia, for any length of time.

Before proceeding to the lamellar epithelium we will allude to the most remarkable vital phenomenon of the tissue, the vibratory or ciliary motion.

Since the vibratory phenomenon outlasts the death of the creature and the separation of the cell from its natural connections for a very unequal length of time, in the individual animal groups, it is of the greatest importance to make a suitable selection. Mammalia and birds, in which the movements of the cilia very rapidly cease, are therefore to be avoided in the first examinations. The naked amphibia, salamanders and frogs, are best adapted; the larger size of their cilia also constitutes a second and not unimportant advantage. Many of the invertebrates, such as the river muscles of the genera unio and anodonta, as well as the genus cyclas, which have on their gills a covering of ciliated cells with splendid long hairs, are exceedingly well qualified for this purpose. By using very powerful lenses one may also obtain a view of an important textural condition in the vibratory cells of the intestine of the river muscle. The cilia are the continuations of fine protoplasma filaments of the cell-body (Eberth, Marchi).

The fluid media are of great importance in the study of the vibratory movements. It is asserted in general that everything which does not affect the cell substance chemically allows the ciliary movement to continue; everything, on the contrary, which alters its molecular condition terminates the movements once for all.

The indifferent natural fluids should therefore be preferred to all others. Blood serum in the first line, also liquor amnii, vitreous fluid, milk, and even urine form excellent fluid media. Iodine serum appears to be very serviceable; bile exerts an unfavorable effect. The addition of pure water increases the activity of the vibrations for a short time, to cause them to cease all the sooner. An alkaline reaction of the fluid media is to be denoted as favorable, acid as unfavorable. Oxygen has an exciting, carbonic acid a paralyzing effect (Kühne). A moderate increase of temperature increases the activity; a higher temperature, which destroys the life of the protoplasma, exerts the same effect on the allied ciliary movement (Roth).

In order to commence the first observations, a piece is to be cut from a membrane covered with ciliated cells (for example, the mucous membrane of the gums, or the pericardium of the frog), serum is to be added, and the membrane folded in such a manner that its cellular surface forms the free border of the fold. In order to avoid pressure, which might dislodge the slippery mucous membrane or force the folds apart, the fragment of a somewhat thick covering-glass is to be laid in the fluid and the preparation covered; or the specimen may be made to adhere to the under surface of the covering-glass and placed in the moist chamber (fig. 64). Blood cells which float in the fluid form a valuable addition (they may be replaced by particles of coal, granules of indigo, and carmine).

If the examination be made with a low power, a movement, a vibration, as the phenomenon has been appropriately named, is recognized at the margin of the fold. The corpuscles will now be seen driven forward in a rapid current, which is always in a definite direction; and if the fold shows hills and valleys, one may see some of the corpuscles driven forward and then suddenly thrown back again. The older observers were led to think of electrical attraction and repulsion. When the phenomenon begins to be paralyzed and the magnifying power is somewhat increased, the movement becomes more distinct and appreciable. The regular and simultaneous vibration of the cilia now appears like an undulating border, like the flaring of a candle, or the rippling of a clear rivulet in the sunlight. If we follow the vibratory movement for a still longer time, and at the same time increase the magnifying power, a moment arrives in which the individual cilia may be distinctly seen to vibrate, but only one direction of the excursion can be recognized at first. Already the blood-corpuscles are driven past more slowly, and we are able to perceive how a cell is driven down into a valley and then, by means of the microscopic whirlpool, it undergoes the above-mentioned repulsion. If the examination be still further prolonged, the number of the individual vibrations becomes less and less; we can now see both of the excursions of the cilia, and a moment soon arrives when small bodies suspended in the water—in our example, the blood cells—present only irregular fluctuating movements in front of the ciliated margin. Finally, the cessation, the expiration of the movements appear. Over a certain space all the cilia are stiff and motionless. There may still be a vibratory movement in the neighborhood for a short time; finally, this ceases also.

It was a beautiful discovery of Virchow's, that the ciliary movement which has but just ceased may be again called to life for a short time. Very dilute solutions of potash and soda are necessary for this purpose.

If the isolated piece of mucous membrane is not too large, one may watch it as it is slowly driven from its position by the united labor of its innumerable cilia.

The ciliary movement may also be examined in still another manner—and this is to be especially recommended for more accurate investigations with powerful lenses. The epithelium is to be separated in shreds by scraping somewhat energetically the surface of the exposed mucous membrane. Here one may at first recognize a few groups of cells engaged in the most active rotatory movements, also isolated disconnected cells with vibrating cilia, etc.

With regard to the number of vibrations which take place in a given space of time, the cilia move so rapidly at first that an estimation having any degree of accuracy is not to be thought of.

It has been assumed that there were a few hundred vibrations to the minute, but this is an uncertain valuation, and is really much too low. Afterwards it becomes more and more easy to count them.

The manner in which the cilia vibrate is by no means always the same. Purkinje and Valentin, who investigated the ciliary movements in the most thorough manner a number of years ago, distinguish four varieties of movements: the hook-like, the funnel-shaped, the oscillating, and the undulatory. The first variety is regarded as being by far the most frequent. According to Engelmann's beautiful investigations: on the contrary, the ciliated cell, when thoroughly unimpaired, only exhibits the undulatory motion. All other forms of vibration are caused by the cilia having become stiff and motionless in certain places.

The examination of the ciliary movement in mammalia and birds requires the rapid preparation of the tissue immediately after the death of the animal, the addition of its blood, iodine serum, and the hot stage. Sometimes, in spite of all haste, one is too late, in other cases the vibration continues to be lively for several minutes. A few cases are known in which, long after death, in the bodies of mammalia which have become quite cold, the liveliest ciliary movements were still perceptible to the astonished eye. I have myself observed such a case, years ago.

Human ciliated cells with well-preserved cilia can only be observed in a cadaver which is quite fresh; those with moving cilia may be obtained under certain circumstances from the living. If one bores with a feather (the beard of which has been cut short) in the upper part of the nose, ciliated cells which are still living may sometimes be found in the mucus which is thus rubbed off. They may be more readily obtained by examining the thin watery secretion from the nasal or respiratory mucous membranes at the commencement of a severe acute catarrh. In such cases, together with regular shaped ciliary cells, abnormal examples will also be found in great numbers, some which are swollen, others which present a more globular form, and within which may be recognized a granular body, a pus cell (fig. 99, *f*, p. 248). (Rindfleisch.)

All of the varieties of simple epithelium which have thus far been mentioned consist of relatively alterable soft cells.

It is different with the stratified flattened epithelium, as met with on many mucous membranes, and most strongly developed as a covering to the external integument. In these cases it is only the deeper layers of younger cells which still retain a similar soft and readily alterable constitution. As it appears, these are to a great extent joined together in a very peculiar manner (Schultze). The surfaces of these membraneless structures (fig. 109, *a*, *b*) are everywhere covered with points, prickles, and ridges which insinuate themselves between those of the neighboring cells "like the bristles of two brushes when pressed together," so that the name stachel and riff cells is quite appropriate.

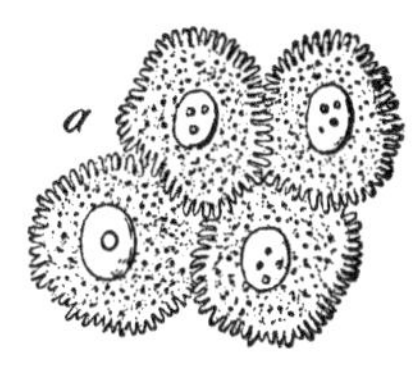

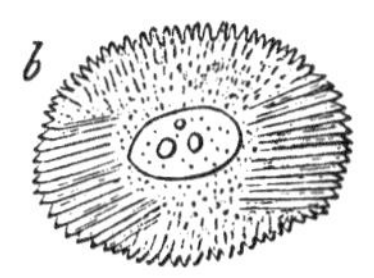

Fig. 109. So-called stachel or riff cells. *a*, from the deeper layers of human epidermis; *b*, cell from a papillary tumor of the human tongue (observed by Schultze).

The older layers of this variety of epithelium show, on the contrary, cells with smoother surfaces, which, in becoming flattened and spread out, are also chemically altered. They consist of a much more resistant modification of albumen; they are cornified, as it is said. The methods of examination are therefore to be modified.

It is self-evident that the various layers, even to the youngest, may be brought to view by scraping, and at the same time one of the ordinary methods of tingeing may be advantageously employed, especially for the youngest cells. By the use of reagents, and especially of a weak acid, we may recognize a considerable power of resistance in the older scale-like epithelial cells, while those which are younger are soon attacked and only their nuclei remain.

Maceration in iodine serum is to be especially recommended for demonstrating and isolating the stachel cells (see above, p. 124).

Drying and hardening in alcohol are employed for obtaining transverse sections through a whole stratum of epithelium. The first method is in general to be preferred for the external integument, the last for the mucous membranes. Weak tingeing with carmine and subsequent washing in water acidulated with acetic acid yields excellent preparations. The cell nuclei may be recognized even in the most superficial layers of the epithelium of mucous membranes, while the quite colorless non-nucleated scales of the cornified epidermis appear in the most beautiful manner above the deeper layers which are stained. Impregnation with silver may also be used here with good results.

There is no medium which renders such good service in the examination of the stratified flattened epithelium as alkalies, especially potash and soda. By the use of these reagents the cells may be made to assume a sometimes lower, sometimes higher degree of distention; they may be isolated, their nuclei destroyed while their membranes are preserved; and finally, they may be entirely dissolved. The use of alkaline solutions is, therefore, of the greatest value in the enumeration of the superimposed layers, as, on the other hand, they reveal the structural conditions of the epithelial cells better than any other method.

A strong solution of potash or soda forms a combination with the substance of the flattened epithelium in question which causes the cell to swell and which eagerly mixes with water, and thus induces an increasing intumescence till the cell is finally dissolved. Concentrated solutions therefore exert a different effect from those which are more diluted, and the quantity of potash especially which a fluid medium contains is of the greatest importance.

Moleschott, who has examined this subject more accurately, has instituted a series of experiments on this point. He made use of the dried tissue.

A strong solution of potash of 35 per cent. induces only a moderate distention; the cells form a very elegant mosaic and their nuclei are preserved. The intercellular or cement substance which unites the cells is gradually dissolved, the cells become isolated and float about in the fluid. The nuclei are preserved even with solutions of 30 per cent., but they are rapidly attacked by those which are weaker, below 20 per cent.

In order to cause a considerable distention of the cells, to the form of elliptical vesicles, the tissue is to be placed in a solution of potash of 30–10 per cent. for the space of about four hours.

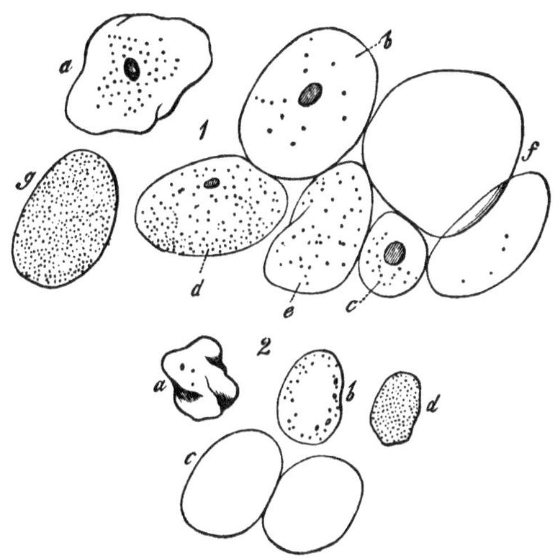

Fig. 110. 1. Epithelial cells: at *a* an unaltered flat cell from the cavity of the mouth; at *b–f* the same variety of cells after treatment with caustic soda, some still with nuclei (*b c d*), some already without nuclei (*e f*); at *g* after the action of soda with the addition of acetic acid. 2. Epidermoid cells: *a*, unaltered; *b*, at the commencement of the soda action; at *c* the more prolonged action of the reagent; *d* with the addition of acetic acid.

If water be added to these swollen cells, they become distended into vesicles which are as transparent as glass, and soon dissolve. Before this, however, the reduction of an albuminous matter (their corneous substance) may be induced in the epithelial cells of the preparation, by over-saturating the fluid with acetic acid. From what has been said, the corresponding illustrations of our fig. 110, which represent the pavement epithelium of the cavity of the mouth (1), as well as that of the external integument (2) under such treatment, must be intelligible.

The cells are gradually dissolved in very weak solutions, from 10–5 per cent., of potash, but weaker solutions, under 5 per cent., exert less effect on this tissue.

Solutions of soda may also be employed with advantage, but they must be more dilute.

A solution of chloride of gold (0.005 per cent.) and its reduction by means of sulphate of iron (p. 165), has recently been recommended by Nathusius for the cornified tissues. Such preparations may afterwards be exposed to the alkalies with advantage.

For the examination of the stratified flattened epithelium, fine vertical sections of the tissue which has been thoroughly hardened in alcohol or chromic acid are to be especially recommended. Tingeing with carmine should also not be neglected.

The cornified cells may be readily mounted in preservative fluids. Glycerine is to be used with stained sections. They often make right handsome preparations when deprived of their water and mounted in Canada balsam.

2. The tissue of the nails. In consequence of their consistence, the nails permit of sections being made in the various directions without further preparation, but their elements are combined in such a manner that one can see nothing but a homogeneous tissue which, in consequence of its brittleness, is permeated by numerous rents and flaws. Reagents which soften and dissolve the intercellular substance are, therefore, indispensable. Sulphuric acid and alkaline solutions have been used. The former, even concentrated, act but slowly when cold, although after a few days epithelial disks may be distinctly recognized. By boiling they make their appearance very rapidly, even in half a minute. The nuclei do not become sufficiently visible by this method.

Solutions of potash and soda act very much better, as was indicated by Kölliker many years ago. One may often obtain very handsome specimens of isolated and distended cells, in which the nuclear formations not unfrequently stand out very beautifully, even without using solutions of known strength. A solution of potash of about 25–27 per cent. appears to be suitable. Weak solutions destroy the nuclei.

A momentary boiling in a dilute solution of soda (about 10 per cent.) also frequently affords very characteristic views. The structure of the nails may be demonstrated in this way almost instantaneously. Fig. 111 shows us the cells of the nails isolated in the latter manner.

As is well known, numerous epithelial new formations occur.

Cysts and encysted tumors are, for the most part, lined with pavement cells. Hypertrophied growths of the skin, indurations, verruca and hornlike excrescences present an arrangement which

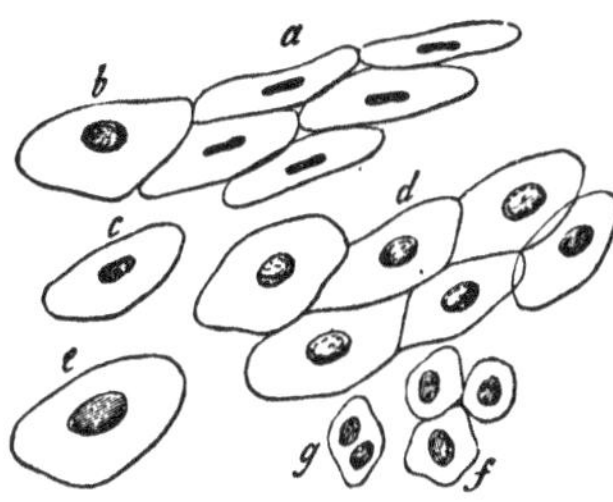

Fig. 111. Tissue of the human nail, in part after the action of the solution of soda. *a*, cells of the most superficial layers in profile; *b*, a cell seen from above; *c*, half profile; *d*, a number of cells which are rendered polyhedral by the pressure of their neighbors; *e*, a cell, the nucleus of which is beginning to disappear; *f*, cells of the deepest layer (stratum mucosum Malpighii); at *g* one of these cells with double nucleus.

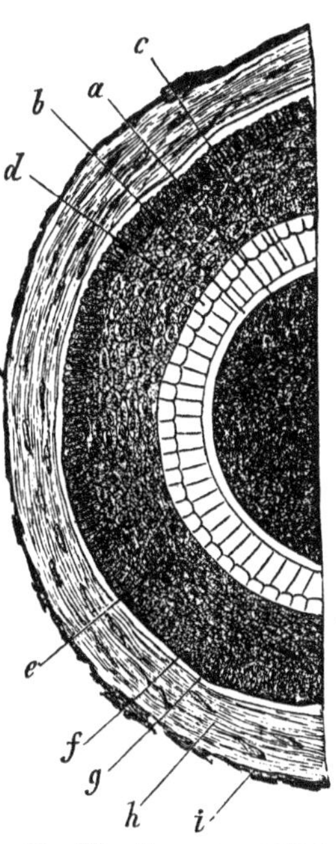

Fig. 112. Transverse section through a human hair and its follicle. *a*, hair; *b*, epidermis of the same; *c*, inner and *d*, outer layer of the inner root sheath; *e*, outer root sheath; *f*, its peripheral layer of elongated cells; *g*, structureless membrane of the hair sac; *h*, its middle, and *i*, external layer.

is similar to the strata of the cornified epidermis and which require analogous methods of examination, such as drying, vertical sections, solutions of potash, etc. The pearl-like tumors (among which are to be reckoned Hassal's concentric bodies of the thymus) and the epithelial cancers or cancroids also have an epithelial character, the first in the form of benign, the latter in the form of malign neoplasms. The preparatory methods are to be adapted to their varying consistence. The fresh tissue may be examined in part as fine sections and picked preparations, in part after the use of hardening agents. Tingeing and the alkalies are also to be employed. Nails undergo but slight changes.

3. Hair and its tissue. We shall assume that the complicated structure of the human hair has already been learned from the text-books on histology.

In order to examine the hair with its sac and the most inferior portion of its bulb, a hair of strong calibre is to be prepared from the skin, or a piece of the skin of the head which has either been dried in the air or hardened by means of alcohol may also be used with advantage. But in making vertical sections, it is necessary to keep as close as possible to the direction in which the hair passes through the skin. Transverse sections through the hair and all its coverings (fig. 112) may be obtained in a similar manner. Moleschott recommends the immersion in his strong acetic acid mixture for a few months.

A hair which has been slowly drawn out from the head may be used for the first examination.

The root will often be found covered with the white substance of the so-called root-sheath, with the exception of its lowermost portion, which has remained with the terminal part of the bulb in the follicle. White hairs are most suitable; blonde are better than dark. Water or glycerine may be used as a medium. Weak powers will then permit of the recognition of the essential structural conditions.

No further preparation with the exception of strong lenses, and at most the use of acetic acid, is necessary for examining the finer structure of the outer root-sheath (figs. 112 *e*, 113 *c*). The inner root-sheath (fig. 112, *c d*) may be obtained either from sections, made parallel with the surface of the skin of hairy parts of the body, or from hairs which have been drawn out after stripping off the outer sheath and removing it from the shaft of the hair. It may also be seen on short transverse sections, made through the bulb of the moistened hair while lying on the slide, and then picked with needles. Although the first few attempts may fail, a little attention and the use of strong lenses will lead to the recognition of the two differently shaped layers of cells (figs. 112, *c d*, 113 *a b*) of the inner sheath. The structure of the shaft and bulb of the hair, as well as the epidermoid covering, may even now be recognized to a certain extent. Reagents, the application of which we will now consider, are necessary for the further penetration into their structure.

Let us commence with the last-mentioned covering of epidermoid cells (figs. 112 *b*, 114 *f*). The action of concentrated sulphuric acid for a few minutes, the importance of which was indicated many years ago by H. Meyer, is an excellent means of separating the cells. The same result may be accomplished with alkalies, though much more slowly. Moleschott praises a potash solution of 4.6 per cent. When this has acted for about forty hours (in the cooler temperature of winter) the cells begin to separate from the shaft of the hair. After three or four days the cells are everywhere isolated in the most beautiful manner. Solutions of soda may also be used.

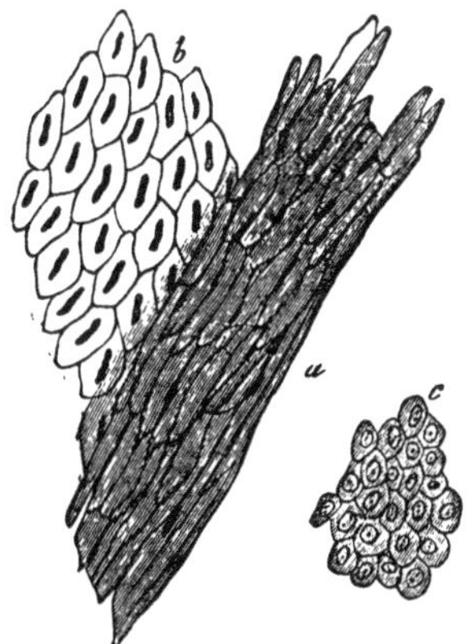

Fig. 113. Cells of the root-sheaths; inner root-sheath with Henle's (*a*) and Huxley's (*b*) layer; *c*, cells of the external sheath.

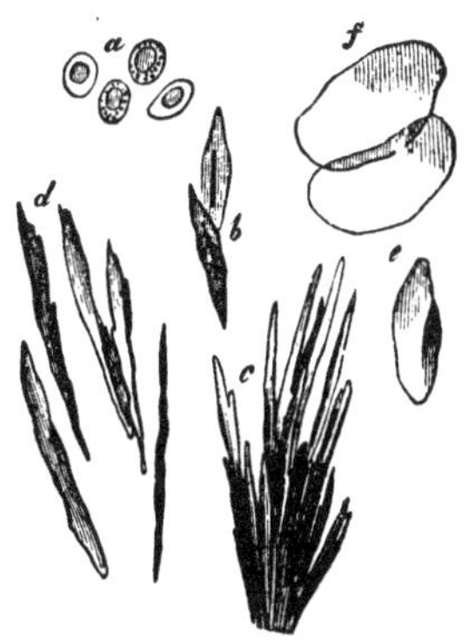

Fig. 114. *a*, cells of the hair-bulb; *b*, from the commencement of the shaft; *c* cortical mass treated with sulphuric acid, and at *d* separated into plates; *e f*, cells of the cuticle of the hair.

The use of concentrated sulphuric acid at a moderate temperature is the best means of demonstrating the cortical layer of the shaft of the hair, and for isolating its peculiar plate-like cells. After several minutes it will be found that the cuticle is commencing to separate, and that the surface of the shaft has become rough and felt-like. After a short interval, especially when the hair is made to roll with a little pressure, the spindle-shaped cells begin to peel off. The inner layers (*b d*) afterwards become separated, until, finally, the medullary substance is exposed.

These plates may also be split off in groups by mechanical means. For this purpose, a dry hair is to be placed on the slide and scraped in the direction from the point towards the root. The scrapings are to be moistened and placed under the microscope (*c*).

Alkalies were long since recommended for rendering visible the shrunken air-vesicles of the medulla (Kölliker). For the medullary cells of the hairs of the beard, and especially blonde hairs, Moleschott praises the maceration, for one or two days, in a 3 per cent. solution of soda. Placing the hair for several days in a 2 per cent. solution of potash, or a longer immersion in one of 4.6 per cent. produces good specimens.

The following process is most to be recommended for obtaining transverse sections through the hair-shaft: A bundle of hairs is to be stuck together with glue or gum-arabic and dried. The sections, obtained by the aid of a sharp knife, are to be softened in hot or cold water. Henle, many years ago, gave another original method. A short time after shaving, the same operation is to be repeated, and the sections of the beard fished out of the lather. Transverse sections may also be made from hairs which are fixed in cork or gutta-percha (Harting, Reichert).

The employment of very dilute acetic acid is very useful for examining the cells of the outer root-sheath. Stronger solutions of potash are used for the cells of the inner root-sheath.

We shall afterwards consider a few of the pathological conditions.

The first rudiments of fœtal hairs are obtained from sections of skin hardened in chromic acid or alcohol. Tingeing with carmine is very useful. The later stages of development are studied in a similar manner.

Hair preparations are to be mounted dry in Canada balsam or in glycerine, according to circumstances.

Section Thirteenth.

CONNECTIVE TISSUE AND CARTILAGE.

Modern histology designates at present with the name of connective substance a series of tissues which are nearly related, although proving different enough in their terminal forms, and which are all, directly or indirectly, interchangeable. They also take their origin from very similar textures, and thus present substantial evidence of being members in a natural series of relationship. Gelatinous tissue, reticular and ordinary connective tissue, fat, cartilage, bone, and dentine are to be enumerated here.

These members also resemble each other in still another physiological regard. They are tissues of low rank, which do not take part in the higher vital processes, but, on the contrary, they form throughout the whole body, in all its parts, a widely expanded framework (although of varying quantity), in the spaces of which are embedded other tissues, such as muscles, nerves, vessels, glandular cells, etc. Virchow deserves the credit of having, by a series of investigations, shown the importance of connective tissue in pathological new formations.

1. As gelatinous tissue, we distinguish soft transparent masses, consisting of round or star-shaped cells (figs. 115, 116), which have between them a considerable quantity of an ordinary homogeneous, slimy intercellular substance. Nearly all of them belong to the fœtal period of life, and concern either transitory organs or are only development

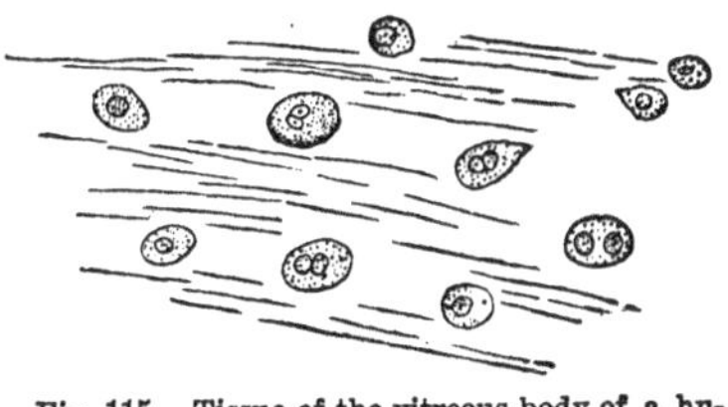

Fig. 115. Tissue of the vitreous body of a human embryo at the fourth month.

stages of the ordinary connective tissue. A single one of these, of a peculiar watery form with stunted cells, persists: the vitreous body of the eye (fig. 115). The extreme softness of all these tissues renders it very difficult to obtain tolerable preparations. At the most, the pale and delicate cells may be studied with a strongly shaded field without further treatment. Hardening media are therefore necessary, and among these chromic acid and bichromate of potash take the first rank. As a rule, a chromic acid solution of 0.5–2 per cent. hardens the tissue to such an extent that sections may be made from it with a sharp razor. With one of the organs which belongs here, the umbilical cord, the drying method may be very suitably employed. Neumann has given a peculiar but very appropriate process for the vitreous humor. It is to be saturated for 1–2 days with the albumen of an egg, and then hardened by a momentary immersion in hot water, after which it is to be placed in alcohol. In this way the organ is not only rendered more consistent, but also darker.

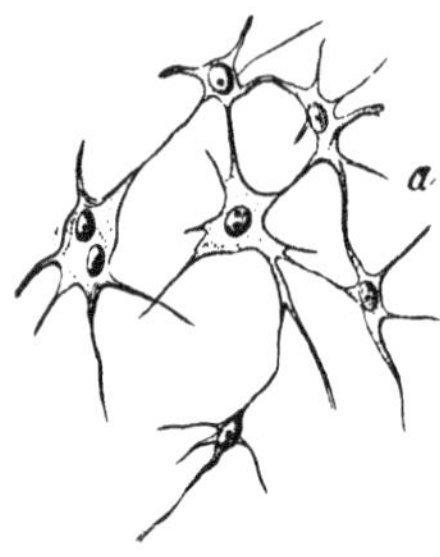

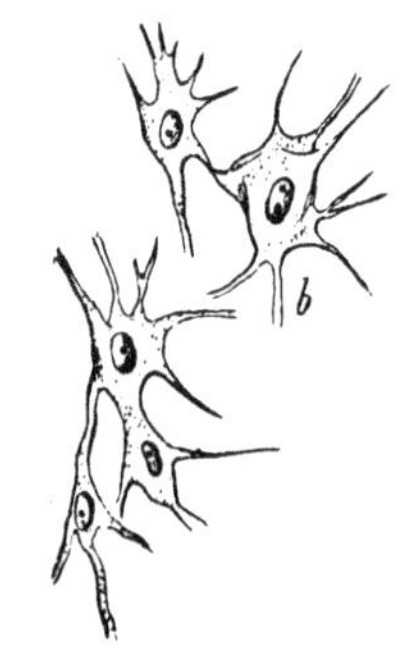

Fig. 116. Cells of the enamel organ of a four months' embryo; at *a* smaller, at *b* larger and more developed star-shaped cells.

In consequence of the delicacy and paleness of the cells of gelatinous tissue, tingeing is very much in place. Excellent preparations of the vitreous fluid are obtained with aniline blue.

Preparations of the gelatinous tissues may be preserved, after tingeing, in glycerine.

2. With the name of reticular connective tissue we designate a reticulated framework made up of star-shaped cells, which no longer harbors in its sometimes wider, sometimes extremely narrow meshes the mucous fluid of the gelatinous tissue; on the contrary, its contents are different. They consist either of lymph-corpuscles—in which case the tissue has recently been called "adenoid" or "cytogenetic" (His, Kölliker)—or of drops of fat (hibernating glands), or of nervous elements (spinal cord, brain and retina). Here, as in the whole connective substance

group, we cannot speak of a sharply demarcated tissue. The reticulated often passes over into ordinary connective tissue, and probably also into gelatinous tissue.

If any tissue of the body is qualified for showing the great value of the more modern methods of investigation, it is just this reticular connective tissue (fig. 117), which for many years has been frequently examined and formerly caused numerous controversies. All the forms in which our tissue occurs in the lymphatic glands, lymphoid follicles of the thymus, spleen, intestinal mucous membrane, etc., are too soft to permit of an examination being made without previous preparation. It is, therefore, indispensable to employ hardening media for several days beforehand. Among these, chromic acid, bichromate of potash, and alcohol take the first rank. When these glandular organs or the intestinal mucous membrane have attained the

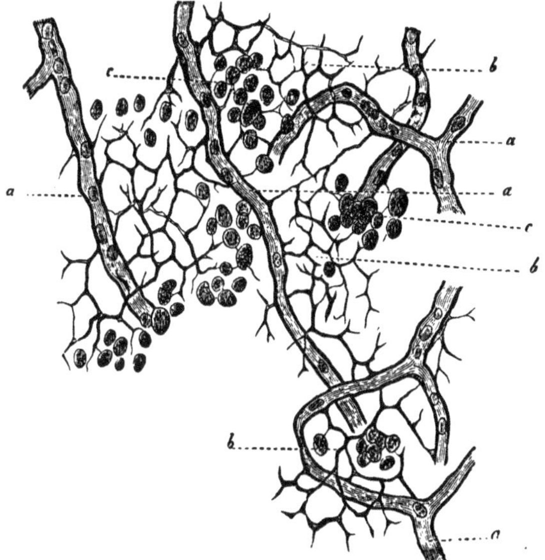

Fig. 117. Reticulated connective tissue from a Peyer's follicle of an old rabbit. *a*, the capillary vessels; *b*, the reticulated connective tissue framework; *c*, lymph corpuscles.

proper degree of hardening, the finest possible sections are to be made from them with the sharp and moistened blade of a razor. These sections are to be carefully brushed, in the manner indicated by His, with a soft camel's-hair brush. Tingeing with carmine and subsequent washing in slightly acidulated water serve for demonstrating the nuclei at the nodular points of the network. They will then be readily seen, especially in young subjects. However, all of the innumerable nodular points do not have a nucleus, as not only the simple processes of the cells, but also the ramifications of these processes become united with each other, so that the cell radii present, together with the nucleated centres, a number of non-nucleated nodular points. Tingeing with carmine will also prevent one from mistaking the stained nucleus for the transverse section of a vertically ascending colorless reticulated fibre. In older animals—and our drawing is taken from such a one—the nuclei may, indeed, be entirely wanting in a few places, and frequently only such as are stunted and shrivelled can be recognized. In conditions of irritation, however, they soon resume their former distended appearance. A larger or smaller residue of the lymph-corpuscles (*c*) will be perceived in the meshes of the tissue, according as the brushing is continued for a longer or shorter period.

In many cases it is not easy to hit upon the proper degree of hardening, and on it, in reality, everything depends. If the preparation is over-hardened it does not permit of the sufficient removal of the lymph-cells; if not hardened to a sufficient degree, the whole frequently falls to pieces after a few strokes with the brush.

For the intestinal mucous membrane and most of the lymphoid organs I prefer alcohol to chromic acid. The pieces, which should not be too large, are to be placed in a considerable quantity of fluid for the first two days. This should consist of alcohol of about 36°, which is to be diluted with an equal part of water. The fluid is then to be replaced by the same alcohol, but without the former addition of water; generally, after from four or five days to a week the tissue is in a proper condition for brushing. Well-hardened pieces may in this way be preserved in weak alcohol for months and years. Very strong alcohol is to be entirely avoided.

If it be desired to use chromic acid, commence with a solution of about 2–5 per thousand, and proceed gradually, changing the fluid to one of 1 per cent. Chromate of potash is to be used in a corresponding quantity (concerning which p. 139 is to be compared).

The reticular connective tissue of the lymphatic glands, Peyer's follicles, and the Malpighian bodies of the spleen, may be recognized with comparative facility. The thymus and the tissue of the pulp of the spleen give more trouble. Its recognition is difficult in the hibernating glands, which I have investigated with Hirzel, and to a still higher degree in the nervous organs, especially in the gray substance of the spinal cord and the brain, likewise the retina of the eye. More dilute solutions of chromic acid than those above mentioned ($\frac{1}{6}$–$\frac{1}{4}$ gr. to 1 ounce) are to be used, and allowed to act for several days. Very strong objectives should also be used. We shall return to this subject in speaking of the organs in question.

Tinged preparations, mounted in dilute glycerine, afford the best specimens for a collection.

3. The examination of fat tissue is simple and causes no trouble, whether a simple form of the same (fig. 118) is concerned, or a pathological new formation, for example a lipoma.

A small piece of the tissue (*a*) is to be picked in the fluid media and then reviewed with a low power. The large, sometimes smoother, sometimes rougher cells will be recognized pressed closely together, often with a polyhedral flattening. At the same time, a number of free globules of fat (*b*) will be met with in consequence of the laceration of the cells which has occurred. The optical properties of both are quite similar. We perceive a pellucid, sometimes slightly yellowish tinged substance, having sharp, dark contours with transmitted light, but with incident light a silver-like lustre and a whitish or yellowish periphery. While the fat-cells are of a definite size, that of the free drops of fat

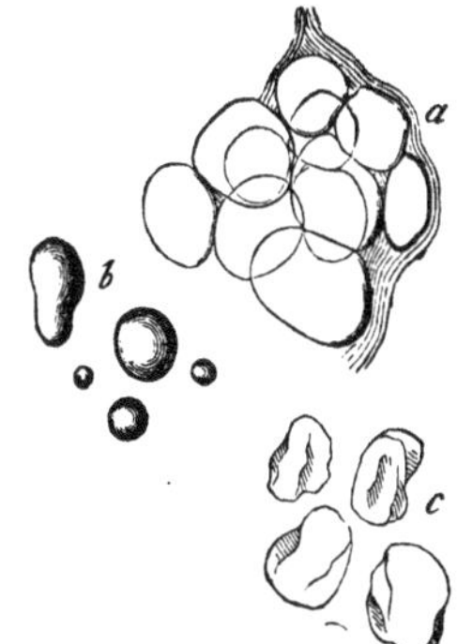

Fig. 118. *a*, human fat-cells completely filled with fat, lying together in groups. *b*, free globules of fat; *c*, empty envelopes.

varies exceedingly. The latter flow together under pressure, but the cells do not.

In order to demonstrate the cell membrane, it is necessary either to rupture the cell, in which case the former remains behind as a collapsed sac (*c*) after the fat has flowed out, or the fat is to be removed by chemical means with alcohol, ether, or benzine, which last was recently recommended by Toldt for this purpose. The nucleus may be demonstrated by tingeing it with carmine in the ordinary way.

The treatment with picric acid and carmine, and the subsequent addition of formate of glycerine affords very handsome specimens (Flemming).

Fig. 119. Human fat-cells containing crystals. *a*, a few needles; *b*, larger cluster; *c*, the cells themselves having these clusters within them; *d*, an ordinary fat-cell free from crystals.

Not unfrequently, a deposition occurs within the fat-cells of crystalline needle-shaped masses (fig. 119, *c*), the same as we have already met with in acid pus. A prolonged immersion in glycerine almost invariably produces such a crystallization within the cell cavity.

In order to study the blood-vessels of fat tissue, injections are to be made with transparent masses, carmine, or Prussian blue, and in making the microscopic examination pure glycerine is to be used as a medium, which last, in consequence of its strong refracting power, is very well adapted for fat-cells in general.

Glycerine (pure or mixed with formic acid, p. 211) is to be used for mounting, or, if the fat tissue is injected, Canada balsam may be advantageously employed. The previously-mentioned solution of arsenious acid (p. 215) has been recommended by Harting.

4. Ordinary connective tissue is very widely diffused and consists, in its developed form, of a fibrous substance which is divisible into bundles and fibrillæ, and in which long or stellate cells, the frequently-mentioned connective-tissue corpuscles, are met with, as are also the various phases of elastic tissue. The

whole lies embedded in an extremely variable quantity of homogeneous basis substance.

If living connective tissue be selected from a suitable place, for instance, in the frog, the thin transparent membrane (to which Kühne called attention) between the crural muscles (fig. 120), and examined with the addition of lymph, one may recog-

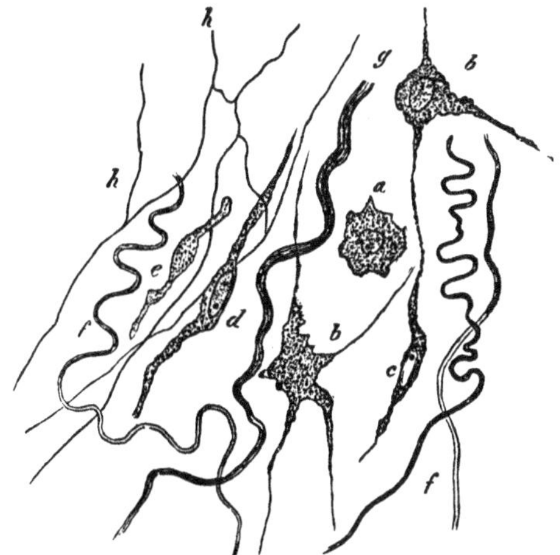

Fig. 120. A piece of living connective tissue from the upper part of the thigh of a frog (the cells are represented somewhat more pressed together than they usually lie). *a*, contracted cell; *b*, radiated, expanded connective-tissue corpuscles, one of them without a visible nucleus; *c*, one with a vesicular nucleus; *d* and *e*, immovable cells; *f*, fibrillæ; *g*, simple bundle of connective tissue; *h*, plexus of elastic fibres.

nize in the pellucid basis-substance the fibrillæ (*f*) and bundles of the connective-tissue fibres (*g*), and also a network of very fine elastic fibres (*h*). The connective-tissue corpuscles then appear as flat, membraneless cells, consisting of a nucleus and fine granular protoplasma. Several varieties of these cells may be noticed (*a* and *b*, *c*, *d* and *e*), and one may at the same time convince one's self that the first two varieties of the connective-tissue corpuscles (*a b c*) have a sluggish, but unmistakable vital contractility, so that the cell *a* gradually changes to such forms as are represented in our figure at *b*. However, as we now know, the shape is even yet not completely described. An uncommonly pale peripheral portion, which may be readily overlooked, gives the whole thing, after its death, the appearance of an oblong, indentated lamella, which is bent over and folded in various ways. This appearance of the connective-tissue cell (noticed many years ago in tendons) is also maintained in the solidified connective tissue of the higher animals (Ranvier, Schweigger-Seidel, and Flemming). Together with these, the remarkable amœboid migratory cells, the emigrated lymph-corpuscles, which we mentioned at page 245, are met with. Accordingly, the *fixed* have been distinguished from the *migratory* cells of the connective tissue.

As the proportion of the cells, fibrillæ, and elastic elements varies considerably, the form of the tissue is necessarily modified in accordance with these several admixtures. The diversity which the twisting and interweaving of its bundles present is not less considerable.

If a piece of dead connective tissue be prepared in a fluid medium with the aid of sharp needles, it may be very readily

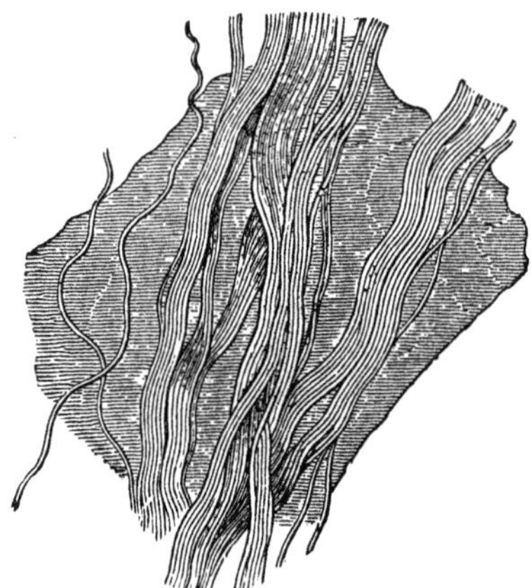

Fig. 121. Connective-tissue bundles (at the left, several isolated fibrillæ) with very abundant homogeneous interstitial substance.

separated into the above-mentioned strings or bundles (fig. 121). The bundles themselves show a striation running parallel with their long axes, and may be separated, in accordance with the latter, into finer strings, and, finally, into extremely thin, homogeneous, more or less undulating fibres or threads, the so-called primitive fibrillæ.

While in former times the anatomists adopted a fibrillated

condition of the connective tissue as a simple expression of this very easily made observation, Reichert, in the middle of the fifth decade of this century, pronounced these fibres to be artificial products, and the longitudinal striations to be the optical expression of the folding and wrinkling of a thoroughly homogeneous membrane.

Controversies were waged for a long time concerning the latter acceptation. It was only at a later period, when the method of isolation by chemical means had been discovered, that the preëxistence of these fibrillæ (which the reader already recognizes from fig. 120), could be proved in the most indubitable manner. If the connective tissue be treated alternately with reagents which cause it to swell and to shrink, the finest fibres appear in the most beautiful manner (Henle). Further observations were then made by Rollett.

If a portion of human tendon be immersed for a week or more in lime-water, and then a bundle of it placed on the slide, one may readily succeed, by inserting the preparing needles into the middle of the same, in drawing it out into longitudinal fibres of greater or lesser calibre, which cross each other at acute angles. All attempts at spreading out the tissue as a homogeneous membrane in accordance with Reichert's views are unsuccessful, and cause it to split up into fibrillæ. Baryta-water produces the same effect, but in a much shorter time—from 4 to 6 hours.

For the microscopical investigation it is necessary to remove the hydrate of lime or baryta, either by washing for a long time in water, or by the addition of so much acetic acid as just suffices to neutralize the lime or baryta. The lime or baryta water dissolves an albuminoid body, which is evidently the cement substance of the fibrillæ.

Although one series of connective-tissue textures behave similarly in this regard, others present a deviation. The corium may serve as an example. This is separated by the same treatment into stronger, apparently entirely homogeneous fibres which can only be split up into the longitudinally arranged fibrillæ after a long-continued maceration (from 10–12 days) in lime-water.

According to Rollett's observations, the fasciculi of the sclerotica, the aponeuroses, the fibrous capsular ligaments, the dura mater, and the interosseous ligaments, are formed after the type of the tendons.

The examination of connective tissue by polarized light also speaks for the presence of the fibrillæ. They are positively double refracting, and the optical axis lies in the longitudinal direction of the fibrillæ. None of the reagents which maintain the fibrous appearance of the connective tissue alter the optical nature of the same to any considerable extent. On the contrary, methods of treatment which render the connective tissue apparently homogeneous, also change the double refraction very much (W. Müller).

The same arrangement as in the external skin is found in the conjunctiva, the subcutaneous cellular tissue, the submucous tissue of the intestinal canal, and the tunica adventitia of the vessels.

The interweaving of the connective-tissue bundles, and the entire arrangement of a connective-tissue part, may be recognized in dried preparations by making coarse sections from them and simply softening these in water. Tingeing with carmine may also be suitably employed.

A finely punctated substance may be discovered in the transverse sections of the fasciculi, which many microscopists have declared to be the transverse sections of the connective-tissue fibrillæ, as, for example, in the tendons.

In order to demonstrate the cellular and elastic elements which occur between the fibrillæ, reagents have been used for many years which cause the fibrillæ to swell, and at the same time lower their refractive power to such an extent that it becomes equal to that of the water which is added. Thus is caused an apparent dissolution of the fibrillæ, and the remaining elements of the connective tissue are rendered prominent, although the cells have undergone extensive modifications and disfigurations.

This manner of action has been longest known in connection with acetic acid, but other organic acids may also be employed. Pyro-acetic acid has been frequently used for this purpose, sometimes undiluted, sometimes diluted with an equal volume of water. Mineral acids, such as nitric and muriatic acids, in a condition of extreme dilution, exert a similar effect. The latter, of 0.1 per cent., behaves the same as acetic acid.

It is only necessary to neutralize the acid with ammonia to cause the fibrillæ to reappear.

The connective tissue fibres also undergo the same swelling in alkalies as in these acids. As was the case with epithelium, the supplementary addition of water produces a rapid dissolution of the fibres.

In still another, much more conservative manner, namely, by employing a fluid medium of strong refractive power, the embedded structures may be recognized in the unswollen connective tissue. Glycerine is of the highest value in this regard.

The swelling of the connective-tissue from the above-mentioned action of acids may give rise to peculiar appearances. In many portions of the body the connective-tissue bundles are invested like a sheath by a condensed substance. Now this expands to a much slighter degree, and consequently it is not unfrequently torn across and then more and more pressed together by the contained mass, which protrudes with a certain force, until finally, in consequence of the strong compression, it has assumed the form of a ring. This ring is very similar in appearance to that of an elastic fibre running in a circular direction, for which it has also frequently been mistaken.

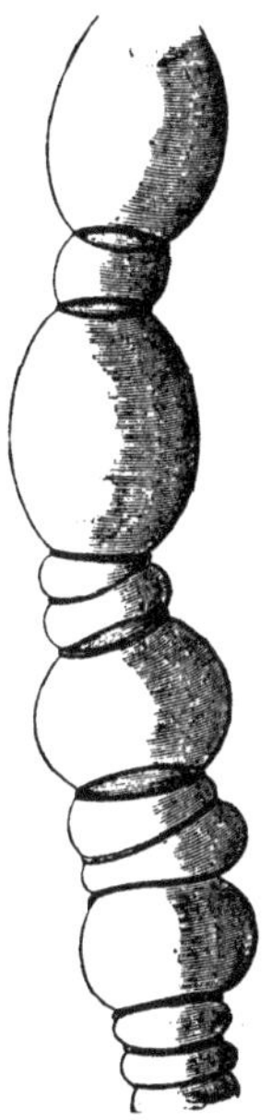

Fig. 122. A connective-tissue bundle from the base of the human brain, treated with acetic acid.

But enough about the fibrillæ. Let us inquire into the methods of examining the cells.

A thin strip of interstitial connective tissue may be cut from the living body and, after being moistened with lymph and enclosed in the moist chamber, examined. The appearances presented are instructive; but the wrinkling of such a lamella is a fatal circumstance, as every observer has experienced.

We are therefore indebted to Ranvier, a French histologist, for the discovery of a new method. Artificial œdema is to be produced by injecting the tissue. Iodine serum or a weak solution of the chromate of potash, for instance, may be injected into the subcutaneous or intermuscular connective tissue of a frog. A beautiful preparation may be obtained by rapidly placing a fine section of this gelatinous infiltrated mass on the slide under the covering-glass. A weak solution of nitrate of silver (0.1 per cent.) is in a still higher degree qualified as an injecting fluid, as by this means the pale borders of the connective-tissue cells are covered with a granular precipitate and thus rendered more distinct. Masses which become hardened, such as solutions of gelatine, are much preferable. Flemming, imitating Ranvier's process, made use of glycerine-gelatine (p. 211), to which half its volume of the above-mentioned solution of nitrate of silver was added. The infiltrated mass was then cut out and exposed to a freezing temperature. Thin sections were then made and washed out, then colored for 6–12 hours in picric acid, and, after being again washed out in water, mounted in glycerine (plain or containing formic acid).

Formerly, the cellular elements were isolated by dissolving the intercellular substance. This readily succeeds, as the latter becomes changed into gelatine.

As is well known, connective tissue may be converted into gelatine by treating it for a variable length of time with boiling water. This may in part stand in connection with structural conditions, as soft connective tissue usually dissolves more rapidly than that which is more firmly united.

This action is, however, altogether too energetic for histological purposes. The softer connective tissue, at least, may be dissolved by other means in a much more conservative manner. After softening it for about a day in very slightly acidulated water, it may be dissolved in 24 hours by moderately warming the water to 35–40° C. At the muscular tissue we shall again have to speak of this procedure, which deserves a more extended application.

It would lead us too far to describe these artificially changed connective-tissue cells in this place. For the rest we refer to the two adjoining figures, 123 and 124, which were drawn from alcoholic preparations.

Cohnheim has very properly recommended the impregnation of connective tissue with gold.

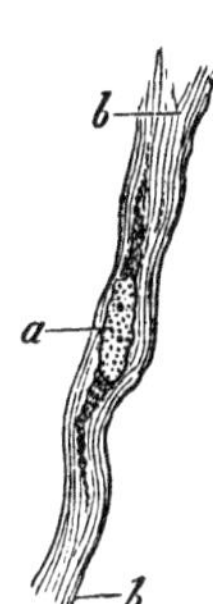

Fig. 123. Spindle cell, from the tendon of a hog's embryo eight inches long. *a*, cell with protoplasma; *b*, connective-tissue fibrillæ.

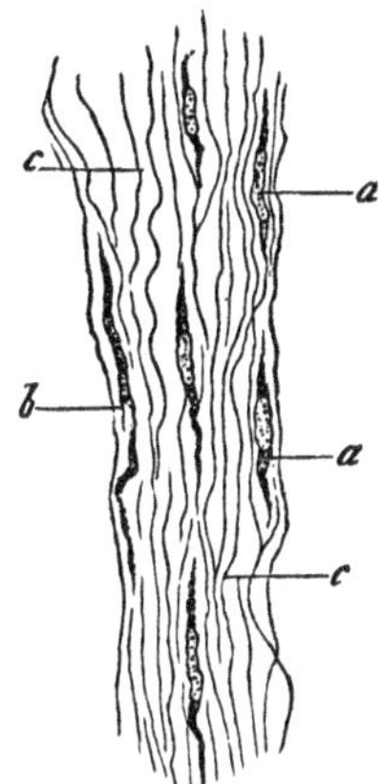

Fig. 124. Soft connective tissue from the vicinity of the tendo Achillis of a human embryo of 2 months. *a*, spindle cells; *b*, a very elongated one; *c*, intercellular substance with fibrillæ.

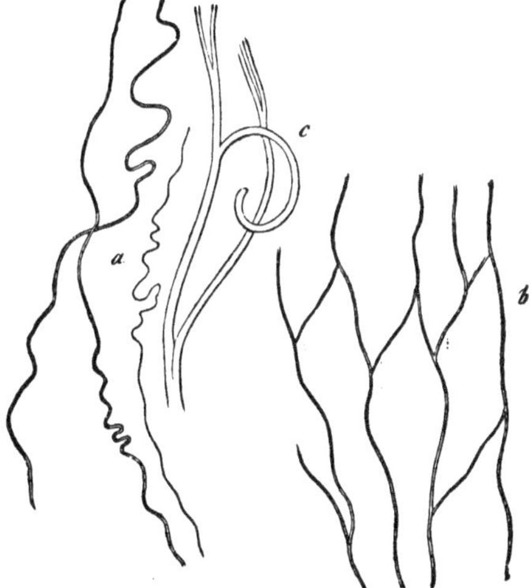

Fig. 125. Various forms of human elastic fibres. *a*, unbranched, *b* and *c*, ramified.

The great majority of the so-called elastic fibres (fig. 125) are undoubtedly of a solid nature, and all attempts to stain them with carmine fail. The great power of resistance which they present renders their examination comparatively easy and simple.

Parts which are very rich in elastic tissue require a somewhat more careful preparation. In this way the great extensibility of the finest fibres (*a*) will be noticed; at the same time it will be seen that these fibres may assume the strangest convolutions in the swollen connective tissue. Thicker elastic filaments prove to be much less extensible, and frequently present themselves as fragments (*c*).

The fasciculi of a tendon of the external skin or of the subcutaneous cellular tissue may be employed for the first examination of connective tissue. Do not spare the trouble of unravelling a very small piece in water or an indifferent fluid. For demonstrating the connective-tissue corpuscles, it is customary to use reagents which cause the tissue to swell, especially acetic acid. Tingeing with carmine is also useful. Ranvier has also made us acquainted with a useful method of examining tendinous tissues.

The extremely thin fibres from the tails of small mammalia, such as young rats, mice, and moles are used. If the last caudal vertebræ be torn from their attachments, a considerable length of the tendons remains connected with the separated parts. Their ends are to be fastened to the slide with sealing-wax, carmine is then to be employed, together with the usual supplementary treatment with acetic acid; glycerine containing formic acid is finally added.

Elastic fibres become prominent after treatment with acids and alkalies. An opportunity of studying them is presented in the subcutaneous cellular tissue, the corium, and the ligamentum nuchæ of the mammalia. A more suitable object for learning the great diversity in the manner of appearance of elastic tissue can scarcely be found than the wall of a large artery from one of the larger mammalia, the various layers of which are to be separated with forceps and scalpel.

Embryonic connective tissue (and many of the pathological new formations of our tissue, at a similar stage of organization and consistence, are to be enumerated here) is to be examined partly fresh in indifferent fluids, partly after the production of an œdema, and finally, although not so good, in preparations hardened by chromic acid or chromate of potash. In order to decide what is here present as elastic tissue (fig. 126), the use of alkalies, preferably boiling for a short time in a 10–15 per cent. solution of potash, should not be neglected, as the connective tissue corpuscles (*A a*) are in this way made to disappear, but not the elastic fibres (*B c*). Both elements react alike with acetic acid

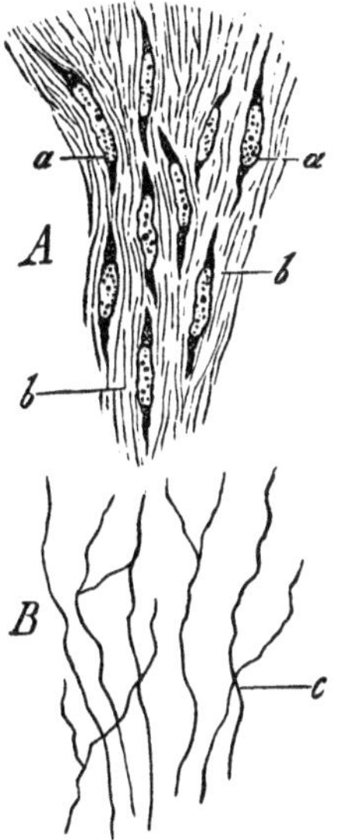

Fig. 126. From the ligamentum nuchæ of a hog's embryo, 8 inches long. *A* side view; *a*, spindle cells in a fibrous basis substance *b* (alcohol preparation); *B*, the elastic fibres brought out by boiling in a solution of potash.

At the commencement of this section we alluded to the importance, which is indeed very great, of connective tissue in pathological formative processes. We are prevented by the narrow limits of our little book from making more than a few remarks on this subject, which is now being so completely revolutionized.

Although, until within a few years, it was generally considered that pathological growths consisting of lymphoid cells were formed by the division of the normal connective-tissue cells, the emigration of the former elements from the blood passages has become prominent, in consequence of the Weller-Cohnheim theory. It is certain that this plays an important though not exclusive rôle, as, according to Stricker's experience, a complete absence of participation cannot be ascribed to the neighboring connective-tissue corpuscles. At all events, in such difficult matters one should avoid springing with inconsiderate haste from one extreme to the other.

Such collections of cells may again disappear, the mass melting down and becoming "pus," in the old sense of the word. They may also become organized, that is, form new connective tissue accompanied by the growth of vessels in them, whereby the migratory cells are transformed into connective-tissue corpuscles and an interstitial substance which is split up into bands and fibres. Parts which have been divided are reunited in this manner, and then one speaks of cicatricial tissue. Loss of substance, occasioned by suppuration or ulcerative destruction, undergoes essentially the same restorative process. Luxurious growths of this unripe tissue, which is overloaded with lymphoid cells, constitute the so-called granulations.

Hypertrophic connective-tissue formations are frequently found as a result of a continued distention of a part with blood, the so-called congestive and inflammatory processes, but likewise without any cause, spontaneous, as it is said. Thickening of the various integuments, the corium, the fibrous and serous membranes, etc., are to be enumerated here; likewise interstitial growths between muscles, nerves, glands, etc. In these cases the microscopic examination shows an increase in the number of the cells and of the intercellular substance.

The various tumors consist either entirely of connective tissue, or, together with other elements, have at least a connective-tissue framework. The forms in which they appear differ extremely. In many we meet with an entirely undeveloped structure, resembling granulation tissue or that of lymphatic glands, as, for instance, in syphilitic tumors and in tubercle. Others, the polymorphous group of the sarcomata, constitute a transition to a more highly organized variety of our tissue. The latter belong, for the most part, to the *fibroid* or cellular tissue tumors. The lipomata are formed of connective tissue with collections of fat-cells, a pathological fat tissue. New formations of gelatinous tissue also occur under various circumstances and constitute the myxomata.

The carcinomata or cancerous tumors, those enigmatical and most dangerous new formations of the body, are embedded, at least in the normal connective-tissue textures, and show, in conformity therewith, a framework consisting of a connective-tissue intercellular substance. In the sometimes larger, sometimes smaller spaces of this framework lie embedded cells which may, in certain cases, resemble those of pavement epithelium, but generally, however, they present a character which does not thoroughly correspond to that of any of the normal cells, although they may have taken their origin from glandular or epithelial cells. These "cancer cells" are capable of an unlimited, exuberant multiplication. It is customary to distinguish between certain forms of carcinoma. A tumor is generally called scirrhus (Faserkrebs) when there are only small collections of cells embedded in a firmly interwoven connective-tissue framework, so that the tumor possesses throughout the character of hardness and firmness. Inversely, one speaks of medullary carcinoma where large aggregations of cells occur in spaces of considerable size, the whole having a softer consistence, and the groups of cells forming masses of a butter-and-cream-like nature. If the cells have the appearance (but not the group-like arrangement) of pavement epithelial cells, we have one form of epithelial cancer, while the others have cylindrical cells, in both cases certainly descendants from the epithelial and glandular cells. If the substance of the framework presents a strongly marked fungous (alveolar) structure, and in the numerous spaces lie cells which are undergoing the colloid transformation, we have the alveolar or colloid cancer of the pathological anatomists. It is well known that sharp boundaries between these various forms of carcinoma do not exist, that they frequently pass into each other, and that in one and the same tumor some localities may be more of one character, and others more of another.

Finally, let us inquire into the method of examining these abnormal connective-tissue structures. They are substantially the same as those which we have mentioned for the normal tissues. Sometimes one procedure, sometimes another will be necessary, according to the exceedingly variable consistence. In a fresh condition, and by the employment of truly indifferent media, we may obtain satisfactory views of the cells and their transformations by picking, by scraping the cut surfaces, etc. Hardening methods (chromic acid, chromate of potash, and alcohol) are generally resorted to in order to ascertain the further arrangement. It is very well to place small pieces of such tumors, and if possible while still warm, in a considerable quantity of absolute alcohol. One may then proceed, even after a few hours, to the preparation of thin sections (Waldeyer). Even here, tingeing with carmine shows many things very handsomely; brushing leads to the isolation of the framework substances. Fine sections also constitute the most important means of ascertaining the relations of the so important region of demarcation between the normal and diseased connective tissue.

It will be necessary to mount most preparations of connective tissue in fluid, if they are to be kept as permanent specimens. The first of Pacini's fluids (p. 213), also a solution of sublimate (1), salt (2), and water (100) may be employed. Another mixture, consisting of sublimate (1), acetic acid (3), and water (300) is also well adapted for preserving, although the action of the acid makes itself felt. Glycerine media will be resorted to, as a rule. If an untinged preparation is to be mounted, the glycerine is to be diluted with a larger quantity of water, so that the former may not become too transparent. Stained specimens permit of the use of a more concentrated glycerine. The latter preparations, for example a cornea, the section of a tendon or of a scirrhus, deprived of their water by means of absolute alcohol, not unfrequently present a very fine appearance when mounted in Canada balsam.

5. The examination of the tissue of cartilage is very simple, as it has a degree of consistence which permits of thin sections being made without any further preparation. Cartilage which has been hardened in alcohol and chromic acid also affords very characteristic and good specimens.

Notwithstanding its consistence, cartilage is a tissue which requires foresight in the employment of fluid media if one desires to have a view of the unaltered texture. Even ordinary water has a strongly alterative action on the cartilage cells, especially those of young animals.

There are three varieties of cartilage to be distinguished: the hyaline, with a homogeneous interstitial substance (fig. 127); the fibrous or reticular, with a basis substance which is split up into bands (fig. 128); and, finally, the connective-tissue cartilage (fig. 129), in which a few cartilage cells are found between the bundles of connective tissue.

For the first examination a fœtal cartilage may be used, the fine sections from which, in consequence of their transparency, require a certain shading of the field. A bone which is com-

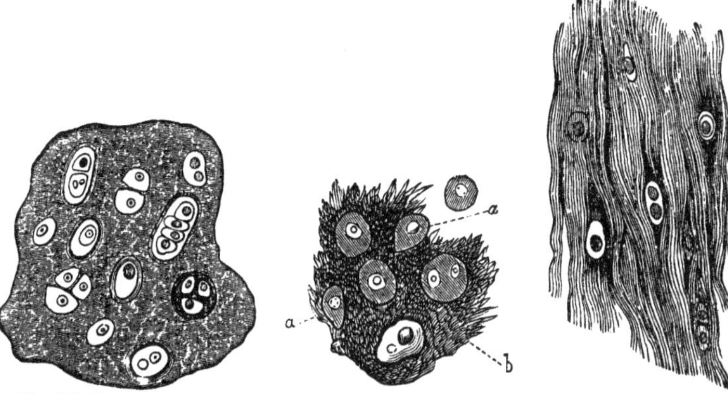

Fig. 127. Hyaline cartilage.

Fig. 128. Reticular cartilage.

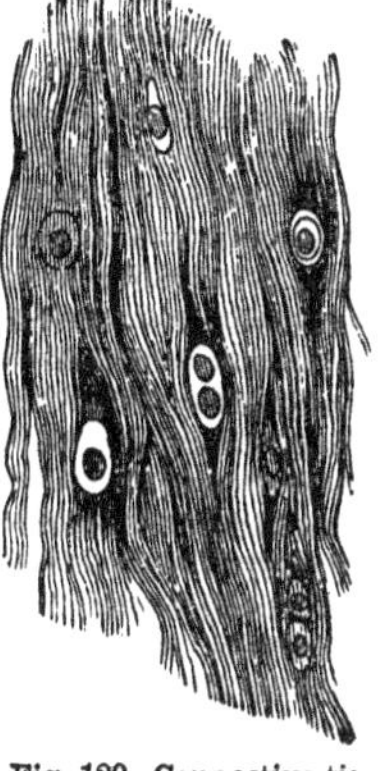

Fig. 129. Connective-tissue cartilage.

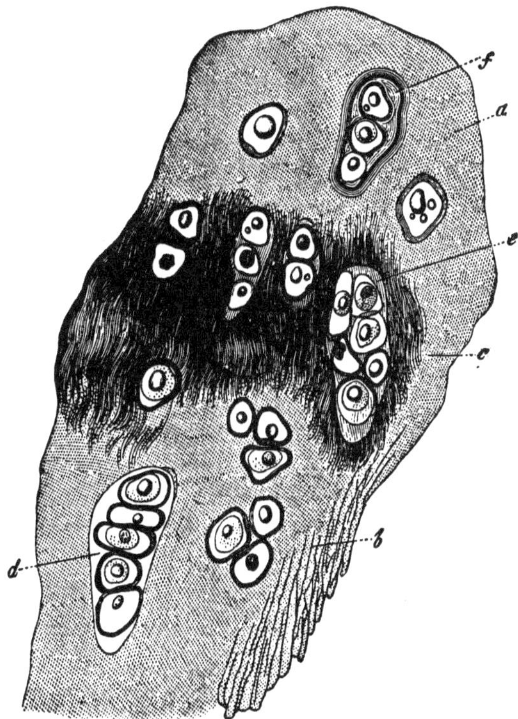

Fig. 130. Costal cartilage of an old man. *a*, homogeneous; *b*, split up into bands; *c*, fibrous interstitial substance; *d*, *e*, large mother-cells; *f*, a mother-cell with considerably thickened capsule.

mencing to ossify may be used for studying the formation of the daughter-cells. These cell formations may be met with, of an elegant appearance, close to the calcified tissue. The articular cartilages of adults form very suitable objects, and the costal cartilages of older men (fig. 130) are especially useful for investigating the textural changes which occur in cartilage which is undergoing senile degeneration. Together with ordinary semi-transparent places (*a*) in the section, others will be discovered which appear more opaque with transmitted light, and with incident light they have a peculiar asbestos-like lustre. The transformation of the interstitial substance into a system of delicate fibres (*c*) running in a parallel and straight direction may also be seen; one may also meet with large, often colossal mother-cells (*d e*), with whole generations of daughter-cells, to which Donders called attention many years ago. Such a costal cartilage affords an excellent opportunity for studying the various stages of thickening of the capsules of the cartilage cells (*f*).

Calcified cartilage tissue requires various methods of treatment, according to the quantity of calcareous molecules embedded in it. If the latter be but scanty, an ordinary watery medium suffices. If the calcification be more extensive, glycerine or Beale's mixture of alcohol and soda is to be used, in consequence of its stronger refractive power. A stage of calcification soon arrives, however, in which even these reagents are no longer capable of rendering the opaque, dark preparations transparent. In this case a method practised by H. Müller is to be especially recommended. The cartilage is to be immersed for a considerable time in a strong chromic acid (1–2 per cent.), the action of which may be assisted by the addition of a few drops of muriatic acid. After the solution of the lime molecules, the preparation is rendered very intelligible by the addition of glycerine. We shall soon see, when speaking of the process of ossification, how important this very method is for the recognition of extremely difficult relations.

The epiglottis or the cartilage of the ear is to be selected for the first examination of reticulated cartilage. In consequence of the opacity of the basis substance the section cannot be made too thin. At the borders of such a preparation one not unfrequently meets with a few of the cartilage cells projecting more or less above the interstitial substance. Glycerine again constitutes a very suitable medium.

The examination of connective-tissue cartilage requires the same methods as connective tissue. The cartilages of the eyelids are to be recommended for the first examination.

The intervertebral ligaments are to be selected for demonstrating the varieties of cartilage in a small space near each other.

The polarizing microscope teaches us that cartilage likewise belongs to the double refracting tissues. We are not as yet sufficiently enlightened with regard to the direction of the optical axis.

Recently, by means of energetic reagents, the apparently homogeneous substance of hyaline cartilage has been completely reduced to a system of thick rings or areæ surrounding the individual cells or groups of cells; and in this manner the origin of these basis substances from the cellular elements has been proved beyond all doubt (Heidenhain, Broder).

In order to obtain this important appearance (fig. 131), the cartilage may be digested in water, at a temperature of from 35 to 50° C.; it may be exposed to the action of diluted sulphuric acid (1:25), or the familiar mixture of nitric acid and chlorate of potash may be used. We would especially recommend the latter, particularly the combination of 80 ccm. of nitric acid of 1.16 sp. wt. with an equal quantity of distilled water, and the addition, at an ordinary temperature, of sufficient chlorate of potash to saturate. After a few days the desired reduction will be obtained, and the appearances will be rendered very beautiful by tingeing with aniline-red or carmine. According to Landois' experience, these areæ are even rendered distinct by tingeing with fuchsine sections of cartilage which have been deprived of their water by means of alcohol. However, even without any artificial interference, the central por-

Fig. 131. Thyroid cartilage of the swine. The basis substance is divided into cell-districts by means of chlorate of potash and nitric acid.

tion of the cartilage of the ensiform process of the rabbit usually presents the same appearance of the basis substance (Remak).

There are various means for dissolving the interstitial substance of cartilage. This is effected by immersion for several hours in a concentrated solution of potash. The same object may be accomplished by immersion for four hours in sulphuric acid containing an atom of hydrate water, and the subsequent addition of water. The means most commonly used, however, is a long-continued boiling in water. While the cartilages of small embryos undergo this solution after several hours even at a moderate temperature, the older tissues require, with the access of air, to be boiled for 12, 18, sometimes 24 to 48 hours. If the cartilage which is being treated in this manner be examined at each stage of its decomposition, one may recognize the obstinacy with which the cartilage cells themselves resist the boiling temperature, and that in none of their parts do they contain any gelatine-producing substance. Even when the entire basis substance is dissolved one may meet with numerous cells floating in the fluid.

The capsules of the cartilage cells also resist the effects of the boiling water more energetically than the interstitial substance, so that the chondrogenous matter of the latter is in no wise to be regarded as being the same as that of the capsules. The substance of reticular cartilage shows the extraordinary insolubility of the so-called elastic tissue.

Pathological cartilage tissue is not a rare occurrence. It appears as an inflammatory new formation in chronic arthritis and in the formation of callus; but, as a rule, such cartilaginous tissue makes its appearance in the form of tumors, the so-called enchondromata. The textural conditions of such cartilaginous tumors present different appearances, the same as in the normal tissue. Thus the basis substance may appear homogeneous (and this is predominantly the case), in places it may form an elastic framework, or, finally, it may have a connective-tissue character; not unfrequently one may meet with all three of these varieties of cartilage in the various parts of one and the same enchondroma.

It would be superfluous to enter further into the consideration of the methods of examination; they are the same as for the normal tissue.

Various fluids have been recommended for preserving preparations of cartilage. Even distilled or camphor water renders good service. Glycerine strongly diluted with water (2 parts water, 1 part glycerine) also acts advantageously, at least in many cases. Harting sometimes employs creosote water (p. 215), sometimes a solution of sublimate (1 part to 2–500 water). I have also used the latter fluid with success. Furthermore, the sublimate in combination with phosphoric acid (sublimate 1, phosphoric acid 1, and water 30) has also been recommended (p. 214). Cartilage strongly tinged with carmine and deprived of its water by means of absolute alcohol may be mounted in Canada balsam with advantage.

Section Fourteenth.

BONES AND TEETH.

We consider these two members of the connective-substance group in a special chapter because they require peculiar methods of investigation, in consequence of their hardness and density.

There are two kinds of preparatory treatment of the bones and teeth, according as one desires to preserve these parts with their inorganic elements or deprived of the same. Let us first speak of the latter.

Various acids are used for decalcifying, such as muriatic and nitric acids, also chromic acid, the latter partly pure, partly mixed with muriatic acid in order to obtain a more energetic action. Small pieces of bones or teeth, placed in muriatic or nitric acid, lose their earthy salts in a few days if the fluid is changed several times; chromic acid requires a longer time. One should always select high degrees of dilution (about 5 per cent. of muriatic acid), and not begrudge a few days more, or even a whole week, if one desires to spare the tissue. Chromic acid, in combination with a few drops of hydrochloric acid (p. 132), deserves to be most recommended. The immersed object will be seen to become gradually paler and more flexible, and finally to resemble cartilage in appearance and consistence. The action of the acid is now to be discontinued, and the bones or teeth carefully washed out with water. The parts which have thus been decalcified, or—to use a badly selected expression—the bone and tooth cartilages then permit of the same methods of examination as cartilage tissue proper. This method is most to be recommended for all investigations where it is necessary to obtain a large series of views with economy of time and labor. Dried specimens may be decalcified in this manner as well as those which are fresh, taken directly from the body. Bones in the latter condition, treated with chromic acid, show at the same time the substance which fills their canals and cavities, the medulla; pyroligneous acid accomplishes the same.

Decalcified in the same manner, the texture of the dentine and also of the cement may be well recognized, but not that of the enamel, in consequence of the considerable amount of mineral elements which it contains.

Naturally, more energetic measures are necessary if it be desired to isolate the walls of the canaliculi and lacunæ together with the cell remnants, that is, the bone corpuscles of the bone and the dentinal tubes of the dentine.

Virchow showed how to liberate the bone corpuscles in this way, years ago. A thin piece is to be removed from a fresh bone and either simply macerated in muriatic acid or boiled, in a decalcified condition, in distilled water, or (which is preferable) in a solution of soda. A period then arrives in which the tissue assumes a pulp-like softness. Preparations which are now removed (fig. 132) show us, especially if a slight pressure be made on the covering-glass, the bone-corpuscles together with their processes and nuclei protruding from the dissolved basis substance. Occasionally, some of these may be entirely isolated in this way (*a c d*). That they have undergone considerable changes in consequence of this energetic treatment is sufficiently obvious.

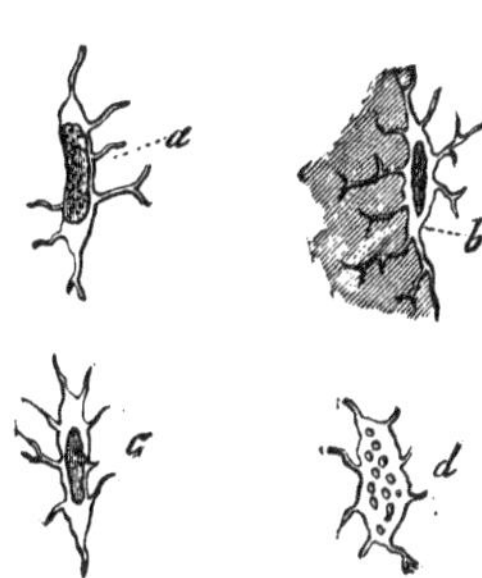

Fig. 132. Remains of the bone-corpuscles with their boundary layer, from the decalcified shaft of the femur, after boiling in a solution of soda. *a b c*, corpuscles containing nuclei (at *b* an adherent residue of the basis substance); *d*, a bone-corpuscle with a crumbled nucleus.

Förster has made us acquainted with another method of isolation by means of strong nitric acid. Thin pieces of the dried bone or tooth are to be placed in concentrated or but slightly diluted nitric acid, to which a little glycerine is added. The desired effect is obtained after a series of hours, sometimes not till the following day. Even bones in which all the soft parts are destroyed yield a similar appearance with the same treatment (Neumann).

A maceration in strong muriatic acid, likewise a protracted

boiling of the piece of decalcified bone in a Papin's digester, also causes the destruction of the interstitial substance and the isolation of the bone-corpuscles with their systems of processes. Thin scales of bone fall to pieces after the action of diluted solutions of potash or soda for half a day or several hours.

To demonstrate the bone-cells proper (fig. 133), however, very thin scales of fresh bone, and especially such as are carefully tinged with carmine, are necessary. These (*b*), surrounded by the already mentioned elastic boundary layer of the basis substance (*a*), represent the bone-corpuscles of the foregoing woodcut.

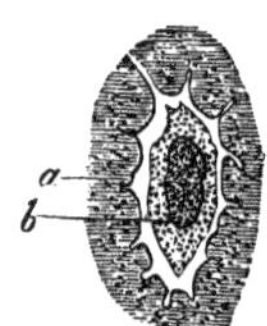

Fig. 133. Bone cells from the fresh ethmoid bone of the mouse, tinged with carmine. *a*, limiting layer; *b*, cell.

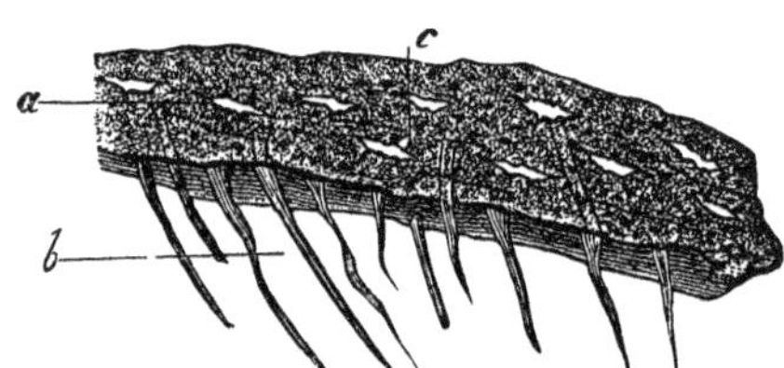

Fig. 134. The Sharpey's fibres, *b*, of a periosteal lamella of the human tibia; *a c*, lacunæ.

The impregnation with gold has also been recommended for the recognition of the bone-cells. The thin bones of the cranium of the water salamander, after an immersion of from one hour to an hour and a half in a 1 per cent. solution, and a subsequent reduction in acidulated water, yield good specimens in from one day to a day and a half. The adherent soft parts may be scraped from the bone while it is in the gold solution. Even fragments of large bones permit of this treatment (Joseph).

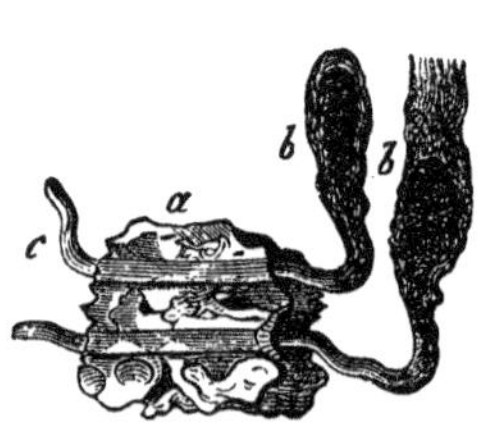

Fig. 135. Two dentinal cells, *b*, which pass with their processes through a portion of the dentinal canals at *a*, and protrude from the fragment of dentine at *c*; after Beale.

The decalcified bones of man and the mammalia are also to be used for demonstrating the so-called Sharpey's fibres (undeveloped connective-tissue fibres), fig. 134.

The walls of the dentinal tubes of dentine may also be isolated by similar methods. In fragments of fresh teeth one may see these tubes occupied by a system of soft fibres (fig. 135, *c*), which latter represent the processes of the dentinal cells of the tooth-pulp, or the so-called odontoblasts (*b*) (Tomes).

An entirely different procedure is necessary for the examination of the calcareous tissue of the bones and teeth. Thin plates must be sawed from the bone and ground on a stone till they have become as thin as paper, and have acquired the transparency necessary for their examination. The whole procedure requires time, is troublesome, and therefore, as a rule, it is shunned by microscopists. However, with a little perseverance, one may obtain excellent and uninjured preparations.

The desired object may be accomplished in various ways, and there are numerous methods extant for producing sections of bones and teeth. We will here communicate to the reader a procedure which leads to the production of very beautiful specimens, and the outlines of which were given a few years ago by Reinicke.

A fine saw, the blade of which is made from a watch-spring and is held by screws, is employed for sawing out the plates of bones or teeth. The bone or tooth is to be firmly secured in a vice. Brittle objects which are liable to splinter are to be previously wrapped in paper.

After sawing out the plate, it is to be ground on a small rotary grindstone, the handle of which is to be turned by the left hand while the plate is pressed by the fingers of the right hand on one of the flat surfaces of the stone. The stone is to be kept moistened by means of a trough placed under it containing water. If this tolerably inexpensive apparatus is not at hand, the first excess may also be removed with a file.

In order to obtain a smooth surface, the thin preparation is now to be placed on a fine flat whetstone, such as is used for sharpening razors; in this way, held by the fingers, it may be further ground on both surfaces. This may also be accomplished between two such whetstones, and indeed more rapidly. Small objects are to be previously cemented on to a glass plate with Canada balsam; ether serves best for loosening the bone and for removing the remains of the balsam. The bone may also be very conveniently cemented with red sealing-wax, and the adequate thinness of the section may finally be recognized by the liveliness with which the red shines through it. The sealing-wax is to be dissolved with strong alcohol. The preparation is then to be cleaned in water, either with a camel's-hair brush or a soft tooth-brush, and dried. If the grain of the whetstone is sufficiently fine, nothing further is necessary. If one desires to obtain a better polish, one may employ a glass plate or a piece of soft leather which is nailed to a flat strip of wood, and which is smeared with tripoli or some other polishing powder. A beautiful polish may also be produced in a short time with fine emery paper. A preparation obtained in this way—for example, a transverse section (fig. 136)—presents a charming appearance. The various general or fundamental lamellæ (*a d b*) may be recognized passing throughout the entire bone, and the transverse sections of the Haversian canals with their special lamellæ (*c*) surrounding them, as well as the innumerable so conspicuous lacunæ with their canaliculi (*e*) may also be seen.

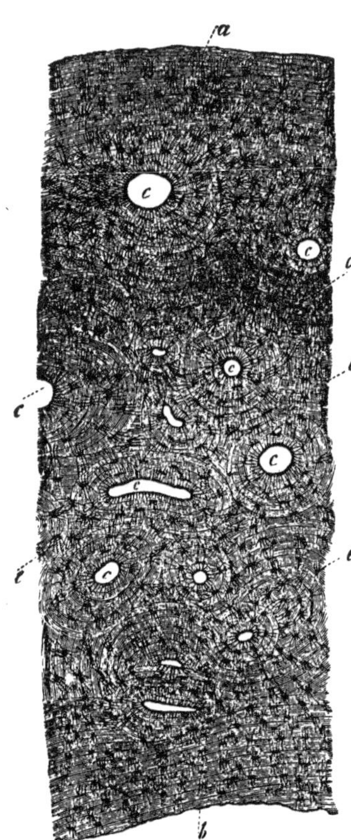

Fig. 136. Transverse section of a human metacarpal bone. *a*, external, *b*, internal surface; *d*, interstitial lamellæ; *c*, Haversian canals in transverse section with their lamellar systems; *e*, lacunæ and canaliculi containing air.

To obtain these appearances, however, the latter system of canals must be dry and filled with air. A section which is sufficiently thin presents this appearance without any addition, and in this condition it may enter the collection as a permanent preparation. Very handsome preparations are obtained by melting the sections into a resinous substance. Ordinary fresh Canada balsam is not adapted for this purpose, as, in consequence of the slowness with which it hardens, the air escapes more or less completely from the section. In order to obtain a good medium for mounting, proceed in the following manner:—A quantity of fresh Canada balsam is to be poured into a watch-glass, and placed, with a bell-glass over it, on a warm stove for several days, until the Canada balsam has become quite hard and solid. With this, and by strongly heating the glass slide, the sections of bones and teeth may be mounted without displacing the air, especially if the preparations are exposed to the cold immediately afterwards.

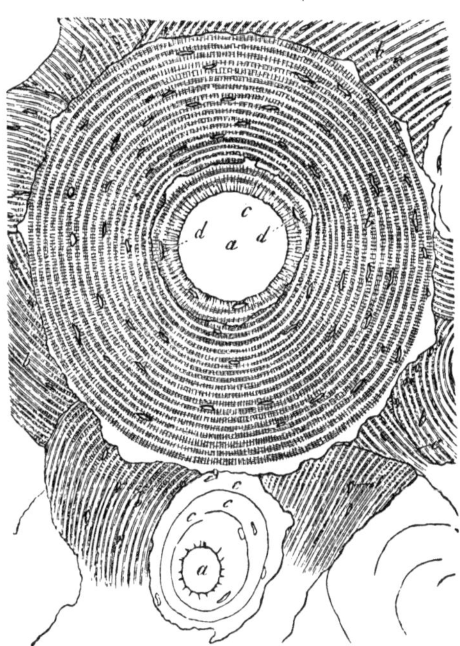

Fig. 137. Portion of a transverse section of the shaft of the humerus, treated with oil of turpentine. *a*, Haversian canals; *b*, their lamellar systems; *c*, newly-deposited osseous substance; *d*, lacunæ.

If, on the contrary, it be desired to bring the lacunæ and canaliculi filled with fluid into view, in the form of cavities, oil of turpentine is to be used in the examination; and for mounting permanently, fresh fluid Canada balsam. Tingeing with carmine may precede as a useful accessory (fig. 137).

In order to fill the blood-vessels, which is not very easy, the gelatine may be injected by a large vessel (in small animals), or by the nutritive artery (in larger creatures).

Injected bones, decalcified slowly and conservatively in dilute chromic acid, may be examined and preserved in Canada balsam or in glycerine. For such purposes one should select a durable coloring material. Bones injected with soluble Prussian blue have afforded me very handsome preparations. It is advisable to brush out the cavities a little.

We are indebted to Gerlach for a method of filling the cavities of the lacunæ and canaliculi with coloring material, and thus demonstrating their hollow character in the most perceptible manner. A transparent coloring material, and one of the smaller long bones which has been sufficiently macerated and deprived of its fat, should be used. A hole is to be made in the epiphysis for the reception of the canule, and the entire surface of the bone is to be covered with shellac, so that the injection fluid may not escape from the apertures of the Haversian canals.

In order to recognize the double refraction of the uniaxal negative bone, undecalcified sections are to be sawed out as accurately as possible in the transverse or vertical direction. They should be neither too thin nor too thick, but should be rendered strongly transparent by means of Canada balsam or turpentine. If we have a suitable transverse section, in which the diameter of the Haversian canals is perpendicular to the long axis of the bone, we may recognize, by polarized light, a regular and elegant cross, which is not changed by rotation. However, only a minority of the bone sections fulfil these requirements sufficiently. Very beautiful appearances are obtained by the intercalation of suitable films of selenite or mica. The reader will find further details in Valentine's work.

In investigating carious teeth, they may be broken to pieces in a vice, as recommended by Neumann, and then sections made from the portions which have become brown and deprived of their lime salts. But if one desires to follow the transition from healthy to diseased tissue more closely, the previous decalcification is to be recommended. Tingeing with carmine and iodine also renders good service.

It is much more troublesome to prepare the enamel of the teeth than bones and dentine for examination. It is preferable

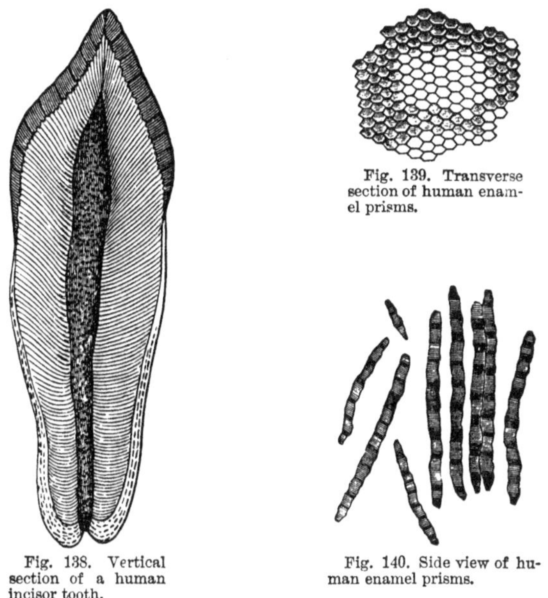

Fig. 138. Vertical section of a human incisor tooth.

Fig. 139. Transverse section of human enamel prisms.

Fig. 140. Side view of human enamel prisms.

to use only young teeth in a fresh condition; much care should be taken in sawing, and still more in grinding. Dried teeth may be rendered serviceable again by soaking them for several days in water. Longitudinal and transverse sections of the enamel prisms (figs. 139, 140) may be recognized in good specimens. The transverse lines of the enamel may be best seen by moistening the object with muriatic acid. Developing teeth are to be used for isolating the enamel prisms.

The dental pulps are to be examined in fresh teeth; they are to be liberated by breaking the teeth; in a vice or by the blow of a hammer. Teeth carefully decalcified by means of chromic acid and then hardened in alcohol also afford very good views especially in transverse sections. We shall speak of the nerves further on.

With regard to the so difficult and complicated histological relations of the development of the teeth, we must refer to the text-books. As material for examination, may be selected embryos from the third to the sixth month of fœtal life, likewise those of the mammalia, as for example of the hog, or, among the carnivora, of the dog and cat; they should be immersed in chromic acid. The new-born may also be used with advantage. It is preferable to immerse only the jaw. The finest specimens are obtained by a very gradual decalcification, occupying several weeks, by means of chromic acid solutions of 0.1–0.3 per cent., which must be frequently changed. A 5 per cent. solution of the officinal nitric acid is also very highly praised for this purpose by Boll. Fine sections are to be made with a razor, in various directions, through the jaw thus softened, and examined in glycerine. Canada balsam is to be recommended for mounting permanent preparations after tingeing with carmine.

The examination of developing bone is not less difficult. Although twenty years ago, in consequence of the incompleteness of the methods of investigation at that period, the osteogenetic process could scarcely be found out, we have succeeded more recently, by the aid of improved methods, in unfolding the chief points, at least, of the textural conditions which here occur.

The various portions of the skeleton are divided into such as are preformed in a cartilaginous state, and others in which such a cartilaginous preformation cannot be recognized. We have finally ascertained, through modern researches, that in the former the cartilage does not become transformed into bone substance, as was considered to be the case at a former period; but rather that the cartilaginous tissue disappears in consequence of the formation of vessels and the deposition of bone salts, and that in the spaces formed by its dissolution the bone substance appears as a secondary, newly formed tissue.

Cartilage which is destined to give place in this manner to the bone substance is seen to be permeated by canals filled with small cells, and in which the development of blood-vessels takes place. This observation is in general easy to make in a few sections of fœtal bone cartilage. If, as is at present the custom the embryos of man and the mammalial animals which have been immersed in chromic acid are made use of, the blood-cells will not unfrequently be recognized in glycerine preparations as a reddish-brown mass filling these newly developed vessels. The so-called points of ossification then make their appearance; that is, those places in the cartilaginous skeleton where numerous calcareous granules lie embedded in the interstitial substance (fig. 141 *a*), and where the solution and melting down

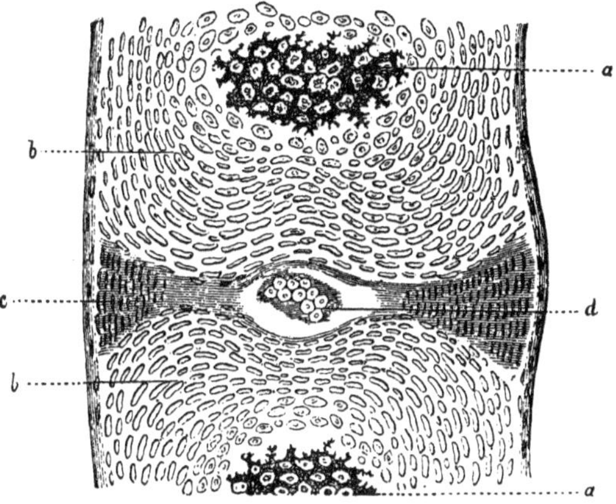

Fig. 141. The last dorsal and first lumbar vertebra of a human fœtus of ten weeks in vertical section. *a*, calcified, *b*, soft cartilage; *c*, oblong cells at the periphery of the developing symphysis; *d*, remains of the chorda dorsalis becoming transformed into the gelatinous nucleus of the vertebral symphysis.

of the cartilage, which soon commences, begins. Chromic acid preparations are also exceedingly well adapted for this purpose, as, after the decalcification, the places in question still remain recognizable by their cloudy appearance and the irregular constitution of the interstitial substance; but it is only by the use of glycerine that they obtain such a degree of transparency that the processes in question can be investigated in all their details.

By the aid of similar methods—and we here most urgently recommend tingeing with carmine—the later stages of the process are also to be followed (figs. 142 and 143), such as the more and more preponderating formation of cavities in the cartilaginous skeleton (*a b d f*) in consequence of the continued melting down of the tissue of the cartilage, the progressing calcification of the cartilage at the periphery, the formation of daughter-cells, etc., concerning which the text-books on histology are to be consulted.

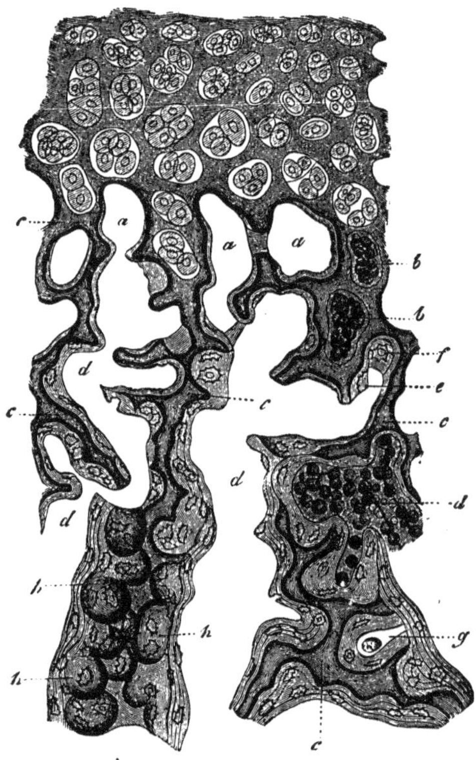

Fig. 142. A phalangeal epiphysis of the calf, cut perpendicularly through its ossifying border. At the upper part the cartilage with its irregular capsules containing daughter-cells. *a*, Smaller medullary spaces penetrating the cartilage, in part without visible entrance; *b*, the same containing cartilage marrow-cells; *c*, remains of the calcified cartilage; *d*, larger medullary spaces, on the walls of which the newly-formed, partly thin and unstratified, partly thicker and lamellated bone tissue is deposited; *e*, a developing bone-cell; *f*, an opened cartilage capsule with an embedded bone-cell; *g*, a partially filled cavity, covered externally with bone substance and containing a marrow-cell; *h*, numerous apparently closed cartilage capsules with bone-cells.

If the sections are brushed out a little, the newly-formed bone substance may be noticed in the form of a homogeneous layer (figs. 142 *d d*, 143 *b*) covering the walls of the cavities, together with the young bone-cells (*e*), which are at first thin, soft, and unstratified, afterwards thicker, stratified, and diffusely calcified in the outer layers.

The recognition of the origin of the bone-cells requires a more accurate examination, and a careful analysis of the cell formations which occupy the cavities.

These (figs. 142 *b b* and 144 *a*), generally regarded as derivatives from the daughter-cells of the perishing cartilage tissue, are seen by the naked eye as a soft, reddish mass, and appear in the form of lymph-cells as roundish, small, and granulated, with a simple or double nucleus. Many assume spindle and stellate shapes (*c c*) to become transformed into connective-tissue cells, others form capillary vessels, others again, with a proportionate increase, probably become at a later period enlarged in size and transformed into the globular fat-cells of the bone-marrow (*d e*). If attention be directed to the periphery of these cellu-

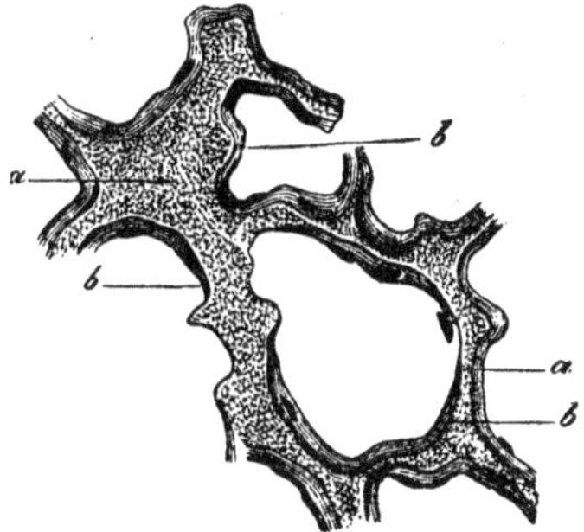

Fig. 143. Transverse section from the upper portion of the femur of a human embryo at the 11th week. *a*, Remains of cartilage; *b*, covering of osteoid tissue.

Fig. 144. Cartilage marrow-cells. *a*, from the humerus of a 5-months human fœtus; *b*, from the same bone of the new-born; *c*, star-shaped cells of the former melting down into fibrous formations; *d*, formation of the fat-cells of the marrow; *e*, a cell filled with fat.

lar contents, especially in thin sections which have been slightly and cautiously brushed, a layer of peculiar cells will be seen closely pressed together, which differ somewhat from the ordinary marrow-cells and remind one of epithelium (fig. 145 *c*). From these, the "osteoblasts" of Gegenbaur, the separation of the basis substance of the bone tissue takes place in an outward direction, and some of these cells, advancing beyond the crowded ranks, sink into this tissue (*g*) to grow in a radiated manner and become bone-cells (*f*).

Fig. 142, *d*, *e*, shows such cells commencing to assume the stellate form; some of them are already entirely surrounded by a homogeneous interstitial substance, others have only a portion of their surface (that which is directed outwards) covered in this manner.

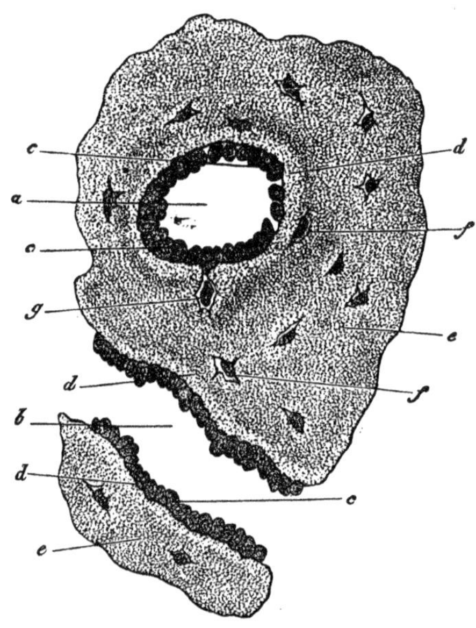

Fig. 145. Transverse section from the femur of a human embryo of about 11 weeks. *a*, a transverse, and *b*, a longitudinally divided medullary canal; *c*, osteoblasts; *d*, the more transparent, younger; *e*, the older bone substance; *f*, lacunæ with the cells; *g*, cell still united to the osteoblast.

The continued formation of new cavities in the remaining portions of the cartilage causes the opening of numerous cartilage capsules. These spaces are also soon occupied by bone-cells and intercellular masses. If the entrance of a cavity filled in this manner with young bone substance (fig. 142 *f*) be recognized, the appearance is easy to comprehend. This aperture is, however, much more frequently not to be seen (*h h*), and then it makes an impression as if there were bone corpuscles lying in the interior of unopened cartilage capsules. Earlier observers had frequently met with such appearances in their investigations, and were thus led to the erroneous conclusion that (after the manner of the formation of the porous canals in plants) the cell remains of the unevenly thickening cartilage capsule became transformed into bone corpuscles. Very instructive examples of these apertures in the cartilage capsules may be obtained from the comparison of a series of consecutive transverse sections (Müller). However, the question as to whether the bone-cells occur in ruptured cartilage capsules only, and not in those which still remain closed, is one which is not, as yet, definitely solved.

The later phases, the increasing deposition of new bone-lamellæ, and the final melting down of the last remains of the cartilage (figs. 142 *c*, 143 *a*), may also be investigated by the aid of the above-mentioned methods. Tingeing with carmine should be invariably employed when it is necessary to distinguish the already diffusely calcified older bone tissue from that which is quite young and still soft. The soft (osteogenous) bone substance readily assumes a lively red color, while the older, calcified (osteoid) takes up the coloring matter less readily and much more slowly, even in cases where a considerable portion of the bone salts have been removed by means of chromic acid. This method is also excellent for the inverted process, for the normal as well as the pathological decalcification and melting down of bone-tissue.

To demonstrate the manner of growth of fœtal or young bones, longitudinal and transverse sections should be made from those which have been decalcified. In the former we see the growth in length taking place at the expense of the cartilaginous, articular portion, and showing the same structural changes which we have just mentioned in discussing the primary formation of bone.

Transverse sections which are to be tinged with carmine deserve the preference, as a rule, for studying the manner in which bones increase in thickness. This takes place by the formation of a new osteogenous tissue (fig. 146 *e*) from the connective tissue of the periosteum (*a b*) with the help of a similar layer of osteoblasts (*c*). To these is due the elegant and regular structure which the bone assumes after the melting down of the primary, irregularly deposited osteoid substance.

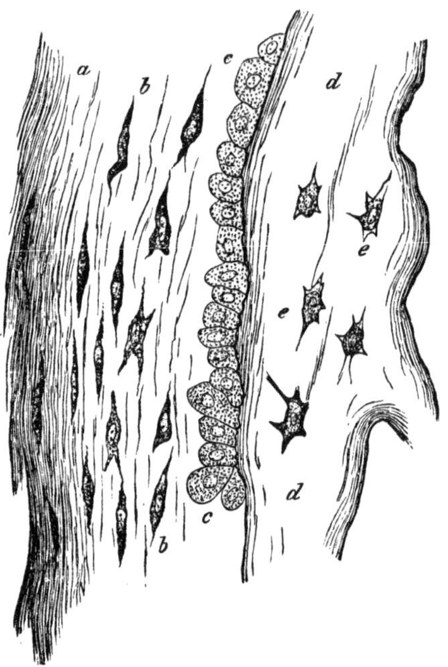

Fig. 146. Formation of secondary bone substance. Longitudinal section through the femur of an older fœtus of the sheep. *a*, the inner surface of the periosteum, consisting of connective tissue; *b*, the younger, or Ollier's layer of the periosteum; *c*, the stratum of osteoblasts; *d*, newly-formed bone tissue; *e*, bone cavities and cells.

The bones of the second group, originating from connective-tissue substance, without a cartilaginous preformation, coincide very nearly with the periosteal mode of growth and require the same methods for their examination. Decalcification by means of chromic acid and subsequent tingeing with carmine has afforded me the best specimens. Pyroligneous acid, as recommended by Billroth, may also be employed with advantage, as a few trials have shown me.

If one fortunately succeeds in injecting the embryos destined for such examinations with transparent masses, many things will be better recognized in this case, as in the investigation of all osteogenetic processes, than when the blood-vessels are not filled. The use of warm water, in the manner indicated at p. 280, to cause the interstitial substance to assume a soft pulpy condition, deserves a further trial, as it will probably in this case, as in that of ripe bones, render the cells sharper and more distinct.

For the examination of the bone-marrow, the preparatory hardening methods with chromic acid, bichromate of potash, and Müller's fluid may be employed. The fresh tissue, with indifferent fluid media, is also to be recommended. Here, for instance, in tadpoles, the vital changes of form of the bone-marrow cells may be readily appreciated (Bizzozero). In this way numerous lymphoid cells which are undergoing transformation into red blood corpuscles may be seen in the red bone-marrow of mammalial animals. This source of the latter cells was not noticed till quite recently (Bizzozero, Neumann). We have already mentioned this circumstance, in a cursory manner, at p. 230. The idea that our cells pass through the thin walls into the vessels of the bone-marrow is obvious.

With regard to the ossification of permanent cartilage which occurs in the later periods of life, such as those of the ribs and many of those of the larynx, this may, as a rule, be considered as a calcification of the cartilage, which is the same process that takes place on a more extensive scale in the fœtal skeleton, and never ceases entirely at any period of life. As in the embryo, so also in the aged, the calcified cartilaginous tissue may be absorbed, and osteogenous substance deposited in the cavities thus formed.

The examination of rachitic bones constitutes an interesting study, completing that of the normal fœtal bone formation. The appearances naturally vary, according to the grade of the disease, the attempts at restoration which have taken place, etc. The individual parts of a bone also present numerous variations.

An insufficient, occasionally almost entirely wanting calcification of the cartilages, the persistence of considerable portions of the fœtal cartilage, together with peculiar transformations of its capsules, and an osteogenous substance which is sometimes inadequately, sometimes not at all impregnated with bone earths may, in general, be regarded as the chief anomalies.

In the rachitic cartilaginous skeleton the medullary cartilage cells are met with the same as in normal bones, as also a similar rupture of the cartilage capsules and the deposition of bone-cells with their interstitial substance. Even in the medullary spaces, anomalies of form and extent may be seen. They frequently advance beyond the calcifying border of the cartilage and even to a considerable distance into the unaltered portion of the latter. Very deceptive appearances are presented by the capsules in that portion of the cartilage which remains; in consequence of the thickening of their walls, the residue of their contents may be recognized as star-shaped bodies. Appearances are caused in this manner which bear great resemblance to bone corpuscles and which, in fact, can scarcely be distinguished from many ruptured capsules in which true bone corpuscles are embedded, unless the point of entrance happens to lie in the plane of the section. Hence we shall readily comprehend that until within a few years rachitic bones were regarded as affording the most certain proof of the transformation of cartilage cells into bone corpuscles, and were accepted as true paradigms of the process of ossification. In reality, however, they constitute very insidious and deceitful objects.

These few remarks must suffice, in consequence of the narrow limits of our little work. The investigations of Bruch, Kölliker, Virchow, and Müller are to be consulted for further details.

Fresh bones or those preserved in alcohol may be selected for examination. Those which have been dried also occasionally afford very handsome specimens. Müller found the employment of weak solutions of chromic acid, with the subsequent addition of glycerine, very useful in these cases.

In consequence of the exuberant vitality of bones, new formations of osteogenous tissue, physiological as well as pathological, are of very extensive occurrence. In both cases the point of origin of the new bone tissue may be from the periosteum and the endosteum, that is, the connective-tissue layer which lines the medullary cavity. The former is, however, much more frequently the case, and Ollier's interesting experiments show that the living periosteum is not deprived of its bone-producing power by transplantation to a remote part of the body.

A beautiful, accurately-investigated example of this double origin is afforded by the reunion of fractured bones, the so-called callus formation. If the examination be made with the aid of the methods at present used for the normal osteogenesis, the newly-formed osteogenous substance which has originated in the periosteum may be seen surrounding the ends of the bones like a ring. Beneath the periosteum, which is here thickened and swollen, appear the various layers of osteogenous tissue which are formed from it. In man these strata have, as a rule, a connective-tissue character, much less frequently that of cartilage (while in mammalial animals, under similar conditions, there is a more plentiful production of cartilage). Secondly, there is a reunion of the bone tissue beneath the endosteum. The latter also swells up and produces new osteogenous tissue, which spreads through the medullary cavity and plugs it up.

Where there is a greater loss of substance in a bone, the regeneration takes place from the periosteum.

Other new formations of bone tissue, such as the hypertrophies or hyperostoses, the inflammatory productions of the same and the bone tumors originate in part and chiefly from the periosteum, in part from the connective tissue of the medullary cavity.

Hyperostosis is, in reality, exactly the same occurrence which is met with in the increase in thickness of young bones, and, in suitable transverse sections, it presents very similar appearances. The local, more or less salient new formations of bone substance of this kind, which are not separated from the ordinary tissue by any line of demarcation, constitute the compact exostoses. Next to these come the tumors which are composed of a denser bone tissue. They show, in part, the ordinary compact texture; in many cases they are of a more spongy nature; in others, finally, they have an ivory-like hardness, in consequence of the slight development of medullary canals. The osteophytes have a sponge-like arrangement.

The cases hitherto mentioned have placed before the reader bone tissue formed from the periosteum. In the so-called scleroses of bones we meet with the new formation of osteogenous tissue proceeding from the medullary cavities and the medullary canals. Among the osteosarcomæ, the central ones are developed from the large medullary cavity, the peripheral ones from the periosteum. They show, in the main, only isolated globular and flake-like masses of bone tissue, without vessels or medullary canals.

The new formation of osteogenous substance in soft tissues, independent of preëxisting bone, has certainly been very much exaggerated in favor of the modern connective substance theory. In most cases there is only calcified connective tissue with indented corpuscles. However, the production of true bone substance in connective tissue parts does take place, although rarely. The stratified arrangement of the basis substance, and the radiated bone corpuscles, connected with each other by their branches in a reticular manner, secure one against mistaking the one for the other.

The resorption of the previously decalcified bone tissue constitutes the opposite process. In normal life the melting down of the bone substance occurs very extensively in growing young bones. Only call to mind the formation of the larger medullary cavities in the fœtus, and the Haversian spaces of later periods! The anatomical events which take place here are the increase of the medullary cells and the enlargement of the medullary spaces, the crumbling down and fatty degeneration of the bone-cells together with the decalcification of the neighboring osteoid substance and the subsequent dissolution of the same. Hereby the liquefying bone tissue frequently shows excavated borders, as if gnawed out. If such a condition occurs at a later period as an abnormal process, we have the so-called osteoporosis. Osteo-malacia also presents us with a similar increase of medullary cells and medullary spaces, with impoverishment of the osteoid substance in bone earths, and the dissolution of the same. In reality the same process occurs in the formation of granulations. While, however, in this case the interstitial substance of the granulating medullary cells still presents a certain firmness, similar to the ordinary consistence of fœtal bone tissue, in other cases there may be a liquefaction of the interstitial mass. The cells which are suspended in such a fluid are then called pus corpuscles, and the process itself is called caries. The latter, corresponding to the two localities of the osteogenesis, may either occur within the bone, in its medullary cavities, or externally in the canals which are filled with bone marrow by the periosteum. Thus the microscope here teaches in a very beautiful manner how normal and pathological processes may pass over into each other.

Decalcified bone substance is said by many histologists to be capable of transformation into ordinary connective tissue. According to our views, this is incorrect. This substance is incapable of any future development; it simply undergoes liquefaction, sooner or later.

Finally, if inquiry be made as to the methods of examining diseased bones, reference is to be made to what has already been remarked. They are the same as for the normal tissue. Dried bones are less to be recommended than moist ones, which may be decalcified by means of chromic acid with the addition of a little muriatic acid, and which may be subsequently hardened again, according to circumstances, in strong alcohol. Bones which are strongly impoverished in bone earths may be examined fresh, or as alcohol preparations, without the use of acids. As we have already remarked, the decalcified tissue distinguishes itself in a very beautiful manner from that which is still calcified by the readiness with which it imbibes carmine.

Section Fifteenth.

MUSCLES AND NERVES.

MUSCLES and nerves, in consequence of their softness, require very different accessories than the hard tissues which we have just left.

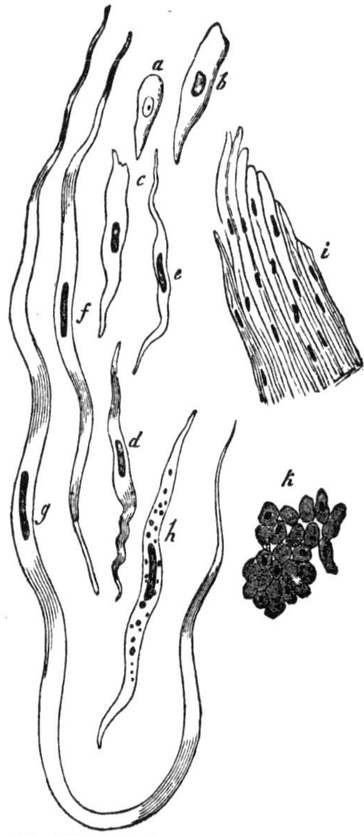

Fig. 147. Smooth muscular tissue. *a*, the fœtal developing cell from the stomach of a pig; *b*, a somewhat more advanced cell; *c–h*, various forms of the contractile fibre-cells from the mature body; *i*, bundle of smooth muscle; *k*, transverse section of the latter.

The muscular tissue of man and the vertebrate animals consists of two forms of fibres, the smooth and the transversely striated.

The latter muscles show as their elementary formation a generally undivided, more rarely branched fibrilla, which is marked with fine and closely arranged transverse lines (the so-called primitive bundle), while the smooth muscles are formed of spindle-shaped, linearly arranged cells. With this difference in structure is also associated differences in their manner of action. The smooth muscles of man always act involuntarily and slowly; the transversely striated muscles, on the contrary, are obedient in their rapid contraction to the impulses of the will. But the heart, a transversely striated muscle, also contracts involuntarily like the smooth tissue, but rapidly.

The examination of smooth muscles (fig. 147) is in general difficult.

This very tissue shows how important the employment of suitable reagents is for the recognition of many textural conditions.

For a long time histologists regarded the elements of the smooth muscles as flat bands (*i*) containing nuclei placed behind each other, and, in fact, the older methods of investigation yielded nothing further. It was not till about 1847 that the piercing eye of Kölliker succeeded in resolving these bands into long spindle-shaped cells arranged in rows, with a columnar-shaped nucleus (*c–h*). Since that time the elements of the smooth muscles bear the name of contractile fibre-cells.

Formerly acetic acid was generally used in studying the smooth muscles. Boiling the tissues (Henle), or hardening them in alcohol, also affords serviceable preparations, especially with subsequent tingeing with carmine.

More recently, however, we have become acquainted with more conservative methods.

For the first examination the frog may be selected, the urinary bladder and lungs of which afford good objects; the smaller arteries of the frog are also to be recommended. To isolate single fibres without reagents, take the walls of the intestines.

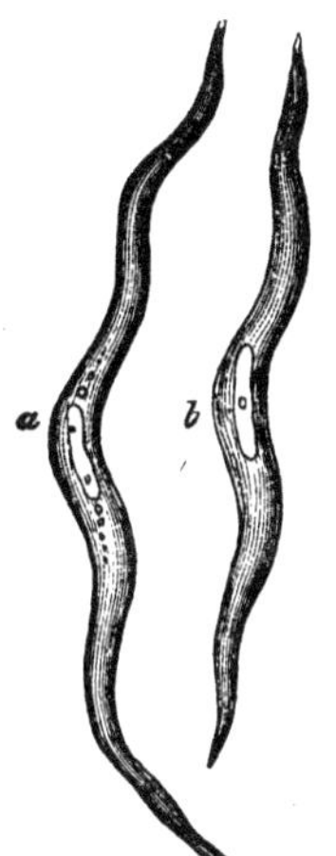

Fig. 148. Elements of the smooth muscles of the rabbit.

The more minute arrangement is to be examined either with the addition of an indifferent fluid, such as blood- and iodine-serum, or by the application of reagents. Impregnation with gold (0.1 per cent.) may be used, but decidedly more may be accomplished by macerating for one or two days in a very weak chromic acid solution of 0.01–0.05 per cent.

The two last-mentioned methods also show us the nucleolus (fig. 148) single or multiple (Frankenhäuser, Arnold, Schwalbe). It was formerly overlooked in the tissue which was altered by acetic acid. The nucleolus is occasionally remarkable, however, even in the fresh cell.

Impregnation with silver is also well adapted for the recognition of delicate strata of organic muscles; for example, in the villi and the mucous membrane of the small intestine (His); likewise chloride of palladium (F. E. Schulze) and picric acid (Schwarz), which color yellow.

Drying, and thereupon tingeing with carmine, and the action of acetic acid, were formerly employed for obtaining transverse sections of bundles of smooth muscles. The preparatory hardening by means of alcohol, chromic acid, or bichromate of potash, appears to be more suitable. The most conservative treatment consists, however, in the freezing method, with a subsequent addition of serum (Arnold), or of a 0.5 per cent. solution of common salt (Schwalbe). For this purpose select the walls of the stomach or intestines of a frog or mammalial animal, the urinary bladder of a dog (Schwalbe), or a section may be made in a vertical direction through the coats of a larger artery. The two umbilical arteries also afford handsome specimens by this method. Thus (fig. 147 *k*) the transverse sections of the fibre-cells will be seen, some more round, some more polyhedral in shape, and in many of them the transverse section of the nuclei may also be recognized, and one will be readily convinced that the contractile fibre-cells are by no means flattened structures, but rather spindle-shaped.

For isolating the cells we have three especially good methods at present in use.

1. The maceration in nitric acid of 20 per cent., with which Reichert and Paulsen have made us acquainted. The first effect is to render the tissue darker and more yellow; after 24 hours the separation of the bundles into contractile fibre-cells commences, and after three days the latter readily fall apart, especially with a little shaking. At the same time a peculiar transversely wrinkled or folded appearance occurs in the elements of the smooth muscles.

Muriatic acid of 20 per cent. also exerts a similar effect.

2. Diluted acetic acid.

This has at all times played an important rôle in the examination of the tissue with which we are at present occupied, and was also extensively made use of by Kölliker in his investigations. Its value lies, firstly, as we have already remarked, in the rapidity with which it renders the so characteristic nuclear formation visible; then, by making the connective tissue transparent, it causes the bundles of smooth muscles themselves to become prominent. Solutions of 2–5 per cent. are to be used.

3. Treatment with 30–35 per cent. solutions of potash.

If the demonstration of the nucleus be renounced, the solutions of potash of the strength mentioned, or one of 32.5 per cent., constitute the best means of isolating and demonstrating the contractile fibre-cells. After an action of 15, 20–30 minutes, numerous examples of the latter may be obtained, which are frequently of an undulating, serpentine form.

The maceration in iodine-serum, or the already mentioned extremely diluted chromic acid, also leads to the isolation of the muscular elements.

The destruction of smooth muscular tissue by fatty degeneration of the cells is a not infrequent event, as a normal occurrence (uterus) as well as a pathological one; likewise the new formation of the tissue from the pre-existing. The latter phenomenon, however, requires a more careful study.

The transversely striated muscles (fig. 149) offer much more remunerative objects. The more important elements make their appearance readily and beautifully, and only the resolution of certain very delicate textural conditions leads to a difficult domain lying at the limits of our present instruments.

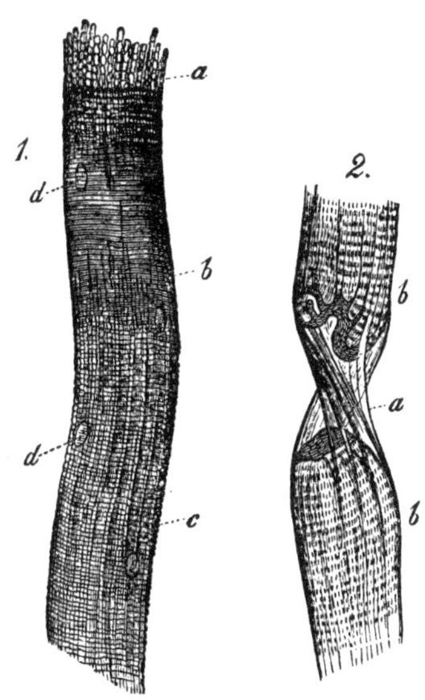

Fig. 149. 1. Transversely striated muscular fibrillæ; *a*, so-called primitive fibres; *b* and *c*, transverse and longitudinal lines; *d*, nuclei. 2. A muscular fibrilla, the sarcous portion, *b b*, of which is torn in two and shows the empty primitive sheath at *a*.

If we desire to bring the fibres of the transversely striated muscular tissue to view, in as unaltered a form as possible, the frog is to be especially recommended. The animal is to be decapitated, and the well-known cutaneous thoracic muscle, or one of the flat muscles running from the os hyoides to the lower jaw, immediately cut out, avoiding all straining and tearing. These, with the addition of blood-serum or some other indifferent fluid, will afford us excellent views of the familiar longitudinally and transversely striated fibrillæ (compare figs. 149, 1; 150, 6). We obtain similar appearances in the living creature bv select-

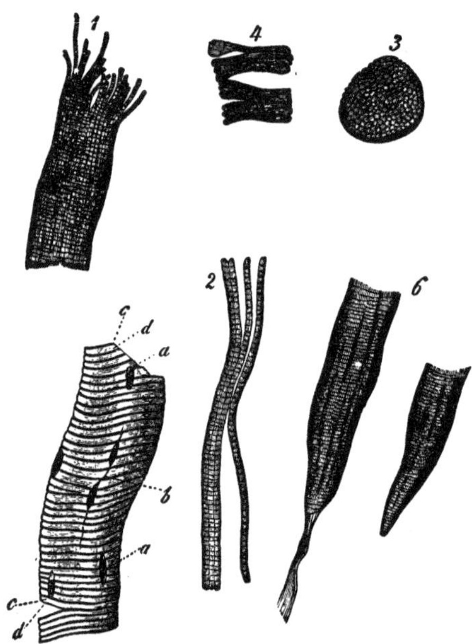

Fig. 150. 1, Muscular filament with so-called primitive fibrillæ and distinct transverse lines; 2, isolated fibrillæ more strongly magnified; 3, the sarcous portions united as a disk; 4, the commencing separation of the disks; 5, muscular filament after longer maceration in muriatic acid; *a* and *b*, nuclei; *c* and *d*, brighter and darker zones; 6, two pointed fibrillæ of the biceps brachii, already ending in the course of the muscle.

ing the tail of the frog's larva; young fishes just hatched also afford admirable objects. If entire freshness be disregarded, a muscular fasciculus from the body of any vertebrate animal may be used several hours after death. A small portion of tissue, carefully picked apart with needles, always affords good specimens, and shows us the fibrillæ varying in their diameters and markings.

To recognize the nuclei, use a weak acid (diluted acetic acid, muriatic acid of 0.1 per cent., etc.). They will then be discovered in the form of oval bodies (figs. 149, 1, *d*; 151, *c*). A remainder of original cell-substance (protoplasma) surrounds the nucleus and extends beyond both its poles in a spindle-shaped prolongation (muscle corpuscles).

We do not see the sarcolemma, or the primitive sheath of the muscular filament in the ordinary method of examination, as this envelope closely surrounds the contractile contents. One may succeed in recognizing them in various ways. The muscles of the pisciform amphibiæ, the proteus and axolotl, which have been in alcohol for some time, afford, without further preparation, a very good view of the loose and expanded envelope. If the greater portion of the various albuminous substances are dissolved by a longer-continued maceration in 0.1 per cent. muriatic acid, one may perceive the softened contents projecting from a surrounding sheath at the divided extremities of the muscular fibres. The sarcolemma may be completely isolated, as Kühne has taught us, by means of a somewhat complicated process. For this purpose the muscular fasciculus of the frog is to be macerated for a day in water with 0.01 per cent. of sulphuric acid of 1.83 sp. wt., and then freed from its connective tissue by digestion in water at 35–40° C., which likewise requires 24 hours. The fibre is now to be exposed for a day to the action of muriatic acid of 0.1 per cent.

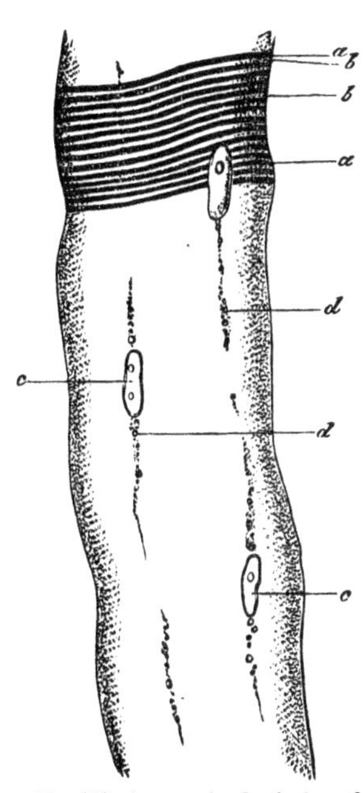

Fig. 151. A muscular fasciculus of the frog by 800-fold enlargement. *a*, dark zones with sarcous elements; *b*, bright zones; *c*, nuclei; *d*, interstitial granules. (Alcohol preparation.)

However, we possess still other accessories, by means of which we are able to bring the sarcolemma into view instantaneously. If a bundle of muscular fibres be drawn with a pair of sharp forceps out of one of the muscles at the upper part of the thigh of a freshly decapitated frog, one will soon be able, by the addition of water, to recognize numerous separations of the primitive sheath from the contractile contents, as a result of the energetic imbibition. At first there are small, limpid pouches; these soon grow larger and larger under the eye of the observer, neighboring ones flow together, and the vesicular, elevated portions of the sarcolemma are separated by annular constrictions from those which still remain attached.

Other muscles may also afford us the desired result, if in their preparation we subject the several fibres to a strong tension and tearing. In some of them the contractile contents are torn in two; while over this place the more extensible sarcolemma remains intact. Such an appearance is seen in the muscular fibre fig. 149, 2, *a*.

The drying method served for a long time to show the relation of the several muscular fibres with each other, as well as the formation of the bundles of muscles and of the entire muscle. Thin sections which were again softened, and especially such as were soaked in an ammoniacal solution of carmine, and subsequently treated for a few minutes with very dilute acetic acid, then presented the frequently mentioned and characteristic appearance of fig. 152 *a*. One may recognize, at the same time, in the muscles of man and the mammalian animals, the manner in which the nuclear formation is imbedded in the periphery of the contractile substance, and lying against the inner surface of the primitive sheath (*e*). In the muscular fibres of the heart, on the contrary, nuclei also occur in the more central parts, a condition which seems to predominate in the lower vertebrated animals.

The freezing method presents very much better results, however (Cohnheim). By the aid of the highest magnifying powers one may then recognize groups of sarcous elements as a mosaic of small, dull areolations of various forms (fig. 153 *a*), and surrounding these groups a lattice-work of transparent glistening lines (*c*).

To demonstrate the branched muscular fibres, as they occur in the heart and tongue, a 30–35 per cent. solution of potash may be used for the former organ, while tongues are to be immersed fresh in pyroligneous acid, or they may be exposed to the action of this reagent after hardening in alcohol or chromic acid. The value of the pyro-acetic acid (or of a dilute acetic acid) naturally consists in its property of rendering the connective tissue transparent.

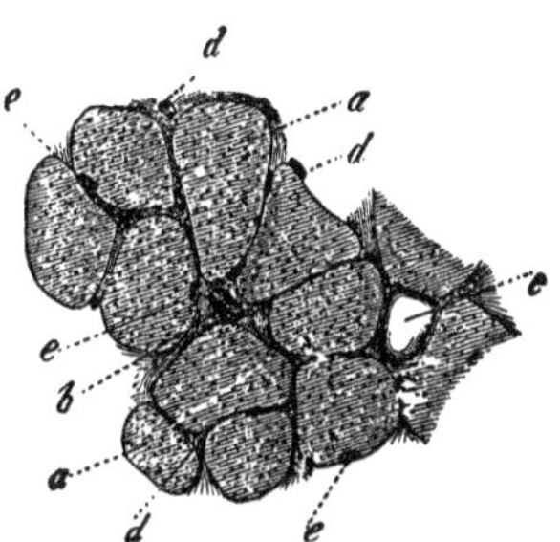

Fig. 152. Transverse section through a bundle of the biceps brachii of man. *a*, the muscular fibres; *b*, transverse section of a vessel; *c*, a fat-cell lying in a large connective-tissue interstice; *d*, capillaries cut across; *e*, nuclei of the muscular fibres.

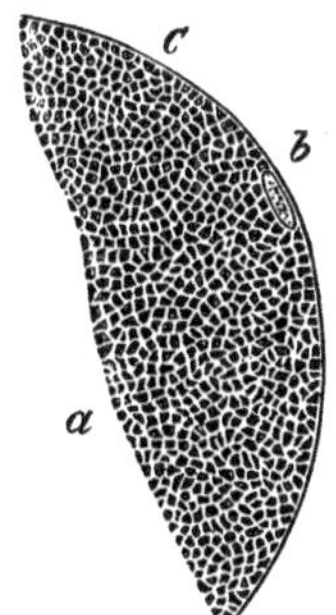

Fig. 153. Transverse section through a frozen muscle of the frog. *a*, groups of sarcous elements; *b*, a nucleus; *c*, transparent transverse connecting medium (Querbindemittel).

The isolation of the muscular fibres in their entire length is necessary for a number of purposes of investigation. In this way we recognize the course of the fibres in a muscular fasciculus, the divisions of the same as phenomena of growth, and the increase in the number of fibres in the enlargement of the muscle, etc. We have the choice of several methods for this purpose.

1. The mixture of chlorate of potash and nitric acid may be employed in various degrees of concentration. We are indebted to Kühne for a suitable procedure. The bottom of a beaker is to be covered with crystals of the chlorate of potash, slightly moistened with distilled water, and four times the volume of pure concentrated nitric acid added. After considerable stirring, the fresh muscle (of a frog) is to be placed at the

bottom of the vessel and buried beneath the crystals of potash by means of a glass rod. In about half an hour the muscle is to be removed from the mixture and placed in water in an ordinary test tube. This is to be strongly shaken, and then, in fortunate cases, the tissue separates completely into fibrillæ. If this separation does not succeed the first time, the muscle is to be replaced in the mixture and exposed to the same procedure at intervals of five minutes.

By this means exquisite specimens are obtained, and the nuclei appear very finely in the slightly browned sarcous substance.

The method given by Wittich for using this mixture is also very judicious. That is, boiling in chlorate of potash and nitric acid strongly diluted with water (water 200 ccm., nitric acid 1 ccm., and chlorate of potash, 1 grain).

2. The same may be accomplished by maceration for twenty-four hours in 0.01 per cent. sulphuric acid and the subsequent treatment for one day with warm water, as recommended above (p. 317) for the demonstration of the sarcolemma.

3. After the example of Rollett, the muscle may be placed, without any addition of water, in a small glass tube which may be hermetically sealed by the heat of a lamp and then warmed to 120°–140° C. on a sand-bath for ten minutes. The tube is then to be broken and the muscle agitated in warm water.

4. A strong, but no longer fuming muriatic acid (p. 129) may also be employed with advantage. After an action of several hours the interstitial connective tissue is likewise found to be dissolved.

5. Finally, a potash solution of about 35 per cent. also forms a very good accessory. A sometimes slighter, sometimes larger portion of the muscular fibres will always be found isolated, after an action of from fifteen to thirty minutes.

The high value of reagents is not more striking in any question with regard to the structure of muscular tissue than that of the relation cf the muscular fibre to the tendon.

Until recently it could only be stated as a true expression of what had been observed, that there was no boundary to be discovered between the contractile substance of the fibrillæ (fig.

154 *a*) and the connective-tissue fibrous portion of the tendon (*b*), whether the muscle be inserted into the tendon in a straight line or at an oblique angle. It was therefore extremely probable that the sarcous portion, as well as the sarcolemma, were continued directly into the tissue of the tendon. This continu-

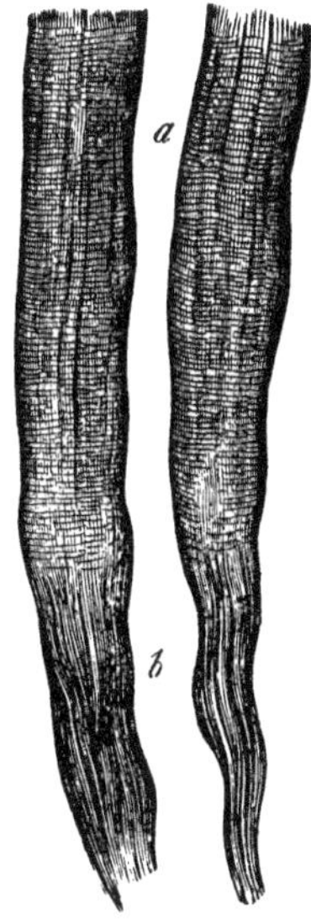

Fig. 154. Two muscular fibrillæ (*a*) with the apparent continuation into the connective-tissue bundles of the tendon (*b*).

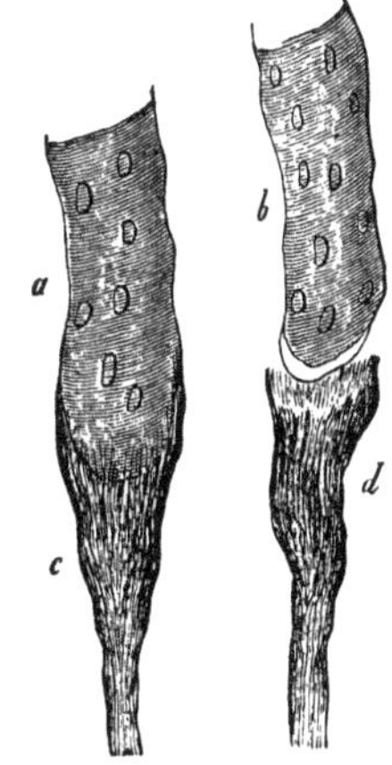

Fig. 155. Two muscular fibrillæ (*a b*) after treatment with solution of potash. The one still in connection with the tendon (*c*), the other separated from the same (*d*).

ity of the contractile substance and the connective tissue was certainly somewhat strange, and we might say inconvenient.

At the present day we must all admit the error, even though we formerly defended this theory, since Weismann has discovered a medium in the 35 per cent. solution of potash, which decides in a beautiful and certain manner the long-contested structural relation.

After 10, 20–30 minutes the muscular fasciculus presents the appearance shown in fig. 155 *a*, *b*. The apparent continuity has disappeared. The former, covered by the sarcolemma, is separated by a sharp line of demarcation from the tendinous fasciculus (*c*). In many specimens they are even seen disconnected from their tendons (*d*), especially when a slight pressure has been made. There is therefore no longer any doubt that the fasciculi of the muscles and tendons are only "cemented" together in the firmest manner. It is this substance, this "tissue cement" which the potash solution has dissolved, which holds them together.

While it was formerly supposed that every transversely striated fibre continued throughout the entire length of its muscle, more recently numerous exceptions to this have been observed; that is, muscular fibres which terminate in a point or some other form at a greater or lesser distance from the tendinous extremity (Rollett, Weber, Herzig, and Biesiadecky). Such fasciculi (fig. 150, 6) have their connection with the tendon, to a certain extent, in the interstitial connective tissue. For these investigations, which are easy to make, fresh as well as boiled muscles may be immersed for twenty-four hours in glycerine, or the solution of potash mentioned may be employed.

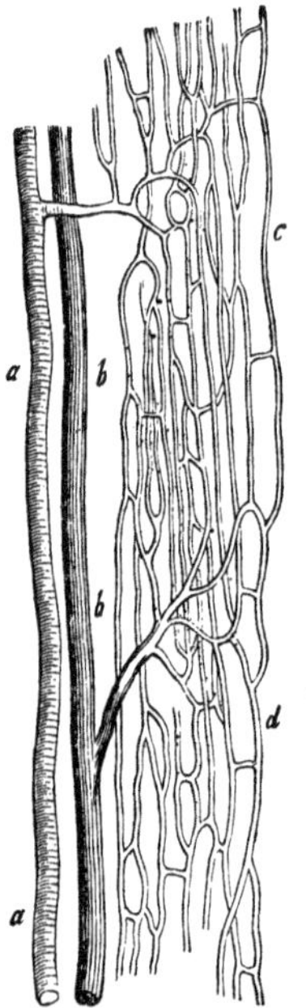

Fig. 156. Vascular network of a transversely striated muscle; *a*, artery; *b*, vein; *c d*, the capillary network.

To see the extended capillary network of the muscular tissue (fig. 156), they should be injected with transparent masses, with carmine or Prussian blue. Thin, flat muscles taken from a frog which has been drowned in alcohol, and placed on the microscopic slide, will bring to view the capillary system filled with blood in the most beautiful manner; and with a little contraction of the muscular fibres, the delicate windings of the capillary vessels may be readily recognized.

Concerning the nerves of the muscles, reference is to be made to one of the following pages.

The structural conditions which have thus far been mentioned of the transversely striated muscular tissue are all, as we have remarked, relatively easy to examine, and, with the requisite methods, they afford a suitable study for the beginner. It is otherwise with regard to the subtle question concerning the constitution of the contractile substance which they contain, the "sarcous substance."

The muscular fibrillæ (fig. 157, 1) show a double marking, which, however, is subject to considerable variation as to sharpness and distinctness. We recognize, coming to the surface of the fleshy substance, sometimes over a long space, sometimes only for a short distance, and then disappearing in it, a fine longitudinal marking (*c*), extending throughout the entire thickness of the former; and secondly, a likewise very fine, transverse linear marking (*b*), which also may be followed through the entire substance of the muscle. In many fibrillæ the latter is alone present; in other specimens the longitudinal lines predominate, occasionally to exclusiveness, and fine bands and fibres (*a*) may project from the cut extremities. It was especially the latter cases which, in former times, led the microscopists to the acceptation of a further composition of the muscular fibrillæ out of the finest fibres, the so-called "primitive fibrillæ" (fig. 150, 1, 2). The transverse lines were then generally referred to a knotty condition of these elementary fibrillæ resembling a string of pearls.

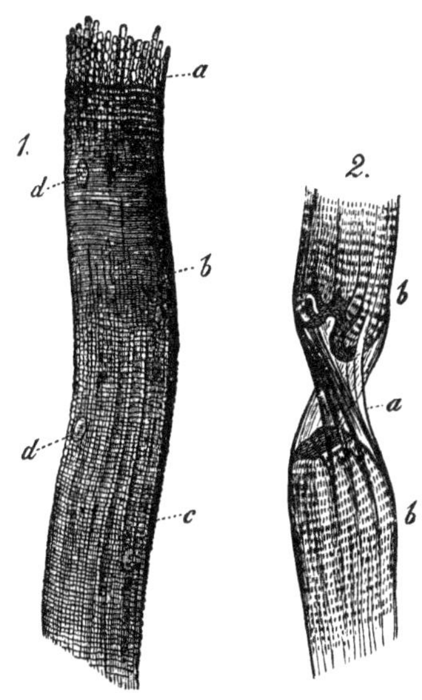

Fig. 157. 1. Transversely striated muscular fibrillæ; *a*, so-called primitive fibres; *b* and *c*, transverse and longitudinal lines; *d*, nuclei. 2, a muscular fibrilla, the sarcous portion, *b b*, of which is torn in two, and shows the empty primitive sheath at *a*.

At the present day this theory still finds its defenders, and even among renowned investigators, although the so much improved optical accessories of the period by no means decide in their favor.

In other specimens the transverse markings come out sharper and more distinct (fig. 158, 6). If the longitudinal lines are wanting, one might even here imagine that the muscular fasciculus was composed of disks or plates arranged over each other in layers. The appearances become still more deceptive where the transverse lines stand farther apart than is the rule,

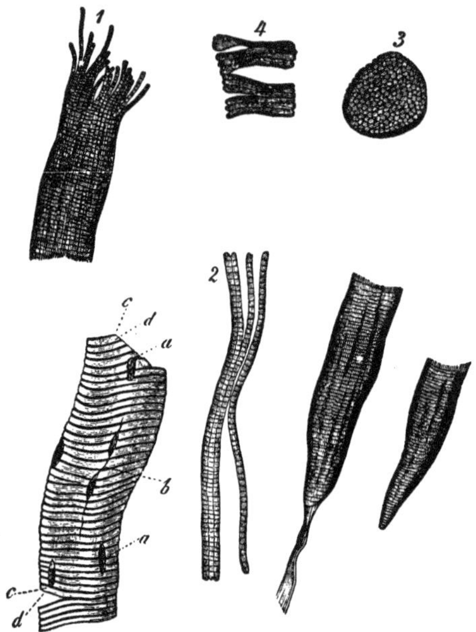

Fig. 158. 1, Muscular filament with so-called primitive fibrillæ and distinct transverse lines; 2, isolated fibrillæ more strongly magnified; 3, the sarcous portions united as a disk; 4, the commencing separation of the disks; 5, muscular filament after longer maceration in muriatic acid; *a* and *b*, nuclei; *c* and *d*, brighter and darker zones; 6, two pointed fibrillæ of the biceps brachii, already ending in the course of the muscle.

and where the border or periphery of the fibrilla has indentations corresponding to the striations.

The theory originated by the English histologist Bowman, and further improved by some of his countrymen, finds at present the most adherents. According to it, the contents of the muscular fasciculi consist of small molecular corpuscles, the so-called fleshy particles or "sarcous elements," which are held together by a homogeneous connecting medium, which is in reality twofold, and not exactly alike chemically. According, now, as the one or the other of these two connecting media predominates, we see the sarcous elements united either longitudinally or laterally; in the first case, the appearance of fibrillæ (1, 2) results; in the latter, that of transverse lines (1) increasing to transverse plates (4, 5).

The sarcous elements in the muscular fibres of man and the mammalia are entirely too small to enable us to state anything with certainty in regard to their form, although by the use of very strong magnifying powers they may be shown with satis-

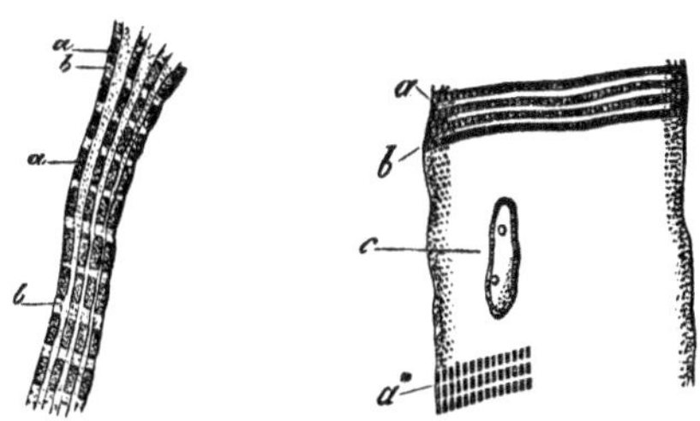

Fig. 159. Two muscular fibrillæ, from the proteus, 1, and the hog, 2, magnified 1000 times (the first from an alcohol preparation, the latter treated with acetic acid of 0.01 per cent.). *a*, sarcous elements; *b* bright longitudinal connecting medium. At *a** the sarcous elements are further apart, and the transverse connecting medium is visible. *c*, nucleus.

factory distinctness (fig. 159, 2, *a a**). The muscles of the lamprey and of the pisciform amphibiæ, on the contrary, have prismatic sarcous elements of considerable size, so that in alcoholic specimens of the proteus (fig. 159, 1) the recognition of these prisms (*a*) 0.00075‴ in size, and of the more transparent longitudinal connecting medium (*b*) becomes very easy. Furthermore, the muscles of the house-fly form very beautiful objects; their prismatic sarcous elements distinctly assume an oblique position during contraction (Amici). Similar sarcous elements are frequently to be met with in insects; their mean longitudinal diameter may be assumed to be about 0.0015‴ (Schönn). The crab was also very properly recommended for this observation years ago (Häckel), as may be proved by any one who has one of Hartnack's immersion systems, No. 10 or 11.

Without mentioning that the various appearances of the muscular fibrillæ may be readily explained by what has been stated, this theory has received still further important supports, partly chemical, partly optical in their nature, through the labors of German investigators.

Firstly, we have a series of reagents which attack the longitudinal connecting medium more or less, while the transverse remains unaltered, or is only subsequently affected.

Very dilute acids are here accorded the first rank. Thus acetic acid of from 0.5–1 per cent. causes in a short time the longitudinal lines to disappear, and the transverse striations to become distinct in the swollen muscular fibres. Other acids, as, for example, diluted phosphoric acid, cause similar effects. The finest appearances are, however, afforded by the strongly diluted muriatic acid of 0.5, 0.1–0.05 per cent.

After several hours one may perceive not only the most distinct transverse lines (fig. 158, 5), but also a regular breaking-up of the muscular fibrillæ into transverse disks (4). The same effect is also produced by the gastric juice by means of its free acids. Vomited pieces of meat often present similar extremely elegant appearances. Older muriatic acid preparations show the molecular decomposition more and more, till at last a mucilaginous granular mass issues from the opening in the sarcolemma.

Concentrated muriatic acid causes the muscle to shrink, whereby sharp transverse markings also, not infrequently, make their appearance.

However, not only solutions of acids, but also those of many salts of the alkalies and alkaline earths, such as those of the carbonate of potash, the chloride of calcium, and the chloride of barium, present excellent media for rendering transverse plates visible in most shrinking muscular fibres. The transverse lines are gradually rendered very sharp, and there is frequently a distinct breaking-up into transverse plates, especially from the action of the carbonate of potash.

On the other hand, we have become acquainted with a series of other reagents which first attack the transverse connecting medium of the sarcous elements, then dissolve it, and are thus capable of producing the separation of the muscular filament into the so-called primitive fibrillæ.

Among these may be enumerated the maceration of the muscle in cold water, the boiling in the same, an immersion in absolute alcohol, in diluted alcohol, in diluted solutions of chloride of mercury, chromic acid, and chromate of potash.

The latter, after acting for about a day, may in favorable cases produce appearances such as are represented in fig. 160; the muscular filament becomes separated, like a string, into long crooked threads.

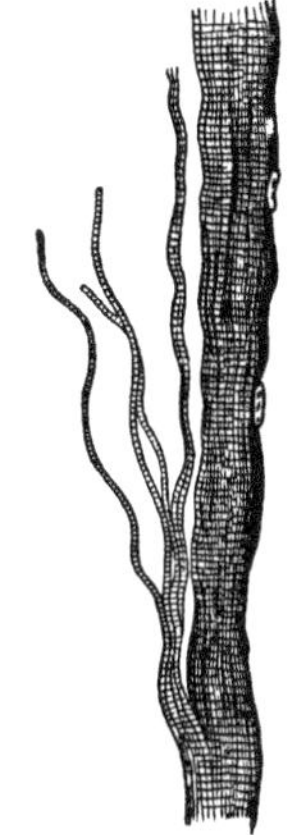

Fig. 160. A muscular filament after treatment for twenty-four hours with chromate of potash.

If such a fibre be examined with very strong objectives, one may distinctly recognize that it is composed of alternating darker and brighter zones (the sarcous elements and the longitudinal connecting medium). Frequently a delicate transverse line may also be seen passing through the middle of the more transparent intervening substance; probably the place where, in the division of the plates, the longitudinal connecting substance usually separates.

We cannot omit to mention that investigations have lately been made public by Krause and Hensen, which are said to prove a further more complicated composition of the transversely striated muscular fibre. The object and narrow limits of our book do not allow us to enter into this domain, which is at present so obscure.

The various substances of the muscular fibre are still further distinguished, as Brücke ascertained, by dissimilar optical properties. The substance of the sarcous elements consists of a double refracting material, while the longitudinal connecting medium only refracts simply. Even with crossed nicols, one may recognize the bright and dark zones interchanging in a beautiful manner; still finer appearances are produced by the intercalation of a selenite or mica plate. According to the observations of this scientist, the muscular fibre is positively uniaxal, and the optical axis coincides with the long axis of the image. The muscles of insects, deprived of their water by means of alcohol, and mounted in Canada balsam, may be employed for these investigations, the correct interpretation of which has since been questioned by Valentin and Rouget. Smooth muscles consist, according to Valentin, of a doubly refracting substance.

The changes which occur in the transversely striated muscle

during its contraction, as well as those which take place after death during the rigor mortis, deserve a more accurate study with the aid of our improved optical accessories. A few communications have recently been made concerning the contraction of the smooth tissue.

The larva of frogs, or the embryos of the hen or of the mammaliæ, may be used either fresh or hardened in alcohol or chromic acid, for the study of the fœtal muscles and of their manner of origin. The methods of investigation consist in the preparation of fine sections, tearing with needles, tingeing (glycerine-carmine, hæmatoxyline), and the application of weak acids.

Fatty degenerated muscles and those streaked with fat are to be examined either fresh or from chromic acid preparations. The latter, in which the connective tissue between the muscular fibres is changed into fat tissue, that is, into a series of fat-cells

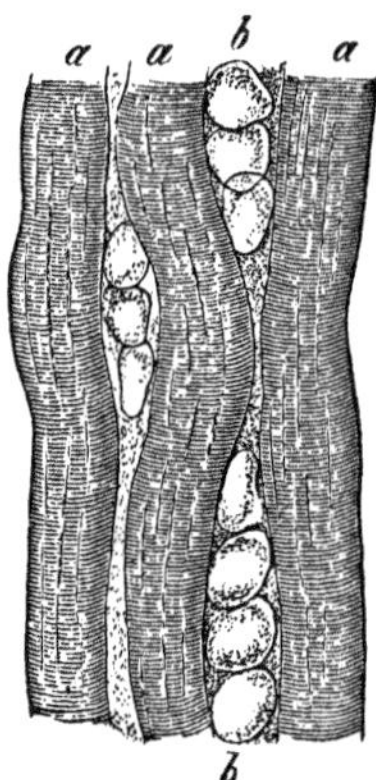

Fig. 161. Human muscle streaked with fat-cells. *a*, muscular fibres; *b*, rows of fat cells.

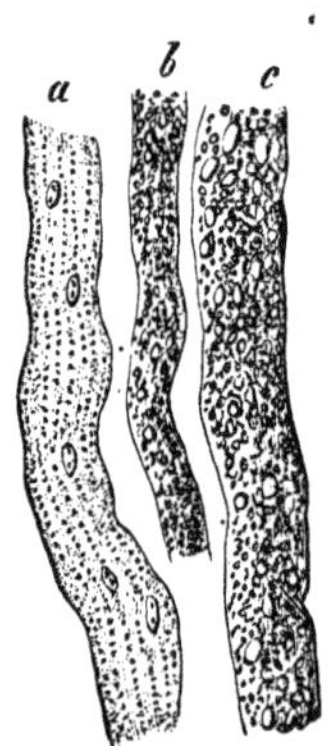

Fig. 162. Fatty degenerated human muscular fibres. *a*, slighter; *b*, increased; *c*, highest degree.

—a condition which also occurs with high degrees of obesity and over-feeding—is shown in our fig. 161. Fig. 162 represents the former condition, in which fat molecules are formed within the sarcolemma at the expense of the sarcous elements, which undergo fatty degeneration.

The inflammatory changes of the muscles, with their increase of cells and the typhous transformations so beautifully described by Zenker a number of years ago, also require similar methods of treatment.

We should render ourselves responsible for a defect if we should pass over in silence a subject which has recently awakened the greatest interest among physicians and laymen; we refer to the occurrence of trichinæ in the transversely striated muscular tissue.

The trichina spiralis, these small forms of nematodes, are, as is well known, eaten with the flesh of the hog and, after a few days, they arrive at a condition of sexual ripeness in the human intestinal canal, so that we now meet with examples of somewhat larger females (measuring more than 1‴) and smaller males (intestinal trichinæ). About a week after the transplantation they produce a multitude of very small living young ones, which, after perforating the walls of the intestines, find their way into the muscles. Here they force themselves through the sarcolemma into the fibres of this tissue, in which they increase considerably in size, so that they may obtain a longitudinal dimension of from $\frac{1}{3}$ to $\frac{1}{2}$‴ (muscular trichinæ).

All the transversely striated muscles, with the exception of the heart, serve as a location for these small parasites, whose number may not unfrequently become extraordinarily large, in consequence of repeated immigrations. Nevertheless, the muscles of the jaw and neck and the diaphragm are distinguished as favorite localities. The tendinous extremity of the muscles — obviously because there is here a mechanical impediment to further emigration — usually shows the greatest abundance of the dangerous guests.

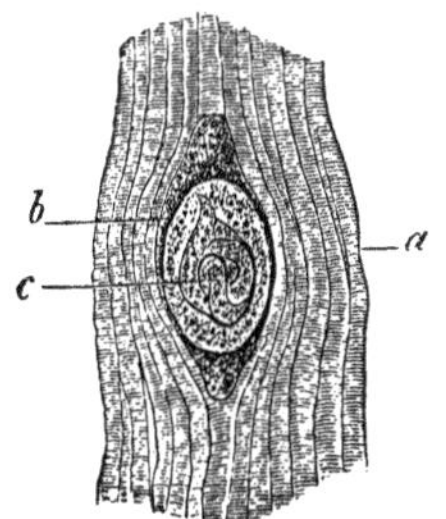

Fig. 163. Incapsulated trichina in man. *a*, muscular fibres; *b*, capsule; *c*, worm.

The little worm devours a portion of the fleshy mass beneath the sarcolemma of the muscular fibre, where it gradually rolls itself up in a spiral manner. Around this a capsule (fig. 163) is gradually formed, which requires months for its completion. While this is taking place, we see the muscular corpuscles of the neighborhood increasing luxuriously and forming a more compact internal investing layer, to which the thickening sarcolemma is also associated as an external layer. The form and size of the capsules vary; we meet with oval (more rarely cask-shaped), spindle, and lemon shaped forms, generally with considerably thickened extremities. The length is usually 0.2, 0.3, 0.5‴. The calcification of the capsule, which commences in the internal portions, begins late (scarcely before the expiration of a year). This process, with its further development, renders the whole thing visible to the unaided eye as a white point, which was not the case with its earlier phases. The trichinæ were first discovered many years ago, in just this latter condition, in which the parasite may preserve its extremely tenacious life for many years in the calcified capsule.

The examination of muscles infected with trichinæ is very easy. Thin sections made in the direction of the fibres, with or without picking apart, will show the presence of the worms when examined in the ordinary fluid media, with the addition of acetic acid or alkalies. A magnifying power of about 40 diameters suffices for the first examination; for more accurate investigation one of 150 or 200 should be used.* In invalids, where trichiniasis is suspected, fragments of muscle may be removed from the body with a small harpoon-shaped instrument. For the microscopical examination of the flesh of the hog, a number of very thin, as large as possible sections should be taken from the muscles of various parts of the body, but especially from the chief localities of the parasites.

Only injected muscles, or those intended for polarized light, are to be mounted in Canada balsam for preservation. The other preparations require to be put up in a moist condition,

* This object may be accomplished with ordinary and therefore cheaper microscopes, with which the several enlargements mentioned in the text may be obtained. With this optical apparatus, the weaker power should show distinctly the larger scales of the lepisma saccharinum (p. 63), while the stronger combination should afford a satisfactory image of the smaller forms of the scales of this insect, with their longitudinal and oblique lines. Some of the modern "trichina microscopes" fulfil these requirements in a satisfactory manner, but a quantity of the most miserable trash has also been put in circulation.

for which purpose glycerine diluted with water stands in the first line; in this tissues, especially those tinged with carmine, keep for years in a beautiful manner.

The elements of the nervous system are marked by exceedingly variable qualities, so that in their investigation numerous precautionary measures become necessary.

We distinguish, as is taught by every hand-book, white and gray substance. The former consists exclusively of one of the two elements of tubes or fibres, called nerve-tubes, nerve-fibres, primitive fibres of the nervous system. In the gray substance we meet with the second element, together with a sometimes slighter, sometimes greater quantity of the nerve-fibres. This second element, the ganglion-body, ganglion-cell, or nerve-cell, is generally a large cellular structure with a vesicular-shaped nucleus. The other constituents consist of connective substance in various stages of development, and blood-vessels.

In order to see the nerve-tubes, which consist of an albuminous central fibre, the so-called axis cylinder; of a peculiar substance surrounding this, the nerve-medulla or the medullary sheath; and of a very fine envelope surrounding the whole and holding it together, the primitive sheath or the sheath of Schwann, in as unaltered a condition as possible, we cannot proceed too rapidly, and must, at the same time, avoid almost all preparatory manipulation. For this reason, there are but few parts of the bodies of vertebrated animals which afford suitable objects. The cornea of a small mammalial animal just killed, for example, that of a rabbit or of a mouse, may be examined on the warm stage without the addition of any fluid medium. The cornea should be incised at its margin. A very fine form of nerve-fibres will here be met with. The frog affords better preparations; its transparent eyelid shows larger tubes, isolated or lying together in bundles. The tails of their larvæ permit the observation to be made on the living creature.

Quite fresh, unaltered nerve-fibres should present the appearance of entirely homogeneous, opaline, cylindrical fibres, in which there is no trace of any further composition to be recognized. The addition of iodine serum is here to be recommended.

If we take a nerve from the body of an animal recently killed, and pick it in water with needles, notwithstanding the greatest rapidity of manipulation, it is no longer possible to obtain the natural condition; on the contrary, its appearance is more or less changed; there is an alteration of the medullary sheath which it has been agreed to call a coagulation.

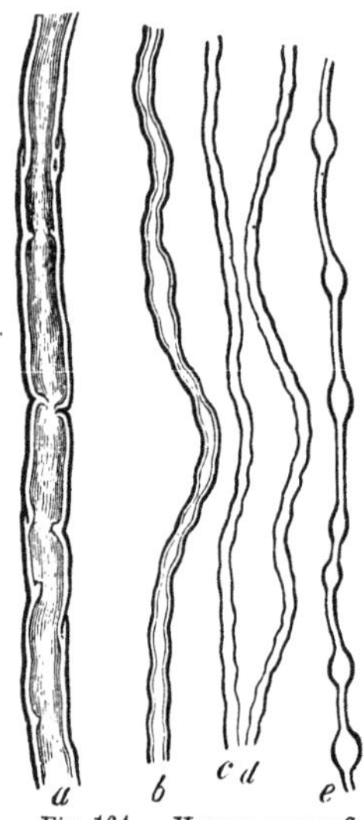

Fig. 164. Human nerve-fibres. *a*, broad; *b*, medium breadth; *c d e*, fine.

Our fig. 164 may represent the commencement of this coagulation. In the beginning it gives the nerve-fibre a darker contour. Soon, however, we see a thin peripheral layer coagulated and separated by a second, more internal and finer line from the central portion of the medulla, which is not, as yet, drawn within the sphere of these changes. But to present these "double contours" the nerve-tubes must have a certain thickness (*a b*). If the transverse diameter falls below a certain size, the tubes then and afterwards appear with only simple contours (*c d e*), but at the same time they readily assume a peculiar appearance,—they become "varicose," as it is called.

Further changes render the coagulated peripheral layer broader, and frequently show an irregularity of the inner contours. The process may here become stationary; the coagulated portion protects the internal, still uncoagulated medulla to a certain extent. But generally this is also drawn into the sphere of the changes; the previously homogeneous appearance is lost; a few lumpy formations make their appearance in it and increase in number and size; not unfrequently the whole becomes a granular, crumbling mass. But all nerve-tubes do not behave in exactly the same way; we may meet with them lying close beside each other, showing various phases of coagulation.

We have not yet perceived anything of the axis-cylinder and the delicate envelope.

The so-called nerve-medulla is a mixture of peculiar substances of cerebrine and lecithine, with a very changeable body belonging to the albuminous group. We shall therefore understand why reagents, such as strong alcohol, concentrated chromic acid, a solution of corrosive sublimate, and many others which have a coagulating effect on albumen, should also produce the more advanced phases of coagulation almost instantaneously.

For the same reason it is unnecessary to state that such a coagulated nerve-medulla, by the addition of alkaline solutions, such as those of potash or soda, again assumes a more fluid and homogeneous condition, and exudes from the cut ends of the nerve-tubes in the shape of double-contoured, fat-like drops and filaments.

If strong pressure be made with the covering-glass on the nerve-tubes, treated in this manner with alkalies, the medulla may be forced out of many of them, and in this way the empty, homogeneous, extremely delicate primitive sheath may be seen. If careful search be made among the nerve-fibres, which have been isolated by picking a nerve-trunk apart, a few will be met with in which their contents have been somewhat displaced, in consequence of the pulling and the pressure of the preparing-needle; and the sheath, which is generally collapsed, may also be recognized for a short distance.

It was frequently denied, at a former epoch, that the axis-cylinder was an integral constituent of the nerve-tubes, and this was quite proper, for, with the accessories then employed, it could only be brought to view in an isolated condition. At the present day it is a small matter to demonstrate these fibres in all nerve-tubes; and we have the choice between several methods.

For the demonstration of the same, Schulze's reagent, the mixture of chlorate of potash and nitric acid may be employed (Budge and Uechtritz). Chloroform renders good service (Waldeyer), and collodion is very excellent (Pflüger). A fresh nerve is to be picked apart on the slide, without any addition of fluid. A large drop of collodion is then to be added, the covering-glass placed over it, and the examination made immediately.

The nerve-tubes rapidly become more and more pale, and, instead of the dark medulla, only a few granules are to be seen enveloped by the distinct primitive sheath. This becomes contracted, and thus frequently shows a series of extremely characteristic invaginations. The axis-cylinder appears in each tube as a pale fibre. During the contraction of the nerve-fibre it frequently appears to be too long; and not unfrequently, under the eye of the observer, it shoves itself through the axis towards the periphery, and protrudes as a fibre from the cut extremity (fig. 165 *c*).

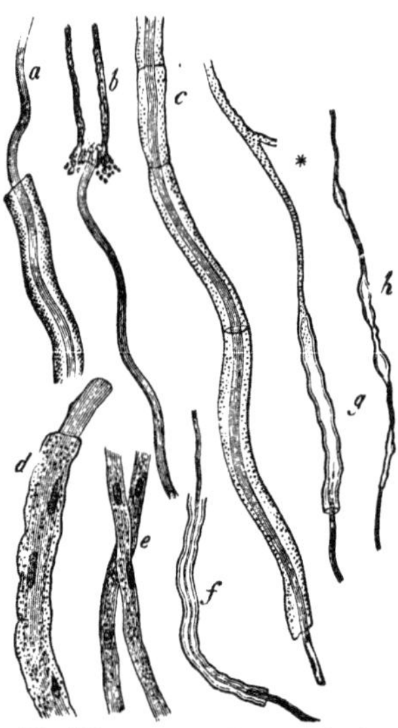

Fig. 165. *a–c*, nerve-fibres of the frog treated with absolute alcohol (*a*), chromate of potash (*b*), and collodion (*c*), all showing the axis-cylinder; *d*, non-medullated fibre from the lamprey; *e*, non-medullated nerve-fibres from the olfactory nerve of the calf; *f*, *g*, *h*, nerve-tubes of a fine form with axis-cylinders; that of *g* becomes at * a branch of a ganglion-cell.

This interesting appearance may be followed in this manner for a short time, but it soon undergoes further changes, and often becomes entirely unserviceable in a quarter of an hour.

I afterwards found in aniline red of the above-mentioned (p. 156) strength a new accessory for the demonstration of the axis-cylinder in fresh medullated tubes. Frog's nerves, picked apart and placed in the solution, show in from 4 to 12 hours the beautifully reddened axis-cylinders glistening through the fatty enveloping mass.

Still other methods permit of the recognition of the axis-cylinder in a beautiful manner. Thus, after a longer treatment with strong alcohol or ether, it may be rendered visible in the tubes deprived in this way of their fat. A solution of sublimate and Moleschott's acetic acid mixture also afford good specimens. Chromic acid preparations (or those obtained by means of chromate of potash) present very beautiful appearances. Long, hardened filaments frequently project from the cut extremities (*b f*).

Impregnation with various metals has also been used recently for demonstrating the axis-cylinder. Nitrate of silver either gives it a uniform dark color, or causes it to assume peculiar transversely striated appearance, reminding one of muscular fibre (Frommann, Grandry). The chloride of gold, recommended by Cohnheim (if successfully employed), shows the axis-cylinder shining bright-red through the dark-red medullary substance; it afterwards appears blackened. Osmic acid, on the contrary, very soon blackens the nerve-medulla, while the axis-cylinder remains colorless or only slightly browned (M. Schultze), so that we possess in our reagent an excellent accessory for deciding the presence or absence of the medullary sheath of the peripheral ramifications of the nerves.

We have finally to mention the recognition of the axis-cylinder in transverse sections of previously hardened nerve-trunks; this is also of particular interest in still another regard. If a human or mammalial nerve be immersed for a short time, first in chromic acid solution of 0.2, then in one of 0.5 per cent., it will attain such a consistence as to permit the finest transverse sections to be made with a sharp razor. These, tinged with carmine, are to be deprived of their water by means of absolute alcohol, and, after soaking in turpentine, mounted in Canada balsam. Then, the medulla having become transparent, the axis-cylinder may be recognized as a small reddened circle, surrounded by transparent medulla, which forms a single or multiple circle around the axis-cylinder (a condition to which attention was called several years ago by Lister and Turner, and which it has not as yet been possible to explain), and finally the whole is found to be surrounded by the simple contour of the transversely divided primitive sheath.

Formerly, the axis-cylinder was generally regarded as a homogeneous structure, although there has never been any want of manifold testimony to its more complicated formation. Newer, more conservative methods show that it is with great probability composed of the finest fibres, the axis fibrillæ of Waldeyer or the primitive fibrillæ of M. Schultze (fig. 166). The white substance of the brain and spinal cord serves best for their recognition in medullated nerve-fibres. The fresh object may be examined in blood-serum with very strong magnifying powers, but it is preferable to macerate for a day or more in iodine-serum. Osmic acid ($\frac{1}{2}$–$\frac{1}{8}$ per cent.) renders

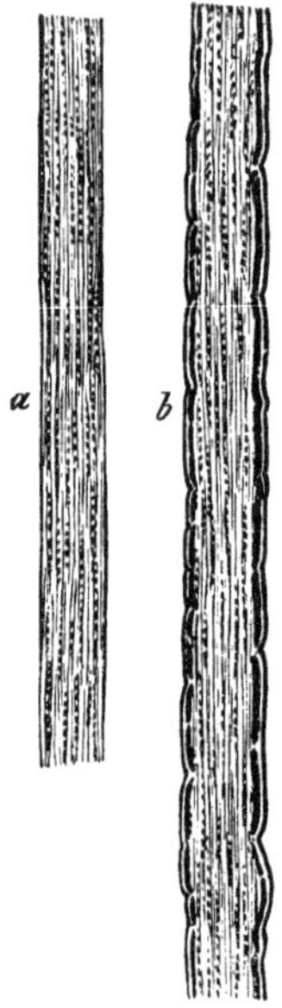

Fig. 166. Fibrillated arrangement of the axis-cylinder after Schultze. *a*, a thick axis-cylinder from the spinal cord of the ox; *b*, nerve-fibre from the brain of the torpedo.

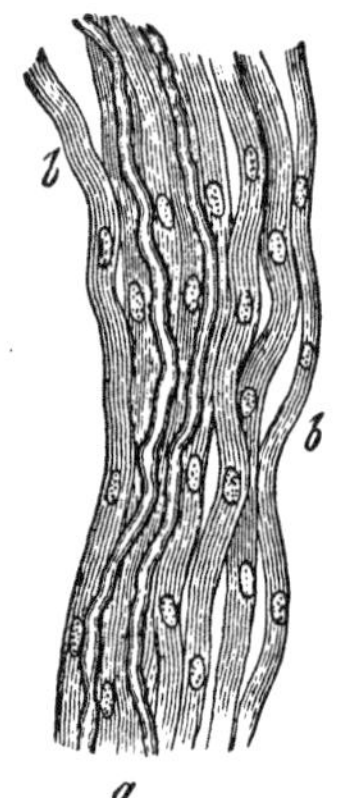

Fig. 167. Sympathetic nerve-branch. Two medullated nerve-tubes (*a*), surrounded by numerous fibres of Remak (*b*).

excellent service. After a short action the axis cylinder becomes sufficiently hardened without any granular opacities, and shows the longitudinal markings very distinctly, especially when freed from the medullary sheath (Schultze); although this sheath does not form any obstacle to the recognition of the axis-cylinder.

Medullated tubes are not shown by all the nerve-trunks in man and the mammalia, however. The fibres of the olfactory nerve (fig. 165 *e*) all appear pale and nucleated, and by proper treatment may be resolved into a bundle of the finest primitive fibrillæ. In the ramifications of the sympathetic nervous system of man and the higher vertebrata, there also occurs intermingled with medullated nerve-tubes a system of pale, nucleated fibres, which bear the name of Remak's fibres, after their discoverer, Remak (fig. 167 *b*). Considerable controversy has arisen as to their nature, whether nervous or of connective tissue; but at the present time there is no further doubt as to their nervous constitution. Indeed, in the earlier embryonic periods, all nerve-tubes appear pale, non-medullated and nucleated. Finally, in the lower vertebrata, all the nerve-tubes may remain through life at this stage of development, as, for instance, in the lamprey, of which such a nerve is represented by our fig. 165 *d*.

For the examination of these pale, nucleated fibres, the fresh tissue may be employed, with picking and perhaps the addition of a weak acid. A longer immersion in very dilute acetic acid (about 20–50 ccm. water, with a few drops of hydrated acetic acid) is preferable. A maceration in weak solutions of chromic acid and of chromate of potash, of the degrees of concentration given by Schultze (comp. above p. 131), also produces very beautiful specimens. Chloride of palladium was also recommended by Bidder. One of the ordinary tingeing methods may be employed for demonstrating the nuclei.

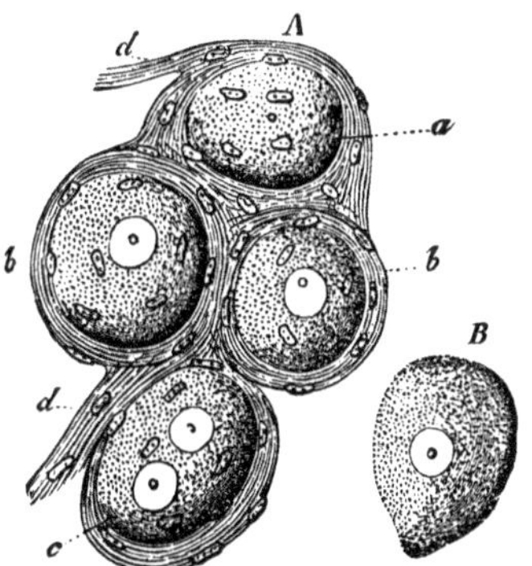

Fig. 168. Ganglion-cells of the mammalia. *A*, cells with connective-tissue envelopes, from which spring Remak's fibres *d d*; *a*, a cell without a nucleus; *b*, two single nucleated ones; and *c*, one with two nuclei; *B*, a ganglion body without an envelope.

The examination of the nerve-fibres in polarized light shows the interesting fact of a double refracting positive sheath, and a likewise double refracting but negative medulla. The long axis of the primitive fibres and the optical axis coincide. Valentine, to whom we are indebted for this interesting result, remarks that we may thus, with the aid of the polarizing apparatus, distinguish the medullated from the non-medullated nerve-tubes.

We have now to speak of the examination of the second

elementary form of the nervous system, the ganglion cells (Figs. 168, 169.)

These appear as cells of considerable size, although subject to great variation in this regard, with large, round, vesicular nuclei and a rather thick, very finely granular, sometimes colorless, sometimes pigmented cell-body, which usually hardens into a thin rind at its periphery. Accessory envelopes are found covering these ganglion bodies in the peripheral ganglia, and are either (as is generally the case in the lower vertebrates) a homogeneous membrane or a thicker, nucleated, connective-tissue substance, which shows numerous nuclei imbedded in it, and not unfrequently runs out into thread-shaped processes presenting the appearance of Remak's fibres.

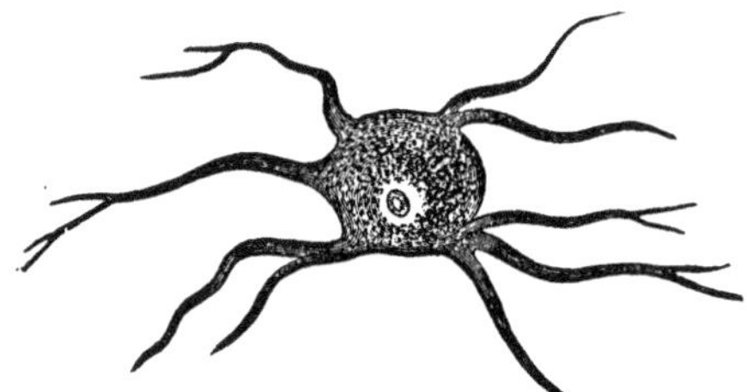

Fig. 169. A multipolar ganglion-cell from the gray substance of the human brain.

An epithelium-like lining on the inner surface of these envelopes is of interest. Nitrate of silver, or the method of impregnating with gold indicated by Gerlach (p. 165), may be employed for demonstrating the latter.

The first incomplete view of the ganglion bodies may be obtained either by selecting a small ganglion, for example, a spinal ganglion of a frog or a mouse, and carefully picking it apart with sharp-pointed needles, with the addition of an indifferent fluid medium, or a thin section from a larger fresh ganglion may be subjected to the same treatment.

Naturally, by this procedure, numerous divisions of the connection take place, and the absence of a sufficient insight into the arrangement of the whole is felt. To obtain these, places should be selected iu small creatures where microscopic ganglionic swellings occur on fine nerve-branches, which may be viewed in their totality without preparatory manipulation. For this purpose the frog stands in the first line. The small embedded ganglia, frequently consisting of only a few cells, which may be recognized on the cardiac nerves in the septum of the ventricles, or on the branches of the sympathetic system, afford admirable specimens. Schwalbe praises the spinal ganglia of the lizard. A very dilute acetic acid may here be used with advantage. Strongly diluted phosphoric acid has also been recommended for this purpose; likewise (although less suitable) very weak solutions of potash and soda (Schwalbe).

The relation of the nerve-fibres to the ganglion bodies is of great importance. As is known, the opinions of investigators have undergone great changes in this regard of late years, and even at the present time we are far from meeting with coinciding or even similar views.

Although it was at first considered that there was but a simple juxtaposition of both elementary forms in a ganglion (Valentin), connections of the ganglion-cells with the nerve-tubes were afterwards frequently observed (Wagner, Robin, Bidder, and others), and the doctrine of bipolar, multipolar, unipolar, and apolar ganglion cells founded. This is not the place to test the correctness of these various opinions, and we must refer on this point to the text-books on histology.

The several animal groups are of very unequal service for ascertaining the origins of such fibres by means of picking. The scanty admixture of a soft, loose connective tissue with the nervous elements of a ganglion facilitates the acquisition of this knowledge very much. A more plentiful admixture of a firmer, interweaved connective-tissue formation either renders the isolation difficult to a high degree, or makes it entirely impossible. The cartilaginous fishes (rays), therefore, form extremely favorable objects in the former regard, and many osseous fishes are, at least, serviceable. The bodies of the naked amphibiæ are less suitable, and the ganglia of man, the mammalia, and birds are scarcely to be mastered with the preparing needles.

Suitable ganglia, for example, those of the trigeminus, vagus, and the spinal nerves of the pike and the burbot (Gadus lota) may be picked apart either fresh or, which may not be called unserviceable, a few hours, 10–15, after death. A preparatory maceration for a day in dilute chromic acid (0.1–0.5 per cent.) may also be employed. We also recommend the trial of a method indicated by J. Arnold, which affords good results, at least with the frog. The ganglion is to be placed for four or five minutes in acetic acid of 0.3–0.2 per cent., and then for twelve to forty-eight hours in a 0.02–0.01 per cent. solution of chromic acid. The preparatory treatment with a very dilute solution of chloride of gold (0.005 per cent.) has also been used (Bidder). Notwithstanding every precaution, numerous fractures and lacerations are unavoidable.

Hardening in chromic acid or in chromate of potash may also be used with the higher vertebrates. In such cases one should begin with weak solutions of the acid, of 0.2–0.5 per

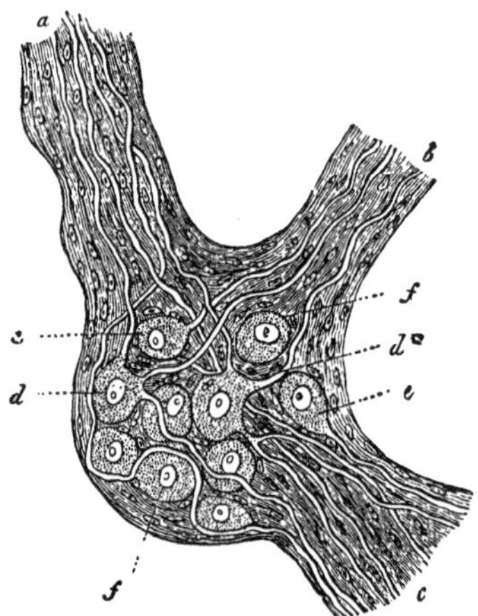

Fig. 170. A sympathetic ganglion of a mammalial animal. The medullated nerve-tubes of the three trunks, *a b c*, pass through a profuse, nucleated fibrous tissue (Remak's); *d*, multipolar ganglion-cells; at *d**, one with a dividing nerve-fibre; *e*, unipolar, *f*, apolar cells.

cent., change them frequently, and gradually increase the concentration. The chromate of potash is employed in corresponding quantity (comp. p. 139). The ganglia thus hardened permit of very thin sections being made with a sharp razor, which are to be examined in glycerine diluted with water. One will thus be able to recognize appearances, for instance, in a sympathetic ganglion of a mammalial animal, which come near to our fig. 170, which is indeed drawn somewhat diagrammatically. Multipolar cells (*d d*) are, as it seems, of very frequent occurrence in the sympathetic ganglia of the mammalia, in contradistinction to those of the lower vertebrata, in which bipolar and unipolar constitute the rule. The ganglion-cells of the sympathetic of the rabbit and the Guinea-pig appear to have two nuclei.

More recently we have become acquainted with other suitable methods. These sections of chromic acid preparations may be placed for 12–24 hours in a solution of osmic acid (1 per cent.), in which the nerves become blackened. The solution of the chloride of palladium (1:500) is still better, however, because it hardens and colors at the same time. Even after 24 hours (if the fluid has been changed in the mean time) the ganglion may show a blackish-gray color and be ready for preparation. If the cut surface is still yellow, a further action till a following day is then sufficient. Very instructive appearances result, as the connective tissue is pale, the ganglion-cells yellow-brown, and the nerve-fibres blackish (Schwalbe).

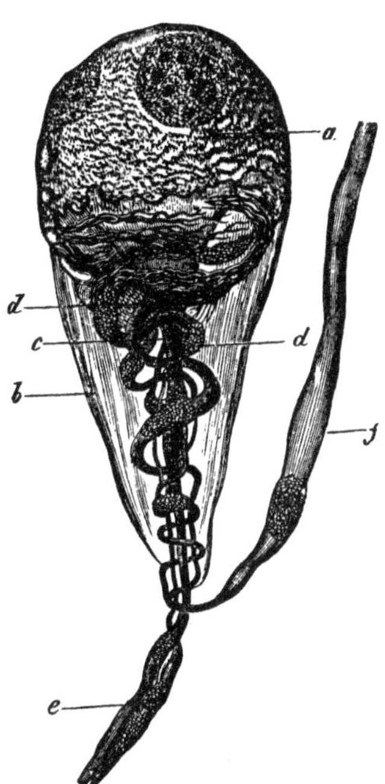

Fig. 171. Ganglion-cell from the sympathetic of the hyla or green-tree frog (after Beale). *a*, cell-body; *b*, sheath; *c*, straight nerve-fibre; and *d*, spiral fibre; continuation of the former, *e*, and of the latter, *f*.

These hardened ganglia may be examined in still another way. The sections are to be stained, then deprived of their water by means of absolute alcohol, and oil of turpentine added. If the brain of a small mammal, a rabbit, or a Guinea-pig, be thoroughly injected from the arch of the aorta with carmine gelatine, the Gasserian ganglion, after delicate staining with carmine, affords excellent specimens of this kind.

Within a short time another interesting structural condition has been observed in the ganglion-cells of the sympathetic of the frog (fig. 171). From the cell (*a*)—and from the central portion of its body—arises a straight fibre (*c e*) (axis-cylinder), in which a nuclear formation is not unfrequently remarked. These are surrounded by one or several spiral fibres, which likewise present nuclei (*d*). They originate from the surface of the cell-body.

This was the condition which Beale found in glycerine preparations tinged with carmine. Arnold, an able investigator, who made use of the method* mentioned at p. 340, states that both varieties of fibres arise from the nucleolus of the ganglion-cell. I could not convince myself of this, and am inclined to regard Beale's spiral fibre as elastic. Nevertheless, by this the possibility is not to be denied that, in the bipolar ganglion-cells, where both nerve-fibres arise close to each other, the one may surround the other with convolutions.

Remarkable ganglionic apparatuses of microscopical fineness have recently been discovered in the walls of the abdominal viscera.

* In a second article the author communicates new and more complicated methods for the investigation of these ganglion-cells. To isolate the spiral fibres in the greatest possible length, immerse them in 5 ccm. of nitric acid of from 0.01 to 0.02 per cent. Even after five to ten minutes the structure of the ganglion-cell becomes clear. After an immersion of from twelve to twenty-four hours, however, these fibres may be followed very far into the nerve-trunk, and be seen to become true nerve-fibres. Chloride of gold also colors both varieties of fibres, the straight as well as the spiral. Immerse the tissue in 4–5 ccm. of a 0.02–0.05 per cent. mixture of 1 per cent. acetic acid and chloride of gold and potassa. When the first traces of a violet color make their appearance (after about three to four hours), the main trunk of the sympathetic is to be placed in 10 ccm. of acetic acid of the above-mentioned strength. The color has become intense in from four to five days, and the connective tissue remains light and loose. A microscopic preparation, to which acidulated glycerine has been added, is now to be placed on a white surface, and exposed to the action of day or sun light for the further reduction of the gold. After four to five days the straight nerve-fibre has become bright red; the thickest of the spiral fibres also present the same appearance, while the finer ones only gain an intense color on the eighth or tenth day.

To these belong the ganglia in the submucous connective tissue of the digestive apparatus (fig. 172), discovered by Meissner,

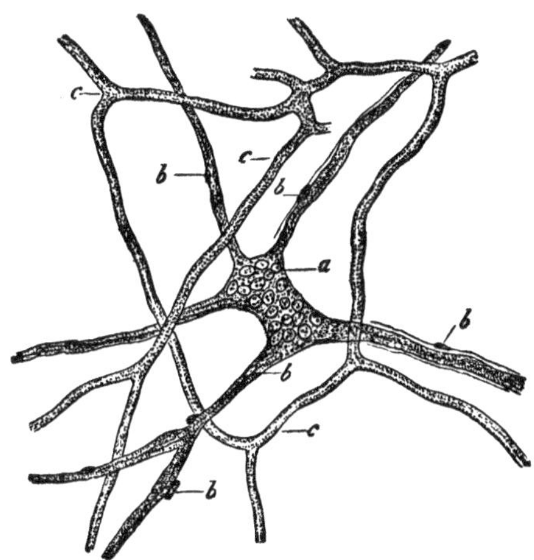

Fig. 172. A ganglion from the submucous tissue of the intestinal canal of a child ten days old (pyroacetic acid preparation). *a*, ganglion; *b*, its radiating nerve branches; *c*, injected capillary network.

and then investigated by Remak, Manz, Kollmann, Billroth, and others. Likewise the plexus myentericus, pointed out by Auerbach, a highly developed ganglionic network between the two muscular layers of the intestinal canal.

The examination of these submucous nerve-ganglia has for the most part been made with the aid of maceration in pyroacetic acid. But many observers (Billroth, for example) have committed the fault of allowing this reagent to act in much too energetic a manner, and were therefore able to describe artefactions only. Portions not too large, taken from the fresh body, are to be placed in purified pyroligneous acid diluted with one or several times its volume of water. After one, two, or three days the examination should be attempted on vertical sections, or the isolated submucous tissue (likewise surface sections of the latter made with the scissors), in order to recognize the horizontal ramifications.

A certain attention is always necessary in this case, because exactly the proper degree of maceration must be employed in the investigation, and an excessive action of the pyroacetic acid soon follows. The pyroacetic acid may also be replaced with very dilute acetic acid. One may also succeed, for example, in the new-born child, in demonstrating this ganglionic plexus

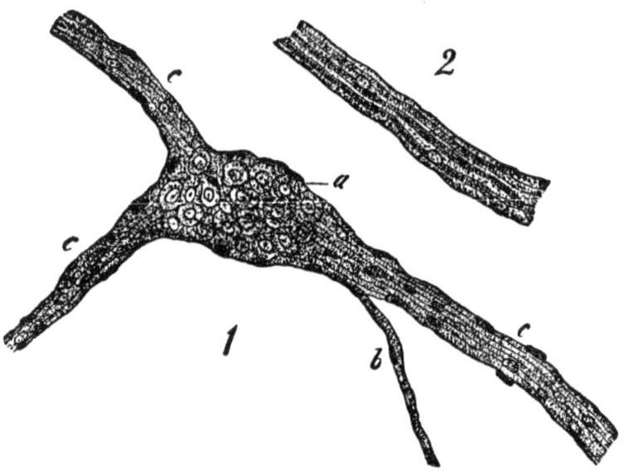

Fig. 173, 1. A large ganglion from the small intestine of a child 10 days old. *a*, the ganglion, with the ganglion cells; *b c*, efferent nerve-branches, with pale nucleated fibres in a fresh condition. 2. Such a nerve-trunk from a boy 5 years old, with three pale primitive fibres, treated with pyroacetic acid.

(fig. 173, 1) in the fresh intestinal canal, with the cells (*a*) and the pale nerve-fibres (*b c*). Fine vertical sections may be employed, or (which has proved to be more suitable) a portion of the intestine may be well stretched, and the muscular layer and the mucous membrane carefully dissected from both sides, so that the submucous connective tissue remains isolated. Even without any further addition, one may, with some pains, discover a few ganglia, but as soon as the connective tissue has been rendered transparent, by means of very dilute acetic acid, the whole arrangement may be readily seen. Simple chromic acid preparations also frequently afford good specimens, at least in vertical sections.

The ganglionic plexus mentioned has been, in an almost incomprehensible manner, asserted to be a capillary network. All doubt may be removed by injecting a portion of intestine with Prussian blue or sulphate of baryta, and immersing it in pyroacetic acid (fig. 172, *c*).

The plexus myentericus (fig. 174) of the larger mammalia

and of man is only to be recognized with difficulty and pains, in consequence of the thickness of the muscular coat. Macerations in dilute pyroacetic acid or acetic acid, also appear to constitute the best accessories. The demonstration succeeds very

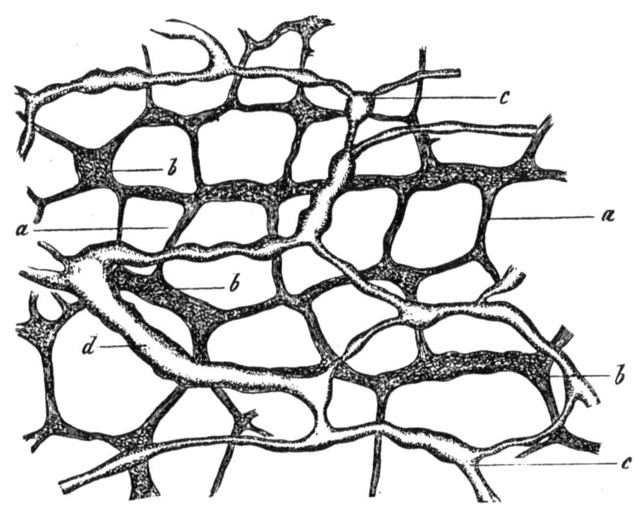

Fig. 174. Plexus myentericus, from the small intestine of the Guinea-pig. *a*, nervous plexus; *b*, ganglia; *c*, lymphatics.

readily, on the contrary, with smaller animals, such as rabbits, but especially with Guinea-pigs, rats, and mice. A portion of the small intestine, still better of the colon of a Guinea-pig, immersed in purified pyroacetic acid diluted with several times its volume of water (20 to 15 per cent.), will after 24 hours (or even sooner) have acquired such a swollen and macerated condition as that the mucous membrane may be readily removed. If the thin muscular and serous layers be now immersed in watery glycerine, and placed under the microscope with a low magnifying power, a surface view of the whole elegant nervous apparatus (*a b*) may be at once obtained. Here, as with the ganglia of the submucous layer, care should be taken to have the action of the pyroacetic acid too weak rather than too strong, as otherwise the appearance of the cells and nerve-fibres is entirely altered. In all these cases the acid contained in the pyroacetic acid solutions should be more accurately determined by the method of titrition.

The structure of the central organs of the nervous system, the spinal cord and brain, is so complicated and obscure, and at the same time the subject of such frequent controversy, that it would lead us far beyond the limits of this book if we were to enter at all into the details of these textural conditions. We therefore limit ourselves chiefly to the description of the methods of investigation generally employed at present.

These may be divided into two series: first, such as are intended for isolating the elementary structures; and then the others, which are to harden the central organs to such a degree as that thin sections may be conveniently made from them, and thus an appreciation of the entire arrangement obtained. It is hardly necessary to remark that a thorough promotion of our knowledge requires the combination of both these methods of investigation.

The older investigators frequently attempted to examine the ganglion-cells and nerve-fibres in picked preparations of portions of brains and spinal cords, which were as fresh as possible, as well as of those which were not so fresh. The delicate nervous structures are too intimately united by the connective-tissue framework, however, for one to hope to find more than their fragments. And, in fact, we have more recently discovered much better and more productive methods. The highly diluted solutions of chromic acid, and of the bichromate of potash, recommended by Schultze, constitute accessories of the first rank, as they exert a partly macerating, partly hardening effect on the various elements of these organs, without producing any deeper textural changes.

The reader would deceive himself, however, if he were to regard the successful application of the solutions as a relatively easy affair. Even after following certain directions, selecting only fresh, preferably still warm organs, and especially those of the larger mammalia, such as the ox and calf; furthermore, not to place too large pieces in a relatively small quantity of fluid; there are, nevertheless, many difficulties still remaining. Next arises the problem of hitting upon the proper degree of concentration of these fluids, and this, lying within pretty narrow limits, requires careful experiment, as further differences present themselves according to the warmth, nature, and age of the animal. Solutions which, to the ounce of water, contain more than 0.1–0.125 gr. of the chromic acid, or more than 2 grains of its potash salt, are absolutely objectionable. Even much higher degrees of dilution are frequently employed with advantage.

Let us listen to Deiters, the most competent of the modern investigators in this department of technology. He recommends the immersion at first in a solution of chromate of potash, which contains 0.5 gr. to the ounce, till the second day, whereby the desired result is not unfrequently obtained. If the latter has not yet taken place, or if it be desired to preserve the preparation for a few days longer, this solution may be doubled in strength for an additional day, and then increased to 2 grains for another twenty-four hours. Not unfrequently weaker than half-grain solutions are to be preferred. Thus, one may commence with 0.125 and 0.25 gr., and then terminate with 0.5 gr.; or solutions of chromic acid may be first employed, and then its salt, whereby greater looseness of the preparation is obtained. The strength in which the chromic acid is employed is from 0.033, 0.05 to 0.1 gr. to the ounce. Two days are to be allowed to pass without changing the fluid; on the third day it is to be renewed. Now, occasionally even sooner, a very good degree of maceration for many parts is obtained. For the combination of both these methods it is recommended, after using the chromic acid for two days, to place the piece in a 0.5 gr. solution of chromate of potash, then on the following day in one of 1 gr., afterwards perhaps increasing the strength to 2 grs. After this, to obtain a considerably macerated condition of the framework substance, such objects may be advantageously exposed to the action of extremely dilute solutions of the alkalies. A drop of a 28 per cent. solution of caustic potash may be added to the ounce of water; after having been exposed to this for an hour, the preparation is to be removed and washed (in dilute chromic acid, for instance), then replaced in the solution of chromate of potash, which should contain at first 0.5, the following day 1, afterwards perhaps 2 grs. to the ounce.

The simpler of these methods, or a combination of them (it

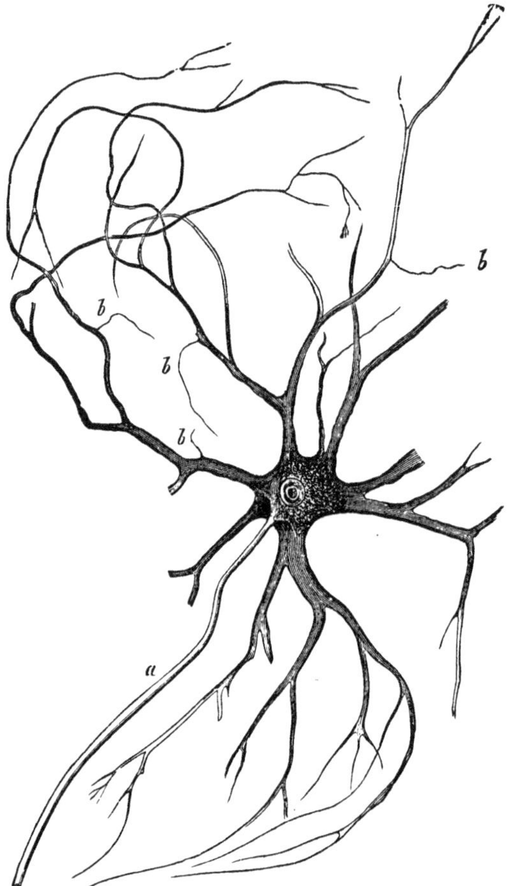

Fig. 175. Multipolar ganglion-cell from the anterior horn of the spinal cord (of the ox), with the axis-cylinder process (*a*), and the branched protoplasma processes, from which at *b* the finest filaments arise (after Deiters).

is better to employ several of them simultaneously), will, together with many failures, also afford suitable objects for examination, although they only continue to be serviceable for a few days. It is best to remove a small portion of the tissue with the point of a knife, and pick it apart as carefully as possible.

With such accessories Deiters succeeded in making a remarkable discovery concerning the multipolar ganglion-cells of the central organs. They (fig. 175) have two kinds of processes. The greater proportion of the latter are only continuations of the same protoplasma-like substance which is presented by the body of the ganglion-cell. These processes, the "protoplasma processes" of Deiters, divide into manifold ramifications, until at last they disappear as the finest terminal branches in the supporting substance. At the first glance an exceedingly long process (*a*) may be distinguished from the protoplasma processes, which arises either from the cell-body itself or from one of the broadest of the former processes; it never presents any ramifications, and is afterwards covered by a medullary sheath. Deiters has called it the "axis-cylinder process." Finally, one may also recognize, passing off at right angles from the protoplasma processes of our multipolar ganglion-cells, extremely fine filaments (*b*, *b*), which the above-mentioned investigator believes to be a second system of extremely fine axis-cylinders.

Still another method has recently been recommended by Gerlach, an investigator who has accomplished much for microscopical technology, for isolating these ganglia and a very fine nervous network (that is, their protoplasma processes) connected with them (of which, according to his views, the gray substance of the spinal cord consists). Thin longitudinal sections are to be made with a razor, preferably through the region of the anterior horns of the quite fresh and still warm spinal cord of a mammalial animal. These are to be placed for 2–3 days in very weak solutions of the bichromate of ammonia (0.01–0.02 per cent.). They are then to be placed in a likewise extremely dilute ammoniacal solution of carmine, which produces the necessary tinge in about a day. Those portions which are thinnest and best tinged are then to be carefully picked apart.

Still another complication of the structure of these ganglion-cells of the central organs has been observed. According to

Schultze's investigations, both varieties of the processes of the central ganglion-cells (fig. 176) present a fibrillated structure; this is most distinct in the axis-cylinder (*a*), however, while in the protoplasma processes (*b*) there is a greater quantity of

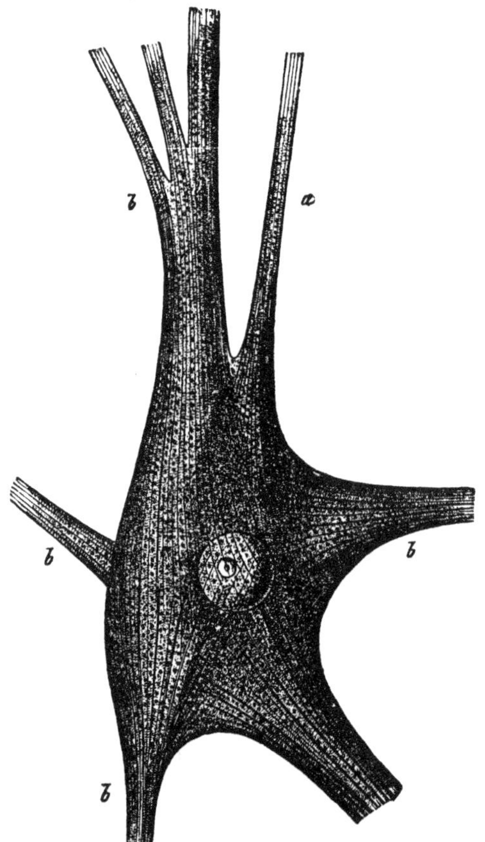

Fig. 176. Ganglion-cell from the anterior horn of the spinal cord of the ox, after Schultze. *a*, axis-cylinder; *b*, cell-processes.

a granular intervening substance. The "primitive fibrillæ" (p. 335) may be followed into the body of the ganglion-cell, where they are seen to have a complicated course. This arrangement, first noticed by Remak, may be observed without difficulty in fresh specimens simply moistened with serum, or in osmic acid preparations.

Frommann asserts that he has ascertained, by treatment with nitrate of silver, that these fibrillæ originate in the nucleoli and are surrounded like a sheath by tubes which proceed from the nuclei. Arnold also informs us of similar results. He used serum or chromic acid (0.01 per cent.) and chromate of potash (0.02–0.05 per cent.) as fluid media.

These points will have to be decided by future investigations. It was supposed, many years ago, that the nerve-fibres originated in the nucleolus and nucleus of the ganglion-cell (Harless, Axmann, Lieberkühn, Wagener).

The art of giving the substance of the brain and spinal cord a consistence suitable for making sections has been possessed for many years.

Alcohol, the solutions of chromic acid and of the bichromate of potash, and ammonia are used for hardening. Whether the one or the other of these fluids is used, only pieces of the brain and spinal cord which are quite fresh and carefully removed from the recently killed animal and deprived of their envelopes, should ever be immersed in them. The pieces should not be too large. If their volume be too large, the exterior may acquire a good consistence, but the central portions will, on the contrary, remain soft or even become decomposed. I would recommend, as a useful method, to allow such pieces to float in a tall glass cylinder, suspended by means of a silk thread to a hook in the glass stopper.

Among the reagents mentioned, alcohol takes the lowest place. Preference has therefore, for many years, been given to solutions of chromic acid and the chromate of potash. Here we must also censure the custom of estimating the strength of these solutions only by their color. Now and then the proper degree of concentration may be obtained in this way; but in many cases one will be deceived and fail in obtaining the desired result, which might have been accomplished with less trouble by making an accurate solution.

Now, what degree of concentration should be given to such solutions? Here it must be remembered that the fluids used by former observers were generally much too strong, so that a considerable shrinking of the tissues took place, and not unfrequently the whole became entirely too brittle to permit of a section being made. A chromic acid solution of 1 per cent. is certainly too strong for commencing the hardening. I have obtained good results with mammalial animals as well as with cold-blooded vertebrates, such as fishes and frogs, when I commenced the hardening with solutions of 0.2 per cent.; then changed the chromic acid after a few days, replaced it by a stronger solution, and thus finally arriving at those of 1 per cent. Chromate of potash is to be used in the corresponding strength of 2–6 per cent. (comp. p. 139). Deiters employed the following method for hardening the brain and spinal cord: He first immerses the preparation for one or two weeks in a solution of the chromate of potash (15 grains to the ounce of water). Then (when a uniform saturation has taken place and the hardening has commenced) the preparation is placed either immediately or after washing out the potash salt, in a solution of chromic acid which contains 2 grains to the ounce and may be increased to 3 grains.

Gerlach recommends the use of a 1–2 per cent. solution of the bichromate of ammonia for at least fifteen or twenty days, occasionally for five weeks.

It is impossible to state anything definite with regard to the time which is generally required for the hardening. Chromic acid salts act more slowly, the free acids more rapidly. I have often found that the spinal cord of small animals has become sufficiently firm in a week in these solutions of the free acid. As a rule, a space of three to four weeks, not unfrequently still longer—six weeks or more, is necessary. There is great variation, however, in this regard. Reissner is therefore correct in asserting that the central parts, especially the spinal cord of the various kinds of animals, also show differences in regard to the time which is requisite to harden them. It is generally stated, it is true, that the hardening takes place more rapidly with small animals than with those which are larger. This is, however, by no means universally the case, as, according to his experience, the spinal cord of the calf becomes hard in weaker solutions than does that of the rabbit, the mouse, or the rat.

To ascertain the proper degree of consistence, nothing further is necessary than to make a trial section with the razor from time to time. The hardening should have progressed in exactly such a manner as that with the moistened blade, a very thin layer may be removed conveniently and without crumbling. If the tissue crumbles it is already overhardened; if the consistence is insufficient, only thick sections are possible. In the latter case a further immersion is necessary; in the former, the procedure is a failure.

If one has been so successful as to have obtained the proper consistence, the hardened object is to be placed, after having been washed, in weak watery alcohol. It may be preserved in this for a long time without undergoing any changes, to serve for future investigations.

With some practice and a good razor one soon learns to make marvellously thin sections. In doing this, the object is to be held with the points of the first three fingers of the left hand, and care is to be taken to keep the preparation and the blade sufficiently moistened with alcohol. It is impossible to hold very small objects, such as the spinal cord of a mouse, for instance, with the points of the fingers. They may be placed in a larger animal body, the spinal cord of a larger animal for instance, or in a piece of elder pith; or an embedding method (p. 117) may be employed.

Sections made in all directions are naturally necessary, in order to construe from the various preparations the structure of such central parts as the spinal cord, for example. Transverse sections are first made, then longitudinal ones, of which vertical and horizontal longitudinal as well as oblique (that is, such as are made, for instance, from the right posterior cornua towards the left anterior cornua) sections are of importance. Those which are made oblique to the long axis of the cord appear to be of less importance.

For most examinations it is advantageous to tinge the sections thus obtained. Staining with carmine is nowadays generally employed in these cases.

Here, as with all delicate tissues, I employ a solution of carmine obtained with a minimum of ammonia, which is to be considerably diluted with water, and then an equal volume of glycerine added. The sections are to be washed out in very dilute alcohol, to free them from any chromic acid which may be present, and then placed in the solution to acquire the desired redness, for which two, four, eight to twelve hours is necessary, according to the concentration of the coloring medium.

The object is next to be placed for a short time in clean water to wash out the superfluous carmine, then in water very slightly acidulated with a few drops of acetic acid, or in dilute alcohol thus acidulated. The diffuse redness disappears, and the remaining carmine is then fixed to the cells, nuclei, and axis-cylinders. If individual differences occur with regard to the capability of imbibition of the tissue elements of the brain and spinal cord, the epithelium, ganglion bodies, axis-cylinders, and the nuclei of the connective-tissue framework are to be designated as those parts which become especially impregnated with the coloring matter.

Preparations thus treated may be examined in a moist condition. A solution of the chloride of calcium has been recommended for rendering them more transparent (Schröder van der Kolk). I must acknowledge, with Reissner, that I have accomplished nothing with this. Glycerine renders better and satisfactory service in this case.

The preparation may be rendered more permanently transparent, however, by immersing it, after it has been carefully and cautiously deprived of its water, in oil of turpentine or Canada balsam. This method is the most popular at present, and also affords the most beautiful and most durable preparations for a collection. We refer for it to page 207 of our book.*

* We here add several other methods:—

1. Years ago Lockhart Clarke, who was afterwards imitated by Lenhossek, made use of the following process: The fresh spinal cord is to be hardened in alcohol; the first day it is to be placed in such as is diluted with an equal volume of water, then left in pure alcohol until thin sections are possible, a condition which is generally obtained, in the cooler portion of the year, in from 5–6 days. The sections are then to be immersed for 1–2 hours in the mixture

If it be desired to tinge blue, the soluble aniline blue of the strength previously mentioned (see p. 157) affords, after somewhat energetic action, handsome and permanent preparations; the hæmatoxylin (p. 158) is still finer. Osmic acid, which acts in the manner indicated at page 163, not only on such sections, but likewise on such as have been tinged with carmine also, promises to be of importance (M. Schultze).

Several years ago Gerlach praised the treatment with chloride of gold and potash (p. 165) as an excellent means of rendering the course of the nerve-fibres in the spinal cord visible.

Fine transverse sections are made from the spinal cord after it has been hardened by immersion for 3–6 weeks in a solution of the bichromate of ammonia. The sections are to be placed for 10 or 12 hours in a 0.01 per cent. solution of the gold salt to which a little acetic or muriatic acid has been added. Then (when the white substance has gained a pale lilac color, and the gray presents but a slight tinge), the section is to be washed with a to-and-fro movement, continued for several minutes in very dilute hydrochloric acid (1 : 2–3000). Gerlach now places it for about 10 minutes in a mixture of 1 part muriatic acid and 1000 parts alcohol of 60 per cent., and afterwards, finally, for a few minutes in absolute alcohol. Oil of cloves serves to render the section transparent, and it is then mounted in Canada balsam, which terminates this somewhat complicated process. To render the ganglion-cells visible, it is necessary, before immersing the preparation in the gold salt, to employ one of the other methods of metallic impregnation for several hours, such as that with chloride of palladium (p. 165), or, which the author

---

mentioned at page 142, consisting of 1 part acetic acid and 3 parts alcohol. In this, not only the nerves and fibrous elements are rendered more sharply prominent, but the gray substance is also rendered quite transparent.

2. J. Dean, to whom we are indebted for a very superior work on the central organ, hardens in alcohol or chromic acid, and colors the washed sections moderately in glycerine-carmine, in which they remain from 4–8 hours, according to the intensity of color desired. Alcohol, oil of turpentine, and Canada balsam are then employed. According to Dean, thick copal varnish is often preferable to the balsam for the recognition of fine details. Dean also praises Clarke's method highly, and, we will add, very appropriately, if the mixture is allowed to act on tinged preparations.

prefers, to make use of a very dilute solution of a metallic salt which has not been previously employed in histology, the nitrate of uranium.

With industry and perseverance, and by the aid of the methods of preparation which have been given, one may succeed in recognizing the more important textural relations of the spinal cord (those of the brain with greater difficulty), whereby, as was remarked, the examinations should commence with transverse sections. At the same time one will also recognize the great difficulty of accurately describing the textures of the central organs—a difficulty which is founded in part on the nature of the object; in part, also, on the methods of investigation, which still remain insufficient. Many observers have certainly exaggerated the results of their investigations very much, frequently drawing very false conclusions from fragmentary, isolated appearances. Other investigators have, however, become infected with an excessive skepticism. Attempts have even been made to disprove the reticulated commissural communication of the large multipolar ganglionic bodies in the anterior cornua of the spinal cord, as also the continuation of some of their processes into nerve-fibres of the anterior motor root. In our opinion these textural conditions may be observed with entire certainty, although only with difficulty and on very rare occasions.

To obtain injected preparations of the brain and spinal cord, one should proceed in the following manner. One of the smaller mammalia, a rat, a Guinea-pig, a rabbit, or a cat may be selected, and the canula inserted into the commencement of the aorta, after this vessel has been ligated below the carotids and the subclavians. The injection readily succeeds on the fresh body (with some loss of injecting fluid, it is true) with cautious management of the syringe. But to judge of the correct moment for terminating the procedure is somewhat difficult. If a white rat or an Albino rabbit has been used, the complete injection of the eyeball affords a criterion. To inject the upper half of the spinal cord, the aorta is to be ligated as it passes through the diaphragm, and the procedure completed as above mentioned. Deep-red carmine-gelatine constitutes the best injection fluid. Alcohol is used for hardening, and a blue tincture for staining the sections. Prussian blue is to be preferred, if it be desired to accomplish the former with chromic acid.

The blood-vessels in the central organs are loosely enveloped by a connective-tissue adventitia, and, according to the interesting testimony of His, the lymph flows in the intervening space which is thus formed. This vaginal space may be readily injected by means of the puncturing method. A similar vaginal formation may also be seen around the blood-vessels of the pia mater (perivascular space of His).

A further difficulty in the investigation of the central organ of the nervous system is found in distinguishing the connective-tissue framework (neuroglia) from the nervous elements. Although the tacit presumption, that everything which is present in the brain and spinal cord must also be of a nervous nature, was accepted for many years, the extensive occurrence of a connective-tissue substance, in which the nervous-tissue elements are imbedded, was afterwards rightly maintained by Bidder and his followers, although with some exaggeration.

There is repeated in the brain and spinal cord, therefore, one of those undeveloped, reticular connective substances, such as have been frequently observed of late in the human body—one of those reticulations and frameworks with cell-bodies in some of the nodal points.

In the white substance, this is of a more compact structure, and appears in transverse sections as a network with isolated nuclei and round apertures for the reception of the nerves (fig. 177).

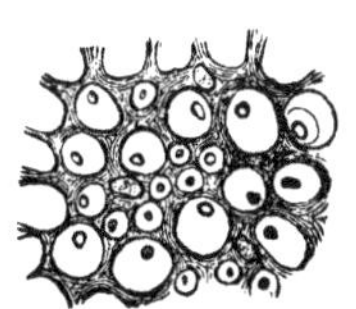

Fig. 177. The connective-tissue framework from the posterior column of the human spinal cord.

The reticular connective tissue in the cortical layer of the white substance passes uninterruptedly into the pia mater, where it is more richly developed and the reticulations become very much finer.

The reticular framework of the gray substance of the spinal cord likewise appears to be extraordinarily delicate, and frequently the meshes are extremely fine. This also appears extremely developed and with distinctly radiated connective-tissue cells in the interior of the so-called central thread of ependyma.

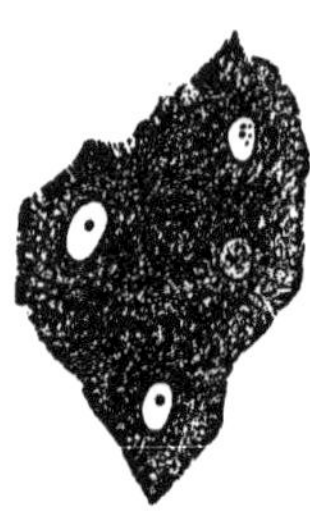

Fig. 178. Porous tissue of the gray substance of the human cerebellum; obtained with extremely dilute chromic acid.

Such a supporting substance certainly occurs in the brain also, although not so well known (fig. 178). In the gray substance of the cortex, the network, which has nuclei in its nodal points, assumes an infinite fineness and delicacy of its fibres and meshes, so that its existence has frequently been totally denied.

Methods of maceration such as were recommended by Deiters (p. 347) serve for the recognition of these framework substances, which are also extremely important for the pathologists. The action of nitrate of silver on segments of the fresh tissue, with a subsequent addition of glycerine, has also been used with success (Frommann); likewise osmic acid.

Gerlach recommends the two methods of treatment with chloride of gold and potash and carmine, mentioned above (p. 355), for distinguishing the neuroglia of the gray substance from the very fine network of nerve-fibres which also occurs there. Only the nervous elements, but not those of the connective tissue, become colored. For the diagnosis of nervous and connective-tissue cells in them, a suitable reagent is still wanting.

Fig. 179. Amyloid corpuscles from the human brain.

Tumor-like new formations of the framework substance mentioned occur in the central organs and in the retina. They have been called glioma (Virchow).

A deposit of peculiar structures, frequently discussed of late, the so-called amyloid bodies, corpuscula amylacea (Fig. 179), takes place in this framework substance after death, as a result of decomposition, and even during life in abnormal conditions.

These, varying in size, occur as globular, ovoid, or double-loaf-shaped structures, in which a distinctly stratified appearance may frequently be recognized under the microscope. In these cases they remind one very much of amylon granules, for which they have also been mistaken. Their reaction may be that of amylon, taking a blue color with a solution of iodine. Others, on the contrary, assume a violet hue from the action of iodine and sulphuric acid (see p. 136), and put one in mind of cellulose.

In addition, let it be remarked that similarly reacting masses may also make their appearance in many other parts of the body, and an amyloid degeneration has consequently been recently assumed.

As we are at present referring to chemical matters, let us also at the same time mention the so-called myeline. It appears under the microscope in the form of double contoured drops and lump-like masses, and is also not limited to the nervous system.

Our fig. 180, in its lower half, may afford us a representation of this optical peculiarity of that substance. The upper portion

Fig. 180. Crystals of cholesterine and deposits of myeline.

of the drawing is occupied by crystals of cholesterine, a peculiar substance, widely spread throughout the animal body (which has also been more recently discovered by Beneke and Kolbe in plants). This cholesterine forms an element of the nerve-substance; it also occurs in the blood, though in small quantity, more abundantly in the bile (and especially in the biliary calculi); likewise in most of the other animal fluids, with the exception of the urine. Finally, it occurs in pathological fluids and tumors, and has the signification of a product of decomposition.

It occurs in very delicate, thin, rhomboidal tables (with acute angles of 79° 30′, also of 87° 30′, and of only 57° 20′), and is thus for the most part readily recognizable. It likewise shows certain characteristic reactions. If a mixture of 5 parts of sulphuric acid (of 1.85 sp. wt.) and 1 part of water be added to the crystals of our substance under the microscope, a peculiar change of colors takes place. The borders of the tablets become carmine red, then, with a commencing dissolution into drops, violet. If iodine and sulphuric acid be applied, pure cholesterine assumes a blue; impure, a violet, reddish, or variegated color. The whole affords a beautiful microscopical image, but is, as a rule, without any practical value, as the shape of the crystals is for the most part entirely sufficient for the recognition of cholesterine.

We have, finally, to mention the methods of investigation which are used at present for the recognition of the terminations of the nerves.

These vary exceedingly, according to the constitution of the part to be investigated, as, together with the examination of the freshest possible and the altered elements, there is also an innumerable quantity of methods employed which vary with the parts of the body.

Let us commence with the manner of termination of the motor nerves, especially those in the transversely striated muscles.

The tissue taken from an animal which has just been killed is to be first employed, as, immediately before the rigor mortis takes place, the muscular filaments present a considerable degree of transparency, which soon gives place to a more cloudy condition. In such investigations the object is to be examined either without any fluid medium, and covered with a thin glass scale only (which may, at the most, be very carefully pressed upon, to give a smooth surface), or with the addition of indifferent fluids. Only certain, especially thin membranous muscles, are adapted for such investigations.

The orbital muscles of small mammalial animals, and among

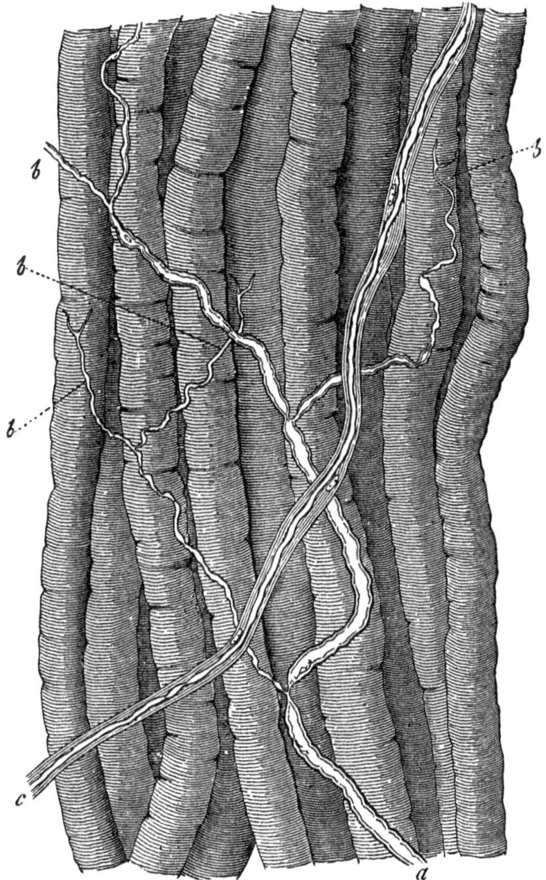

Fig. 181. Radiation of the nerves in the voluntary muscles of the frog. A nerve-fibre, *a*, without a neurilemma, with repeated subdivisions into several smaller branches, *b*, *b*; *c*, a nerve-fibre with a neurilemma of the simplest kind, without division.

these the retractor bulbi (of the cat), also the psoas muscle of these animals, furthermore the flat muscles which pass from the hyoid bone to the lower jaw of the frog, and the cutaneous thoracic muscle of this animal, the very short muscles in the tail of the lizard, etc., may be advantageously employed.

One may also, without trouble, succeed in obtaining appearances like our fig. 181, with suitable preparations, from the frog. The division of the dark-bordered primitive fibres into medullated branches, and their continued splitting up into finer dark ones, may be followed, until at last fine terminal branches appear to end in the muscular filaments. It was believed for many years, in fact, that we had thus penetrated to the ultimate terminal branches.

A series of investigations lately undertaken shows that these earlier views are at all events untenable, and that the nerve-distribution takes place beyond these alleged terminal branches. The results of the observations instituted by Kühne, Margo, Kölliker, Rouget, Krause, Engelmann, and others do not, however, agree. Still, after unprejudiced examination, it can no longer be doubted that the nerve perforates the sarcolemma (whereby its neurilemma becomes continuous with the latter), and terminates beneath it in a nucleated, fine, granular, lamellated substance. The latter, however, pass at their borders and inner surfaces uninterruptedly into the sarcous substance of the muscular filaments (Rouget, Engelmann).

Our fig. 182 shows the terminal structures in question, which have been suitably designated by the name of "terminal plates," from the psoas of the Guinea-pig, at the left in profile, at the right as seen from above. In mammalia, in which they are well developed, the terminal plates have a magnitude varying in the mean between 0.0177 and 0.0267‴. The number of their nuclei fluctuates between 4, 6, 10, and 20.

The terminal plates become more and more simplified in the lower vertebrates.

As has been shown by more recent investigations (Kühne, Engelmann), however, these terminal plates do not represent the entire arrangement. Suitable profile views show that the axis-cylinder becomes divided and spread out into a tree-shaped figure in the external portion of the terminal plate. Beneath it, "like a sole," lies the granular nucleated mass.

Most of the accessories which have thus far been employed are intended to render the whole muscle, or at least a part of it, as transparent as possible, so that the distribution of the nerve-fibres may be better followed than in the unaltered

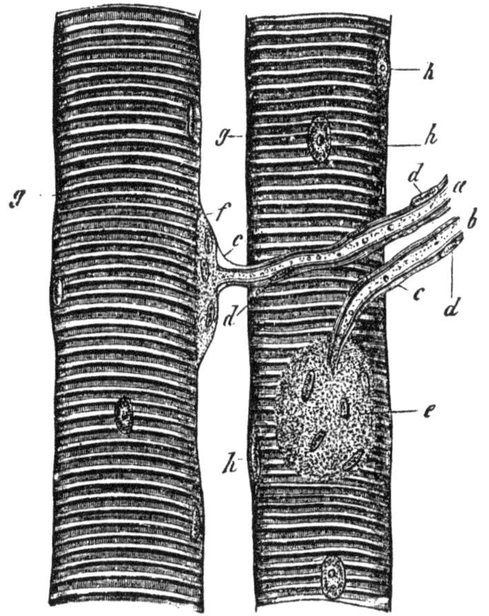

Fig. 182. Two muscular filaments from the psoas of the Guinea-pig. *a b*, the primitive fibres and their continuation into the two terminal plates *e f*; *c*, neurilemma continuous with the sarcolemma *g g*; *h*, muscular nuclei.

tissue; and secondly, to isolate the transversely striated muscular filaments with as little injury as possible, and subject them to examination freed from their interstitial connective tissue.

The alkalies are unserviceable for the former purpose, but various acids, highly diluted, are, on the contrary, very good.

Kölliker recommends the dilution of 8, 12–16 drops of acidum acet. concent. of the Bavarian Pharmacopœia, of 1.045 sp. wt., with water to 100 ccm., and to immerse in it the cutaneous thoracic muscle of the frog for 1½–2 hours, after which time it is said to become transparent like glass. I have accomplished the same result with 1–2 drops of hydrated acetic acid to 50 ccm.

of water. Highly diluted acetic acid also proves very serviceable for the muscles of other animals (Engelmann). I would also give this acid the first rank for such purposes. Muscles which have thus been rendered transparent may be preserved for some time in 1–2 per cent. acetic acid.

Muriatic acid of 0.1 per cent. is likewise a suitable reagent. It produces a similar condition of the muscle in from eight to twelve hours at the ordinary temperature of the room.

The action for twenty-four hours of nitric acid, of the same concentration as the hydrochloric acid, is also serviceable.

Cohnheim recommends the treatment with nitrate of silver, likewise with chloride of gold, in which Krause agrees.

The still living muscular fibres, fortunately isolated by mechanical means, often give the most characteristic appearances.

We have received from Kühne a good method for the further isolation of the muscular filaments (naturally, with as little injury as possible).

He was able, by means of the mixture of nitric acid and chlorate of potash, mentioned above at the muscular tissue, p. 319, to isolate the muscular filaments with the adherent nerve-fibres very handsomely; but it was impossible to ascertain the further distribution of the latter. On the contrary, the treatment which we have also mentioned, with extremely dilute sulphuric acid, and the subsequent digestion in water, constitutes a very suitable process.

Krause recommends, furthermore, the immersion of the muscles for several days in a 33 per cent. solution of acetic acid, and then the isolation of the filaments by careful picking from the swollen connective tissue. He also obtained good preparations by placing them in a 2 per cent. solution of the chromate of potash, with the subsequent action of 25 per cent. acetic acid. He furthermore praises solutions of sublimate of 0.3–0.5 per cent., and the subsequent treatment with the same acid; and finally, sulphuric acid of 0.1 per cent.

The so readily decomposable structures of warm-blooded animals are less to be recommended than those of the scaly amphibiæ for seeing the arborescent distribution of the nerve-fibres in the terminal plates. A lizard or a ring-snake, killed twenty-four hours previously by destroying the central nervous system, presents, with the addition of a 0.5 per cent. solution of chloride of sodium, very characteristic appearances (Engelmann).

It is much more difficult to follow the terminal forms of the nerve-fibres in the smooth muscles than in the transversely striated tissue, and our knowledge is therefore very uncertain on this point. The broad ligaments of the uterus of the rabbit (Frankenhäuser), likewise the urinary bladder and the small arteries of the frog (Klebs, Arnold), are at present considered the most suitable objects for examination. Highly diluted acetic and chromic acids are to be tried here. Klebs recommends a 5 per cent. solution of cane-sugar to which sulphurous acid is added, and the subsequent immersion in phosphate of soda. Frankenhäuser advises highly diluted chromic acid, $\frac{1}{37}$–$\frac{1}{50}$ per cent. and also acetic acid of 20 per cent. Finally, we are in-

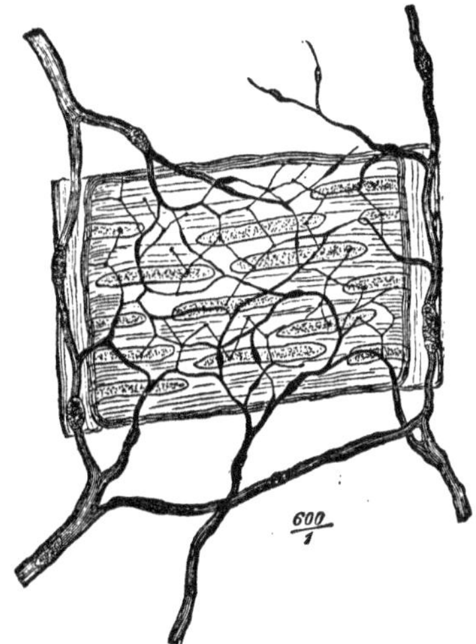

Fig. 183. Alleged nerve-termination in the muscular coat of a small artery of the frog, after Arnold.

debted to Arnold for very accurate directions. The object is to be placed for two to four minutes in 4 ccm. of acetic acid of 0.5–1 per cent., and then for half an hour in the same quantity of chromic acid of 0.01 per cent. This investigator found the gold method, and likewise the treatment of transverse sections of frozen muscles with chloride of gold and chromic acid solutions, advantageous.

According to the investigations of Frankenhäuser and Arnold, however, the termination is very peculiar. These nerves form manifold networks. A secondary plexus of this kind lies close to the muscular layer (fig. 183). It consists of fine, pale, nucleated filaments. Still finer fibres spring from it to form a new network with small meshes, the extremely fine terminal fibrillæ of which are said to end in the nucleoli of the contractile fibre-cells. This condition may be represented to the reader by the above figure. The correctness of this statement has of late, however, again become very questionable. Engelmann, in a re-examination, was unable to see any trace of this manner of termination—and we have been equally unfortunate.

The nerves of the cornea of the eye, which have been formerly and recently frequently examined, present interesting objects to the microscopist.

Soon after entering the cornea at its periphery they lose their medullary sheath, become pale, and form a plexus which permeates the corneal tissue, of very fine fibrillæ with nucleated enlargements at their nodal points. Nerve-fibres pass from this network in two directions: those which proceed in the one direction terminate in the corneal tissue itself; while the others, after perforating the anterior homogeneous limiting layer (Hoyer), find their terminus in the epithelium (Cohnheim).

The cornea removed from an animal immediately after death, is naturally to be employed for examination. One may here, for example, with the cornea of a frog, proceed in the following manner (Kühne): The point of a knife-blade is to be inserted near the sclerotic border, and the aqueous humor drawn out with a pipette; the cornea is to be rapidly separated with a pair of fine sharp scissors and placed on a slide, with the small quantity of aqueous humor from the pipette. The whole is then placed in the previously (p. 103) described moist chamber, to remain for hours under the microscope, and at the same time to gradually reveal not only the course of the nerves, but also many other remarkable things in the most conservative manner.

The procedure mentioned may also be used, with slight modifications, for other animals. The corneæ of the smaller animals, the mouse, the rat, and the squirrel, are generally to be recommended. They are to be removed in connection with a narrow zone of the sclerotic, and it will generally be necessary to make several incisions in the direction of the radii.

If one desires to employ reagents, the highly diluted acetic acid recommended by Kölliker for the nerves of the muscles (p. 363) is advisable (Müller and Saemisch). Even after 10–15 minutes the epithelium may be removed with the forceps; while at least several hours' action of the reagent is necessary for the examination of the nerves. The action of very dilute chromic acid (0.1–0.01 per cent.), to which 0.25 per cent. of chloride of sodium may be added, is also advantageous, at least with the frog (Kühne).

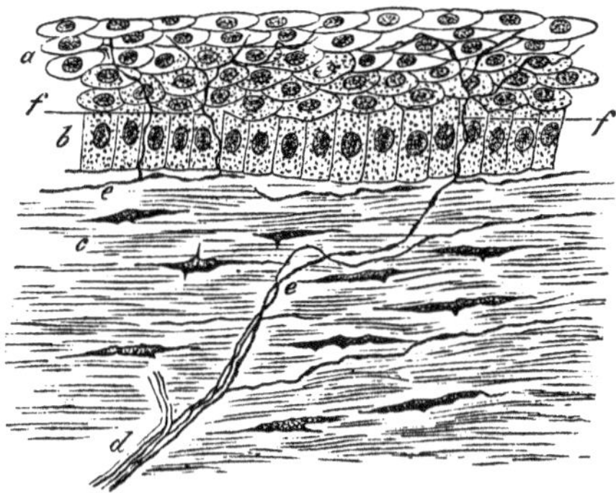

Fig. 184. The cornea of the rabbit in vertical transverse section, after treatment with chloride of gold. *a*, the older, *b*, the young epithelial cells of the anterior surface; *c*, corneal tissue; *d*, a nerve-branch; *e*, finest primitive fibres; *f*, their distribution and termination in the epithelium.

To recognize the penetration of the (extremely fine) nerve-fibres into the epithelium of the conjunctiva, and thus to verify Cohnheim's beautiful discovery, one should resort to chloride of gold (p. 163) and use the eyes of Guinea-pigs and rabbits (fig. 184). The frog's cornea shows its epithelial nerve-distribution even without any reagents, by simply remaining in the moist chamber (Engelmann).

We thus become acquainted in the most certain manner with the penetration and termination of the finest nerve-fibres in an epithelial layer.

Many additional observations of a relative nature have been made in modern times.

Thus, Hensen informs us that in the tail of the frog's larva he saw the terminal branches of the cutaneous nerves terminate in the form of infinitely fine filaments in the nucleoli of the epidermoid cells. Lipmann recently asserted the same with regard to the pavement epithelium of the posterior surface of the frog's cornea. He employed the chloride of gold. These observations (which would prove a parallel case with the contractile fibre-cells) still await confirmation. Langerhans—again with the help of the gold method—found that in the human cutis branches of pale nerve-fibres penetrate between the cells of the rete Malpighii; here they probably enter small radiated cells, the ascending processes of which are said to terminate with slight enlargements beneath the stratum corneum.

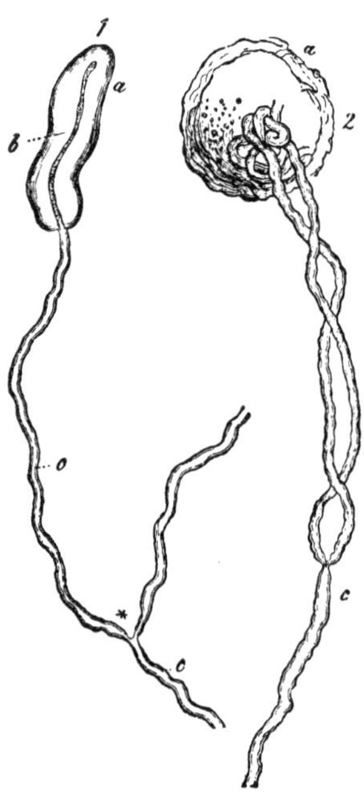

Fig. 185. Terminal knobs—1, from the conjunctiva of the calf.
a, knob; *c*, the medullated nerve-fibre, dividing at *; *b*, its pale terminal portion. 2, from man.

To obtain the first view of the numerous nerves of the dental pulp, one of the large incisor teeth of the rabbit is to be crushed and the examination made in iodine serum. The finest terminal fibrillæ (which probably penetrate a portion of the dental tubes) are difficult to examine. Chloride of gold and osmic acid are of no service (Boll). Solutions of chromic acid are the most useful.

Disregarding the higher nerves of sense at present, we here treat only of the so-called terminal knobs, the tactile and the Pacinian bodies, remarkable terminal structures which have been discovered and more closely investigated within the last few years.

The terminal knobs, for a knowledge of which we are indebted to Krause, are represented by the drawing fig. 185. In mammalial animals, as it is known, they are of an oval form (1, *a*), while in man they are more globular in shape (2, *a*). They belong especially to certain mucous membranes, although they may also occur in the external integument.

The ocular conjunctiva is to be selected for their examination, and it is preferable to use the fresh, warm eye of an animal, such as the calf or hog, that has just been slaughtered. Portions of the conjunctiva are to be carefully freed from the connective tissue which lies beneath it, and examined without any fluid media. With a little perseverance, the structures in question may be recognized in the connective tissue by their bright appearance. An examination of this kind always presents difficulties, however.

Krause has made us acquainted with a good accessory, which is to be especially employed with organs which are no longer fresh, and may therefore be used with those of man. This consists in an immersion in ordinary vinegar, continuing for several days or a week. The tissue is rendered transparent, and the arrangement of the nerves may be observed in an elegant manner. Single nerve-fibres are found to enter the terminal knobs which are now become cloudy and have dark contours. The pale terminal fibres can no longer be perceived by this method, however. The vinegar may be replaced by dilute acetic acid; a mixture of acetic acid and alcohol also renders good service. Finally, tingeing with carmine is to be employed with advantage.

The tactile bodies (fig. 186) which occur in certain portions of the external integument (the volar surfaces of the fingers and toes, the palm of the hand, the sole of the foot, etc.) are embedded in a part of the sensitive papillæ of the cutis, and are also rather difficult objects for investigation. Although, with suitable methods of treatment, one may soon recognize these structures and become convinced of their connective-tissue nature, and likewise that the oblong, transversely arranged bodies on their surface are not nervous structures, the investigation of the end of the nerve-fibre still presents the greatest difficulties.

Various methods of examination have been employed in the investigation of the tactile bodies.

The freshest possible human integument, when stretched,

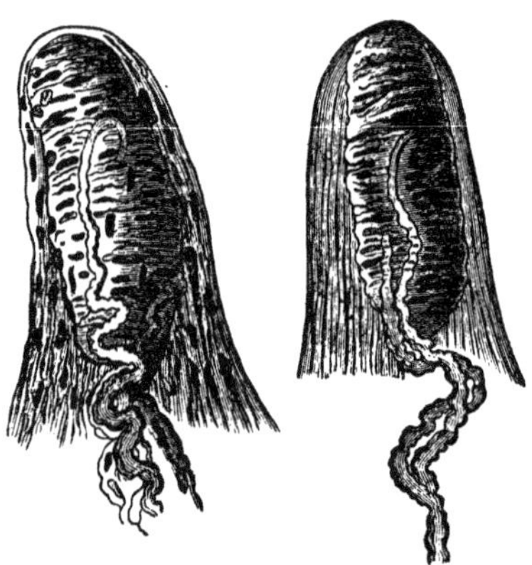

Fig. 186. Two nervous papillæ from the volar surface of the index finger, with the tactile bodies and their nerves.

permits of quite thin vertical sections being made with a sharp knife. These, in consequence of their fibrillated structure, require additional measures to render them transparent. There are two media especially which are used for this purpose; a sometimes more concentrated, sometimes more dilute solution of soda, and the diluted acetic acid. The former affords quite handsome but also very perishable specimens. Thin sections placed in a watch-glass swell considerably after a time, and then permit the epidermoidal layer to be removed. Any fragments of the rete Malpighii which may remain are to be removed by brushing. The examination is to be made with a strongly shaded field, and also, according to circumstances, with the employment of a drop of acetic acid. Other investigators have given the dilute acetic acid entire preference over the solution of soda, and, in fact, it cannot be denied that many details of the tactile bodies, and especially of the course of the nerve tc and in the same, may be more conveniently brought to view by means of this reagent. Such sections, tinged with carmine, present handsome appearances.

Fresh integument, such as that of the point of the finger, may likewise, when carefully dried, present serviceable views in vertical sections, the more so if tingeing with carmine is also employed. Suitable portions of the integument with their natural injection, or injected with Prussian blue and treated in this manner, are adapted for distinguishing the two kinds of sensitive papillæ of the skin from each other. Chromic acid and even alcohol preparations occasionally show very handsome tactile bodies.

Transverse sections of the papillary bodies of the skin, hardened in alcohol or chromic acid, tinged with carmine, cannot be dispensed with.

Gerlach, years ago, made us acquainted with another method. A piece of integument removed from the volar surface of the finger is to be placed for a moment in hot, and then in nearly boiling water. The epidermis is then to be taken off, and any portions of the same which may remain are to be removed with a brush. The piece of skin is then to be hardened for several days in a solution of the chromate of potash. Transverse sections are now to be made from the papillæ with a razor, and placed in water under the microscope. Strong acetic acid serves to render them transparent. The transverse sections of the nerve-fibres may be recognized within the tactile bodies. Injected skin is to be used in order to avoid confounding them with transversely divided capillary vessels.

All these methods of investigation of a former period are unserviceable, however, when the termination of the nerves is concerned. Here the freezing method, in combination with the impregnation with metals, such as chloride of gold, osmic acid, and chloride of palladium, promises the most.

There is still remaining to consider the remarkable Pacinian or Vaterian body (fig. 187), the most complicated form of these terminal bodies of the nerves of sensation.

It is preferable to select for their investigation the mesentery of the cat, in which they at once become evident to the eye, and are isolated by slight preparation. By the application of indifferent fluids they present excellent specimens, in which the structure, the concentric capsules (*b*), the entering nerve (*a*), with the pale terminal filament (*c*) may be readily recognized. Here, also, the latter distinctly shows a combination of primitive, or axis-fibrillæ, as was ascertained by Grandry and confirmed by Schultze, Michelson, and Ciaccio. Previous injection with cold-flowing transparent masses is a good accessory; diluted acetic acid and tingeing may also be employed.

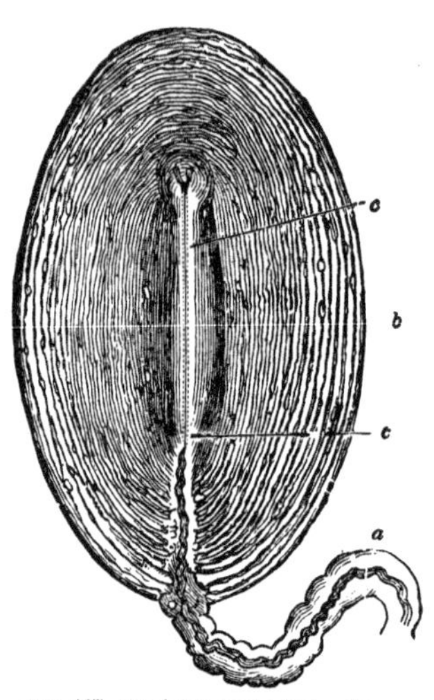

Fig. 187. Pacinian body from the mesentery of a cat. *a*, Nerve-fibre; *b*, the capsules; *c*, the pale-contoured terminal filament of the nerve-tube.

Dilute chromic acid, or corresponding solutions of chromate of potash, may likewise be employed for their preservation and examination. I find the acetic-ple microscope serve for the detachment of the capsules. Impregnation with silver also shows on these the familiar mosaic.

The Pacinian capsules of man are to be obtained without much trouble, by preparing the cutaneous nerves of the palm of the hand and the sole of the foot. The methods of investigation are the same as for those of the cat.

The textural conditions of the nervous system in the fœtus, and the history of the origin of their elements, are, as yet, by no means known with desirable certainty. The embryos of our domestic mammalia or of the hen, which have been immersed, as fresh as possible, in dilute solutions of chromic acid, or of the bichromate of potash, and slowly hardened, are to be employed. Transverse sections of hardened embryos afford very fine review preparations for the spinal cord, the spinal ganglia, etc. The objects tinged with carmine are to be mounted in Canada balsam.

Many characteristic views of the peripheral nerves of the fœtal period may be readily obtained in the fresh larvæ of the frog and salamander. Together with the nitrate of silver and the chloride of gold treatment (Eberth), a very excellent procedure, given by Hensen, may also be employed. The larvæ are to be dipped for 20–50 seconds in a chromic acid solution of from 3–4 per cent., and are then thrown, still living, into spring-water. Then, or after half an hour, the epithelium may be removed from their tails by brushing. Eberth recommends for the same purpose to place the frog's larvæ for a half or a whole hour in a weak solution of nitrate of silver (1 gr. to 5 ounces).

In consequence of the extreme delicacy and alterability of these tissues, however, the most conservative methods are always the best.

Somewhat more strongly hardened embryos afford good preparations relating to the structural conditions of the developing spinal cord and brain. The changes in form of the former, and also of the spinal ganglia, with a progressing development, may be readily perceived. Here, also, chromic acid and bichromate of potash deserve decided preference over alcohol. Transverse sections, with the aid of carmine tingeing, suffice for the first examinations.

Concerning the tunics of the central organ, it is better to examine the arachnoid and pia mater fresh, employing the reagents which are customary for connective-tissue parts.

The numerous capillaries, and the small arterial and venous branches which are associated with them, cause the latter membrane to appear very well adapted for the study of the vessels. In suitable specimens (as well as in mechanically isolated vessels of the nerve-substance) one may readily recognize that the formation of tubercles commences in the adventitia. It was thought that the rudimentary cells, which are found here (the so-called vascular nuclei), increased by proliferation. At the present time, a wandering of the lymphoid cells of the blood into the surrounding layer has become more probable. If suppuration takes place in the pia mater, in consequence of inflammatory irritation, the emigration of the lymph-cells from the blood-current may be most distinctly recognized (Rindfleisch).

The dura mater may be examined fresh, dried, or hardened by means of chromic acid,—methods which are also employed with the neurilemma of the larger nerves. The plexus choroidei scarcely requires special directions; its injection with that of the brain readily succeeds. Here Müller's mixture (p. 139) affords good preparations. The calcareous concretions of the same, the so-called brain-sand (which, as is known, also occurs in the human pineal gland), are to be studied with the employment of acids and media for rendering them transparent, especially glycerine.

The pituitary gland, for which the calf is to be especially recommended (Peremeschko), is to be hardened in chromic acid, Müller's fluid, or alcohol. Thin sections, brushed and tinged with carmine, readily yield the essential appearances.

We have already noticed above, the great difficulties which the investigation of the normal textural conditions of the central organs of the nervous system still present. Hence, we shall comprehend that their numerous pathological alterations are still very inadequately known, and can only be assailed, histologically, with slight results. It is usually accepted that the nervous elements indeed undergo numerous secondary processes of degeneration, such especially as the fatty, and also amyloid and colloid transformations, but that the true new formations proceed from the connective-tissue framework and from the vessels. The correctness of the former doctrine has more recently, however, become doubtful, and the lymphoid migratory cells exert at present a deep and disagreeable influence on the latter. The finer textural conditions of the framework substance are also uncommonly difficult to follow, as the very methods of hardening which are customary for the normal structure frequently render very little service in the domain of pathology, so that often one is able to examine only fresh objects. For connective-tissue formations one should try very weak chromic acid, according to Schulze (p. 131), likewise Müller's fluid, diluted with about an equal volume of water. Finally, the skilful employment of tingeing methods will afford much additional assistance.

Good review preparations are not unfrequently afforded by a

method practised by Billroth,—placing small pieces of brain and spinal cord in powdered carbonate of potash or chloride of calcium for 24 hours. By this means the objects usually gain a degree of consistence, which permits of fine sections being made; these are to be examined in water (or with the addition of glycerine).

To investigate the fatty degeneration of the nerve-fibres, as well as the textural changes which occur in the peripheral portion of a divided nerve, the animal which has been subjected to the above operation is to be examined, after the proper interval of time, either quite fresh or with the employment of the solutions of free chromic acid and of bichromate of potash, which have been recommended for the spinal cord and brain. This is one of the few structural changes of the nervous apparatus which present but slight difficulties to the practised observer.

Section Sixteenth.

VESSELS AND GLANDS.

The methods of investigating the vessels differ somewhat according as the contained mass consists of blood or of lymph; furthermore, they vary considerably in proportion to the size of the vessels. One variety of accessories is therefore necessary in the examination of the capillaries and small vessels; others are required for the investigation of the larger and largest trunks.

The finest canals of the blood-passages are, as is known, the capillaries (fig. 188, *A B*), which are very thin, nucleated, branched membranous tubes. The narrowest capillaries (*A a b B a*), which, however, do not occur in all parts of the human body, are only sufficiently wide to allow of the passage of the blood-cells singly, one after the other. A similar condition recurs in the several groups of vertebrated animals, modified, naturally, by the size of the blood-corpuscles. Frogs and naked amphibiæ have therefore capillaries of much more considerable diameter than are met with in the human body, and the capillaries of these creatures are consequently better adapted for many investigations than our own.

Until recently the purport of the almost universally accepted history of the development of the capillary vessels was that they originated in the blending of the formative cells which, coming together in a single row, opened into each other, so that the cell-cavities became united into a capillary tube, the membranes of the cells were transformed into the wall of the vessel, and the persisting nuclei into the nuclear formations of the latter.

It has been ascertained by the concurrent investigations of several observers (Hoyer, Auerbach, Eberth, and Aeby) that the walls of the capillary vessels are not in reality structureless, but rather that they are formed by the melting together of very thin and flat nucleated cells (the cavity of the capillary is therefore an intercellular passage). The boundary-lines of these cells are only to be rendered visible by means of impregnation with silver, and were previously entirely overlooked.

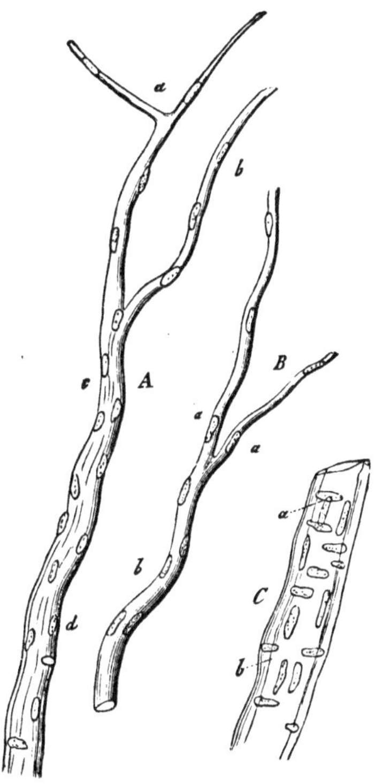

Fig. 188. Small vessels from the pia mater of the human brain. *A*, a vessel whose trunk acquires, at the lower part, *d*, a double membrane, and at its upper part passes over into two fine capillaries, *a* and *b*; *B*, a second one; *C*, a somewhat larger trunk with a double membrane, the inner one, *a*, with longitudinal nuclei, and the external one, *b*, in the central portions of which transverse nuclear formations may be recognized.

This important discovery may be readily verified (figs. 189 and 190). A frog, a mouse, or a Guinea-pig is to be allowed to bleed to death, and then a 0.25 per cent. solution of silver is to be injected into the vessels. The simple immersion of organs, such as the retina or pia mater of mammalial animals or of man, which are deprived of their blood, also leads to the desired result. The object is to be washed in spring water and examined in acidulated glycerine.

More recently increased attention has been paid to a somewhat more complicated arrangement of the capillaries which occurs in the lymphoid organs, the lymphatic glands, the Peye-

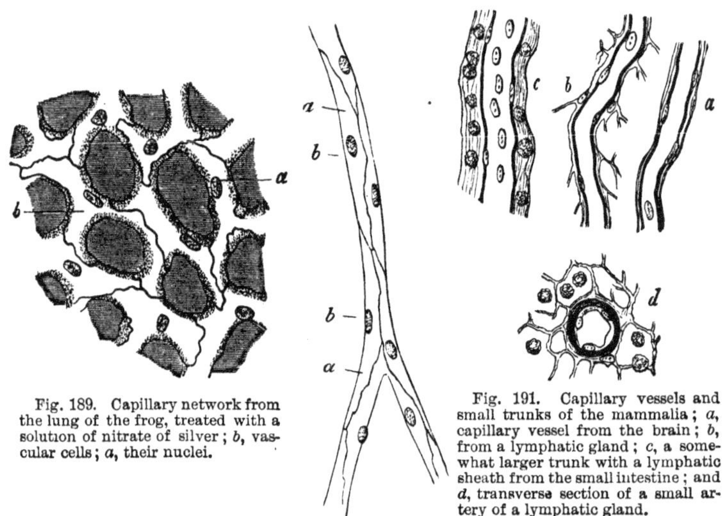

Fig. 189. Capillary network from the lung of the frog, treated with a solution of nitrate of silver; *b*, vascular cells; *a*, their nuclei.

Fig. 191. Capillary vessels and small trunks of the mammalia; *a*, capillary vessel from the brain; *b*, from a lymphatic gland; *c*, a somewhat larger trunk with a lymphatic sheath from the small intestine; and *d*, transverse section of a small artery of a lymphatic gland.

Fig. 190. Capillary vessel from the mesenterium of the Guinea-pig, after the action of a solution of nitrate of silver. *a*, vascular cells; *b*, their nuclei.

rian and solitary follicles, the tonsils, Malpighian bodies of the spleen, the thymus gland, and also in other glands. It consists in the spreading out of the reticular connective tissue of these organs, in the form of a membrane, around the primary walls of the capillaries, and thus forming a second accessory layer, a so-called adventitia capillaris. This may be represented by fig. 191, *b*. Furthermore, microscopic vascular trunks may be surrounded, with a larger intervening space, by a connective-tissue sheath (*a*) whereby the cavity which is thus formed (lymph-sheath) serves for the current of the lymph (*c*).

There are by no means many organs which are adapted for the primary examination of the capillary vessels of man and the mammalia. The most suitable, and therefore, also, the most generally recommended, appear to be the brain, the pia mater of the same, the retina of the adult body, and the lymphoid organs.

To obtain characteristic views from the former parts a small blood-vessel which is barely visible in the gray substance is to be seized and an attempt made to remove it by traction. It is then to be placed in water and freed from any adherent brain-matter by means of the wash-bottle, or, still better, by brushing. In this way will be perceived a trunk with abundant ramifications and numerous capillaries as terminal branches, and not only these finest forms of vessels may be studied, but also a series of transformations into more complicated vessels. The pia mater when picked also affords excellent objects, especially if portions be selected which pass over a furrow between the convolutions of the brain. The capillaries of the retina are to be treated in the same manner as those of the brain-substance.

The organs belonging to the lymphatic system require somewhat greater preparations for the investigation of their capillaries. Either none at all or only extremely unsatisfactory views are to be obtained from the fresh parts. For this reason, preparatory hardening methods (chromic acid, alcohol, etc.) are to be first employed. Thin sections from the tissue which has become more resistant must then be freed from the innumerable lymph-corpuscles which fill the connective-tissue network before the desired delicate capillaries can be brought to view.

Any ordinary preparation serves for the recognition of the nuclei. These are rendered sharper by dilute acetic acid, likewise *by carmine tingeing, which, together with staining with* hæmatoxyline and aniline blue, deserves recommendation. Chloride of gold also affords serviceable specimens.

In organs with a fibrous structure, even where there is a considerable vascularity, we may, as a rule, search in vain for capillaries except we employ especial accessories for rendering them prominent. The empty capillaries disappear most completely in the fibrillary tissues. The familiar action of acetic acid may here be made use of, or, which is to be preferred, the preparation may be first colored with carmine and then subsequently exposed to the action of the acid. Glandular organs, on the contrary, require the application of alkalies if the capillaries are to be rendered prominent after the dissolution of their cells.

After what has just been mentioned it is unnecessary to discuss further the great value of transparent injections with Prussian blue or carmine for rendering the capillary and larger vessels of an organ visible; the silver solution (p. 186) may also be used with advantage. The slight trouble of injecting should, in fact, never be shunned in such investigations, as the structural relations of all organs are usually rendered much more comprehensible as soon as the distended capillaries have afforded a landmark for the eye.

Where it is possible to retain the blood in a vascular district, such natural injections may replace the artificial ones, but the preparations should not be moistened with water.

If a frog or a salamander be at hand, the capillaries of certain parts of the body may be examined with advantage for comparison with the human textural condition. From the first animal (preferably killed by means of ether or chloroform) the lower eyelid or one of the flat, transparent muscles which we have mentioned above, p. 361, is to be taken. The membrana hyaloidea also affords magnificent views.

The larger blood-vessels which follow these no longer show the original simplicity of structure possessed by the capillaries. First appears the original capillary membrane (fig. 188 *C*) metamorphosed with cells and longitudinally arranged nuclei (*a*); then a second layer with transversely arranged smooth muscular elements scattered through it, the nuclei of which are seen at *b*. This is the earliest appearance of a tunica media *seu* muscularis, to which in vessels of a somewhat larger calibre is gradually associated the tunica cellulosa, the external connective-tissue layer. In other vessels the epithelial tube is seen to be surrounded by an elastic inner membrane; this is the first appearance of the tunica serosa. We meet with other trunks (and these have for the most part the character of venous tubes) which present the most internal cellular layer, the elastic inner membrane, and the adventitia; but, on the contrary, no trace of any muscular middle layer can be recognized in them. A complete contrast is presented by the arterial trunks (fig. 192), the muscular middle layers of which are very strongly developed, and its contractile fibre-cells are found packed closely together.

The methods for investigating these vessels are the same as

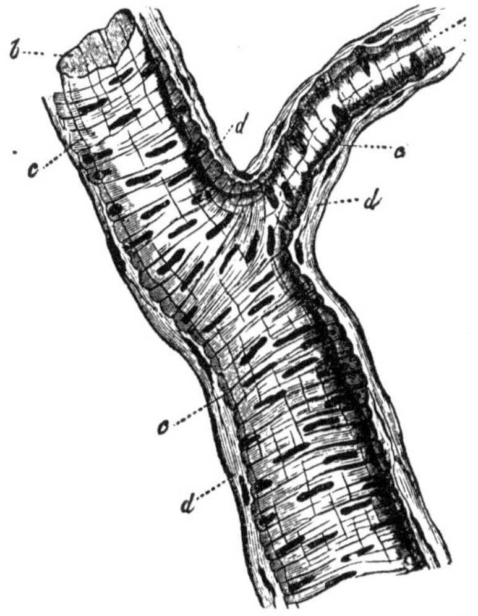

Fig. 192. A small arterial trunk without an epithelial covering; *b*, the homogeneous elastic internal layer; *c*, the middle layer, consisting of transversely arranged fibre cells; *d*, the external connective-tissue covering.

for the capillaries. Even the localities, such as the substance of the brain, the pia mater, and the lymphoid organs frequently remain the same. Together with these the mesenteric arteries may also be used with advantage. Advantageous use may be made of tingeing methods, especially of carmine tingeing with the subsequent action of acetic acid, and also of dilute acids and alkalies. Transparent injections are very useful here, especially for estimating the extremely variable thickness of the walls of small venous and arterial branches.

The epithelium in somewhat larger trunks is to be examined in part in fresh, unaltered specimens, or by means of carmine tingeing and silver impregnation (see page 160).

Tingeing with carmine, the application of potash solutions of 30–35 per cent., and of 20 per cent. nitric acid are to be recommended for the recognition of the muscular layer. The action of nitrate of silver may also be employed for rendering visible the contours of the individual contractile fibre-cells.

The usefulness of still another method here makes itself felt, which is of unsupersedable importance in the examination of larger and the largest vessels—we refer to the preparation of thin sections through their walls. Drying was formerly employed for this purpose. At the present time the embedding method (p. 117) has taken its place.

Preparations hardened in alcohol or chromic acid also afford handsome sections of such vessels (fig. 191 *d*). Beale distends such arterial and venous trunks as much as possible by the energetic injection of uncolored gelatine, and then makes fine sections through the hardened mass. He recommends this method very properly, especially for the demonstration of the contractile fibre-cells. We recommend for this purpose especially the Malpighian bodies of the spleen, the follicles of the lymphatic glands and of the kidney, whereby the slight trouble of a careful brushing should not be avoided.

Vessels whose walls can no longer be seen in their totality with the microscope require the preparation of thin, partly longitudinal, partly transverse sections. The fresh vessel may, without further treatment, be dried or embedded, and then examined with the application of acids and alkalies, or it may be previously boiled in vinegar or dilute acetic acid. The action of 20 per cent. nitric acid, likewise Schulze's treatment with chloride of palladium, and the subsequent tingeing with carmine, as also Schwarze's double tingeing with carmine and picric acid (p. 159) deserve to be recommended.

The various strata of elastic membranes, connective tissue, and muscular layers are thus rendered most distinctly visible. The best views are obtained of the development of the muscular layers in arteries and veins of medium size, and of the retrogression of this tissue in the largest vessels. It will frequently be found that the epithelium is no longer preserved.

A second, and indeed older procedure, consists in opening a vessel while it is in a moist condition, and then with the scalpel and forceps separating, under water, the individual layers successively from within outwards, or from without inwards, to study them with the application of suitable media. By scraping the fresh specimen with the blade of a scalpel, larger or smaller shreds of the epithelial covering may be readily brought to view. The free border of the valve of a vessel not unfrequently presents a fine view of this covering, and, at the same time, a good means of measuring the slight thickness of the same.

For the recognition of the vascular nerves, a network of very fine pale filaments, which occupies the tunica media and the contiguous portions of the external layer, select the mesentery of a frog, treat it with dilute acetic acid, and remove the epithelium by brushing (His).

The arrangement of the various capillary districts according to the magnitude and manner of distribution of the vessels, as well as the size of the tissue-spaces surrounded by the meshes of the network, has occupied the attention of anatomists and physiologists for a long time. Even the charming appearances which successful preparations of this kind unfold under the microscope must exert an attractive influence. Then it is only by the vascularity of an organ that a conclusion can be made as to the quantity of matter which it assimilates, either in its own interest or for the service of other organs (glands). The relative arrangement of the capillaries is of great importance in the mechanism of the circulation.

As the technology of injection has already been mentioned in detail in a previous section (p. 169), we may simply refer to the same. For the study of the capillaries, as has been already remarked, one should only examine objects injected with transparent masses in a moist condition (either quite fresh or after a short immersion in alcohol, and then with a subsequent addition of glycerine), as opaque masses conceal too much, and all dry preparations present a distorted appearance. The simple injecting fluids (the cold flowing Prussian blue of Beale or of Richardson, see p. 183, 184) will suffice for most of the investigations of the capillary networks. If one desires to employ a double injection, Beale's carmine mixture (p. 185) may be added.

It would lead us beyond the limits of this little work were we to mention here more fully the various appearances of the capillary network according to the size of the vessels and meshes, as well as the form of their arrangement. A few remarks may therefore be sufficient, and the text-books on histology recommended for further instruction.

Many parts of the body, as is known, remain entirely without vessels; others are but slightly vascular, and are only permeated at considerable distances by capillary vessels; while in the

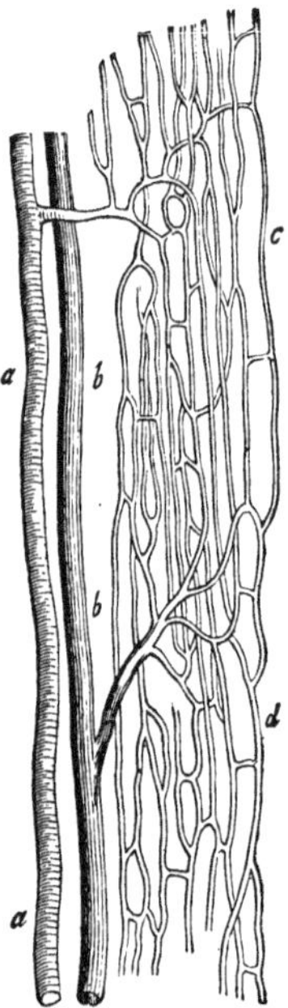

Fig. 193. Vessels of the voluntary, transversely striated muscle. *a*, arterial; *b*, venous branch; *c*, the straight capillary network.

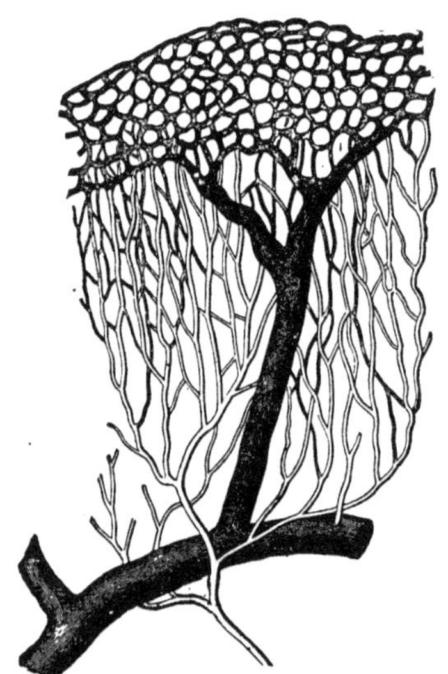

Fig. 194. Vessels from a vertical section of the mucous membrane of the stomach; the fine arterial branch divides into a straight capillary network, which forms circular meshes at the surface of the mucous membrane, and passes over into the thick venous trunk.

extremely vascular organs the capillaries are closely approximated to each other, and the meshes are smaller.

The anatomists have distinguished two fundamental forms of capillary networks, according to the shape of the parenchymatous spaces surrounded by them; namely, 1, the straight, and 2, the circular network.

Both forms arrange themselves according to the shape of the tissue-elements. Circular parts formed of cells or gland-vesicles have a similar shaped, that is, circular network of vessels; while those with a decided fibrous disposition, or parts formed of glandular passages and tubes running in a parallel direction, present the straight capillary network.

Fig. 193 shows the straight capillary network of the transversely striated muscle; fig. 194 the same of the mucous membrane of the stomach surrounding the gastric glands. The latter assumes the form of a circular network at the surface of the mucous membrane, where the gland-ducts terminate with round apertures.

We have already in a previous section (p. 273) learned that

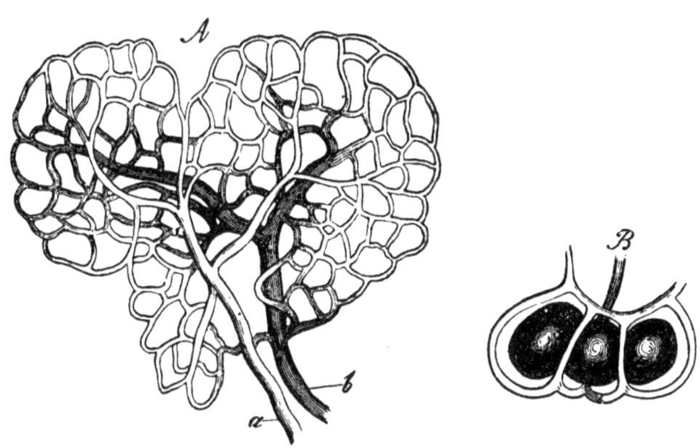

Fig. 195. Vessels of the fat-cells. *A*. Arterial (*a*) and venous (*b*) branches, with the capillaries between them. *B*, the capillaries around three cells.

the lobules of fat-tissue consist of groups of large globular cells. The capillary network, fig. 195, is in accordance therewith. A likewise more circular, but large-meshed and peculiarly formed network of capillaries is seen in the inner layer of the retina (fig. 196). Where small papillary projections appear (external integument and many mucous membranes), we meet with simple capillary loops (fig. 197); where these are larger, with a looped network, as in the intestinal villi. The arrangement of the capillary networks of the human organism is frequently of such a peculiar nature that the practised

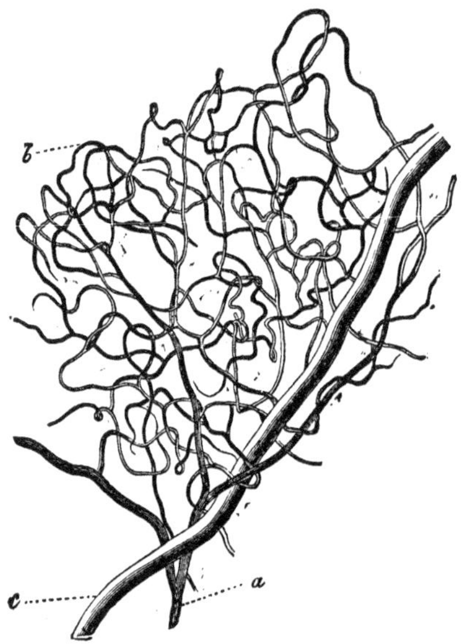

Fig. 196. Vessels of the human retina. *a*, arterial, *c*, venous branch; *b*, the capillaries.

observer can readily recognize, with the greatest certainty, from what part of the body the preparation is obtained.

Fig. 197. Capillary loops of the papillæ of the skin (in others appear the tactile bodies).

The first appearance of the vessels in the embryo, as well as the subsequent formation and transformation of the fœtal vessels, constitute, as is known, a very difficult and therefore still to a great extent deficient section of histology. The more recent discoveries concerning the structure of the walls of the capillaries also make a revision of the earlier statements urgently necessary.

Very young embryos of birds, mammalia, and fishes are to be recommended for the investigation of the origin of vessels. Among these, the embryos of the hen have been employed for many years and stand in the first line, in consequence of the facility with which suitable material may be obtained. The formation of the first capillary reticulation may be observed in the area vasculosa from the end of the first and on the second day of incubation. For this purpose, the germinal membrane is to be cut out beneath the surface of lukewarm water to which a little chloride of sodium and albumen has been added. It is then to be examined either quite fresh with the application of an indifferent fluid, or of a strongly diluted solution of chromic acid; or, which is to be considered as more advantageous for many purposes, after having been hardened in chromic acid or bichromate of potash, it is to be investigated with the aid of glycerine and tingeing methods.

In order to follow the further peripheral formations of vessels one may use the embryos of the hen at a more advanced stage, for example, their allantois; or the embryos of mammalia may be employed. In the latter, the urinary sac, the membrana capsulo-pupillaris and hyaloidea of the eye also afford admirable preparations.

A very convenient object for examination is presented during the early part of the summer by the tail of the frog's larva. As the living animal may be examined either on the Schulze's object-bearer (p. 243), or by means of a strip of moist blotting-paper wrapped round its body without injury, and again placed in the reservoir, it is possible to follow from day to day, in one and the same specimen, the alterations in the vascular formations which take place in accurately designated places. Further assistance is afforded by brushing off the epithelium, as performed by Hensen (p. 373). The application of a dilute solution of nitrate of silver promised important disclosures here. It has presented the unexpected result, that vascular cells cannot be rendered visible in developing capillary tubes (Kölliker, Golubew). Many vascular districts of adult creatures, however, behave in a similar manner; thus, for example, those of the hyaloidea of the frog (Golubew) and of the mammalial liver (Eberth). Darkness, therefore, still prevails.

Pathological changes of the blood-vessels, as is known, are met with often enough. In so far as they consist of a metamorphosis of the structure, they affect the larger trunks, especially of the arteries, much more frequently than the finest arterial and venous terminal branches, or the capillaries lying between them.

In the arterial walls of older persons there are generally found —increasing in frequency with the advance in life—changes of the inner vascular membrane in the form of smaller or larger whiter or yellower spots and plates projecting somewhat above the surface. These are found by the microscopical analysis to consist of collections of fat-molecules. There may afterwards be a softening and breaking down of these fatty degenerated places.

In the atheromatous processes, also, we again meet with the same fatty deposits, but in the deeper strata of the intema contiguous to the muscular coats, after a proliferating thickening of the inner coats of the vessels has taken place as a result of an irritation. Here, also, a softening of the fatty infiltrated place occurs, and the melting process advances at the expense of the remaining layers of the inner coat of the vessel. When a regular atheromatous pulp (which may force its way into the blood-passage) has been formed, its elements are shown by microscopical analysis to consist of fat-molecules, isolated or united in globular conglomerations, crystals of cholesterine, and fragments of tissue. These thickened places in the intema may, however, undergo still another degeneration, a hardening, which may also be combined with the former; they may become calcified, and form hard plates or tablets in the walls of the vessel.

By such atheromatous changes in the arterial walls are caused, at least to a great extent, arterial aneurisms which in part permit of the recognition of all three of the groups of layers, although thickened and metamorphosed in part after the destruction of the intema, or also of the muscular coat, consist either of both coats or of the connective-tissue coat alone, which latter then presents transformations, thickenings of its tissue, etc.,—matters which we cannot consider further at present.

The question arises—How are such abnormities of the arterial walls to be examined?

In general, by means of the same methods with which we have already become acquainted in the examination of the normal structure. By pulling off the individual layers from the fresh specimen, then on horizontal and vertical sections of the walls after hardening in alcohol or chromic acid, or finally by the aid of the embedding and drying methods. The coats, boiled in vinegar and then dried, also afford handsome sections whereby the fat-molecules of the atheromatous mass make their appearance in an elegant manner. Atheromatous pulp is, like pus, etc., to be spread out with water.

The method of examination also remains the same for the pathological metamorphoses of the structure of the veins. As with the arteries, we here omit the dilatations of the veins, and the occlusions from thrombi and emboli (that is, clots which have originated in a remote locality, and, carried forward by the blood, have finally become wedged into a vessel). Only those layers of the walls which contain vessels, especially the adventitia and the middle layer, are at first concerned in the inflammatory processes of the veins. At the same time there is tumefaction, the formation of so-called exudation masses, and an accumulation of pus-corpuscles. The inner coat, which is not immediately concerned in the inflammatory process, is also at last, as a result of these structural changes, drawn into the sphere of the process. It appears cloudy, thickened, rough, and may become separated in shreds.

Such rough inner surfaces of venous as well as arterial vessels frequently receive aggregations of coagulated fibrine from the blood. Consequently we see such deposits on the intema of inflamed veins, as well as on softening atheromatous patches and in the dilated sacks of aneurismal arteries.

Pathological changes of small vessels, microscopic arteries

and veins, escape the notice of the physician much more readily, as will be appreciated, and also cause much slighter effects during life.

In amyloid degeneration of the smaller arteries the middle coat is seen to be the seat of the deposit. The fibre-cells of the muscular coat lose their structure and become transformed into amyloid flakes; and in calcifications, also, the deposit of bone-earth takes place in this contractile element.

The small arteries of the brain-substance occasionally undergo an interesting metamorphosis. In vessels of extreme minuteness, up to those of 0.5‴ in diameter, a tearing of the inner and middle coats takes place; extravasated blood becomes infiltrated under and into the adventitia, and arches this forward, in various ways, into vesicles and knobs. If, at last, the external connective-tissue layer also becomes torn through, apoplectic effusions occur. Should it hold, however, a striking microscopic appearance is unfolded in the gradual metamorphosis and breaking-down of the extravasated blood-corpuscles; so-called granule cells, aggregations of brown and yellow pigment, and their final dissolution may be observed.

Fine microscopic veins and their branches, which are passing over into capillaries, occasionally show similar varicosities of their lumen. Although with the above-mentioned arteries the bulging is caused by the laceration of the coats and the extravasation of the blood, here all three of the coats are uninjured.

Calcareous and fatty degenerations, and likewise deposits of pigment, have been noticed in the capillaries as well as in the smallest arterial and venous branches which are associated with them. Furthermore, emboli of the same, as well as thickenings of their walls, belong to the more interesting occurrences.

Calcifications have thus far been noticed chiefly in the capillaries of the brain; they occur very rarely. Much more frequently, especially in the brain of older persons, fatty degenerations, groups of aggregations of small fat-molecules around the nuclei or in the place of the same, are noticed. This structural metamorphosis is occasionally diffused, in the most extended manner, throughout a whole brain. Deposits of black pigment molecules have been observed in the capillaries of the spleen, liver, and also of the brain, in melanæmia.

Peculiar emboli of the finest arteries and veins from masses of fluid fat have also been noticed more recently in so-called pyæmia (E. Wagner).

We have already mentioned above the normal occurrence of the adventitia of capillaries. Something of the kind is also met with under abnormal circumstances. The capillaries of a part in a condition of inflammatory irritation gradually receive an aggregation of spindle-shaped cells, precisely similar to those which occur in the normal development. Very fine appearances of this kind may be obtained from the inflamed cornea. An aggregation of this undeveloped connective-tissue formation of the gelatinous tissue may also appear as an adventitia around capillaries (Billroth).

In all textural changes of the capillaries, like those of the larger and largest trunks, the greatest attention is to be paid to the nuclear formation of the so-called vascular cells, as it is just these epithelioid cells which rapidly assume a condition of luxurious proliferation, and thus give rise to numerous new formations (Thiersch, Waldeyer, Bubnoff).

The structural changes which have thus far been mentioned of the smaller and smallest vessels coincide exactly with those of the normal body in so far as the methods necessary for their examination are concerned.

A matter which has caused numerous controversies is the manner in which the origin of vessels takes place under pathological conditions.

Such productions of new blood-vessels, as is known, are not rare occurrences, and appear in hypertrophied organs, in neoplasms, in so called pseudo-membranes, and granulations. Quite massive new formations of blood-vessels may be recognized in the so-called vascular tumors. Numerous sack- and knob-shaped expansions of the dilated capillaries are met with in the capillary telangiectasias, especially such as occur in the skin. The methods for examining the tissues of the skin must here be employed. Preparations boiled in vinegar and then dried afford characteristic views.

When such newly formed vessels are examined they show either—and this is generally the case—the character of the capillaries or those of the arteries and veins, while the blood which circulates through them presents nothing especial. Their diameters are either those of the normal condition, or they are increased frequently in the most remarkable manner. At the same time partial dilatations of the walls frequently occur. Knob-shaped pouches are also met with, especially in vascular tumors, which require more accurate investigations.

At a former period, swayed by the theory of spontaneous cell-formation and the exudation doctrine of that time, it was frequently asserted that these pathological vessels (like the blood contained in them) originated independently of those of the neighboring normal tissue, and only subsequently became united with the adjacent vessels.

At the present day we may say that this theory was false, and it has also never failed to be frequently attacked. No new formation of vessels deviating from that of the fœtal body occurs in the domain of pathology. In both cases the new vessels arise by growth from the existing ones.

According to more accurate investigations which have been made, the formation of vessels in a tumor, like a so-called pseudo-membrane, appears to take place but slowly and gradually, and thus to form a striking contrast to the rapidity with which, for instance, an aggregation of pus-cells may take place.

Either the fresh tissue or that hardened in alcohol, chromic acid, etc., is to be employed for examination. The escape of the blood-corpuscles from the newly formed vessels, which readily occurs in such preparations, is a very unfortunate circumstance, and is responsible to a considerable extent for the scanty results which so many investigators have obtained in this direction. If the injection with transparent masses, which is frequently difficult, it is true, succeeds, the whole naturally gains extraordinarily in clearness. Thus Thiersch made the interesting observation in the healing of wounds of the tongue, that at the commencement a system of lacuna-like passages is formed in the granulation-tissue which lead from the loosened arterial walls to similarly constituted veins. The largest proportion of these "plasmatic" canals afterwards disappear, but a portion of them become widened and form blood-conveying vessels, and the cells of their walls are produced by the adjacent tissues. Injections are an indispensable accessory for such studies.

The lymphatics, as is known, show in their large trunks a structure reminding us of that of the veins, and also coincide with them in the abundance of their valves. The latter also continue in the fine branches, and give them a very characteristic knotty appearance. So long as such a constitution can be recognized the walls of these tubes, although at last simplified to a structureless membrane, are formed of a special tissue which differs from the neighboring tissues.

The same methods are employed for the examination of the walls of these vessels as for the arteries and veins. Large trunks may be dissected out, slit up, and examined by pulling off the individual layers, or, after drying, longitudinal and transverse sections may be made. It is best to inject small trunks with pure gelatine by tying in a fine canule; after cooling, thin transverse sections may be made. The finer lymphatics appear at first to form cavities and passages surrounded only by connective tissue. By the application of the dilute solution of nitrate of silver (preferably in the form of an injection), it has been ascertained, however, that these also have a wall consisting of the characteristic vascular cells (fig. 198, *a*). While the latter, in the capillaries of the blood-passages, however, usually presents a certain independence in opposition to the adjacent tissues, the walls of the lymph-canals are intimately blended with the neighboring connective tissue.

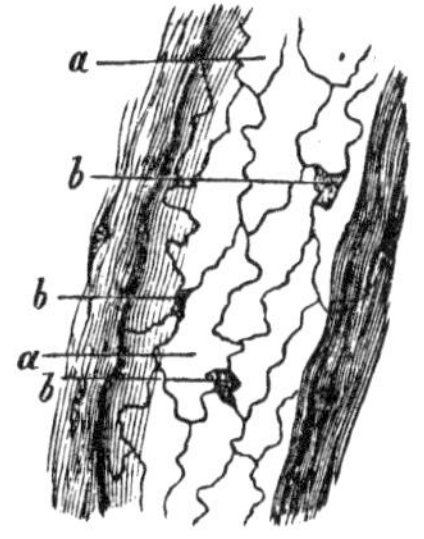

Fig. 198. A lymphatic canal from the large intestine of the Guinea-pig. *a*, vascular cell; *b*, intercalary plate.

In investigating the arrangement of the finer lymphatics of an organ, the injection of cold flowing, transparent masses of Prussian blue and carmine (see above, p. 185) is to be employed, together with the subsequent hardening in alcohol. In doing this the canule is either to be tied in or the method by puncture is to be used.

The injection is sometimes easily accomplished in some parts of the body, occasionally only with the greatest difficulty.

The natural injection, which, for the blood-vessels, may afford to the unpractised a substitute for the artificial injection is, from the nature of the enclosed fluid, of very limited importance for the lymphatics. The lymph proper disappears as a colorless fluid in the tissue of the organ, and the small vessels only glimmer forth from the tissue where a pathological coloring matter, of the bile or of the blood, for example, is retained and mixed with the lymph. The chyle, on the contrary, forms, as is known, a milk-white fluid, in consequence of the larger amount of fat which it contains, and hence its vessels are seen distended in a beautiful manner. Mammalial animals, especially young suckling examples, killed during the digestion of fat (3–5 hours after its reception) afford, therefore, excellent objects for the study of the chyle ducts and vessels, a subject to which we shall return at the investigation of the organs of digestion.

Pathological new formations of lymphatics, especially in tumors, are of frequent occurrence, although this subject, in consequence of the difficulty of its investigation, still represents almost a terra incognita. The younger Krause has lately communicated a few observations concerning them. He succeeded, in scirrhus and medullary sarcoma, in injecting trunks lying in the connective-tissue bands of the framework, and likewise large vessels in myoma of the labia. May these experiments very soon be further extended!

The structure of the lymphatic glands has recently become considerably more comprehensible from the labor of a number of observers.

The extreme softness and the opacity of the fresh organ, caused by the presence of millions of lymph-corpuscles, leads to the employment of hardening methods and brushing.

These methods are the customary ones. Immersion in alcohol, at first in such as is ordinarily used for preparations, which has been diluted with about half its volume of water, leads as a rule, in from 5–8 days, to the desired object, especially if the precaution is observed to change the fluid frequently. The alcohol finally added should no longer become cloudy. If a sufficient consistence is not obtained in this manner, stronger alcohol, and finally that which is almost free from water may be used, and thus, not unfrequently in the middle or towards the end of the second week, preparations will be obtained which are suitable for sections and for brushing. Over-hardening is, however, to be most carefully avoided if the framework substance is to be examined, while strongly indurated alcoholic preparations afford the best specimens for the investigation of the blood- and lymph-vessels of these organs. For many purposes chromic acid deserves the preference to alcohol. Commence with weak solutions and proceed very gradually to stronger ones. The shrivelling which is usually connected with alcoholic preparations may thus be frequently avoided to a considerable degree. Solutions of the bichromate of potash of a corresponding concentration are also very serviceable. All lymphatic glands once hardened by any of these ways may be preserved for a long time in a serviceable condition, and may be used for occasional observations.

Small, fresh lymphatic glands from healthy bodies do not as a rule present any difficulties in hardening. It is otherwise with those which are very voluminous, or no longer fresh, as well as with those which are affected by many varieties of degeneration. Thus, for example, typhus mesenterial glands require much care as a rule, and the object is not always accomplished. The previous injection of the hardening fluid through the blood or lymphatic vessels of the organ to be immersed is a useful accessory with such organs as are difficult to manipulate. Attempts may be made in vain to harden these very glands by immersing them for 8–14 days in alcohol of increasing strength, and finally in that which is almost absolute, success only being obtained afterwards by placing them in chromic acid solutions.

Toldt has recently recommended another procedure which permits of the preparation of the thinnest sections, and thus renders the trouble of brushing unnecessary to a great extent. The fresh glands are to be placed for 3–4 days in very "dilute, wine-yellow" chromic acid. When the hardening has reached the interior it is to be placed for the same length of time in glycerine diluted with an equal part of distilled water.

It would appear almost superfluous to give further directions for the examination of the framework of the alveoli or follicles (fig. 199, *d*) and lymphatic vessels (*e*). The glands of younger animals, or such as are in a swollen condition, are to be employed for the first recognition of the cellular character of the reticular tissue. Among the tingeing methods, that with carmine accomplishes most in these cases. The reagents mentioned at the smooth muscles are employed for the recognition of the fibres of this tissue at and in the septa, especially the treatment with chloride of palladium and the double staining with picric acid and carmine (p. 159).

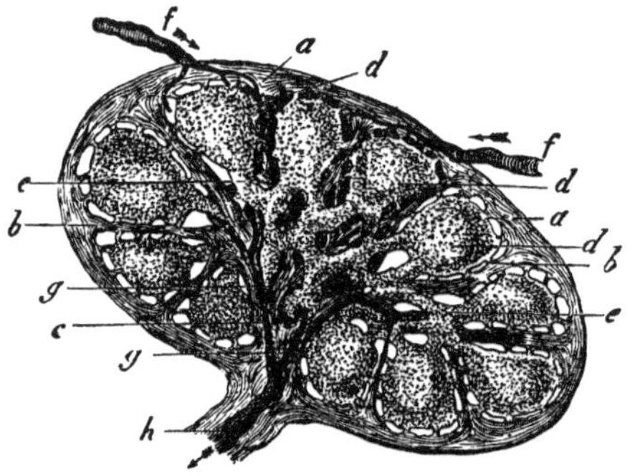

Fig. 199. Section through one of the smaller lymphatic glands, with the current of the lymph —half diagrammatic figure—*a*, the capsule; *b*, septa between the alveoli or follicles of the cortex (*d*); *c*, system of septa of the medullary substance as far as the hilus of the organ; *e*, lymph-tubes of the medulla; *f*, afferent lymphatic currents, which surround the follicles and flow through the spaces of the medulla; *g*, union of the latter into an afferent vessel (*h*) at the hilus of the organ.

The blood-vessels may be injected either from the small arterial branches which enter the gland when the organ is sufficiently voluminous, or, in smaller glands, by the neighboring large trunks; thus, for example, the pancreas Asellii of the smaller mammalia may be injected from the mesenteric arteries and the portal vein. Here the double injection readily succeeds.

A few years ago I gave more accurate directions for the injection of the lymphatics (*f*, *g*, *h*) by the afferent and efferent lymphatic vessels of the glands. In these cases, as a rule, finding the lymphatics usually causes greater difficulty than the subsequent manipulations.

Transparent and—as I will add, relying on recent experience—cold-flowing injection masses should always be used. But not all lymphatic glands are adapted for injecting. As with all injections of lymphatics, fat subjects and bodies already commencing to decompose are to be avoided. Œdematous portions of the body are usually best qualified. A preparatory immersion in water for several hours may also prove advantageous.

If a mammalial animal be employed, the following procedure presents the greatest advantages. The animal is to be killed by a blow on the head, or by strangulation. The ductus thoracicus is to be immediately ligated high up, and the body allowed to lie for 2–6 hours. After this interval the lymphatics are, for the most part, firmly distended, and readily permit of the injection being made in the direction in which their valves open. In injecting the vasa efferentia, on the contrary, it is difficult to overcome the resistance of the valves, and it only succeeds in isolated cases.

The various degrees of distention are here of great importance for the intelligibility of the whole current. At the commencement, therefore, only injections which have been early discontinued should be used, proceeding gradually to the employment of those which have been more prolonged. The injection of a second or even third lymphatic gland by the vasa efferentia of a previously injected gland affords very fine specimens.

It has already been remarked, at page 197, that Hyrtl and Teichmann have facilitated this procedure considerably by means of the puncturing method; and in fact this process accomplishes a great deal for the lymphatic glands. Fine tubes, cautiously introduced beneath the capsules of larger and smaller glands, as a rule, readily fill the investing spaces of the follicles, and from these the passages of the medullary substance. This method is indeed inestimable for the examination of the passages of pathologically metamorphosed lymphatic glands.

It may be practised with the syringe or with the constant pressure.

True glands, belonging in the narrower sense of the word to the lymphatic system, will be met with only in extremely rare cases naturally injected with decomposed hæmatine, in a condition serviceable for the microscopical analysis. On the contrary, animals fed with fat, or the bodies of persons who have died in the act of digesting fat, present a very important and instructive natural injection of the chyle-glands. Take one of the smaller mammalia,—for example, a rabbit or a small dog,—and introduce a considerable quantity of milk through an œsophageal catheter into its stomach. The animal is to be killed after from 4–7 hours, and, as a rule, the most exquisite injection of the entire chylous system will be found.

The recognition, by a finer analysis, of the chyle-fat in the interior of a somewhat more voluminous lymphatic gland is, however, a precarious matter. Fresh sections may be treated with a solution of albumen, as recommended by Brücke. Attempts should be made to render preparations which have been hardened in dilute chromic acid or weak alcohol transparent by means of soda solution. I have never seen any great effects from drying such glands either with or without a previous immersion in boiling water.

Extremely small chyle-glands, especially those consisting of a single follicle, such, for instance, as are found in the abdominal cavity of the rabbit, when examined fresh without further addition, and in the condition of fat assimilation, afford, on the contrary, handsome appearances.

The self-injection of the lymphatic glands has also been employed. Toldt employs for this purpose a very finely granulated aniline blue, precipitated from an alcoholic solution by the addition of water. It may be injected under the skin of the living animal, and being conveyed by the afferent lymphatic vessel, the injection of the neighboring gland may be expected. Or, the lymphatic glands lying in the vicinity of the liver of the dog may be selected.

As, according to Hering's experience, the hepatic lymph of narcotized creatures is rich in red blood-corpuscles which have escaped from the blood-vessels, one may accomplish the object by injecting this aniline mass for 7–8 hours in repeated doses of 12 grammes about every 10–15 minutes, in the vena cruralis of an animal narcotized with opium.

As is known, the human lymphatic glands are subject to numerous structural changes. A part of the latter are to be regarded as senile metamorphoses; others are of a more pathological nature.

Among the former (which may also occur at a relatively early period of life), we must especially adhere to three; namely, the formation of fat-cells, the pigmentation of the lymphatic glands, and the metamorphosis of the framework substance into ordinary connective tissue with the gradual destruction of the entire organ.

The fat-cells take their origin from the connective-tissue corpuscles of the framework of the lymphatic gland, and, as a rule, affect the cortical substance of the gland. It is only in rare cases that they are noticed in the lymph-tubes of the medullary substance. The glandular structure of the lymphatic gland disappears more and more in proportion as groups of fat-cells take the place of single cells.

The pigmentation of the lymphatic glands, as is known, affects the bronchial glands chiefly, and is, after certain periods of life, an almost regular occurrence, although of extremely varying degrees. If this process be followed from its very commencement it will be seen that, at least in most cases, the exciting cause is an inflammatory irritation of neighboring parts of the lungs. The consecutive tumefactions of the lymphatic glands, which are so frequent and with which practical physicians are familiar, are accompanied by very extraordinary enlargements of their finest blood-vessels, so that, for example, almost all the capillaries are found to be dilated to four and even six times their ordinary diameter. In consequence of these expansions, an exudation of the coloring matter of the blood takes place in the bronchial glands (and under certain circumstances in the lymphatic glands of other parts of the body also) so that the gland, saturated with a brownish fluid, assumes a "splenoid" appearance. At the same time lacerations of individual vessels and extravasations are also met with here and there. The molecules of black pigment arise by intermediate stages from the gradual transformation of the coloring matter of the blood. These are seen to be situated, without conformity to any law, partly in the framework substance of the septa and in the walls of the vessels. In many cases it is principally the substance of the follicles which, at least at the commencement, constitutes the seat of the melanosis; in others the condition is reversed, the medulla being attacked.

In this manner then, varying extremely in degree, pigmentations of the bronchial glands take place. Those of slight degree give the organ a sprinkled and spotted black appearance, but in higher degrees cause the black appearance to extend over greater distances, and even throughout the entire thickness of the organ.

While lower phases of this melanosis prove to be relatively indifferent for the organ affected, stronger pigmentations lead to the connective-tissue metamorphosis and atrophy of the lymphatic gland.

Such connective-tissue metamorphoses show bundles of striated and fibrillary tissue, at first isolated and then developed in the most extensive manner at the expense of the reticular framework. The characteristic structure of the organ disappears more and more, and finally, together with the loss of all the lymphatic passages, the whole gland becomes degenerated into a connective-tissue mass. These processes are observed with pigmentations, but also without them. The external lymphatic glands appear to be more subject to this process than those which are situated deeper in the body.

The ordinary methods suffice for the examination of the most prominent structural conditions. The previous injection of the blood-vessels with cold-flowing mixtures should, when possible, be at least attempted.

The true pathological metamorphoses of the lymphatic glands affect in part the framework, in part the lymph-corpuscles, and in part both elements together.

The structural changes of our organ in abdominal typhus are not very easy to follow. In the first, so-called catarrhal period of this disease, a tumefaction of the organ is met with, which consists chiefly of one of the above-mentioned extensive dilatations of the finest blood-vessels. The investing spaces of the lymphatic-gland follicles are enlarged, and in these are found a number of large multinuclear cells (which may also be found, although in smaller quantity, in other irritated conditions). The participation of the framework-substance appears, on the contrary, to be remarkably slight. At later periods these cells break down under fatty degeneration, and produce collections of very unequal extent, of a finely granular substance, the medullary typhous matter. Not unfrequently these form local softenings, into the sphere of which are drawn the adjacent tissues, the framework with the blood-vessels. In favorable cases, the finely granular substance is again removed by the efferent lymph-current.

A similar process, only much more sluggish in its progress, is met with in tuberculous and scrofulous lymphatic glands.

Here, together with the breaking down of the framework substance, the same degeneration also appears—a fine, molecular, fatty, waterless matter, with interposed, shrunken lymph-corpuscles. These "cheesy metamorphosed" masses may have various destinies allotted to them; they may be reabsorbed, become indurated and calcified, or soften and give rise to the formation of a fistulous passage.

In other pathological conditions the participation of the framework substance is more extensive. Thus, in secondary inflammatory conditions of our organ, the meshes of the network are seen to gradually become narrower, the trabeculæ increase in size, and distinct nuclei are again formed in the nodal points. In the more voluminous organs, where the expansions of the capillaries already mentioned may be recognized, there may be a gradual obliteration of the textural differences of the septa of the medullary and cortical substance. The lymphatic passages disappear, and the organ has become incapable of performing its functions. The later appearances of such lymphatic glands vary considerably, however. An interesting structural condition is sometimes presented in such cases, by the enormous thickenings of the capillary walls which are caused by the aggregation of spindle-cells.

Allied structural conditions are presented by the hypertrophies of the lymphatic glands. Here the capsule, the septa, and, at last, the medullary substance also, become transformed into a reticular tissue, which is uniform throughout the entire organ, and encloses numerous lymph-cells. This transformation of the capsule enables one to appreciate how the adjacent connective substance may be drawn into the sphere of the same metamorphosis, and a fusing together of the neighboring lymphatic glands take place. The reticular framework is either like the normal, or the meshes are seen to be narrower. In other cases, the fibres become much more strongly developed, so that a coarse-banded framework, like that of a carcinoma, may be formed. In the latter processes the large nucleated cancer-cells are met with in the meshes, varying in form and arrangement.

Formerly, the inflexible trabecular framework, which permeates the lymphatic channels (investing spaces), appeared to constitute the chief starting-point of the metamorphosis in question, as the cancer-cells arise in its nodal points, and its trabeculæ become the stroma of the carcinoma. At the present day, an original immigration of the first cancer-cells through the vas afferens into the investing spaces has become more probable. The glandular tissue becomes slowly and gradually atrophied.

A newer and more successful method of investigating these diseased lymphatic glands consists in their injection, and in the study of their lymphatic channels by the aid of the puncturing method. In such an organ, so long as there is a simple tumefaction, whereby those enormous distentions of the capillary blood-vessels are frequently met with, the lymphatic passages are all permeable. If the metamorphosis of the gland progresses further, as in typhus, and a breaking down of the lymph-corpuscles into the fine granular "typhus substance" occurs, such places become stopped up; the channels of hypertrophied lymphatic glands likewise become impermeable to a great extent. These are a few of the results which the author of this work has, thus far, obtained by suitable injections.

Many dark points still exist with regard to the origin of the lymphatic glands and lymphatic vessels of the fœtal body. Many years ago we became acquainted, through Kölliker, with interesting lymphatic vessels in the tail of the frog's larva. These run near the capillary blood-vessels, and appear as delicate, twig-like, ramified canals, without the reticular intercommunications of those vessels, and are characterized by the numerous small pouches into which their delicate walls are expanded. They contain a colorless fluid, almost free from cells, and it is certain that they are without an epithelial lining. Aggregations of neighboring spindle-cells are frequently met with on the membrane of the vessel.

We now turn to the methods of investigating glandular tissue.

Three elements participate in the formation of a gland, or —when its volume is greater and the structure more complicated—of its subdivisions. A transparent, apparently structureless membrane (membrana propria) constitutes the framework, and thus determines the form of the organ or part of the organ; strata of cellular elements (gland-cells) cover the inner surface of the latter and play an important rôle in the formation of secretions. Finally, the external surface of the structureless membrane is covered by a reticulation of capillary vessels, from the contents of which the matters for secretion are taken in the form of watery solutions.

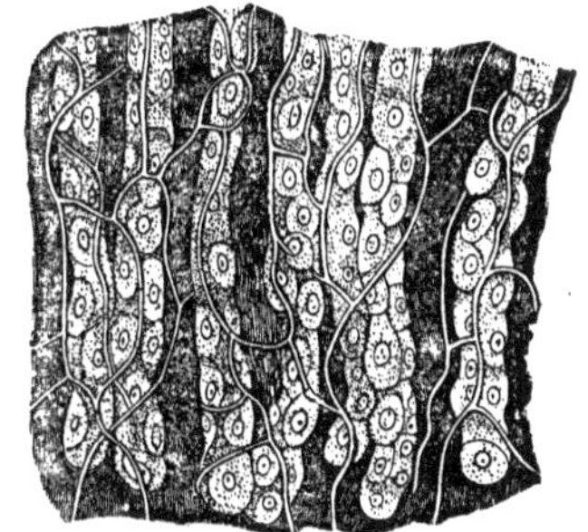

Fig. 200. Gastric glands of the dog, with cells and capillary vessels.

Our fig. 200, which presents the lower halves of long, simple, tubular glands from the gastric mucous membrane, may afford us a representation of this condition. The fine contours of the sinuous, blind-sack-like tubes present the optical expression of the membrana propria; nucleated fine granular cells form the contents, and a capillary network, spread out in the tubular form, surrounds the individual organ with elegant incurvations.

None of the glandular organs of the human body are without capillaries and gland-cells. This is not the case, however, with the membrana propria. It may be absent, and, indeed, under manifold conditions. In the first place we see that the fine membrane which is present in the earliest period of life is blended with neighboring parts, as in the liver. Or, the same has been absent from the commencement, and a firmly woven, connective-tissue limiting-wall encloses the aggregation of cells at all periods of life. The latter condition is shown, in addition to other organs soon to be mentioned, by the glands of Lieberkühn, a tubular form, very similar to the gastric glands of the stomach, which are placed close together in the mucous membrane throughout the intestinal canal. Finally, we have learned through more recent researches that a wicker-work of flattened, multi-radiated connective-tissue cells (p. 274) becomes visible in this membrana propria (fig. 201). This condition has been especially noticed in racemose glands (salivary, lachrymal, and lacteal glands), but also in the tubular glands of the mucous membrane of the stomach. Macerating methods and sections through hardened preparations are employed in these cases. An entire muster-roll of methods has been recommended: vinegar; the 33 per cent. solution of potash; the maceration for several days in iodine-serum, and then, subsequently for 24 hours, in a chromic acid solution of $\frac{1}{32}$ per cent. (or, chromate of potash $\frac{1}{10}$ per cent.); the immersion in osmic acid of $\frac{1}{2}$ per cent.; and the hardening by means of alcohol or the bi chromate of potash, with subsequent carmine tingeing.

Fig. 201. Plexus of star-shaped connective-tissue cells from the membrana propria, isolated by maceration. From the submaxillary gland of the dog, after Boll.

However, the numerous glands of the human body are of such manifold natures, according to their size, their complexity, and their entire structure, that the example made use of above can in no wise suffice for their comprehension.

Together with the simple tubular glands which we have

already become familiar with in the gastric glands of the stomach, other more complicated ones occur in which the lower cæcal extremity, with or without dividing, forms a number of coil-shaped convolutions. These organs have been provided with the appropriate name of convoluted glands. The most extended and familiar example of them is presented by the sudoriparous glands of the skin (fig. 202, *a*, *b*). The network

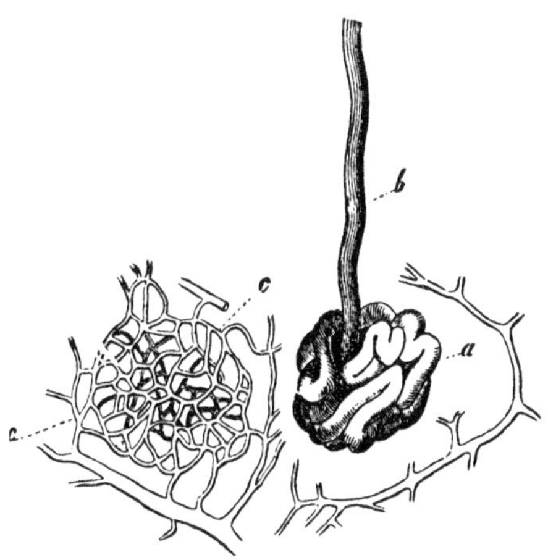

Fig. 202. A human sudoriparous gland. *a*, the coil, surrounded by the commencement of venous vessels; *b*, the excretory duct; *c*, the basket-like capillary plexus surrounding the coil, and the arterial trunk.

of vessels which encircles the coil becomes a sort of wicker-work with rounded meshes (*c*). The kidney and the testicle, two large, voluminous organs of the body, present much longer cylindrical tubes with divisions and reticular intercommunications. Fig. 203 represents these glandular tubes of the kidney, the so-called uriniferous tubes (1, 2).

Another form of glands, the racemose, is very widely diffused.

Roundish sacks (gland-vesicles), which are smaller or larger, longer or shorter, have their outlets associated together in groups. Such groups of sacks (gland-lobules) are again united by short ducts and by elongations of the membrana propria, and thus, sometimes in a slightly, sometimes more considerably (fig. 204), and sometimes extremely complicated manner, the racemose gland is formed. The changes which are here to be observed, and the manner in which the efferent canal-work gradually advances to a more compound texture—these, as well as many other particulars, must be learned by reference to the text-books on histology.

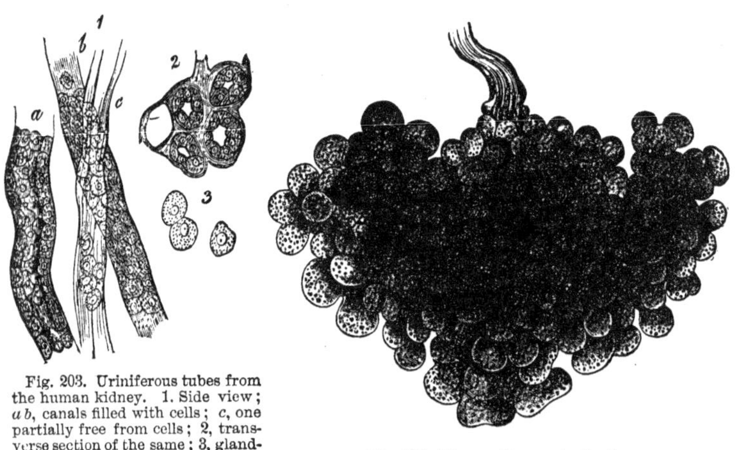

Fig. 203. Uriniferous tubes from the human kidney. 1. Side view; *a b*, canals filled with cells; *c*, one partially free from cells; 2, transverse section of the same; 3, gland-cells.

Fig. 204. Human Brunner's glands.

Notwithstanding many subordinate variations, there is one and the same fundamental design present in all, and easy to be observed from the mucous follicles, which are to be called almost microscopical, to the most voluminous examples, such as the salivary glands and the pancreas.

The gland-cells (which we have still to mention more especially) present numerous variations according to the nature of the actual secretion; the capillary network surrounding them, on the contrary, always exhibits circular meshes (fig. 205).

There is still a third form of glandular organs; such, namely, in which the membrana propria constitutes a roundish vesicle, closed on all sides, with cells in its interior, and surrounded externally with a network of capillaries; the entire organ being composed of a large or larger proportion of such vesicles embedded in a connective-tissue groundwork.

The ovary (fig. 206) represents the latter arrangement. Its

Fig. 205. Vascular network of the rabbit's pancreas.

gland-vesicles, called Graafian follicles (*c*, *d*), contain, together with small roundish gland-cells, a large globular cell, the ovum. The latter, by the rupture of the (rather complicated) wall, hav-

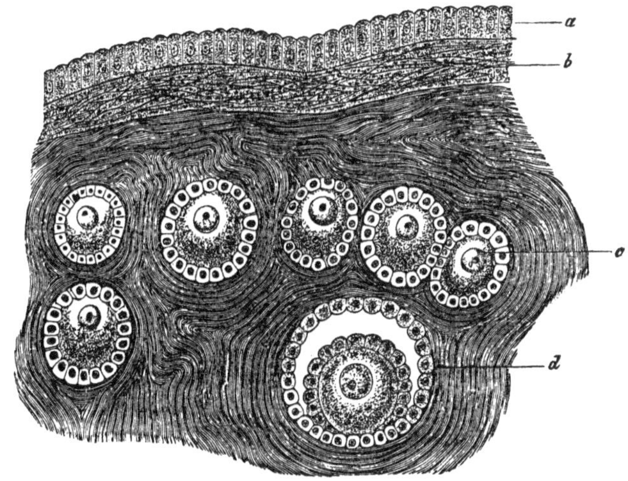

Fig. 206. Ovary of the rabbit. *a*, epithelium (serosa); *b*, cortical or external fibrous layer; *c*, youngest follicle; *d*, a somewhat more developed older one.

ing arrived at the end of its existence, undergoes a process of cicatrization as the corpus luteum, as it is called.

Still another variety of such glands with closed vesicles has been assumed. The capsules are said to form a secretion in their interior from the elements of the blood, and when the secretion is perfected, it is consigned to the blood and lymphatic vessels for removal. This is a very unsatisfactory explanation of the dilemma, arising from the experience that such a dehiscence as is exhibited by the ovary is never observed in the organs in question.

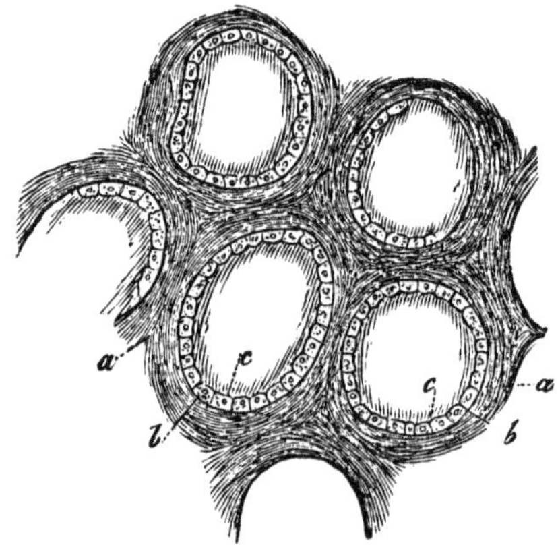

Fig. 207. Thyroid gland of the child. *a*, connective-tissue framework; *b*, capsules; *c*, their gland-cells.

They were formerly rather liberal in the acceptation of such organs, so-called " blood vascular glands." At present we have learned to separate many of them as belonging to the lymphatic glands, or, at least, as being nearly related to them; such as the thymus, the spleen, the Peyerian and solitary follicles of the intestines, the tonsils, and the conjunctival follicles. Only a limited number of enigmatical structures, especially the thyroid gland (fig. 207), the supra-renal capsules, and the hypophysis cerebri still find a place here.

The alleged membrana propria (fig. 207 *b*) which former observers believed that they saw on these structures, appears in reality, however, not to exist. We believe that its presence, at least in the hypophysis cerebri, the supra-renal capsules, and the thyroid gland must be denied. The predecessors were deceived, in consequence of inefficient methods of examination, by the compactly arranged connective-tissue parietes.

Finally, the cellular constituents of our organ are of greater importance. The gland-cells proceed, as we have learned in the most certain manner from Remak's admirable investigations, from the fœtal epithelial layers, the so-called horn and intestino-glandular lamellæ, and present originally partly solid cell-growths, partly hollow diverticula. In accordance with this, much of their character remains closely allied, in all their processes of life, to the nature of epithelium, and, even in the excretory ducts of the glands, the continual transition into the adjacent epithelial tissues may be observed.

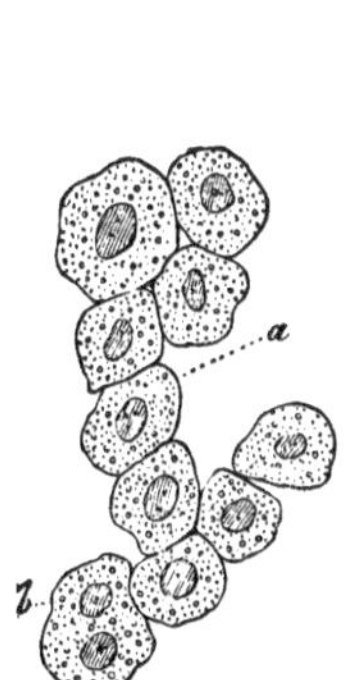

Fig. 208. Human liver-cells, with a single nucleus at *a*; one with two nuclei at *b*.

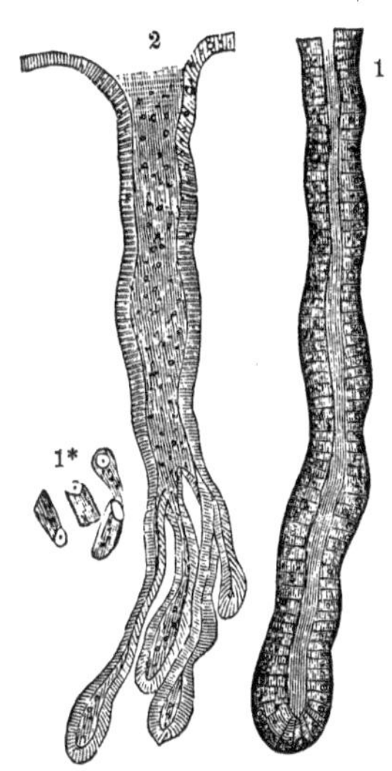

Fig. 209. So-called mucous gastric glands. 1. From the cardiac portion of the hog's stomach; *a*, the cylindrical cells (at 1* isolated); *b*, the lumen. 2. From the pylorus of the dog.

The cells which we meet with in the various glands are in part circular, in part flattened, and in part cylindrical nucleated cells (fig. 208 and 209). As a rule, especially with a certain amplitude of the passages, these cells clothe the inner wall after the manner of epithelium (fig. 209), so that a lumen still remains, and only narrow passages, such, for instance, as those of the liver, are found to be filled up by several cells placed one behind the other. In consequence of mismanagement in the preparation, as well as of cadaverous decomposition, these aggregated gland-cells generally become detached and frequently fill the entire cavity of the gland with fragmentary structures, and even free nuclei and molecules.

The gland-cells display their relationship to the epithelial structures, at least partially, in still another, and indeed physiological manner; namely, by a certain transitoriness of their existence, and by their falling off from the glandular walls. Although the duration of their life varies to a greater extent, and though many gland-cells, such as those of the liver and of the renal passages, are of a more persistent nature, so that the formation and excretion of certain secretory elements is repeated by them for a longer time, there are also, on the other hand, many examples of a more rapid separation. At every act of gastric digestion numerous cells of the gastric glands become separated from their parent tissues and cover the inner surface of the stomach, at least in certain mammalia, with a thick mucous covering. Other glands which prepare a fatty secretion present, as a physiological process, the fatty degeneration of the cells, and in this way innumerable quantities of the latter are destroyed. In this manner, by the destruction of innumerable cells, is formed the secretion of the sebaceous follicles, many sudoriparous and Meibomian glands, and likewise of the lacteal glands.

An example of this physiological cell destruction may be represented by fig. 210, the ovoid vesicle of a sebaceous gland.

This appears at *a* to be lined by stratified layers of rounded cells in which the fat-molecules are to be recognized, sometimes in smaller, sometimes in larger quantities. Other cells (*b*), containing a larger quantity of fat, are already separated from the parental tissues, and, already in part undergoing dissolution, fill the cavity of the gland vesicle. In this way is explained the occurrence of free masses of fat in the lower educting portion of the latter; in this way, also, is formed the smegma cutaneum. The various cells of this form of gland are seen at *B*, *a–f*, more highly magnified.

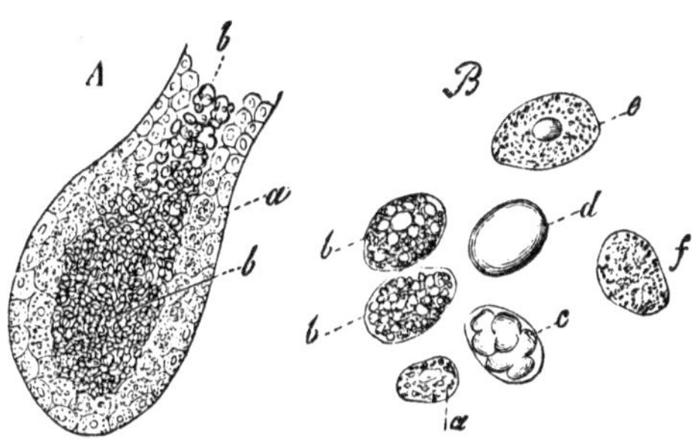

Fig. 210. Human sebaceous follicle. *A*, gland-vesicle with the cells resting on the wall at *a*, and separated and overloaded with fat at *b*; *B*, *a–f*, several of these gland-cells.

In examining the gland-cells (the investigation of which in the living condition has, unfortunately, been almost entirely neglected), the most conservative treatment is necessary. Sections made through an entirely fresh part yield to a knife-blade which is moved or scraped over it, masses which, spread out in an indifferent fluid, will often afford satisfactory examples of the cells in question. Glands will occasionally be met with, which, in a condition of vital warmth, are so solid that with a very sharp and moistened razor very thin sections may be made from them. When these sections are examined in indifferent media, such as iodine-serum or extremely dilute chromic acid, they show the position of these cells, and also permit of their isolation by proper picking. Generally, however, such procedures fail in consequence of the softness of the structure. We therefore recommend the freezing method as the best procedure at present in use for this purpose. Hardening methods have been used for a long time for the examination of cells in situ. Drying the organ is not to be recommended, as deeper alterations of the cells and the subsequent separation of many of these structures can scarcely be avoided. It is better to employ a solution of chromic acid or of the bichromate of potash of gradually increasing strength, by means of which excellent preparations may be obtained.

The immersion of small pieces of the glands, removed from the body immediately after death, in a large quantity of absolute alcohol, is an excellent procedure, and the proper consistence is obtained in a few hours.

Tingeing the gland-cells may be best accomplished with glycerine-carmine, Ranvier's mixture of picric acid and carmine, or a solution of hæmatoxyline.

It is scarcely necessary to mention that the application of numerous chemical reagents is necessary for the recognition of the contents of these cells, as well as that the freshest possible tissue is to be used for this purpose.

Solutions of the alkalies, ammonia, potash, and soda, are most to be recommended for the demonstration of the membrana propria of the glands.

There are various methods for us to select from for the investigation of the relations of the last-named membrane, as well as of the entire structure of the glands. Drying, with the subsequent action of alkalies on the moistened sections, is to be employed with advantage for many parts, as, for example, for the glands of the skin and eyelids. If one desires to study the organs which are embedded in mucous membranes, boiling the piece in question in vinegar and then drying it is to be recommended. It is also well to use this preparatory treatment with vinegar for the skin, the lacteal glands, and likewise for the kidneys.

In the moist condition we can often obtain a sufficient hardening by means of pyroligneous acid, and, as in the previously mentioned process, in consequence of the connective tissue being rendered transparent, we obtain good views from thin sections.

The three above-mentioned, so frequently employed fluids—alcohol, solutions of chromic acid, and of the bichromate of potash—appear, however, to be more important, and, in fact, they generally suffice for the glandular tissues. Where brushing is unnecessary, the tissues may be energetically hardened with stronger degrees of concentration. But if one desires to make use of the procedure just mentioned—and it is of the greatest value for the recognition of the framework of the glands, the vessels, incidental muscles, etc.—the matter should not be overdone. Notwithstanding every precaution, however, many variations will still be met with. Sections of the kidney and of the testicle, and surface sections of the gastric mucous membrane are generally easy to brush; it is difficult, on the contrary, to obtain good preparations from the liver. The chloride of palladium is to be used for recognizing the muscular elements in glands.

The fine blood-vessels which encircle the glands are generally concealed by the cellular contents of the glands, and even after the most careful brushing are only very imperfectly brought to view. The artificial injection with transparent masses, a light blue, should not therefore be omitted. This procedure naturally varies considerably according to the individual organs.

Injections are employed in still another way for glands, naturally only the more voluminous ones; namely, for filling their cavities. Cold-flowing masses (either, and preferably, purely watery ones, or at most mixed with glycerine, but not alcohol), entirely fresh organs, and great care are necessary if such experiments are to be successful. The employment of a constant pressure is decidedly preferable to the syringe for this purpose.

By these means, a plexus of very fine canals, the "gland-capillaries," surrounded by an extremely delicate sheath, has frequently been recognized in an interesting manner, lying between the secreting cells and surrounding each one of them. The presence of this network in the liver has been known for some time. We shall have to mention these biliary capillaries hereafter. Within a few years they have also been discovered in racemose glands, as in the pancreas (Langerhans, Saviotti, Gianuzzi), in the salivary glands (Pflüger and Ewald), in the lachrymal gland (Ball), and in the lacteal gland (Gianuzzi and Falaschi). Our fig. 211 may represent these remarkable canals, which here run partly between the cells themselves, partly on the surface between them and the membrana propria.

For the investigation of fœtal glands, embryos hardened in absolute alcohol or chromic acid are to be chosen, and sections made through them in various directions. The separated skin, as well as mucous membranes, often present very good surface views. The origin of the membrana propria, whether it takes place from the cell-aggregation by a process of separation, or

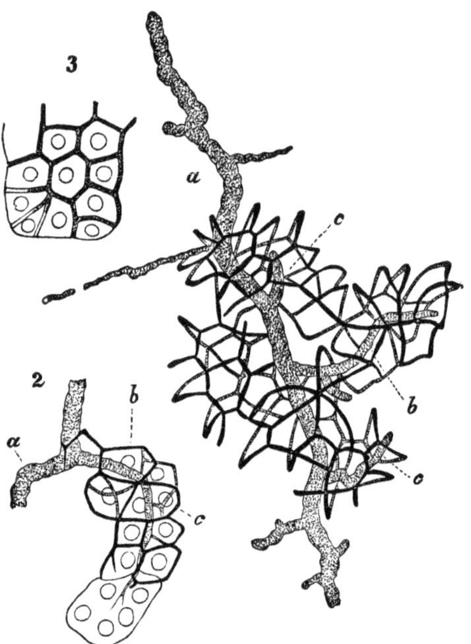

Fig. 211. Glandular canals from the pancreas of the rabbit, injected with Brücke's Prussian blue; after Saviotti. 1 and 2: *a*, larger excretory duct; *b*, that of an acinus; *c*, finest capillary duct. 3, an acinus with cells and only partially filled gland-capillaries.

is formed from the neighboring tissues in consequence of an aggregation of the same, requires more accurate investigation than has thus far been devoted to it.

We would add a few words in closing, touching the pathological conditions of the glandular tissues.

The gland-cells (corresponding to their epithelial nature) present the phenomena of hypertrophy and degeneration, but scarcely that of a transformation into other tissues. This takes place, as a rule, more from the connective-tissue framework substance which permeates the organ, to which, perhaps, the so-called membrana propria of the gland is to be reckoned throughout.

Hypertrophies of a gland show, as a rule, an increase in number of the secreting cells, which we at present ascribe to a more active process of division; although the existing cells may also increase in size, and thus cause an increase of volume. Both conditions are found, for example (frequently enough combined), in hypertrophied livers.

We have already mentioned above the accumulation of fat in the interior of the cells we are at present considering. For many glandular organs it constitutes an entirely normal occurrence. In others such a destruction of the cells is an abnormal phenomenon, a process of degeneration. Pigmentations of the gland-cells are more rare; amyloid degenerations occur in these structures, at least in many cases, while, as a rule, they affect the vessels and the connective-tissue portions of the gland.

Colloid degenerations occur in certain glands at least, and affect their cells quite extensively, especially in the thyroid (fig. 212).

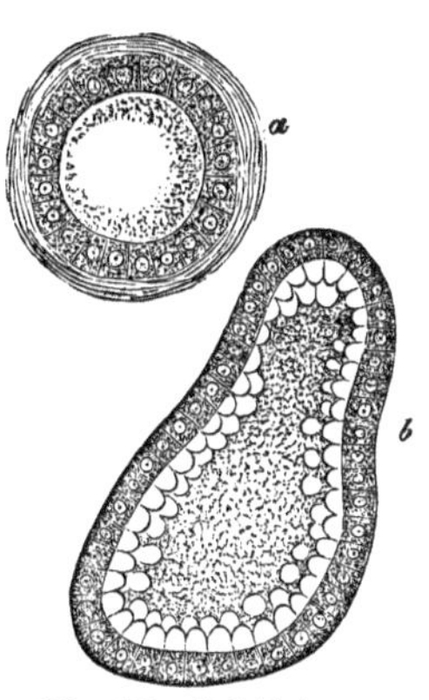

Fig. 212. Colloid degeneration of the gland-vesicles of the thyroid. *a*, from the rabbit; *b*, from the calf, at the beginning.

Swelling of the connective tissue, increase of the interstitial substance, distention of the connective-tissue corpuscles and the division of their nuclei, are met with in conditions of simple inflammatory irritation. A more persistent increase of the connective-tissue of the gland may lead to the destruction of the gland-cells in the compressed cavities. That tuberculous and typhous degenerations, and carcinomatous new formations in glandular organs, also take their origin from the connective tissue, has hitherto been the general modern acceptation. Our present knowledge concerning the structural changes of the liver and kidneys may constitute an important starting-point for subsequent investigations of smaller glandular organs.

Cysts, according to experience, frequently originate from glandular passages when the secretion, in consequence of obstruction to its egress, accumulates more and more and distends the duct.

The new formation of glandular tissue and of entire glandular organs is likewise not an unfrequent occurrence. The former is seen in hypertrophied structures. Entire glands occur in mucous polypi. We also meet with tubular and sebaceous glands, together with hairs, teeth, etc., in cysts of the ovary.

There are no special methods of investigation to be mentioned here.

Section Seventeenth

DIGESTIVE ORGANS.

THE study of the digestive apparatus, its parietes, the glands connected with it, and the substances which it contains, constitutes an extensive section of microscopic investigation. In consequence of the decomposition which so readily takes place in them, most human bodies appear but little adapted for this purpose, so that for many textural conditions, recourse is more advantageously had to a recently killed mammalial animal. The bodies of new-born children are still better adapted.

The lips present a transition of the tissue of the external integument to that of the mucous membrane, as well in its epithelial as in its fibrous layers. The finer structure of the same is to be examined either in dried preparations (also previously boiled in vinegar), or those which have been hardened by means of alcohol or chromic acid. The small sebaceous follicles, which were discovered in them a few years ago, may be recognized without much difficulty by the application of acetic acid.

In the oral and pharyngeal cavities are presented for examination the mucous membrane with the small glands belonging to it, the teeth (already described), the tongue, the tonsils and lingual follicles, and finally the salivary glands as well as the secretion of the oral cavity, the saliva.

In order to accomplish the injection, which is so necessary, of this commencing portion of the digestive tract, we would recommend the use of the smaller mammalial animals and the insertion of the canule in the arch of the aorta, as mentioned above (p. 356), for the brain. A complete injection of the oral cavity, the tongue, and the pharynx may be readily obtained in this way. A blue deserves the preference on account of the subsequent carmine tingeing.

The mucous membrane, with its papillæ, vessels, nerves, and glands, may be reviewed in vertical sections of fresh preparations made as thin as possible, and then rendered still more transparent by means of solutions of soda, or dilute acetic acid. It is always a tedious affair, however, to obtain these from such a soft and slippery tissue, so that naturally the customary hardening methods are also extensively employed.

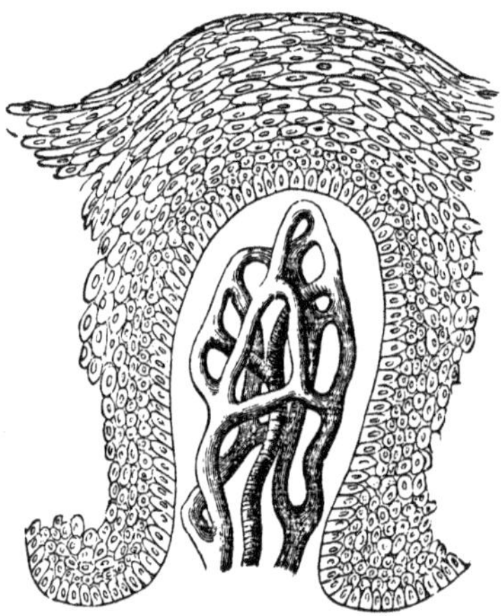

Fig. 213. Injected papilla from the gum of a child.

The mucous membrane and numerous conical or filamentous papillæ, covered by the thickly stratified pavement epithelium, may be recognized with facility in good alcoholic preparations (fig. 213). The numerous racemose or mucous glands of the oral cavity are rendered apparent by the application of this acid or, still better, after the use of alkaline solutions. Their vesicles frequently appear considerably elongated (Puky, Akos), and their gland-cells cylindrical. The gingival mucous membrane of the rabbit constitutes a beautiful object for this purpose.

In order to recognize the general arrangement of the nerves, the gradual hardening with weak solutions of chromic acid or chromate of potash, with the subsequent employment of a very dilute acetic acid, is to be recommended. An immersion of the fresh tissue in the acetic-acid water (1–2 drops of acetic-acid hydrate to 50 ccm.), mentioned at the examination of the muscular nerves, affords, after 12–24 hours, very suitable specimens, especially with the lower vertebrate animals. Pyroligneous acid has been frequently employed here. Osmic acid and chloride of gold are to be tried for more accurate studies.

The examination of the tongue requires various methods, according to the portion of the structure of the complicated organ which one desires to investigate.

To follow the general arrangement of the muscles, one may use tongues which have been for a long time in alcohol, also fresh ones, which must be boiled in water, however, until they have become quite soft. To obtain finer sections, hardening in alcohol or the freezing method may be employed. Thin sections afford beautiful specimens after being tinged with carmine and washed in acetic-acid water, or stained with hæmatoxyline, or with carmine and picric acid, after Schwarz's method, and likewise by the direct application of acetic acid or dilute solutions of soda. The tongues of the smaller mammalia deserve the preference to those of larger animals, and likewise those of embryos to those of older creatures.

Considerable attention has been paid for some time to the divisions of the filaments of the lingual muscles. They may be readily discovered in the lower amphibia, frogs, tritons, etc., by means of the usual maceration in dilute pyro-acetic acid; an immersion in a very dilute solution of chromic acid is also to be recommended. The strong muriatic acid (see p. 129) has been subsequently employed for this purpose, and the divided filaments have also been perceived in the human tongue (Rippmann). The connection of the muscular fibres which ascend into the papillæ of the frog's tongue with the connective-tissue corpuscles, which was observed by Billroth and corroborated by Key, is to be followed in pyro-acetic acid preparations.

The mucous membrane of the human tongue, with its pavement epithelium, does not require any special methods. The epithelial processes of the papillæ filiformes are often quite long, and their formation out of individual cells may be recognized after the application of alkalies.

The nerve-terminations of the tongue will be mentioned further below at the organs of sense.

In larger animals, also, the injection of the blood-vessels does not present any difficulties. The familiar puncturing method serves for the lymphatics and the lymph-passages in general, which are quite abundant in the tongue, and form cul-de-sac-like axile canals in the papillæ filiformes.

Glycerine is adapted for mounting permanent preparations, or, after tingeing, Canada balsam. Excellent preparations are obtained by the latter method, which permit of the recognition of many histological details, not only of the commencement, but also of the entire digestive tract.

Of late years the tongues of mammalial animals have also become important and profitable objects for experimental pathologists. On them Wywodzoff and Thiersch have studied the healing process of wounds. The injection of the blood-vessels with gelatine cannot be dispensed with in these cases. To render the tissue of the organ injected with carmine visible, Thiersch made use of a modified silver impregnation. He dissolved 1 part of nitrate of silver in 5,000 parts of alcohol, and placed the sections in it for 5 minutes, with repeated shaking. They are then to be shaken for a few seconds in a saturated alcoholic solution of chloride of sodium, after which they are to be exposed to the light. They now show their histological details beautifully, and may be mounted permanently in a resinous substance.

We may rapidly pass over the methods of examining the tonsils (fig. 214) and the lingual follicles, for they are the same as

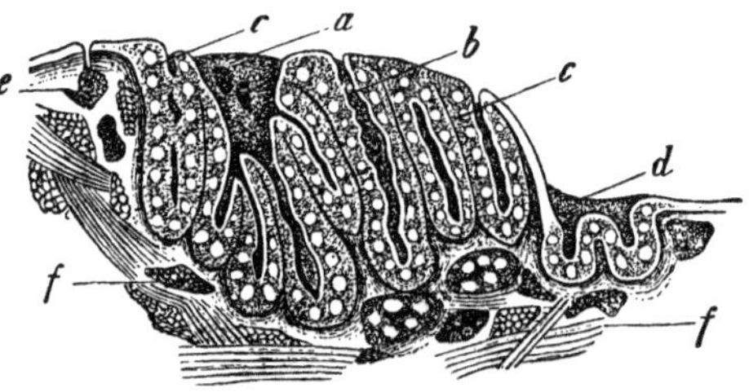

Fig. 214. Tonsil of the adult (after Schmidt). *a*, larger excretory duct; *b*, more simple one; *c*, lymphoid parietal layer, with follicles; *d*, lobule, reminding one of a lingual follicle; *e*, superficial, *f*, deeper mucous follicle.

for other lymphoid organs. Here also chromic acid, bichromate of potash, and alcohol are employed as hardening media. Thin sections, cautiously brushed and tinged, readily permit of the recognition of the structure. In consequence of the numerous diseases of the tonsils, however, the precaution should be observed to use the bodies of new-born or small children;

likewise the younger specimens among mammalial animals. Of these, I would especially recommend dogs, pigs, and calves. The puncturing method, cautiously performed beneath the investing tissues, fills the numerous lymphatics in the calf and ox without difficulty; with somewhat more trouble in the dog; but, according to previous experience, extremely seldom in a satisfactory manner in the hog.

The glandular follicles of the tongue are difficult to inject; the recognition of their structure is, on the contrary, relatively easy.

To obtain the salivary corpuscles which exude from the cavities of the tonsils, pressure should be cautiously made on a tonsil immediately removed from a calf which has just been killed. A thick, glairy mucus, containing a number of these cells, will then make its appearance.

The salivary glands have been frequently examined of late. A whole series of methods of investigation has been given. Heidenhain and Pflüger employ absolute alcohol for hardening, and a subsequent conservative tingeing with carmine. Picked preparations may be made from fine sections of the entirely fresh organ, by adding the gland secretion, humor aqueus, iodine-serum, or a very dilute chromic-acid solution (0.02–0.04 per cent.), with a slight addition of the previously mentioned fluid.

Heidenhain recommends for maceration iodine-serum, chromic acid ($\frac{1}{32}-\frac{1}{2}$ gr. to the oz.), or bichromate of potash ($\frac{1}{4}$–15 gr.). Acidulated water (0.02 per cent. glacial acetic acid), with the subsequent immersion in chromic acid ($\frac{1}{16}$ gr. to the ounce), also affords good preparations.

Pflüger employs the action of iodine-serum for from 4–6 days, either alone or with the subsequent immersion in chromic acid of 0.02 per cent. It is furthermore very good to place the organ (the submaxillary gland of the rabbit) in a small glass vessel, and to add 4 to 8 drops of the chromic-acid solution, and, after an hour, when the gland has become hardened by swelling, to make fine sections from it for the purpose of picking. The 33 per cent. solution of potash also deserves to be employed. Osmic acid serves for the nerve termination.

Krause employed molybdenate of ammonia, with the subsequent use of tannic or pyrogallic acid (p. 158). Finally, Ranvier used for macerating, dilute—for hardening, concentrated picric acid, and for the latter preparations tingeing with picro-carmine (p. 160). The injection of the blood-vessels of the sub-maxillary gland of the dog, for instance, is not difficult. For the recognition of the lymph-passages Gianuzzi recommends exciting a condition of œdema in the same organ. The natural injection may be employed here, the gland being ligated at the hilus, and removed with the preservation of the capsule, and hardened for a few days in a solution of chromate of potash, and then placed in alcohol. Or the gland may be removed, and then carefully injected by the artery with colored gelatine, the aperture of the vein being left open. It is then to be hung in alcohol for a few days, in order that the capsule may be rendered more firm; and, finally, a puncture is to be made near the

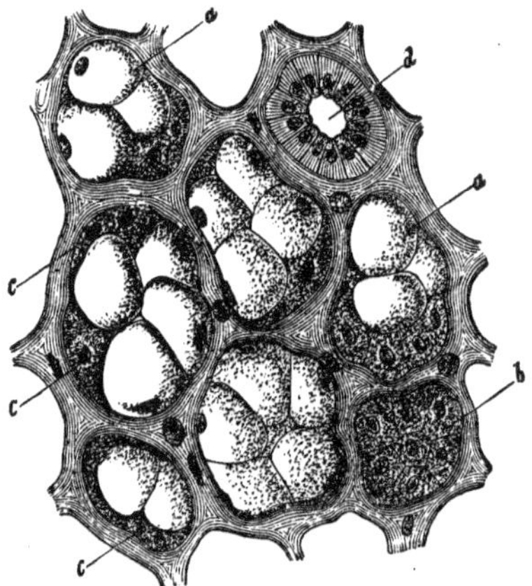

Fig. 215. The submaxillary gland of the dog. *a*, mucous cells; *b*, protoplasma cells; *c*, Gianuzzi's crescent; *d*, transverse section of an excretory duct, with the peculiar cylindrical epithelium.

artery, at the place where it sinks into the gland-tissue at the hilus.

The submaxillary glands of many of the mammalial animals, such as those of the dog and the cat (but not those of the rabbit), are mucous glands. In the quiescent organ (fig. 215) one may recognize, together with granular cells containing protoplasma (*b*), which frequently present a small and compressed semilunar structure (*c*) at the border of the gland-vesicle (the "crescent" of Gianuzzi), other gland-cells (*a*) which are larger and of a hyaline transparency, whose contents are not reddened by carmine, and which prove to be mucous. Our figure shows still another peculiar condition which is very easy to recognize, a delicate longitudinal striation of the cylindrical epithelial cells in the excretory duct (*d*).

When Heidenhain had induced an increased secretion in the submaxillary gland of the dog by a prolonged irritation of the nerves, an entirely different appearance presented itself (fig. 216). The mucous cells had nearly all disappeared, and in their place the acinus was filled by granular structures alone, which were rich in protoplasma (*a*).

If it be desired to attempt the injection of the canal system (p. 413), cold-flowing blue, without alcohol, is the best injection fluid.

Finally, the condition of the oral cavity and the fluids which it contains also require a short notice. The latter consist of the mixture of mucus and the secretions of the numerous glands which open into this cavity; namely, the secretion of the salivary glands. By hawking and coughing, the secretory products of

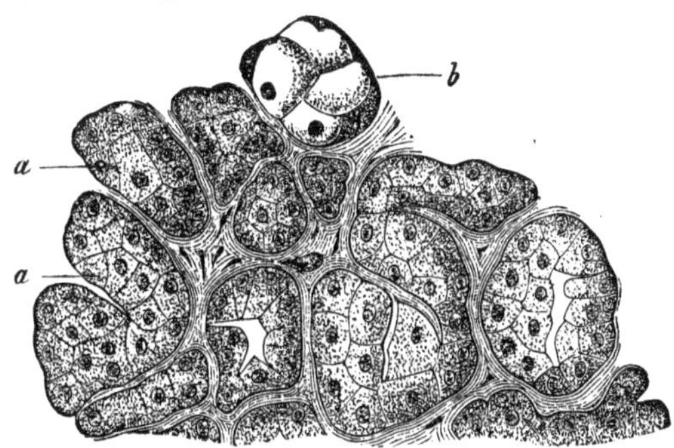

Fig. 216. The same submaxillary gland after prolonged irritation of the nerves, after Heidenhain. *a*, protoplasma cells; *b*, remains of the mucous cells.

the air-passages, and by vomiting, arrested contents of the stomach, as well as remains of food and particles of dust, may be associated with these essential integrants. On examining the parietes of the oral cavity, especially the papillæ filiformes on the back of the tongue (Fig. 217) and the gums at the base of the crown of the teeth, they are found to be covered by a sometimes thinner, sometimes thicker, slightly brown, fine granular covering, which contains, together with decomposed animal substances, the filaments and fragments of a lower vegetable organism from the order of the Schizomycetes (Nägeli). This (the Leptothrix buccalis of Robin) consists of a confused mass of extremely fine filaments.

The gastric furred tongue, when in a rough condition, shows a proliferation of the familiar epithelial processes of the papillæ filiformes, or, when the surface is smooth, a covering composed of luxuriant epithelial cells, the above-mentioned vegetable filaments and mucous corpuscles.

Fig. 217. A papilla filiformis with its epithelial processes, over which is spread out the matrix of the Leptothrix buccalis, from which also single filaments are growing out.

The substances in question may be readily obtained from the living body for examination, by scraping with the blade of a knife. To understand the entire arrangement, a fresh body should be used and recourse be had to vertical sections after a previous hardening.

The vegetable organism just mentioned must, in consequence of its frequency, be designated as a normal occurrence. Another vegetable parasite from the group of fungi, Oidium albicans, occurs in thrush (muguet), a very frequent disease of the earlier period of suckling (fig. 218). In the ordinary, slighter grades of the disease, its collections appear as whitish, later as grayish-yellow plates, sometimes more isolated, sometimes confluent, and, in high degrees, almost covering the entire oral cavity, even extending down into the œsophagus. If we place a small portion of this, mixed with water or some alkaline fluid, under the microscope, we see much broader, jointed, fungous filaments (*a*), with spores (*b*) and mycelium, so that it is impossible to confound them with the Leptothrix buccalis, the filaments of which are so fine.

A drop of saliva, placed under the microscope, shows air-bubbles entangled in it, sometimes in smaller, sometimes in larger numbers, then the separated pavement epithelium of the oral cavity which floats about in the fluid, partly hanging together in shreds, partly isolated (fig. 219), and either unaltered

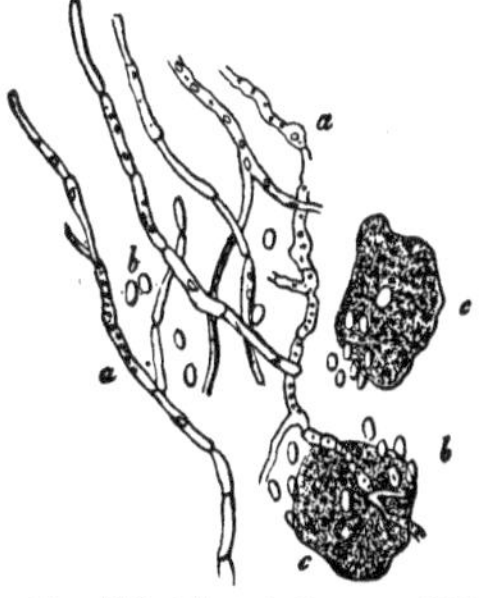

Fig. 218. Thrush fungus, Oidium albicans of the nursing child. *a*, fungous filaments; *b*, spores; *c*, pavement epithelium of the mouth.

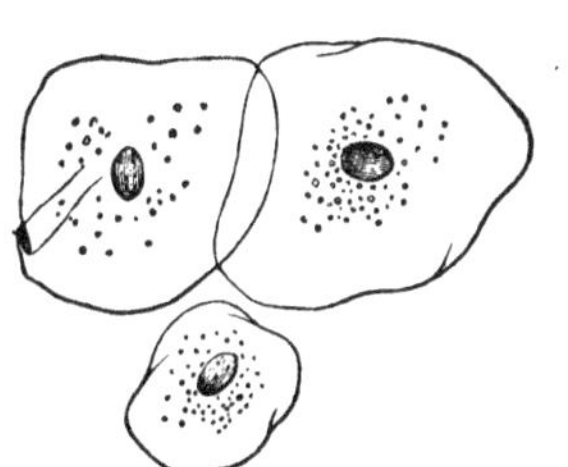
Fig. 219. Pavement epithelium of the oral cavity.

in appearance or having already undergone maceration to a certain extent. Finally, the salivary corpuscles are noticed as an element which, though never absent, still varies in quantity. Fresh, living cells of this kind show, with a higher magnifying power, a distinct dancing movement of the elementary granules which occur in their bodies. Consequently, effete salivary corpuscles, which are undergoing decomposition, no longer present this movement phenomenon.

Filaments of cotton, lint, etc., remains of food—for example, fibres of meat, granules of starch, particles of vegetable tissue, fragments of milk, appearing in the form of fat-globules and drops—form adventitious constituents of the saliva.

The methods for examining the œsophagus are the same as those for the oral cavity, and may therefore be omitted here.

The investigation of the stomach is, on the contrary, of higher importance. In its examination always, when possible, avoid older cadavers and, for many observations, use only the recently killed, not yet cold mammalial animal. Fine sections through the soft tissue are difficult to make, but very easy, on the contrary, through the frozen parietes. On these may be perceived, by the addition of indifferent fluids, the peptic-gastric glands of the mucous membrane, the gland-cells, and finally the cylinder epithelium of their apertures, as well as of the surfaces lying between them. A not too prolonged immersion in a $\frac{1}{4}$–$\frac{1}{8}$ per cent. solution of osmic acid has recently been recommended for these delicate cellular coverings (Ebstein). The addition of dilute alkalies rapidly dissolves these gland-cells, so that the membranes of the tubes only remain. Hardening methods (absolute alcohol, chromic acid, chromate of potash, osmic acid) are necessary for the more accurate study of their arrangement, as well as for that of other elements lying in the tissue of the mucous membrane. Injections readily succeed. Either

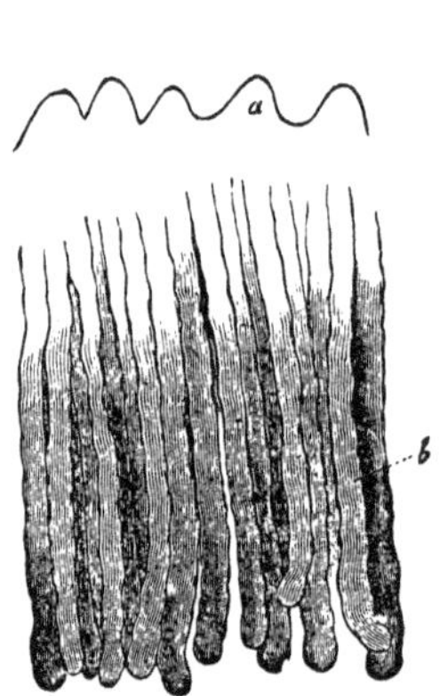

Fig. 220. Vertical section through the mucous membrane of the human stomach. *a*, superficial papillæ; *b*, peptic-gastric glands.

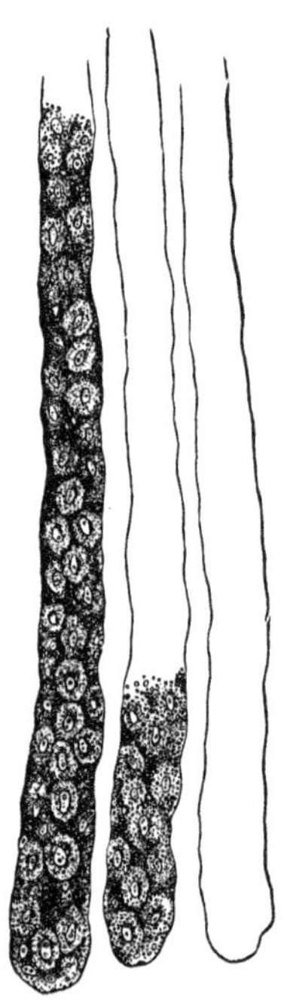

Fig. 221. Three human peptic-gastric glands.

the arteria cœliaca or the vena portarum are to be selected in smaller creatures; in larger animals an arterial branch on the external surface of the stomach is to be used. All attempts to ascertain the presence of a lymphatic apparatus permeating the mucous membrane have been thus far, on the contrary, unsuccessful.

To obtain fine views of the tubular-shaped gastric glands (fig. 220) it is best to prepare thin, vertical sections from the mucous membrane hardened in absolute alcohol; they are to be examined in glycerine, without the addition of any more strongly acting reagent. The simple and more complicated tubular glands, as well as the several varieties of cells which line them, may then be readily recognized. Fine tingeing constitutes an important accessory for further details. We recommend here Heidenhain's directions for carmine and aniline staining (p. 155 and 157). Fine transverse sections are naturally indispensable for other conditions.

One form of the gastric tubes (fig. 221) bears the name of peptic glands. At the present time we can with certainty ascribe the production of the pepsin to these alone. At the first view their compact contents appear as large granular cells (fig. 222).

Recent more accurate investigations (Heidenhain, Rollett), however, show a further composition. There are two forms of the gland-cells to be distinguished. The one, smaller and more transparent, usually appears to line the whole interior of the tube in a coherent layer; the other, larger and more granulated, appears more externally and isolated. The latter is the peptic cell of the writers, called by Heidenhain "*Belegzelle*," by Rollett, "delomorphous" cell. The smaller continuous form is called by the former observer the "*hauptzelle*," by the latter the "adelomorphous" cell. Our figure 223 represents the condition mentioned, from the stomach of the dog (which is especially adapted for these investigations).

A series of statements made by Heidenhain concerning the condition of the peptic-gastric glands in the condition of rest and of activity is extremely interesting. In the fasting animal the tubular glands appear shrunken, their contours are smoother, and their *haupt*-cells are transparent (fig. 223, 1). Several hours after the reception of food the peptic-gastric glands present an entirely different appearance (2, 3). They are swollen, the walls irregularly dilated, the *haupt*-cells are enlarged and rendered cloudy by their finely granular contents. Finally, at a later period (4) shrinking has again taken place, the *haupt*-cells are considerably diminished in size, but are also very rich

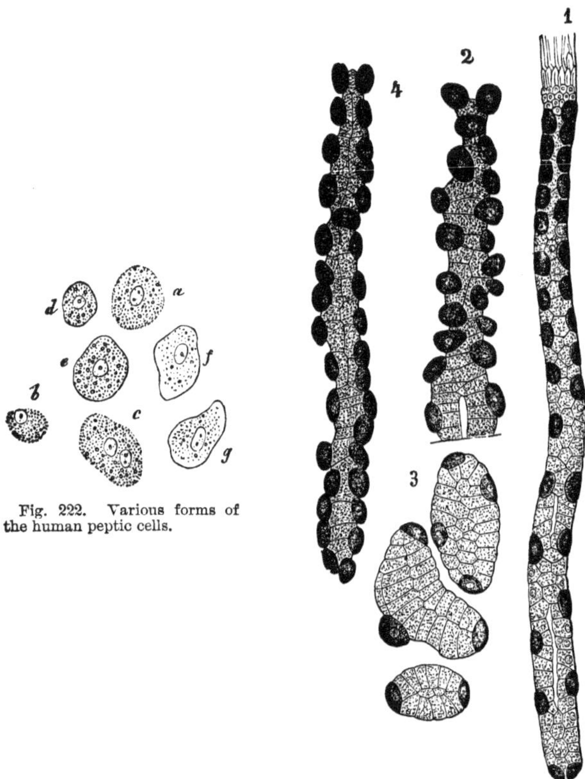

Fig. 222. Various forms of the human peptic cells.

Fig. 223. Peptic-gastric glands of the dog, after Heidenhain, the peptic cells darkened by means of aniline blue. 1, the gland of the fasting animal; 2, portion of a swollen one in the first period of digestion; 3, transverse and oblique sections of the same; 4, tubular gland at the end of digestion.

in granular matter. Their susceptibility to staining is conformable therewith.

If the thick mucous coating which usually occurs on the inner surfaces of the stomach of herbivorous animals, especially the rodents, be examined it will be found to contain a considerable number of the gland-cells in question, part of which appear quite unchanged, part in various stages of decomposition, and thus constitute a surplus of the ferment bodies which are so indispensable for gastric digestion.

Another form of the gland-cells of partly simple, partly branched tubular glands (fig. 224, 1, 2), the so-called gastric mucous glands, is the cylindrical, such as occur in the Lieberkühn's glands of deeper portions of the digestive canal. While, however, the cells of the efferent (occasionally very long) portion of the gland coincide completely with the cylindrical epithelium of the gastric surface, shorter, more granular cells, which are rendered quite cloudy by acetic acid, occur at the fundus of the gland. One is reminded by them of Heidenhain's *haupt*-cells in the peptic-gastric glands. Both varieties of cylindrical

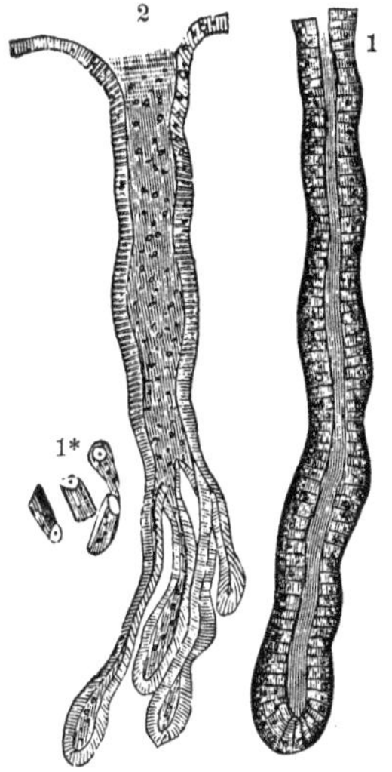

Fig. 224. So-called gastric mucous glands. 1, simpler gland from the hog; *a*, the cylindrical epithelium; *b*, lumen; 1*, isolated cells; 2, compound tubular gland from the dog.

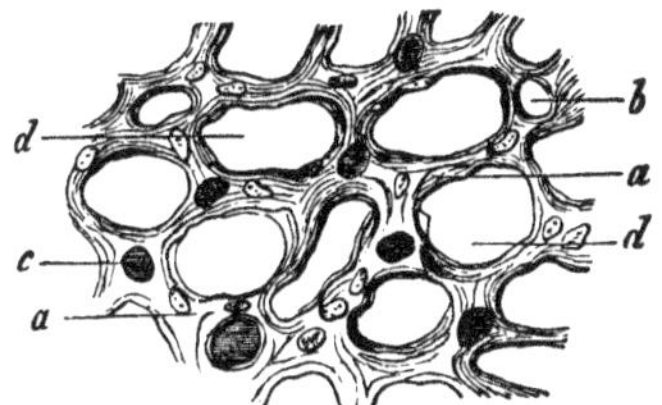

Fig. 225. Transverse section through the gastric mucous membrane of the rabbit. *a*, tissue of the mucous membrane; *b*, transverse sections of empty and injected blood-vessels, *c*; *d*, spaces for the peptic-gastric glands.

cells of the gastric mucous glands also act differently with regard to the above-mentioned methods of staining with carmine and aniline blue. The proper glandular cell-elements at the fundus of the tube appear rich in granules during gastric digestion or gastric irritation, and poor in granules in the fasting animal (Ebstein). Unfortunately, no agreement has yet been obtained in the experiments concerning the fermentative properties of these cells (1*).

Horizontal sections, when brushed a little, show the ordinary fibrous connective tissue of the mucous membrane between the glands (fig. 225). It is as a rule entirely free from lymph-corpuscles. From existing statements of accurate observers it is not to be doubted, however, that they may under certain circumstances obtain a more reticular character in man, and may produce lymph-cells. The frequent occurrence in many persons of scattered lymphoid-follicles, the so-called lenticular glands, in and beneath the gastric mucous membrane, is also an argument in favor of this metamorphosis of the tissue of the mucous membrane.

For the recognition of the muscular tunic of the mucous membrane, vertical sections from the fresh mucous membrane may be acted on for 10–20 minutes by the 30–35 per cent. solution of potash, or thin sections may be made from good alcoholic preparations and stained with carmine (with the subsequent action of acetic acid). Schulze's chloride of palladium method with carmine tingeing, and Schwarz's double staining with carmine and picric acid, deserve recommendation here, as well as for the entire digestive apparatus. The immersion of the fresh gastric mucous membrane in very dilute acetic acid or pyroligneous acid also deserves to be mentioned. These two fluids constitute the most important accessories for the investigation of the gastric nerves which contain small ganglia. They may be readily recognized in the submucous tissue, but, having entered the mucous membrane proper, they escape further observation.

Pathological changes of the walls of the stomach are of rather frequent occurrence.

As a result of chronic catarrh, as well as after small hemorrhagic effusions, the mucous membrane not unfrequently assumes a slate color over larger or smaller places, and the microscope shows an embedment of black-pigment molecules. In slighter degrees of the disease the gastric glands are found to be well preserved; although they often appear distended by large masses of cells, the contents of the latter being opaque (Förster). In such conditions the mucous membrane is not unfrequently found to have an uneven "mammillated" surface, which is dependent partly upon lymphoid follicles, partly on a local hypertrophy of the mucous membrane and its glands, and occasionally also upon a development of fat-lobules in the submucous tissue. Higher degrees may assume the form of polypous protuberances. A new formation of smooth muscular tissue may also take place from the muscular tunic at the pylorus, which then produces an annular constriction of the latter, and has formerly been frequently erroneously regarded as a gastric carcinoma. Vertical sections from the hardened tissue would, in such cases, show the disposition without difficulty.

The microscopic examination of vomited matters has, thus far, yielded only relatively slight results for the purposes of the practical physician.

Among them (fig. 226) appear, first, the constituents of the food which has been taken. These are naturally of the most manifold varieties, and appear partly unchanged, partly slightly altered, partly commencing to decompose in consequence of the action of the lukewarm gastric fluid, or in various stages of digestion from the fermentative action of the gastric juice. At the same time it should not be forgotten to take into account the textural changes which have already been caused in the elements of the food by its preparation.

Fig. 226. Elementary constituents of vomited masses.
a, peptic cells; *b*, cylindrical epithelium; *c*, mucous corpuscles; *d*, pavement-cells of the oral cavity; *e*, sarcena ventriculi; *f*, Cryptococcus cerevisiæ: *g*, starch bodies; *h*, fat-drops; *i*, muscular filament.

Thus we meet with various conditions of the granules of starch (*g*), which, as is known, have a dissimilar appearance according to the several varieties of the starch (rye, wheat, barley, peas, potatoes). The addition of iodine (p. 136) serves for their recognition, if the observer should ever be in doubt. Furthermore, we meet with the greatest variety of vegetable cells, spiral fibres, and other structures of vegetable origin.

If we proceed to the examination of the animal food we find molecules and drops of fat (*h*) coming from milk and fat-tissue;

furthermore, connective tissue parts with hyaline interstitial substance, but unchanged cells, and the likewise unaltered elastic fibres. Muscular fibres (*i*), in consequence of our mode of life, constitute a very general constituent of vomited masses of food. These, in consequence of the free gastric juice, frequently appear in the stage of transformation which we have already mentioned (p. 326), as the effect of the 0.1 per cent. solution of hydrochloric acid, that is, with distinct transverse lines and the separation into plates or disks. Pieces of cartilage will be more rarely found in matters vomited by man, and still less frequently a fragment of bone. While the certain recognition of these constituents suffices for the practitioner, their metamorphoses present an interesting phenomenon for the histologist and the physiologist. It is also very desirable that a systematic study might be made of the action of the gastric juice on the various animal tissues—a research which could be readily instituted with artificially prepared gastric juice.

With these constituents of the food which has been taken are also associated, as admixtures of very unequal quantity, the separated epithelium of the digestive canal—pavement-cells of the œsophagus, and the parts lying higher (*d*), cylindrical cells of the gastric mucous membrane (*b*); likewise the cellular elements of the mucous and tubular glands (*a*), frequently, however, visible only in fragments; and finally, the mucous corpuscles having a granular appearance (*c*).

Pathological conditions of the organ in question may naturally associate new elements with the vomited matters.

In the watery, opalescent, and generally sour fluid which is vomited in so-called pyrosis, we recognize principally epithelial cells and mucous (salivary) corpuscles. Green vomit does not show anything special by microscopic examination. The color is, as is known, due to the coloring matter of the bile.

Large numbers of mucous corpuscles, together with separated pavement epithelium from the oral and nasal cavities, may also be recognized in the rice-water-like substances which are vomited in Asiatic cholera. One notices, on the contrary, but few other cells, such as those of the gastric glands and cylindrical epithelium.

In the brown and black coffee-ground-like masses, such as occur in certain diseases, gastric hemorrhages, gastric carcinoma, and yellow fever, the color is caused by decomposed blood and hæmatine. Here one meets with partly more normal, partly changed blood-cells, lumps of decomposed blood, epithelial and other cells, which appear saturated with hæmatine, and colored brown.

Masses vomited during abnormal fermentative processes of the gastric cavity show interesting microscopical phenomena.

In fermenting fluids, as well as in bread, a fungus, the Cryptococcus cerevisiæ, consisting of oval cells (fig. 226 *f*), occurs. From our mode of life, we frequently receive this with our food, without any injurious effects. Under certain circumstances, however, a very extraordinary increase of these cells takes place in the stomach, and the discharged matters contain large numbers of them.

Another more interesting parasite, though more obscure as to its natural history, is the sarcena ventriculi (*e*), discovered many years ago by J. Goodsir. This—very probably a form of Schizomycetes—consists of regularly united, cubical aggregations of roundish cells. The latter are found united in series of 4, 8, 16, 32. Definite disturbances of the gastric functions do not coincide with the occurrence of the sarcina, so that they are of no pathological importance.

The above-mentioned thrush-fungus of the nursing child (fig. 218), in higher degrees of the disease also occurs in large quantities in the stomach, as would be naturally expected from swallowing the masses of fungus.

The methods of examining the intestinal canal are, for the most part, the same as have been mentioned for the stomach.

Concerning the cylindrical epithelium of the intestines, and the seam which is permeated by porous canals, the essentials have already been mentioned at page 256. Nevertheless, we must here mention a structural condition which has recently been more accurately investigated. Together with the ordinary cylindrical cells (fig. 227 *b*), others (*a*), which were characterized by varying contents, a difference of form, and above all by the absence of a cell-membrane at the upper free extremity, were discovered long ago at more or less regular distances from each other, and varying in number. The structures in question resemble sometimes a pear, sometimes a wide-bellied drinking glass. F. E. Schulze met with them throughout the entire intestinal canal, and in its tubular glands in the vertebrate animals, in the passages of the lungs, and in creatures living in water (fishes and amphibia), on the skin. He has given them the name of "*Becherzellen*" (cup-shaped cells), and declares them to be mucous-secreting structures.

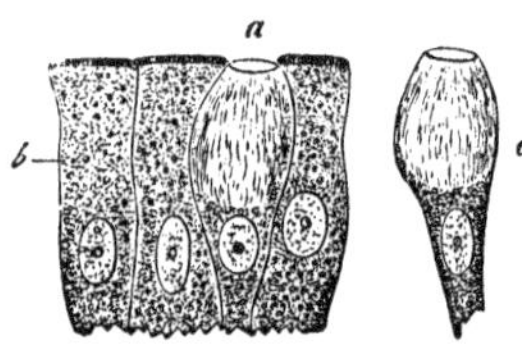

Fig. 227. Cells of the epithelium of the human intestinal villi, treated with Müller's fluid (after Schulze). *a*, Becher-cells; *b*, cylindrical epithelium.

A recently killed animal is to be used for their investigation, and the examination made either immediately, with the addition of indifferent media, such as iodine-serum, or after immersion for several days in Müller's fluid. Nitrate of silver has also been employed.

Mucous- and pus-corpuscles penetrate into the interior of the cylindrical epithelial cells, as probably do also in the rabbit the still so enigmatical Psorosperms (Klebs, Frey, and others), and not only into the cylindrical cells of the small intestines, but also into those of Lieberkühn's glands as well as those of the biliary passages.

The resorption of the chyle-fat by the cylindrical cells of the intestinal villi may be observed in fresh and hardened specimens. The handsomest appearances may be obtained in the smaller mammalia by the previously mentioned injection of milk. Seldom, and only by a rare chance, can the body of a human being who has suddenly died during the digestion of fat be obtained. The examination must naturally be made as soon as possible, as the decomposition which takes place so rapidly in the digestive canal obliterates the delicate textural relations. Older cadavers are entirely useless, as the fine chyle molecules in the intestinal villi usually flow together into large drops, and nothing remains of the cylindrical epithelium.

The contents of the Lieberkühn's glands also show beautifully and distinctly in quite fresh intestines, by the addition of indifferent fluids; likewise in alcohol and chromic acid preparations. Their cylindrical gland-cells (between which, as Schulze saw, cup-shaped cells occur) are also quite perishable, so that one often meets with only the finely granular, nucleated contents of the tubular glands as an artefact.

Hardening methods are to be employed for all the remaining structural conditions. The drying process was formerly employed, in consequence of the poverty of the technique of that period. There is one investigation, the study of the Brunner's glands (fig. 228), for which we would still recommend this procedure, modified by previously boiling the tissue in vinegar. In

Fig. 228. Human Brunner's gland.

fact, handsome preparations may thus be obtained, and the marvellously elegant ramifications of the efferent passages may often be followed in the interior of the body of the racemose gland, especially in thin vertical sections. Nevertheless, at the present day the same purpose is also fulfilled by hardening with chromic acid, chromate of potash, and especially with absolute alcohol, methods which, together with that of freezing, constitute the most important accessories for the remaining structures of the intestinal canal. With them, one may even recognize, the same as in the mucous glands of the oral cavity, the oblong form of the acini and the cylindrical form of the cells of the Brunner's glands (Schlemmer).

Tingeing and brushing may be added according to necessity for the further study of the intestines.

The tissue of the mucous membrane (fig. 229) is differently constituted from that of the stomach.

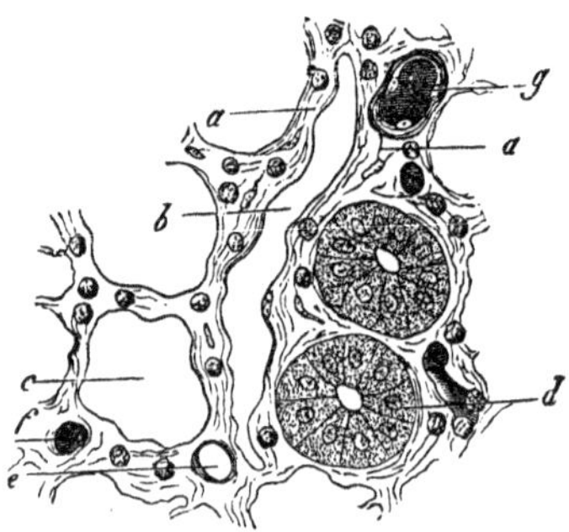

Fig. 229. From the small intestine of the rabbit. *a*, tissue of the mucous membrane; *b*, lymph-canal; *c*, spaces for the Lieberkühnian glands; *d*, transverse section of the Lieberkühnian glands filled with cells.

In the latter organ we meet with ordinary fibrous connective tissue. A looser reticulated substance, with nuclei in individual nodal points, has here taken its place. Lymphoid cells (*a*) lie embedded in the meshes, and are especially numerous in the small intestine. We have here, therefore, a variety of the reticular lymph-cell producing connective substance, which is similar to the framework substance of the lymphatic glands (comp. p. 270). The tissue of the intestinal mucous membrane, however, bears a character of irregularity and mutability which we do not meet with in the lymphatic glands, at least under normal conditions. This tissue becomes condensed into a more homogeneous membranous layer around the glandular tubes at the surface of the intestinal villi, and also forms the limiting layer of the lymphatic canals which pass through the mucous membrane. In places, especially toward the surfaces of the larger blood-vessels and lymphatics, the tissue of the mucous membrane may assume a different appearance, and may even permit of the recognition of the wavy fibrous bundles of the ordinary connective tissue. On the other side, however, as will soon be shown, the tissue in question passes continuously over into the regular reticulated framework of the solitary and Peyerian follicles.

In conformity with this is a textural condition which is interesting for the nature of connective tissue in general. Within a certain space we perceive, at slight distances from each other, the one variety of connective tissue becoming metamorphosed into the other, occurrences which, as is known, pathological histology has so frequently shown to take place temporarily after each other.

The conditions which have just been mentioned are related first of all to the small intestine of man, the mammalia, and birds. The tissue of the mucous membrane of the large intestine appears to be modified more in accordance with the fibrous connective tissue, and is usually poorer in lymph-cells.

Brushing the reticular tissue of these mucous membranes may be accomplished with tolerable facility, and in young creatures the recognition of the nuclear formations is not difficult. In those which are older the number of the nuclei is indeed diminished.

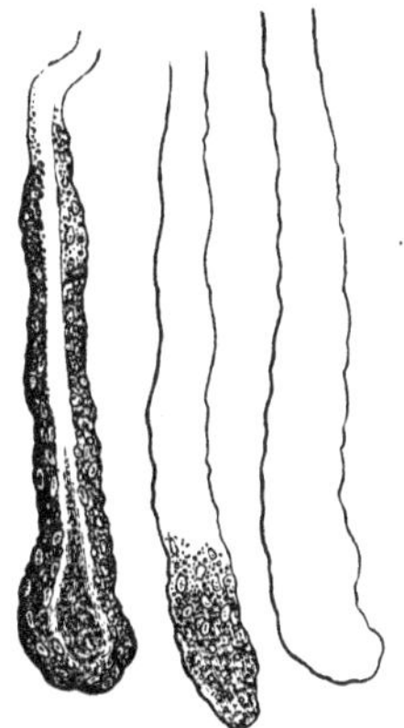

Fig. 230. Lieberkühnian glands of the cat.

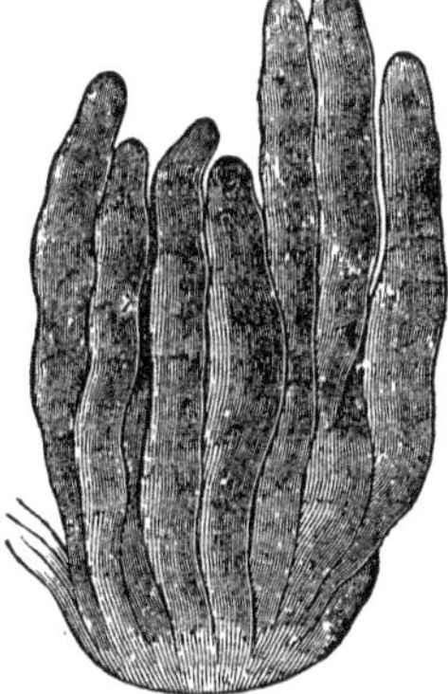

Fig. 231. Tubular glands of the large intestine of the rabbit after treatment with caustic soda.

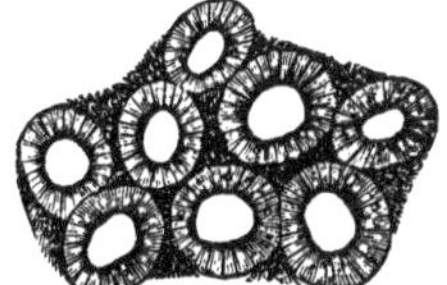

Fig. 232. Apertures of the glands of the large intestine (representing at the same time the transverse sections of deeper lying portions of the glands) from the rabbit.

The Lieberkühnian glands of the small intestine (fig. 230) and the tubular glands of the large intestine (fig. 231), which are quite identical with the former, repeat in their arrangement and frequency the conditions of the stomach. They are to be examined with the same accessories. On thin horizontal sections of freshly immersed parts one may become convinced of the epithelium-like arrangement of the cells, and see how these, conically sloped towards each other, turn their bases outwards and their narrow terminal surfaces towards the axis of the tube (fig. 232). How far they are provided with a special membrana propria to form a line of demarcation between them and the surrounding tissue of the mucous membrane appears still to require a more accurate investigation.

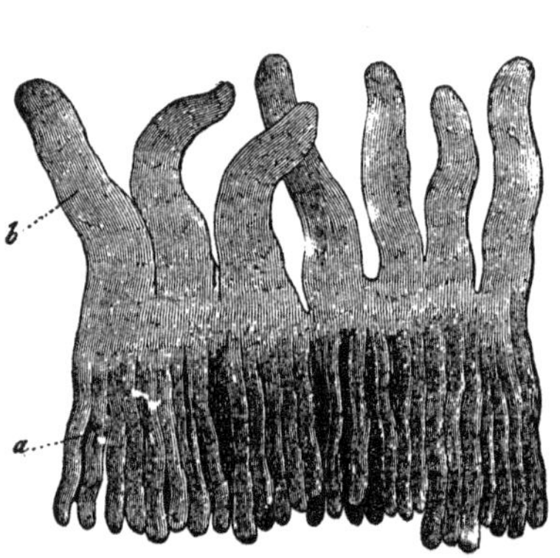

Fig. 233. Small intestine of the cat, in vertical section. *a*, the Lieberkühnian glands; *b*, the intestinal villi.

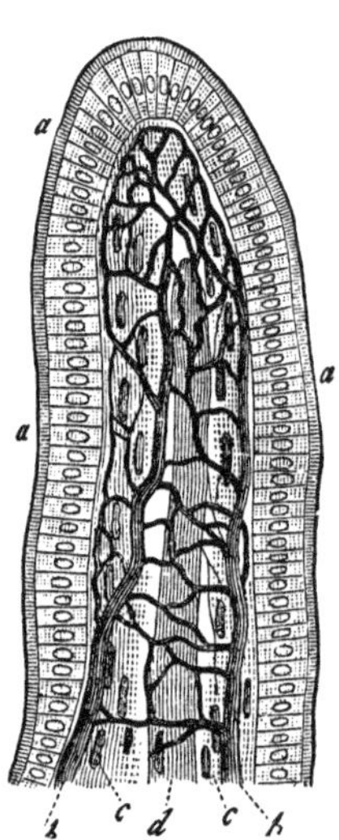

Fig. 234. An intestinal villus. *a*, the cylindrical epithelium with its thickened seam· *b*, capillary network; *c*, smooth muscular-tissue; *d*, central chyle-vessel.

The muscular tunic of the mucous membrane is brought to view by means of the same accessories as were used for that of the stomach.

Peculiar phenomena are presented by the intestinal villi which are met with in the shape of variously-formed projections, pressed closely together in large numbers over the whole surface of the small intestine (Fig. 233, *b*).

Their tissue (fig. 234) bears the same character as that of the remainder of the mucous membrane, and is, as was remarked, membranously thickened at the outer surface, as well as towards the chyle vessel (*d*) which passes through its axis. In birds I succeeded, years ago, in bringing to view a distinctly reticular external surface (as on the surface of a lymphatic gland-follicle) with the greatest certainty. Eberth also found the same in the goose, and was able to recognize a similar condition of the surface of the villi in the mammalia and in man. The intestinal villi of the rat are best adapted for this purpose. Hardening for a month in Müller's eye-fluid has been recommended by that investigator. Longitudinally arranged smooth muscular cells (*c*) also occur embedded in the tissue of the villi, and give these organs their vital contractility, which has been known for a long time and which is so important for the onward movement of the chyle.

Horizontal sections of the villi may be made from well-hardened intestines by means of a very sharp razor with tolerable facility; I find it difficult, on the contrary, to obtain a good vertical section, even from the voluminous villi of larger mammalia, whether the dried or the hardened intestine be employed.

The submucous tissue is to be investigated with the customary methods. The accessories which have been already mentioned (p. 342) serve for the examination of the ganglionic plexuses which occur here (figs. 171, 172).

Their arrangement is to be studied partly on vertical sections, partly on surface views of the submucous tunic from which the muscular and mucous membranes have been separated.

The muscular tunic is to be examined according to the directions previously given for that tissue (p. 313).

The remarkable ganglionic plexus, discovered by Auerbach between the circular and longitudinal muscular layers of the intestine, has also been noticed at the nervous system (p. 343).

Injections of the blood-vessels of the intestinal canal succeed with such relative facility (in the smaller creatures, by the cœliac and mesenteric arteries as well as by the portal vein; in larger ones, by the arterial and venous branches, after the ligation of those which supply the neighboring districts), and afford such a permanent landmark, that they should never be neglected. A capillary network encircles the tubular glands with an abundant, extended, reticular formation in the same manner as in the stomach, so that there, where the surface of the mucous membrane remains smooth, the arrangement is quite the same. Our fig. 235, which presents the capillary network of the gastric mu-

Fig. 235. Semi-diagrammatic figure of the vascular arrangement in the gastric mucous membrane (representing at the same time that of the colon).

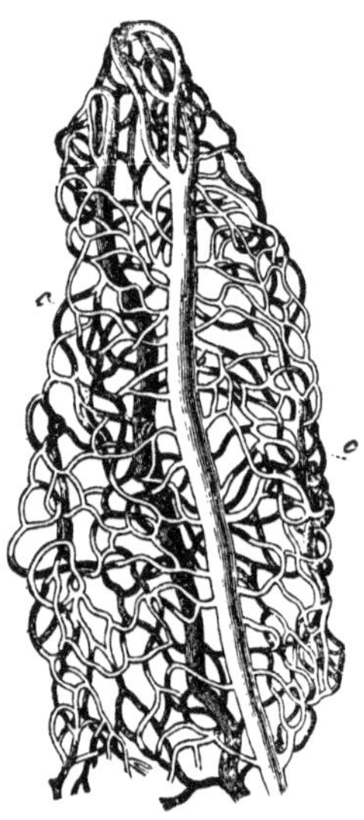

Fig. 236. The vascular network of an intestinal villus of the hare, with the arterial trunk, *b*, the capillary network, *c*, and the venous branches, *a*.

cous membrane in vertical section, may also be regarded as a figurative representation of the blood-vessels in the deeper portions of the colon.

There, however, where projections, papillæ, and villi occur—and this is the case for the entire small intestine, as also, occasionally, for portions of the large intestine—we meet with corresponding modifications of the vascular arrangement. In the intestinal villi especially, the latter become very characteristic and elegant. There is here a so-called looped network, that is, two or more larger trunks pass over into each other in a loop-like manner at the summit of the villus, and are united in their course by an intermediate, more circular meshed network. In larger villi, as our fig. 236 shows, the arrangement may become considerably complicated; in small specimens, those of the mouse, for instance, it remains much more simple.

The capillary network always lies in the peripheral portion of the villus, so that the central portion is occupied by the lacteal vessel which is soon to be described.

The blood readily remains in this vascular district, so that those who shun the trouble of an artificial injection may obtain quite handsome views of the capillaries of the villi from the body of an animal which has been killed several hours previously by strangulation.

The villus-like projections which may make their appearance in the large intestine, such, for instance, as occur in remarkable perfection in the upper portion of the rabbit's colon, have a similar arrangement of the blood-vessels, but differ completely from the glandless villi of the intestines, by being, like the smoothly spread mucous membrane of the colon, permeated by glandular tubes in close apposition.

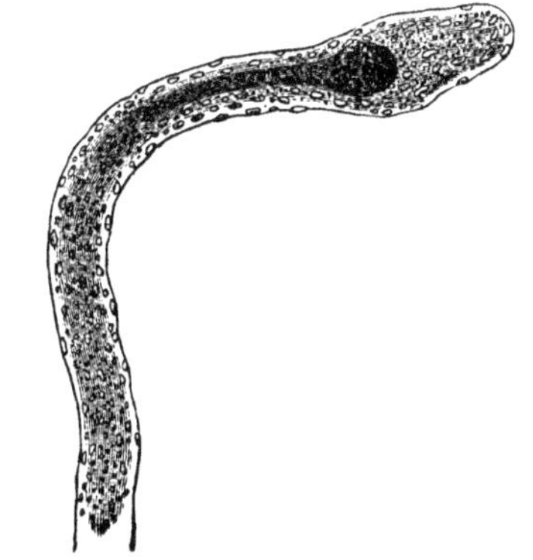

Fig. 237. Intestinal villus of a kid, killed during digestion, with the lacteal vessel in the axis.

Finally, as regards the lymphatics of the intestinal canal, or the so-called lacteals of these parts, much may be recognized, even without injections, in bodies in which the digestion of fat has commenced, and, in fact, in former times, many observers have obtained valuable information in this way. The accumulation of chyle may be observed with facility in the axis of the intestinal villi (fig. 237), and with somewhat greater difficulty the vessels filled with fat of the mucous membrane and sub-mucous layers (p. 394). A suitable medium for rendering such preparations transparent is still wanting, and it is also impossible to keep them in a moist condition for a long time. My attempts, at least, have been entirely frustrated.

The artificial injection by means of the puncturing method was therefore a great improvement, and has, within a few years, increased our knowledge of the lymphatics of the intestinal canal considerably. I believe that I have simplified and facilitated the procedure essentially by the employment of the cold-flowing transparent mixtures.

These injections succeed with greater or lesser facility, and occasionally only with difficulty, according to the frequency and extent of the lymphatic passages and lymphatics with valves which lie in the submucous tissue. The small intestine of the sheep forms a very favorable object, as the submucous layer is occupied, or rather constituted, by a surprisingly large number of very extensive lacteals. The rabbit must also be designated as an animal adapted for these investigations, but the thinness of the intestinal parietes renders the introduction of the fine canules somewhat difficult. The procedure succeeds with less facility, in consequence of the narrowness and greater sparsity of the lymphatics, in the calf and the hog, the dog and the cat; still less in man, although, with some perseverance, one may also succeed with infantile as well as (quite fresh) adult bodies.

With such intestines as are difficult to manage, one may commence with the Peyerian follicles, which are generally easier to inject, and thus from them fill the neighboring portions of the small intestines with their villi. In the sheep and rabbit, on the contrary, where the canule is well introduced, a practised hand almost always succeeds in forcing the mass over more considerable surfaces. Filling the lymphatics of an entire intestine of the sheep, by means of a series of individual injections, of which Teichmann speaks, is, in fact, no great proof of skill.

It would lead us too far, were we here to describe more minutely the relative arrangements of the horizontal lymphatic plexuses in the submucous tissue, the vessels which pass from them into the muscular tunic, as well as the canals which pass up between the tubular glands and frequently reunite in a plexiform manner (fig. 238 *d*). In the intestinal villi, which vary considerably in form and size, being long and thin, and also quite broad and low, there are lacteals of varying diameter with cæcal extremities; in the first case they are single (*a*), in the latter, double (*b*) or manifold (*c*). They may then pass into each other at the extremity of the villus in an arched manner

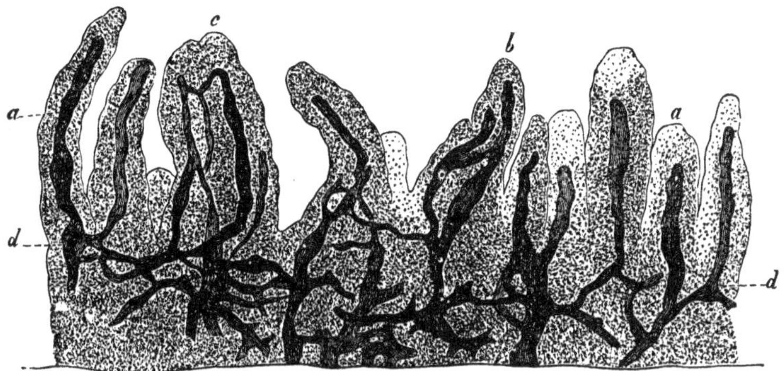

Fig. 238. Vertical section through the human ileum. *a*, intestinal villi with single; *b*, with double; *c*, with triplicate lacteals; *d*, lacteals of the mucous membrane.

(*c*), or still maintain the independent cæcal termination (*b*). Transverse branches occur more frequently in the deeper portions of these more complicated lymphatics.

The injection of the lymphatics in the large intestines, that is, in their mucous membrane, is much more difficult to accomplish. Their occurrence is considerably less frequent, and the entire arrangement is quite variable in the different animals. Horizontal reticulations, passing through the mucous membrane with short knotty vertical passages, and level ramifications passing along the base of the mucous membrane with longer canals, ascending at right angels, etc., also occur. These lymphatics, which have essentially increased our knowledge of the processes of absorption in the intestinal canal, are at present recognized in the ruminantia, the rodentia, and the carnivora. In man (where they are certainly not wanting), the experimental proof of their presence has not, as yet, been adduced.

Have these lymphatics of the intestine a special vascular wall, or are they merely cavities bounded by connective tissue?

Recent investigations leave no further doubt that beneath the serous coverings, and in the muscular layers of the intestinal canal, true "vessels" contain the chyle. Their knotty appearance, caused by the valves, speaks in favor of this, and the walls are also recognizable after the connective tissue has been rendered transparent by means of acetic acid, pyroligneous acid, etc. In part, perhaps in most of the mammalia, this texture is still maintained in the lymphatics of the submucous tissue, while in others there is even here a formation of passages with lacunæ, that is, vessels without an independent vascular wall. Throughout the mucous membrane proper, on the contrary, it is certain that only the latter are present.

They are all, nevertheless, lined with the peculiar vascular cells (see p. 393). These lymphatics are therefore bounded by a very thin but entirely connected epithelium, and this investment is so accurate as to serve the same purpose, at least for the normal condition, as any vascular membrane. Not a granule of the injection mass passes into the adjacent tissue without a rupture. We have frequently injected the small intestines with the finest mixtures under a high degree of pressure, so that the ducts of the intestinal villi, being considerably distended, compressed the spongy tissue of the latter powerfully, but even then not a molecule of the injection mass passed into the tissue. That, on the contrary, an individual immigration into the lymphatics of the relatively gigantic lymph-corpuscles, such as are produced in such abundance by the reticular tissue of the mucous membrane, may occasionally take place is evident. Nevertheless, according to our views, these cells of the intestinal mucous membrane are, under normal conditions, without a future; they arise and disappear in the meshes of the reticular tissue. On the other hand, the possibility cannot be denied, that in morbid processes a more plentiful transmigration into the lymphatic current may take place.

Lymphatic follicles are to be found, although varying in quantity, in the intestinal canals of all of the higher vertebrates and of man. They occur partly isolated or in very small groups, and are then called solitary follicles; they are also, in part, united into larger collections and form the plaques of the Peyerian glands.

The latter structures are most plentiful in the lower portions of the small intestine, but may also—and this is a regular occurrence in many mammalia—be met with in the large intestines. Generally the isolated follicles also show similar conditions.

The structures with which we are occupied, especially the Peyerian glands which are the most thoroughly understood, are embedded in the mucous membrane and the submucous tissue.

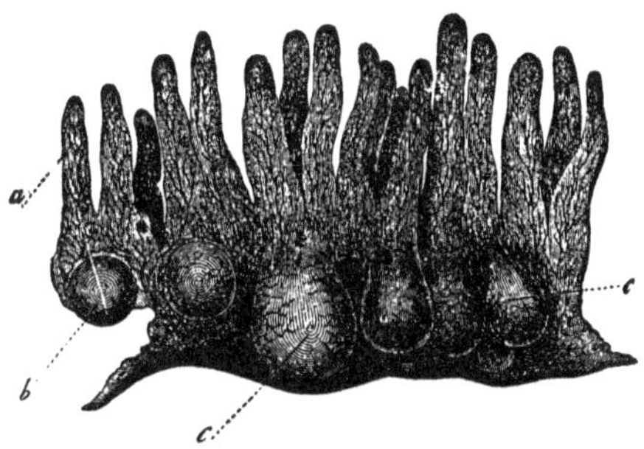

Fig. 239. Vertical section through a fresh Peyer's patch of the ileum from the rabbit. *a*, intestinal villi; *b*, *c*, follicles.

Thus we see in the vertical section of the small Peyer's patch of a rabbit (fig. 239) the bases of these follicles (*b*, *c*) with a globular form in the submucous layer.

Other follicles are much higher and more slender, frequently assuming an appearance resembling the sole of a shoe, and are accompanied by an increased thickness of the mucous membrane and submucous tissue.

At an earlier period, which was poor in methods of investigation, the study of these organs was difficult, so that notwithstanding the interest awakened by the participation of these structures in diseases, especially those of a typhoid nature, our knowledge of them could not be made to progress properly. At the present time, the hardening methods, especially the immersion in alcohol or chromic acid (drying is not so good) are conducive to the purpose. With these must naturally be associated the, in general, not easy (complete) injection of the blood-vessels and the sometimes easier, sometimes more difficult injection of the lymphatics.

The Peyerian follicle (fig. 240) consists, as was remarked, of a sometimes more globular, sometimes more oblong basis portion (*f*) extending freely into the submucous tissue. In many creatures there is a system of connective-tissue partition walls between the basis portions. Secondly, we find the follicle (corresponding to the entire form) projecting freely into the intes-

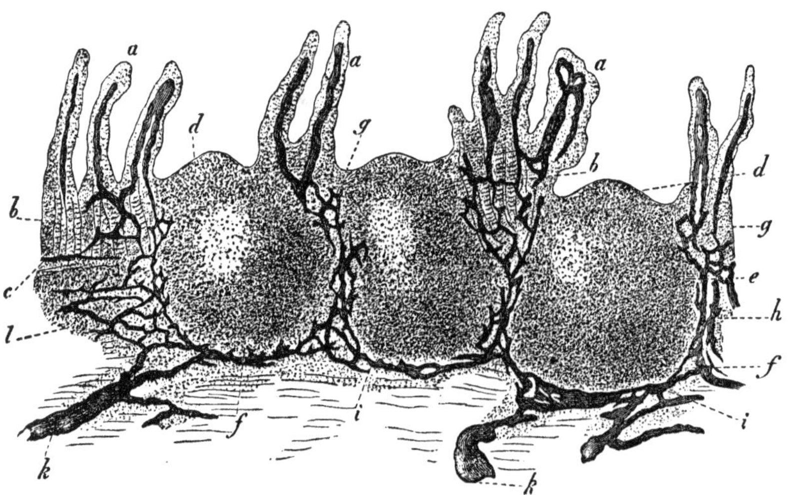

Fig. 240. Vertical section through a human Peyer's patch, with its lymphatics injected. *a*, intestinal villi with their lacteals; *b*, Lieberkühnian glands; *c*, muscular layer of the mucous membrane; *d*, apex of the follicle; *e*, middle zone of the follicle; *f*, basis portion of the follicle; *g*, continuation of the lacteals of the intestinal villi into the mucous membrane proper; *h*, reticular expansion of the lymphatics in the middle zone; *i*, their course at the base of the follicle; *k*, continuation into the lymphatics of the submucous tissue; *l*, follicular tissue in the latter.

tinal canal, with a sometimes higher, sometimes flatter apex (*d*). These, covered with cylindrical epithelium, are surrounded by more or less prominent elevations of the mucous membrane, which are generally furnished with villi (*a a*).

Between the apex and base a middle zone (*e*) remains. In it the demarcation of the two follicular portions is wanting. In vertical and horizontal sections it is seen, on the contrary, how in this middle strata all the follicles of one plaque pass into those of another, and then how this zone is continued uninterruptedly into the adjacent tissue of the mucous membrane (*l*). This is the metamorphosis of the reticular connective tissue of the mucous membrane into the reticular framework of the lymphatic gland follicle, of which we have already spoken on a previous page.

Here also the network of the follicle (fig. 241 *b*) is essentially

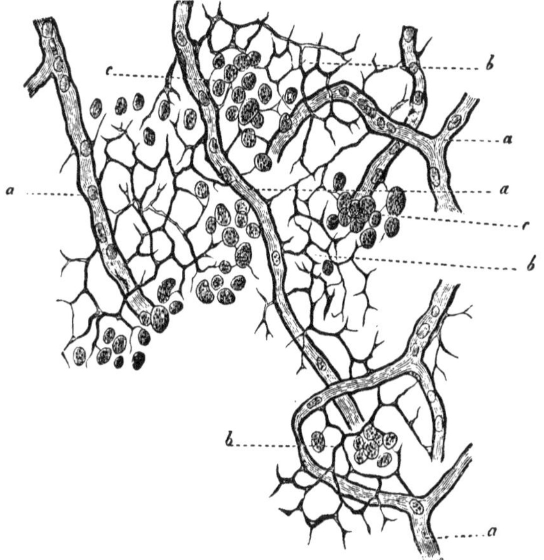

Fig. 241. The tissue of the Peyerian follicle of an adult rabbit, exposed by brushing. *a*, capillary vessels; *b*, reticular framework; *c*, lymph corpuscles.

the same as occurs in the large lymphatic glands; in young bodies it is a cellular reticulation, in older ones it consists more of trabeculæ with shrunken nuclei in individual nodal points. Towards the periphery of the basis portion the tissue assumes a more finely reticulated character (as also occurs towards the investing spaces of the follicles of the lymphatic glands); in the central portions, on the contrary, the meshes are not unfrequently larger.

The blood-vessels of the Peyerian glands have recently been frequently described, so that it must appear superfluous to allude to them more thoroughly again. Only the remark, together with several examples, may here find place, that a non-vascular central portion of the follicle does not, as a normal occurrence, exist. Incomplete injections, it is true, give frequently enough

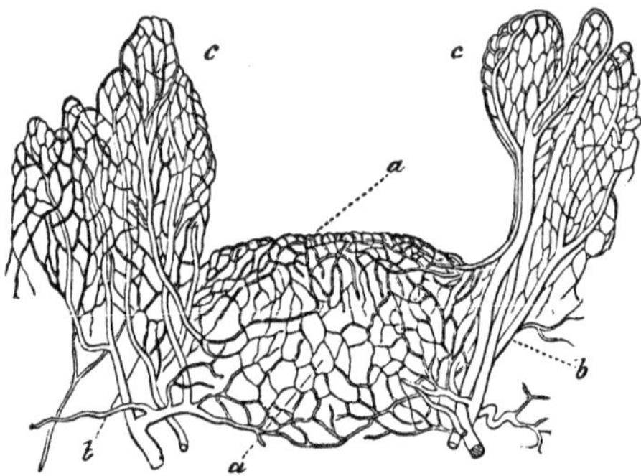

Fig. 242. Vertical section through an injected Peyerian capsule of the rabbit, with the capillary network of the same, *a*, the larger lateral vessels, *b*, and those of the intestinal villi, *c*.

the false image of capillary loops in the internal portions of the follicle. Our figures 242 and 243 represent this arrangement of the vessels in a small Peyerian patch of the rabbit, from a

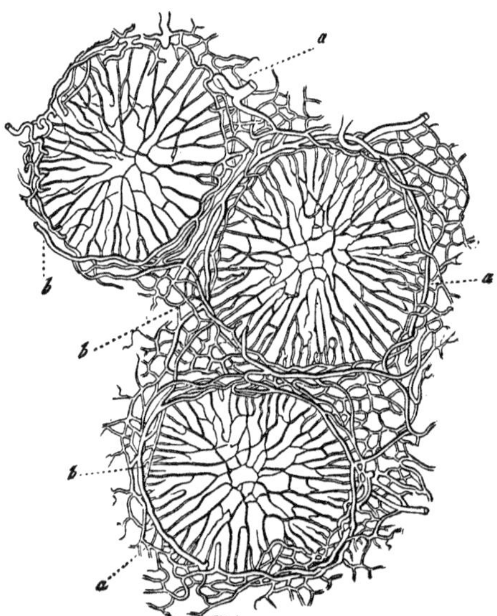

Fig. 243. Transverse section through the equatorial plane of three Peyerian capsules of the same animal. *a*, the capillary network; *b*, the larger annular-shaped vessels.

very complete injection mounted dry. We have also, in addition, accurately re-examined the arrangement in moist specimens from a series of consecutive sections.

Good injections of the lymphatics teach the following:—The lymphatic vessels (fig. 240, *a*, *a*) which return from the intestinal villi (the so-called lacteals) form a reticulum (*g*) around the tubular glands (*b*) which occur in the villous elevations, and this is continuous with a network of reticularly enclosed vessels (*h*) which surrounds the middle zone of each follicle. The latter then open either into a simple investing cavity which surrounds the basis portion like a shell (rabbit, sheep, calf), exactly similar to that of the alveolus, or this is replaced by a network of separated passages and lacunæ encircling the basis of the follicle in a similar manner, so that this portion of the Peyerian follicle (*h*, *i*) appears like a toy-ball around which a thread is wound (as in man, the dog, and the cat). From the latter system of vessels (or from the simple investing space) finally arise the efferent lymphatics of the submucous layer (*k*).

The reader will comprehend that follicles of the latter variety are more difficult to inject than those of the first form with the simple shell-like investing spaces.

The vermiform process, as well as the small and scanty cæcum of many carnivora consists, in a remarkable manner, of only a closely crowded collection of follicles. The processus vermiformis of man and of the rabbit represents, in fact, a Peyerian plaque which, largely extended, forms an entire portion of intestine. Teichmann succeeded in injecting them in man; the injection of the vermiform process of the rabbit is a mere child's play, and the entire organ deserves to be most urgently recommended to any one who desires to study the Peyerian follicles.

Numerous pathological metamorphoses of the intestines become objects of microscopic investigation. The same methods which we have mentioned in the investigation of the normal structures are generally employed. It should be made a rule to obtain the freshest possible objects, as the decomposition which soon commences changes the soft tissues in such a manner as to render them unintelligible. The pathological new formations

29

in the intestinal canal are, in general, the same as in the stomach. Thus we meet with similar pigmentations, connective-tissue productions, lipomata, etc. Carcinomatous tumors occur in the large intestines, especially in the rectum. Tubercles, on the contrary, are met with chiefly in the ileum, less frequently in the jejunum and colon. It is the lymphoid, the solitary, as well as the agminated (Peyerian) follicles of these parts which, like other lymphatic glands, are especially affected by this process. More accurate histological investigations of this metamorphosis with the aid of modern accessories would be in place. Tumefactions of the follicles show themselves conjointly with capillary distentions and cell proliferations. Later, the destruction of numerous lymph-cells takes place and the finely granular so-called tubercle mass is formed. This softens and occasions the formation of ulcers. Usually, the lymphatic glands of the mesentery also take part in this process.

The structural conditions of the follicles in abdominal typhus are very similar in an anatomical point of view. In the first or catarrhal period, the capillaries of the Peyerian follicles are frequently widened to a considerable degree. Large multinuclear lymph-corpuscles are met with here, exactly in the same manner as in the typhoid metamorphosis of the lymphatic glands (p. 400). By means of several injections made at an earlier period, I was able to obtain the conviction, at least, that in this stage the lymphatics of the Peyerian glands are still thoroughly permeable. Later, with the destruction of the cells, the former appear to become stopped up and impermeable. It does not seem to be in place here to speak further of the associated processes of absorption, of the softening of the contents of the follicles, and of the formation of intestinal ulcers and sloughs. The latter masses consist of fine granular matter, nuclei, cells, cell remains, etc. The associated process of cicatrization takes place naturally, by a new formation of connective tissue. Accurate conclusions are not easily obtained in these cases, as I know from my own experience, so that a careful investigation of the cases at hand is very desirable.

Finally, with regard to the methods of preserving microscopical preparations of the digestive canal. The vertical and horizontal sections may be preserved moist, with or without previous staining, either in watery or more concentrated glycerine. If they have been carefully washed before being placed in the latter fluid, they keep well, as a rule, as do also preparations of their vessels and lymphatics injected with transparent masses (carmine, Prussian blue). According to previous experience, the nervous and ganglionic plexuses of the intestinal canal may be best preserved by freeing them from the residue of their acid by washing in distilled water some time before mounting. The method of depriving tinged preparations of their water by means of absolute alcohol, and the subsequent mounting in Canada balsam dissolved in chloroform, must be designated as very serviceable for many of these purposes. Beautiful and durable review preparations for low powers may be obtained in this way. If it be desired to mount thicker masses, as, for instance, a portion of intestinal mucous membrane with the villi erect, glass cells are to be employed. A skilful manipulator will be able to make a handsome preparation with one, even with Canada balsam.

There remains for us to consider, finally, the intestinal contents, and the fæcal masses which are formed from the latter. Although these are not often objects of medical examination, and though disgust deters many observers from the investigation of the latter substances, nevertheless, in consequence of the multiplicity of their elements, they are both very instructive, and not always easy objects of microscopic examination. The alimentary pulp which has been altered by the saliva and the gastric juice, and has left the stomach, has, as you know, received the name of the chyme. In its further progress there become mixed with it the secretions of the liver, of the pancreas, and of the various follicles of the mucous membranes, as well as exfoliated epithelium, gland-cells, and the mucous corpuscles of the intestinal canal; while other matters, such as fat, albuminous bodies, and salts are removed by absorption into the lacteal system. The chyme naturally presents very considerable differences according to the nature of the food; in the carnivora it is different from that of the herbivora.

We omit here the substances which are dissolved in the chyme.

Its elements consist of molecules and drops of fat, altered muscular fibres, portions of connective tissue (in carnivorous animals fragments of cartilage and bone), starch-granules, various vegetable tissues, etc. Fig. 244, which represents the contents

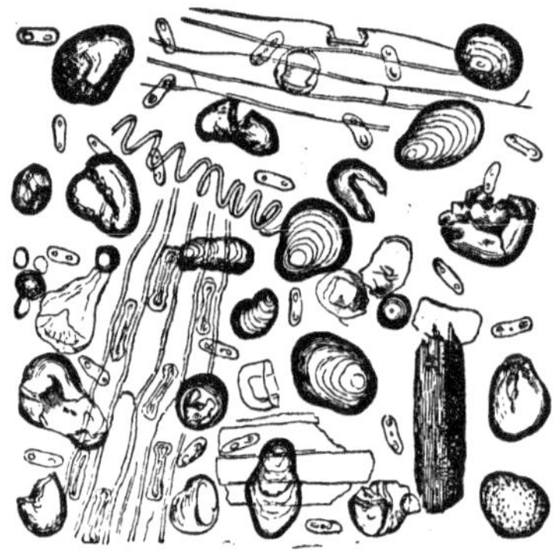

Fig. 244. Contents of the small intestine of a rabbit.

of the small intestine of a rabbit, may give us a conception of the constitution of the chyme after a vegetable diet. In the specimen we meet with starch-granules in various stages of dissolution, in part already changed into empty hollow vesicles, epidermoidal tissue, prosenchyma cells, spiral fibres, etc.

By the onward movement through the large intestine, the mass undergoes further changes. The digestive properties of the so-called intestinal juices make themselves felt; the lymphatics absorb the fluid portions, and, by the transformation of the biliary pigment, as well as by putrefactive decomposition, the masses assume the color and smell of fæces.

Numerous elementary particles of the food, such as filaments of muscular substance, fat-tissue, bundles of connective tissue, elastic fibres, etc., are still to be met with. The muscular fibres are frequently separated into disks, and have a greenish tinge from the biliary pigment. Numerous remains of vegetable alimentary matters, starch granules, spiral vessels, epidermoidal tissue, substances which we have already mentioned in speaking of the contents of the small intestines, show themselves in the human excrements. Remarkable fæcal discharges which cause great solicitude to hypochondriacs, and may also astonish the physician, may frequently be readily demonstrated by the microscopical examination to be merely remains of food.

The human fæces are always very rich in fragments and filaments of the Leptothrix.

With the name of meconium has been designated the dark, pitch-like stools of the new-born child. They contain decomposed bile, separated and decaying epithelium and cells of the intestinal canal, as well as small hairs from the integument which have been swallowed with the amniotic fluid. The meconium is rich in fats, and the ethereal extract forms a deposit of numerous crystals of cholesterine.

Numerous alterations in consistence, color, and constituents are presented by the fæcal masses in disease. The most remarkable stools are found in dysentery, abdominal typhus, and cholera. The alimentary constituents here diminish more and more, as does also, as a rule, the decomposed bile ; the intestinal secretions and the separated cells, on the contrary, increase. Albuminous masses, coagulated fibrine, and blood may be associated with them.

Dysenteric stools contain desquamated cylindrical cells, mucous and pus corpuscles, cell nuclei, gland-cells, fibrinous coagula, blood-cells, and clots of blood.

The peculiar evacuations which occur in abdominal typhus, at the height of the disease, show, together with epithelium, gland-cells, pus-corpuscles, and a fine granular substance with nuclei, which has been regarded as the cast-off ulcerative products of the Peyerian and solitary glands. Not unfrequently, blood-corpuscles also occur in these evacuations.

We mention, finally, the cholera stools. The rice-water-like dejections in this disease contain very large numbers of mucous corpuscles, but, on the contrary, only very little cylindrical epithelium.

Crystalline deposits of the ammonio-phosphate of magnesia (fig. 245) are found in the alkaline fæces of healthy as well as of diseased persons. They present a rhomboidal form, and most frequently appear as three-sided prisms with the two corners corresponding to one side blunted, in the so-called coffin-lid form.

In consequence of the general diffusion of the phosphatic salts of magnesia in the solid and fluid portions of the organism, the double combination we are at present considering forms one of the most frequent occurrences as a result of the development of ammonia.

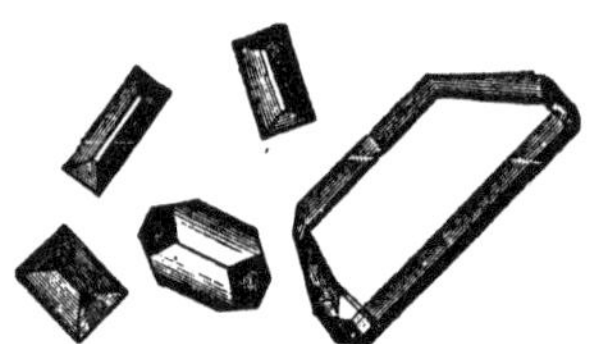

Fig. 245. Crystals of the ammonio-phosphate of magnesia.

Seldom, on the contrary, do we find in the intestinal canal (but even in the stomach, however) crystalline deposits of taurin, a conjugate compound of one of the two biliary acids (fig. 246). Further chemical procedures are necessary, as a rule, for the recognition of this body, as well as of cholesterine.

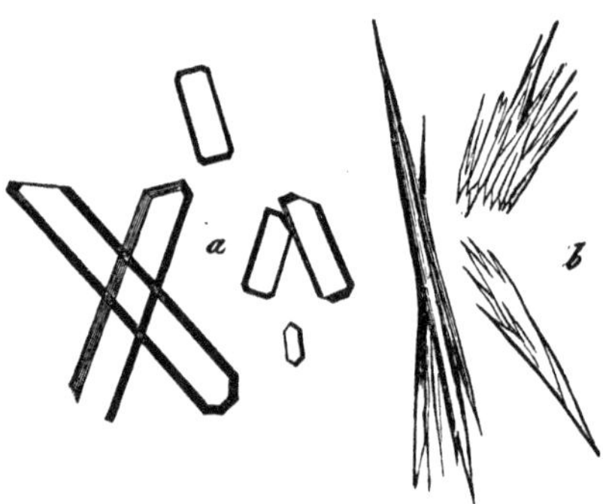

Fig. 246. Crystals of taurin. *a*, completed six-sided prisms; *b*, indefinite sheath-like masses from an impure solution.

We cannot leave the microscopical analysis of the fæces without first mentioning certain of its animal parasites.

A large infusory animalcule, covered on all sides with cilia, the Paramæcium coli of Malmsten, has hitherto had no practical significance. It has been observed a few times in the large intestines of human cadavers, as well as in the stools. The same is also true of the Cercomonas intestinalis, a small creature provided with a whip-like cilia, discovered by Lambl. It has been met with in the hyaline intestinal excretions of children, in intestinal catarrhs, as well as in typhoid and choleraic diseases

(Davaine). Quite fresh intestinal evacuations, or such as have not yet become cold, should be used for their investigation (Ekecrantz).

The microscopical recognition of the ova of the most familiar human helminths is, on the contrary, of much greater practical importance (Davaine, Lambl, Leuckart and others). Leaving out of consideration the trichina, the embryos of which creep out in the maternal body and then perforate the intestinal walls, the ova of the remaining nematodes do not become developed in the human body, but are expelled and appear in the stools; likewise, although merely casually, those of the tapeworms which have been set free by the rupture of a proglottis. It is easy to recognize the ova of the parasites dwelling in the lower portion of the intestine, especially of the Oxyuris vermicularis, numbers of which are presented by every microscopical preparation taken from the external surface of a portion of fæces (Vix). It is more difficult, on the contrary, to discover the ova of the nematodes which live higher up in the intestinal canal, such as the Ascaris, as they do not occur in the mucus which envelopes the solid fæcal masses, but rather in the interior of the latter.

The more solid masses of the fæces are to be spread out with water for the examination, or the coating of slime is to be selected (with Oxyuris). The mucous coating scraped from the intestinal walls with a spatula also presents an abundance of this helminth (Vix).

We give a short résumé of the characteristics of the ova of these helminths (fig. 247).

Tricocephalus dispar (2). Ova double contoured, oval, truncated at both poles, shell and vitellus of a brownish color. Length, 0.0239–0.0257‴; breadth, 0.0111‴.

Ascaris lumbricoides (1). Ova roundish or oval, measuring 0.0363–0.0386‴, next to the largest of all. The shell of the ovum has a double contour and is still covered by the transparent, indentated areole of an albuminous enveloping substance.

Oxyuris vermicularis (3). Ova mostly transparent, double contoured, oval shells (frequently having an asymetrical curvature). Length, 0.0231–0.0248‴; width, 0.0102–0.0115‴.

Distoma hepaticum (4). Eggs oval, very large, yellowish. Length, 0.0572–0.0616‴; breadth, 0.0332–0.0399‴. The anterior pole with the operculum more flattened. Egg-shell double, contents a cell aggregation and vitelline spheres.

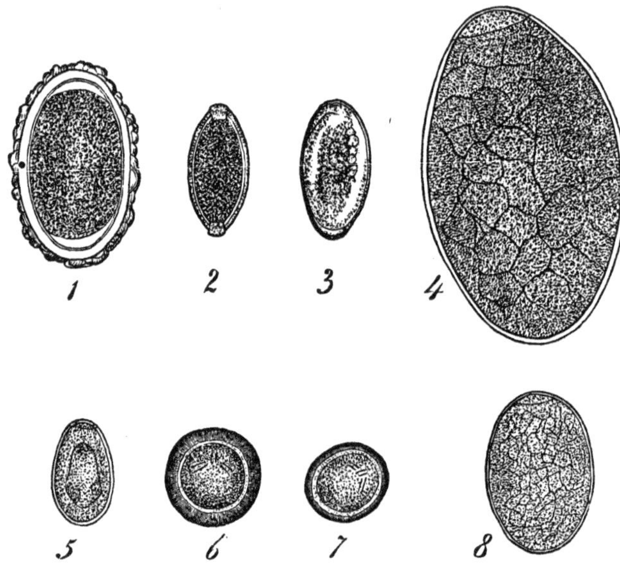

Fig. 247. Ova of the most familiar helminths of man, from a drawing communicated by Prof. Leuckart (all magnified 370 diameters). 1. Ascaris lumbricoides. 2. Tricocephalus dispar. 3. Oxyuris vermicularis. 4. Distoma hepaticum. 5. D. lanceolatum. 6. Tænia mediocanellata. 7. T. solium. 8. Bothriocephalus latus.

Distoma lanceolatum (5). The brown double shelled oval eggs are much smaller, 0.0177–0.0199‴ long, 0.0133‴ broad, are evacuated at a later period than the previous variety, and contain an oval embryo, measuring 0.0115–0.0133‴, with clusters of granules in the posterior parts of its body.

Bothriocephalus latus (8). Eggs oval, averaging 0.0310‴ in length and 0.0199‴ in the middle transverse diameter; they are enveloped by a simple, hard, brown shell, whose anterior pole constitutes a distinctly interrupted, hood-shaped operculum.

Tænia solium. The ova, which develop within the so-called proglottides, present variations in accordance with their age.

The ovum (7) which contains an embryo sometimes shows an oval enveloping layer of albumen and a globular, thick, manifoldly contoured, brownish inner shell 0.0133‴ in diameter, the surfaces of which are covered with closely arranged ciliæ. This contains the spherical embryo, which measures 0.008‴, and is provided with six hooklets. Sometimes the outer layer of substance (which formed the original vitelline layer) is wanting. Undeveloped ova are smaller, globular, at first without the inner envelope, and enclose a vitelline sphere and a special aggregation of embryonic cells.

Tænia mediocanellata. Ova (6) quite similar, but markedly oval and almost regularly provided with the original vitelline membrane. The size and other characteristics of the egg-shell the same as in the previous animal.

Together with these, the presence in the fæces of the familiar hooks of the Tænia and their younger forms, as well as, in the trichina disease, the sexually mature examples of these worms, is applicable for the diagnosis of a helminthic disease.

Section Eighteenth.

THE PANCREAS, LIVER, AND SPLEEN.

We have still remaining the two large glandular organs connected with the intestinal canal, the pancreas and the liver. The spleen is also to be discussed here.

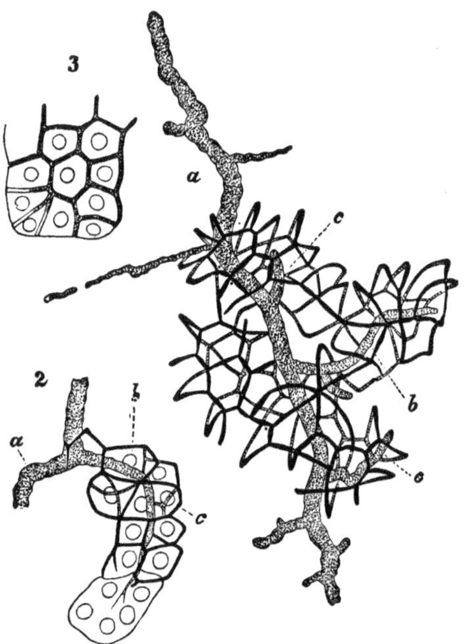

Fig. 248. Glandular canals of the rabbit's pancreas, after Saviotti. *a* Larger excretory duct; *b* that of an acinus; *c* finest capillary ducts.

We can despatch the pancreas with rapidity. The methods for its examination are the same as are ordinarily used for the larger racemose glands. The fresh organ, the usual methods of maceration, portions hardened in alcohol or chromic acid with the addition of the reagents customary for glands, permit of the recognition of its structure. The pancreas of the small rodentia, the mouse, rat, and rabbit, spread out flat, present handsome appearances, while the examination of the human pancreas is very generally found to be somewhat difficult, in consequence of the abundance of fat granules in the gland cells. Injections of the blood-vessels succeed readily; those of the gland-ducts (fig. 248) are to be attempted with cold flowing mixtures, with Brücke's soluble Prussian blue, for example. The syringe may suffice for this purpose, when carefully directed. The constant pressure renders better service for injecting the finest capillary passages (*c*) which run between the gland-cells.

The liver, on the contrary, requires a more accurate discussion, in consequence of its numerous peculiarities. The investigation of this, the most voluminous of all the glands of the body, is in fact difficult, so that some of its structural conditions still remain matters of controversy.

Each of the previously mentioned glandular organs shows the observer at once, together with the parenchyma cells, an investing membrana propria (which, it is true, may be replaced by the adjacent connective-tissue layer). While now the cells of the liver are to be recognized with the greatest facility, the question as to the existence of the membrana propria causes the microscopist great embarrassment.

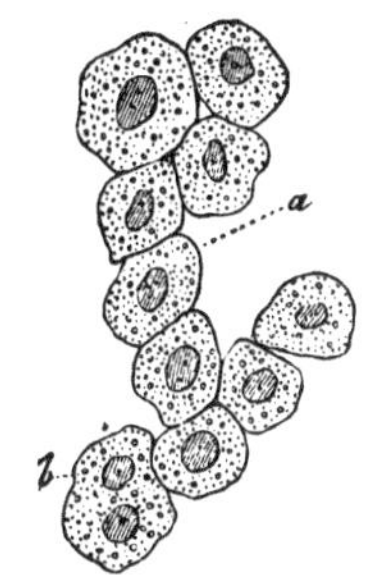

Fig. 249. Human hepatic cells: *a*, with single; *b*, with double nuclei.

The most simple procedure suffices to demonstrate the hepatic cells (fig. 249). If the fresh organ be cut into, and a knife-blade scraped over the cut surface, the brownish mass, diluted with some fluid, presents numerous examples, either single, in series, or in reticulated fragments. The adjacent figure shows the characteristic form, the finely granular cell-contents, very generally intermingled with isolated fat-molecules and the nucleus, two of which not unfrequently lie in one cell-body (according to our present views, a proof of the cell-division). A special cell-membrane cannot, however, be demonstrated on the cells of the liver; its place is occupied much more by a somewhat hardened cortical layer.

The so-called hepatic lobules have been distinguished for a long time. These islets of the substance of the tissue are sometimes brownish red internally with a brownish periphery, sometimes the colors are reversed. In most mammalia they become blended with each other at their peripheries, but are, nevertheless, here and there more distinctly demarcated.

The microscope shows as a cause for such a sharp division of the hepatic lobules a strongly developed connective-tissue boundary layer. The liver of the cat, the sheep, and more especially that of the pig, is of this variety. Many things which are only to be recognized with difficulty in the organ of other animals and of man, appear more distinctly in the last-mentioned animal. The pig's liver has, therefore, very properly been

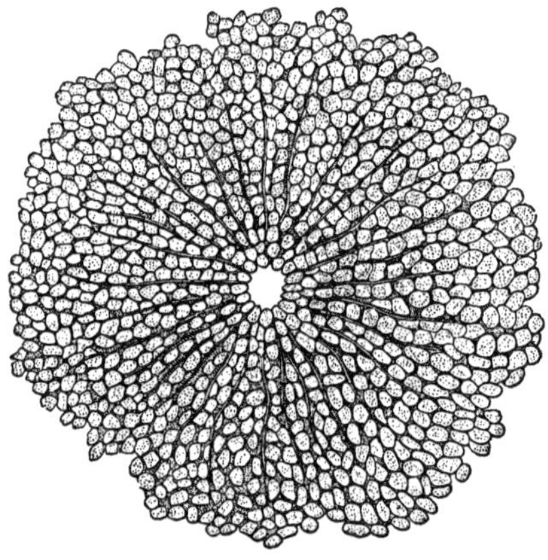

Fig. 250. Transverse section of a human hepatic lobule.

recommended by modern histologists as an extremely suitable object for examination.

With the aid of a sharp scalpel a fine transverse section may be obtained from such a lobule of the quite fresh organ, just beneath the surface, for instance. Valentine's double-knife (p. 111) has been recommended by others for this purpose. It is much better, as we shall have to mention hereafter, to make use of livers hardened in alcohol or chromic acid for the preparation of such specimens. We also recommend the freezing method.

Such a transverse section (fig. 250) shows the columns of the liver-cells or the cell-network arranged in a general radiated manner, and, at the same time, these columns of cells united together in a reticular manner by short transverse rows. In human and mammalian livers the cells of such a network usually lie in single rows, and are only double in places at the nodal points; nevertheless, many variations occur. A system of similar spaces appears in such preparations, for the most part with great distinctness.

If for the purpose of further investigations the blood-vessels be filled with transparent substances (either as a single injection by the vena hepatica or the portal vein, or with two masses by the two veins simultaneously), the radially arranged capillary network appears with surprising beauty, and one is at the same

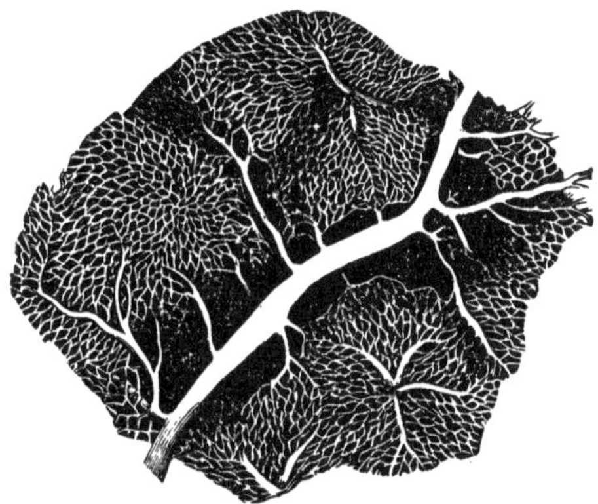

Fig. 251. The injected liver of the rabbit with the branches of the portal and hepatic veins.

time convinced that the origin of the above-mentioned spaces, which were shown by the transverse section of the hepatic lobule, is due to the capillaries of the vascular network, and likewise that the rounded central space (fig. 251) is the transverse section of a branch of the hepatic vein (vena intralobularis of Kiernan).

Fig. 251 may represent to the reader the finer arrangement of the blood-vessels. Several lobules appear to be supplied by each branch of the portal vein with finer ramifications, running in a lateral direction, which are confined to the intervening spaces between the lobules (venæ interlobulares), and in the centre are noticed the branches of the hepatic nervous system. A few branches of the hepatic artery also enter the capillary network at its peripheral portion, so that the injection may be practised by the latter vessel with the same success as by the portal veins.

Even in the fresh condition, the previously-injected liver shows the capillary network occupied by the columns of the hepatic cells, so that two kinds of network, that of the blood-vessels and that of the cellular trabeculæ, are actually thrust into each other.

In well-hardened organs, however, where the razor affords very fine sections, these investigations may be made in a much finer manner. Simple alcohol may be employed, and likewise Clarke's mixture of alcohol and acetic acid (p. 141).

Beale especially recommends the use of alcohol to which a few drops of a solution of soda has been added (p. 142). Such preparations, freed from adherent matters by working, and tinged with carmine or (which is likewise to be highly recommended) hæmatoxylin, afford exactly the appearance as if the cells were embedded quite free in the spaces of the capillary network.

This view was, in fact, maintained for a long time, although the contrary opinion might also have been defended with the same propriety, namely: that a cellular network, enclosed in a homogeneous membrane, was permeated by the reticulated lacunar system of capillary blood-currents.

The modern accessories have led us a considerable step further in this matter.

Fine sections from a liver hardened to the consistence suitable for brushing (I generally employ alcohol for this purpose, at first with considerable water, then with less) permit of the removal of the liver-cells, although only over more limited spaces (fig. 252). In this way there is left a fine and extremely delicate network (*a*) formed of a homogeneous membrane which separates the blood-current from the cell columns. If carmine tingeing be resorted to, the columns of hepatic cells which were not removed by the brushing will appear very beautifully; then, however, one will also recognize in this hyaline

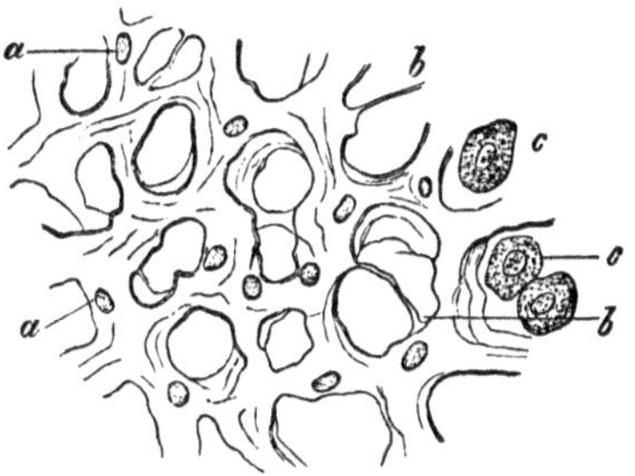

Fig. 252. Framework substance from the liver of the rabbit. *a*, homogeneous membrane with nuclei; *b*, thread-like strip of the latter; *c*, several hepatic cells not removed by the brushing.

membrane of the reticular framework, together with the capillary nuclei, a few small and more rounded nuclei which are, in adult creatures, for the most part shrunken.

If the liver of a new-born child, of a human embryo of the later months, or of a mammalial animal of a corresponding period of life be used, the fine hyaline membrane alluded to appears in places with great distinctness as a double membrane, one of the layers of which corresponds to the capillary walls, while the other limits the cellular network.

According to this, there is no longer any doubt that a thin, often indeed extremely fine layer of homogeneous connective tissue supporting substance (in continuity with the connective tissue which envelops the hepatic lobules), condensed more like a membrane towards the cellular network, constitutes or replaces the long-sought membrana propria of the hepatic cell columns. To it belong as a system of connective-tissue corpuscles, those nuclei which occur more abundantly in the earlier periods of life, and are often surrounded by a distinct cell-body.

While these two membranes, the connective-tissue framework substance and the membrane of the capillary-vessels, appear at first separated, in older creatures they often make the erroneous impression as if they were blended (see below). The beautiful results which Remak made known years ago concerning the manner of formation of the liver may, therefore, be confirmed on the organ of the new-born and the adult. We are indebted in part to Beale, but especially, however, to E. Wagner, for a knowledge of the facts to which allusion has been made.

We come now to the discussion of the biliary ducts. Their branches, provided with a fibrous membrane and a covering of shorter cylindrical epithelium cells, surround the lobules, in parts more continuous, as an extremely delicate circular network (cat, rabbit, Guinea-pig), in part in the form of separated, sinuous, ramified passages; thereby maintaining a course simillar to that of the branches of the portal vein. With careful injections of the ductus hepaticus, these passages (the muscular tissue of which, as Heidenhain has shown, is rendered apparent by treatment with chloride of palladium, 1 : 900) may be recognized with tolerable facility; likewise, after having once observed these canals on fine sections of the hardened organ, with the assistance of brushing and staining. Here and there, the latter procedure will also, occasionally, show still finer passages which run towards the interior of the lobule.

The aid of finer injections of the biliary passages is naturally necessary for the further examination of their structure, and the relation of the ultimate biliary ducts to the cell-columns is to be decided by them. In consequence of the extreme delicacy of the structure of the lobules, and the impediment which the bile accumulated in these canals presents to the injecting fluid, this procedure is difficult, and is also, as a rule, especially with solutions of gelatine, thwarted by rapidly appearing extravasations.

It is only recently that success has been obtained in arriving at a decided result (Budge, Andrejevic, MacGillavry, Frey, Hering, Eberth); namely, in injecting a fine and extremely elegant biliary network which permeates the entire hepatic lobule and surrounds the individual hepatic cells with its meshes. An analogous condition has since been discovered in the racemose glands (p. 413).

For this purpose use the quite fresh liver of an animal which has just been killed, and either the apparatus for constant pressure described at page 189 and represented in fig. 77, or that of Hering. A previous removal of the bile is unnecessary. An aqueous solution of Prussian blue (p. 186, note) serves as an injection fluid which is capable of filling the marvellous network of a lobule by a very moderate pressure (20–25 mm. of

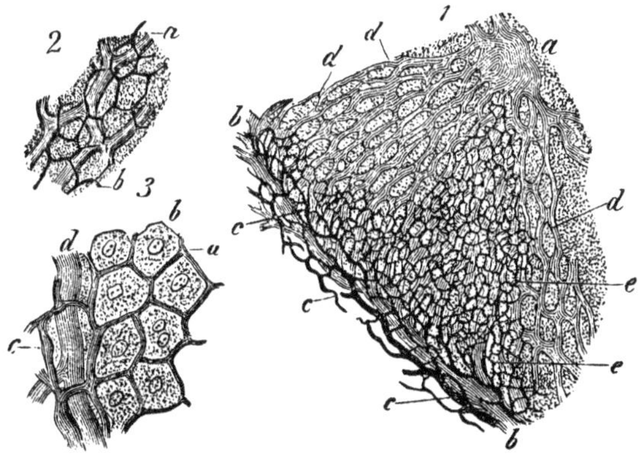

Fig. 253. Biliary capillaries of the rabbit's liver. 1. A part of a lobule. *a*, vena hepatica; *b*, branch of the portal vein; *c*, biliary ducts; *d*, capillaries; *e*, biliary capillaries. 2. The biliary capillaries (*b*) in their relation to the capillary blood-vessels (*a*). 3. The relation of the biliary capillaries to the hepatic cells. *a*, capillaries; *b*, hepatic cells; *c*, biliary ducts; *d*, capillary blood-vessels.

mercury); in other cases only by a cautious increase of the pressure (40–45 mm.). A rounded network of extremely narrow cylindrical tubes, measuring only 0.001–0.0008‴, will then be seen to permeate the entire hepatic lobule. Interwoven with the capillary network of the blood-vessels, it at the same time surrounds the gland-cells with its individual meshes, so that a portion of the surface of each hepatic cell comes into intimate contact with these finest passages, which have been appropriately named "biliary capillaries" (MacGillavry). Our wood-cut, fig. 253, affords the reader a primary representation of this structure; 1 shows the arrangement in the lobule with a low power, 2 shows the biliary capillaries and the capillary blood-vessels, and 3 these, together with the hepatic cells, more strongly magnified.

The recognition of this delicate condition was at first successful only in a few varieties of the mammalial animals. The injection succeeds with tolerable facility in the rabbit; it is more difficult in the dog, the cat, the hedgehog, the calf, and the Guinea-pig. Essentially the same structure was afterwards noticed in the remaining classes of the vertebrate animals (Hyrtl, Hering, Eberth). The injection of indigo carmine in the vein of the living animal (comp. p. 188), which, according to the statements of Chrczonszczewsky and Eberth, is likewise capable of bringing out the network of the biliary capillaries, is also to be recommended for such studies.* What is, however, the more accurate relation of the biliary capillaries to the cells and blood-vessels of the liver?

The coluber natrix (fig. 254, 1) shows, in the most elegant manner, the transversely-divided, finest biliary passages (*c*) surrounded by a wreath of gland-cells (*b*) and separated by these from the capillary vessels (*a*). A similar arrangement is also presented by the liver of the salamander (2).

In the mammalia, on the contrary, the fine system of biliary passages assumes the reticular arrangement represented in fig. 253, in consequence of the extensive development of the lateral branches. Here, now (fig. 254, 3), we see the surface of each hepatic cell (*b*) coming in contact one or more times with the biliary capillaries (*c*). The biliary capillaries and capillary vessels (*a*) are, however, never contiguous to each other, but, rather, a gland-cell or a portion of one always separates the biliary from the blood current. Even in the mammalial animal, therefore, notwithstanding all complications, the old fundamental plan is retained.

If the injection with constant pressure has succeeded—the process should be discontinued as soon as a few lobules on the surface of the liver become slightly blue—the organ may be

---

* Asp showed how the finest biliary passages may be rendered visible, filled with their natural contents. He injected into the ductus chelodocus of a living animal 15 grammes of a saturated solution of gum or tallow. Several days later the creature was killed and the liver hardened in absolute alcohol, chromic acid, or bichromate of potash. The biliary capillaries then appeared as fine gold-yellow shining filaments.

examined fresh. It is more suitable to afterwards fill the blood-vessels with strongly-acidulated gelatine and carmine, and when the liver has cooled, to cut it in pieces and harden it in strong

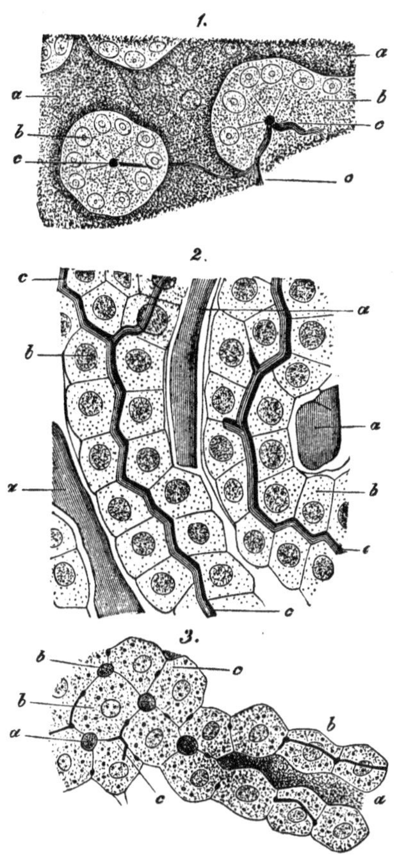

Fig. 254. The finest biliary passages of the liver: 1. of the coluber natrix (after Hering); 2. of the salamander (after Eberth); 3. of the rabbit. *a*, blood-vessels; *b*, hepatic cells; *c*, biliary capillaries.

alcohol to which a few drops of acetic acid has been added. If a weak tingeing with carmine be subsequently employed, the preparations obtained are very handsome and instructive.

If the injection be continued too long, or if too strong a pressure is used, there follows, according to MacGillavry, an extravasation into the lymphatic vessels, into the extremely-developed lymphatic network of the lobules. It is believed, at the first glance, that the capillary blood-vessels have been filled, so deceptive is the appearance; more accurate investigation shows that the injection mass surrounds the blood-vessel like a mantel. The investing lymph-current (which reminds one of a similar condition of the central nervous system, p. 357), therefore, occupies the space intervening between the capillary walls and the connective tissue which surrounds the trabecular cell network after the manner of a membrana propria.

Such extravasations into the lymphatic system, which finally lead to the filling of the interlobular lymphatic passages, readily occur, and have been, here and there, erroneously accepted by earlier experimentalists for successful injections of the biliary passages.

The larger lymphatic canals may be recognized in the vicinity of the lobules. They are regularly arranged and have a partly isolated course, and are partly united into networks of unequal size. Even here these lymphatics begin to encircle, in a reticular manner, the blood-vessels and biliary canals lying between the lobules, which is always the case with the larger branches of the latter vessels. Furthermore, according to Teichmann's statements, the human liver has a single layered network of superficial passages, contained in the peritoneal covering. The meshes are of different sizes and vary in diameter; they are now and then enlarged into considerable lymph-receptacles.

The nerves of the liver come from the plexus cœliacus and consist in part of medullated, in part of Remak's fibres. They have been seen to pass to the vessels, the biliary ducts, and the covering of the organ. Osmic acid has been recommended for their more accurate examination.

The examination of the hepatic secretion of the fresh normal bile shows the microscopist a clear colorless fluid without granules or drops of fat, at the most with a few separated cylindrical cells tinged with coloring matter. The cellular elements of the hepatic substance proper, in contradistinction to many other glands, is entirely wanting in the secretion, so that we still find ourselves in the dark with regard to their destiny and duration of life.

Under more abnormal conditions, sediments are formed in the contents of the gall-bladder. The microscope may show slimy masses, with larger quantities of separated cylindrical epithelium and granulated spherical cells (mucous and pus corpuscles). In bile which has been long retained in the gall-bladder one meets only very rarely with crystals of cholesterine (comp. p. 360); occasionally, on the contrary, one sees deposits of the red biliary coloring matter or bilirubin (cholepyrrhin, biliphæin, bilifulvin). These have, for the most part, amorphous structures, and appear as sausage-shaped, bulbous masses. By treatment with chloroform, larger and more perfect crystals, rhomboidal prisms, needles, and lamina are obtained. The use of sulphuret of carbon is still more advisable. Our fig. 253 shows magnificent crystals of bilirubin, which were obtained in the latter manner by Staedeler from human gall-stones. It has not as yet been definitely decided whether bilirubin and hæmatoidin are similar or only nearly related bodies.

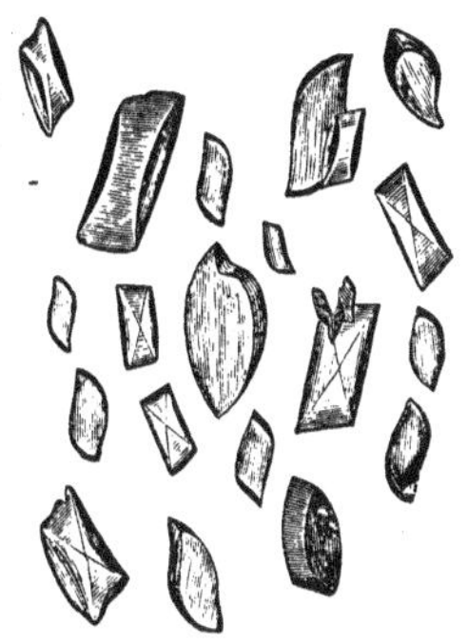

Fig. 255. Crystals of bilirubin. precipitated from sulphuret of carbon.

Pathological changes of the hepatic tissue are frequently met with. Our knowledge of them has recently been especially promoted by a classical work of Frerichs, and the interesting investigations of E. Wagner. Here, as in other glandular organs, we find the cells capable of increase and of manifold changes, but rarely of a transformation into new tissue elements; while here also the new formations may proceed, for the most part, from the small cell-like structures of the connective-tissue framework.

In hypertrophy of the liver we see an enlargement of the existing gland-cells; so that they have gained twice and even three times their normal circumference, and frequently enclose two and sometimes three nuclei. In other cases the microscope shows small, rounded, pale cells, with a large nucleus. These young formations, which have proceeded from the normal hepatic cells, may constitute the greater portion of the parenchyma of the liver, but may also be met with in smaller numbers, together with the large cells mentioned.

A few brown molecules of biliary pigment are met with in the hepatic cells of healthy persons. In obstructed biliary excretion the number of these molecules increases (especially in the cells adjacent to the hepatic vein), or the cell body becomes yellow. The nucleus may also become tinged, and solid, rounded, bulbous or rod-shaped masses of a yellow, brownish-red or greenish color appear in the cell contents. Where the disease has continued for a longer period, concretions of the biliary pigment, frequently in the form of rod-shaped structures, fill the distended biliary capillaries (O. Wyss).

The deposition of fat molecules and drops of fat in the hepatic cells has already been mentioned above. Higher degrees of this process constitute very frequent physiological, as well as pathological occurrences (fig. 256). A fatty or otherwise luxurious diet, combined with deficient bodily exercise, frequently produces such a condition, a so-called fatty liver. It is thus found in the cadavers of quite healthy individuals who have perished suddenly, as well as in nurslings. If cod-liver oil be added to the food of a dog, the hepatic cells of the animal will be filled to a considerable degree with drops of fat, even after a few days, and after eight days they will be quite overloaded with the same. If the cod-liver oil be withheld, this superfluous fat will, after a time, disappear entirely from the cells. The stuffing process produces in geese such a fatty liver, which is highly esteemed by gourmands. The same condition is observed in other cases of a morbid nature, as, with especial frequency, in pulmonary phthisis and dyscrasia potatorum. Locally circumscribed overcharging of the liver with fat also frequently occurs.

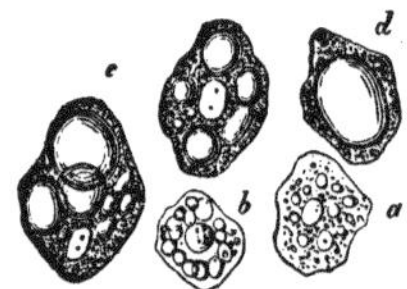

Fig. 256. Cells of the fatty liver.

If we follow the increasing infiltration of the liver-cells with the microscope, we see the, at first, small drops of the molecules of fat become more and more numerous (*a b*), then flow together into a few drops (*c*); finally these also unite into a single drop (*d*).

If such fatty livers be hardened in chromic acid, and the fine sections be rendered transparent with alkalies, the deposition of the fat will be found to proceed in an interesting manner through the cells of the lobule.

Introduced by the portal vein, the fat is first deposited in the cells which belong to this capillary district, that is, in the peripheral portion of the hepatic lobule. The process then advances step by step in a central direction, so that soon only the central cellular trabeculæ, which are adjacent to the hepatic vein, remain free from fat; finally, the deposition of fat also takes place in the latter.

Such a fatty liver will indeed astonish us by the slight quantity of blood which it contains, and will accomplish less for the secretion of bile than the normal organ; but its cells (reminding us of those of fat-tissue) tolerate this fatty deposit well, on the whole, and frequently resume their former condition.

It is otherwise, on the contrary, with the actually fatty degenerated liver. Here, as indeed everywhere, the structure is destroyed by the process of degeneration. Such a metamorphosis is found, for the most part, only in limited portions of the hepatic tissue, in the vicinity of inflammatory foci and tumors.

In a very remarkable, and in its exciting causes still completely enigmatical disease, the acute or yellow atrophy of the liver, there is observed a quick, and often very rapid destruction of the hepatic cells, so that in their place, in cases of a high degree, only a detritus is found, consisting of partly colorless, partly brownish granules, fat-molecules, and drops of fat, as well as crystalline products of decomposition (leucin and tyrosin), which are then partially removed by the urine. The framework of the cellular trabeculæ persists, however, so that it may be readily isolated with the brush; the same is also true of the capillary walls. If, however, the attempt be made to inject the latter, numerous extravasations soon take place, obviously because now, in the place of the former cells, the softened substance of the capillary walls no longer affords any support.

We have just alluded to the crystalline products of decomposition, the occurrence of immense quantities of which in the so-called yellow atrophy was first observed by Frerichs.

In infectious diseases, in typhoid, so-called pyæmic and septic affections, as well as in cases of malignant intermittent fevers, matters occur in the liver, as evidences of an altered assimilation, which in the normal organ are either entirely wanting, or are only present in very much smaller quantity. Among these are to be enumerated a series of crystalline substances which are attributable to organic bases.

Fig. 257. Crystalline forms of tyrosin.

Among these, tyrosin and leucin stand in the first line. Tyrosin (fig. 257) appears in white needles of a silk-like lustre, which occur in part more isolated (*a*), in part, however, united into delicate smaller and larger groups (*b b*). Its reactions may be ascertained from a text-book on zoochemistry.

Leucin is seen in various forms in the examination of the human body. Among these are frequently seen peculiar druses of characteristic appearance, partly small spheres (*a*), partly semispherical structures (*b*), partly aggregations of such masses (*c d*), whereby not unfrequently numerous small, flattened seg-

Fig. 258. Various crystalline masses of leucin.

ments of spheres rest upon a larger spherical body (*e f*). Stratified spheres (*g g*) with smooth borders remind one of starch granules; others have a rough surface. Quite similar druses of fine crystalline needles likewise occur.

Hypoxanthine (or sarcine), a third variety of these products of decomposition, has been much more rarely met with in the diseased liver. With regard to their further properties, we must again refer to the text-books on chemistry. Their combinations with nitric and muriatic acids produce characteristic crystalline forms. Our fig. 259 shows, in its upper half, the appearance of the nitric acid salt, while the lower portion affords a representation of the muriatic salt. The smaller, cucumber-shaped crystals of the nitrate of hypoxanthine are of a particularly characteristic nature.

There is still another nearly related body, xanthine, which constitutes an element of the urine, and is also met with in various organs; it occurs in the healthy and diseased liver, and may here be casually mentioned. Fig. 260 shows the crystalline forms of the combinations with nitric and muriatic acids.

The upper half represents the nitrate of xanthine; the lower portion of the figure is occupied by the characteristic crystals of the nitrates.

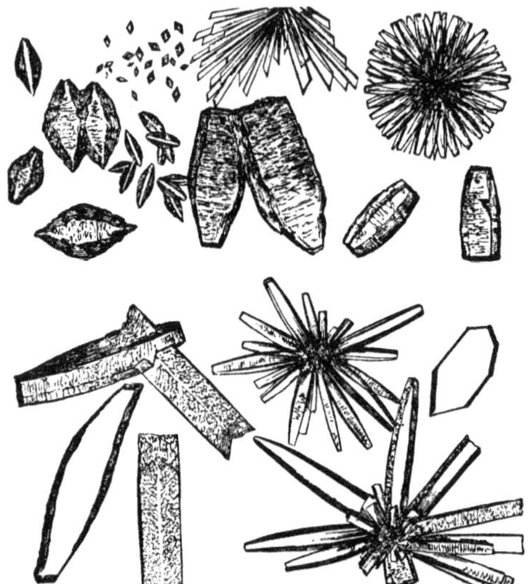

Fig. 259. Crystals of the nitrate and muriate of hypoxanthine.

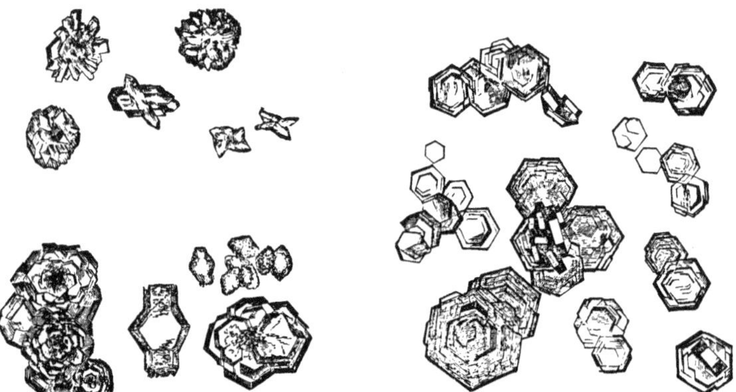

Fig. 260. Crystals of nitrate and muriate of xanthine.

Fig. 261. Crystals of cystine.

Cystine (fig. 261), a product of the decomposition of the body, characterized by the large proportion of sulphur which it contains, has also been observed crystallized in colorless, six-sided laminæ or prisms in the products of decomposition of the liver, in the above-mentioned infectious diseases. It likewise occurs in the normal organ.

While the above-mentioned pathological processes show a transformation of the hepatic cells, in many other diseased conditions of the organ the latter either remain entirely unaffected, or are changed in a secondary manner, and then only subsequently, as, for instance, by compression.

In many cases of malignant intermittent fever an extensive development of melanine has been observed in the tissue of the spleen. Pigmented cells and flake-like bodies, the latter frequently of considerable size, pass through the vena lienalis into the blood of the portal vein, and from here into the vascular district of the liver. If the brown, island-like figures of the lobules, which are often visible to the naked eye, be examined it will be seen that the capillary vessels, and also larger branches which belong to the portal and hepatic veins, are obstructed by these pigmented masses. Similar emboli are also met with in other organs, especially the kidney and the brain. Whether the cerebral symptoms which have been observed in such diseases are to be thereby explained still remains uncertain.

The so-called waxy, lardaceous, or amyloid degeneration of the liver, which, equally and together with that of the spleen and kidney, is not a rare occurrence, does not affect the hepatic cells alone. We have already mentioned this process cursorily at the vascular system (p. 390). The nature of the homogeneous, dull glistening, peculiarly reacting substance was for a long time a subject of controversy, and a definite conclusion has not been arrived at even at the present time. We are now aware, at least, that all the above names are wrong, inasmuch as a product of the metamorphosis of albuminous matters is present, but not fatty substances, or even amylon and cellulose (Kekulé, C. Schmidt). The microscopic examination has shown that the walls of the small arterial branches and of the hepatic capillaries undergo this metamorphosis. The walls of the vessels affected become thickened, stiff, homogeneous, and glistening; thereby a decrease, occasionally an occlusion of the lumen takes place, so that a colorless cylinder results. The normal, fine granular contents of the cell itself disappear more and more, to make place for a homogeneous substance, and the nucleus is gradually destroyed. The cells, which are transformed into flakes, sometimes hang firmly together in the form of consistent, irregular-shaped lamellæ.

We have already mentioned above (p. 136) the peculiar reaction of iodine and sulphuric acid on the substance in question. We will here discuss this more thoroughly by way of example.

The section, which has been made from the fresh hepatic tissue and washed out, is to be placed in a weak aqueous solution of iodine; it is well to continue the immersion for some time, and, to facilitate the saturation, the preparation should be turned over a few times. Even now a characteristic reddish-brown is noticed. The greater portion of this fluid should then be removed and a covering-glass placed over the preparation; concentrated sulphuric acid should then be allowed to flow in from the side as slowly as possible. At very unequal periods, either immediately or after several minutes or hours, or even later, there is either an increase of the red color or a dirty violet, more rarely a blue color produced. Another procedure is, however, more advantageous. Fine sections from preparations hardened in alcohol are to be placed in a glass box with distilled water and 10–20 drops of tincture of iodine added. Then, generally after five minutes (when the coloring of the amyloid substance usually takes place), the preparation is to be washed and again placed in clean water and 3–6 drops of concentrated sulphuric acid added. The characteristic tinge is obtained sometimes rapidly, sometimes only after 2–3 hours, when the ex-examination is to be made with the addition of glycerine. Such objects may be preserved for a sometimes shorter, sometimes longer period; but not, however, according to previous experience, in the form of permanent cabinet preparations.

In tubercle of the liver the ordinary elements are first recognized; nuclei, small cells in the condition of shrinking, together with these, large, flake-like structures with several nuclei. It was formerly considered that these substances originated in the interstitial connective tissue. At the present time, the vascular ramifications are considered to be the points of origin of the tubercles.

A hypertrophy of the connective-tissue framework substance which permeates the liver, with a corresponding transformation of the compressed lobules and gland-cells, is found in the so-called granulated liver, cirrhosis hepatis. The examination may be made in various ways. Sections from the fresh tissue may be picked and treated with reagents, or, which we would prefer, suitably hardened objects may be used. At the commencement of the process it is noticed that the scanty connective tissue which separates the hepatic lobules proliferates considerably; its cells increase and the interstitial substance becomes transformed into a firm, fibrillated substance reminding one of cicatricial tissue. This, by its further increase, compresses the hepatic lobules more and more, so that gradually only island-like remains of the same, with shrivelled, brownish cells, are to be met with. These are in part tinged with hæmatine; they contain in part yellow corpuscles or fatty substances, or finally, amyloid. The membrana propria may hereby be still recognizable, but is likewise finally transformed into connective tissue. The groups and aggregations of brownish molecules which are found embedded in the connective tissue proceed from the remains of disintegrated hepatic cells. The capillaries likewise gradually atrophy, and in the same proportion that the gland-substance disappears, while the interacinous biliary passages often remain permeable for a long time. Injections rarely succeed.

In carcinoma of the liver the connective-tissue framework of the organ is very probably transformed immediately into the framework or stroma of the carcinoma. The origin of the carcinoma cells remains obscure.

We have finally to discuss the spleen. This organ, which still presents so much that is enigmatical concerning its physiology, was also, until within a few years, only imperfectly understood with regard to its structure, and, in fact, numerous accessories are requisite if we would obtain a knowledge which is in any degree satisfactory concerning the latter. The extreme softness, the excessive vascularity of the spleen, and the numerous elastic septal formations render the manipulation very difficult. The latter system of septa (and herein a close parallel with the related lymphatic glands of the creature is shown) is, in large mammalial animals, highly developed and presents a complicated framework, while in small creatures it diminishes more and more, even to an almost complete disappearance. The spleens of the smaller rodents (rabbit, Guinea-pig, squirrel, etc.) form therefore, like the lymphatic glands of these creatures, the most suitable objects for the primary investigation.

One would be very much deceived if one were to expect to find in the fresh spleen, even with the most careful preparation, more than isolated elements, blood-cells, contractile lymph-corpuscles, vascular epithelium, etc. In consequence of the great softness of the organ, there is scarcely an appearance of even fragments of the delicate but developed connective-tissue framework which permeates the whole gland. Injections are also frequently frustrated by the extreme softness of even the freshest spleen. We are here, therefore, admonished to use hardening accessories, and, as the drying methods are not to be thought of, to employ alcohol, chromic acid, and the bichromate of potash.

Assuming that we wish to prepare in this manner the spleen of a small mammalial animal (rabbit, Guinea-pig), the whole organ may be immersed. With the spleens of larger creatures it is judicious to expose only a portion to the influence of the above reagents, and to previously force a stream of the fluid through the blood-vessels with a syringe.

For many purposes alcohol is quite sufficient, especially if it is used diluted at first, and then, after a few days, replaced by alcohol which is stronger. After 6–8 days (occasionally, however, only after a few weeks) the spleen is in a condition suitable for making sections, and has also gained such a consistence that it may be conveniently brushed. Increased hardening no longer permits of the latter important procedure, or only very imperfectly; and, as a rule, nothing further can be done with such spleens. Frequently a spleen is only rendered fit for injecting after an immersion for 24–28 hours in ordinary preparation alcohol. Injected spleens (and here again only transparent, partly solidifying, partly cold-flowing masses should be used) are also, as a rule, to be hardened in alcohol.

For many textural conditions, however, chromic acid renders decidedly better service than alcohol. Portions not too large should be placed in an ample quantity of the fluid, and at first a weak solution of the acid, 0.2–0.1 per cent., is to be used. This is, after several days, to be exchanged for one of twice the strength, and afterwards perhaps for one that is still more concentrated. If test sections be made from time to time with the razor and the brush, good objects will be obtained.

I have seen the finest results, however, from the use of the chromate of potash. If a solution of about 1 per cent. be commenced with, and the concentration slightly increased daily, after several days a period arrives when the organ, which is not yet sufficiently hardened, must be still further hardened by means of alcohol. After a few days further the entire tissue has, with great preservation, gained the proper condition.

The method of further investigation consists in the preparation of thin sections in various directions, which are to be examined partly unbrushed, partly freed from blood- and lymph-corpuscles by means of the brush. An immersion for several hours in pure glycerine is useful, and tingeing with carmine is of the same importance as for the lymphatic organs. The system of septa likewise comes out very beautifully in this way. Acids, the reagents customary for the demonstration of smooth muscular fibres, such as chloride of palladium and the double staining with carmine and picric acid, serve for the recognition of its finer texture.

Nevertheless, although the directions given lead to the hardening of fresh, moderately consistent mammalial spleens, do not think that every human organ can be thereby mastered. The maceration which we meet with in our post-mortems, the often considerable softening which may be found in diseased bodies, not unfrequently render the suitable hardening of the spleen a difficult piece of work, for the termination of which, not only days, but also weeks and months are necessary. Here the weak, watery alcohol is soon to be replaced by that which is stronger; finally, operate with absolute alcohol. Billroth recommends the action of chromic acid in very concentrated solution (even to 20 per cent.) on small portions of spleen, for hardening in typhoid affections of the organ. The framework and the relative arrangement, it is true, become visible in fine sections in this way; the cell metamorphoses and other delicate textural conditions must be followed sooner on the fresh organ or on a portion which is but slightly hardened, for a chromic acid of such strength produces great shrinking.

Preparations of the spleen are to be mounted partly in the ordinary moist way in glycerine, partly dry, whereby, however, absolute alcohol and Canada balsam dissolved in chloroform should always be used. Sections from transparent injections, somewhat strongly tinged with carmine, afford by the latter method very handsome preparations for examination. The system of trabeculæ also appears finest by such treatment.

If now we inquire what information as to the structure of

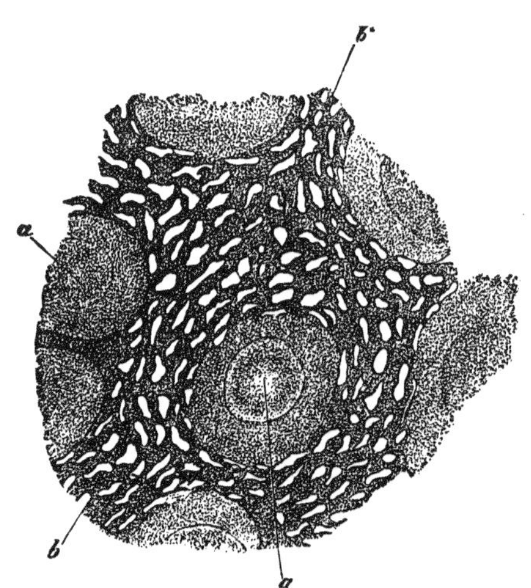

Fig. 262. Transverse section of a rabbit's spleen. *a*, Malpighian corpuscles; *b*, the reticular framework of the pulp, with the spaces filled by the venous blood-current.

the spleen has been obtained by the aid of these accessories, the answer may be given that our organ constitutes a complicated lymphatic gland, in which the lymph-current is replaced by a blood-current, and is therefore a blood-lymph gland, as we might express ourselves in brief.

The Malpighian corpuscles of the spleen (fig. 262 *a*) show the structure of the lymphatic-gland follicles, and, in so far as they do not pass over into tubes or the tissue of the pulp, they have likewise on their surface a more narrow-meshed, reticular border. Nuclei occur in a portion of the nodal points, especially in younger animals. The capillary system presents nothing remarkable, and, with suitable objects, brushing usually succeeds with facility. We would recommend the spleen of the sheep as being very suitable.

In many small creatures (rodentia, for example, the rabbit, Guinea-pig, and marmot) there is also, at some distance from the periphery, a narrow-meshed, concentric layer of reticular connective tissue, the significance of which requires, however, further elucidation.

The pulp (fig. 262 *b*) consists of a system of tubes arising from the Malpighian bodies and united together in a reticular

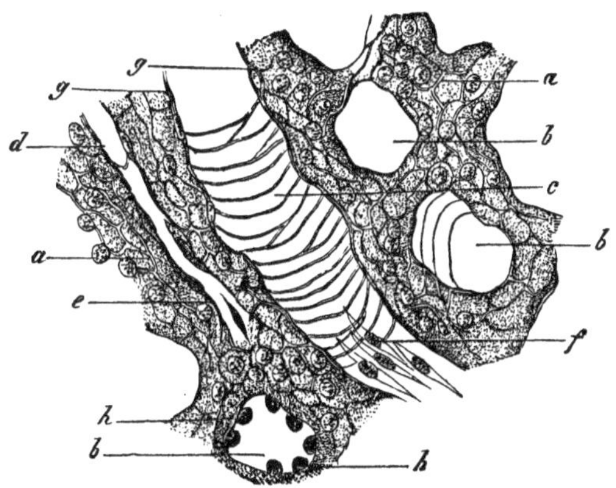

Fig. 263. From the pulp of the human spleen, brushed preparation (combination). *a*, pulp tube with the delicate reticular framework; *b*, transverse section of the cavernous venous canal; *c*, longitudinal section of such a one; *d*, capillary vessel in a pulp tube, dividing up at *e*; *f*, epithelium of the venous canal; *g*, side view of the same; *h*, its transverse section.

manner, which present a much finer and narrower-meshed reticular framework (fig. 263 *a*) and which is much more difficult to isolate. We are indebted to Billroth for its recognition. Permeated by capillaries, it encloses in a sometimes more reticular, sometimes variable form a system of passages which serve for the reception of the venous blood—a discovery to which I was led in the year 1860 by the injection of the human spleen, and which was afterwards also confirmed by Billroth. This system of venous passages reminds one essentially of the cavernous canals which permeate the cortical substance of the larger lymphatic canals and serve for the removal of the lymph.

These passages of the spleen pulp (*c*) are, however, without a membranously thickened wall, inasmuch as the same fine reticular tissue which occurs within the pulp-tubes also encloses the venous current. The passage is also lined by an unstratified epithelium (*f*), which in man has a peculiar spindle shape, and the rounded nuclei of which project into the lumen.

The spaces of the spleen are, as the first examination shows, filled with lymph-cells. As the walls of the fine veins do not appear to be membranously thickened, a wandering of these cells into the venous current, and, with a more considerable increase of the current, a forcible penetration of blood-cells into the pulp-tubes, is conceivable. We therefore see the colored blood-corpuscles in part unchanged, in part in various stages of ruin, and, by no means rarely, free in the tissue of these passages.

Peculiar flake-like structures containing blood-corpuscles and reminding one of cells were described, even many years ago, as being found in the spleen (Kölliker, Ecker, Gerlach, and others). The localities of their occurrence as well as their genesis require renewed investigations, although there is certainly here an admission of an amœboid cell.

With regard to the course of the blood-vessels, it may be said that the greater portion of the arterial trunks can be readily followed in injected preparations; likewise the division of the venous branches. It is also easy to perceive how the capillary vessels of the Malpighian corpuscles are formed by the breaking up of the former. A single or double arterial branch is generally met with on and in the Malpighian corpuscle; veins do not occur here.

The recognition of the capillary blood-vessels in the spleen-pulp as well as their connection with the venous passages is, on the contrary, extremely difficult, and even to the present time there is no agreement of opinions concerning this important structural question. Many investigators assume, after the example of Gray (whose fine monograph is still too little known in Germany), a direct continuation of moderately large capillaries into the venous canals; others believe that they have convinced themselves that a very close network of extremely fine capillary tubes occur here (Key, Stieda). According to our own observations (and we find ourselves in harmony with the most thorough monographist of the organ, with W. Müller), the passage of the arterial splenic blood into the venous roots takes place, on the contrary, in man and the mammalia by wall-less currents. These pass through the reticular framework of the pulp-tubes by using the interstices of the fibres and lymph-cells, we might say, somewhat as the water of a failing brook takes its course between the pebbles of the bed. Our fig. 263 shows a capillary vessel *d* which at *e* is distributed into the network of the pulp, and may represent to the reader the commencement of the intermediate pulp-current. The blood or the injection mass then passes from the spleen pulp through the spaces of the limiting layer (*c*) into the commencement of the veins. Fig. 264 will render this continuation (*b c*) intelligible, and at the same time show that a reticular, scale-like coagulation of the injection mass over the lymph-cells of the pulp-tube explains the pretended capillaries of Key and Stieda.

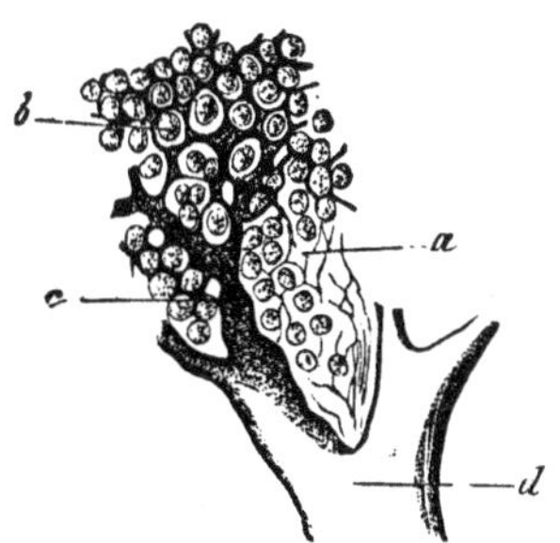

Fig. 264. From the sheep's spleen (double injection). *a*, reticular framework of the pulp; *b*, intermediate pulp-current; *c*, its continuation into the venous roots with incomplete walls; *d*, venous branch.

For the recognition of these important conditions we recommend the injection of a sheep's spleen, as cautiously as possible, but also as completely as possible, with a very intense blue gelatine mass, and to tinge the sections made from the hardened organ with carmine. The comparison with the natural injection is of high value as a control. The organ, hardened in a solution of chromate of potash (1 per cent.) and afterwards in alcohol, shows us, in fine sections treated with glycerine, the uninjured blood-corpuscles in the same places in which we have met with the blue injection fluid (W. Müller).

Lymphatics are, as a rule, very readily recognized in the capsules of large mammalial animals (ox, pig, sheep). Their injection almost never leads into the interior of the organ, and, with the puncturing method, the reticular venous canals are regularly filled. The opinion seems justifiable, therefore, that lymphatics are wanting in the tissue of the spleen (Teichmann, Billroth, Frey). Subsequently, however, Tomsa succeeded in injecting lymphatic vessels in the system of septa of our organ.

The trabecular framework of the human spleen (which arises from the capsule and divides the organ into innumerable irregular compartments) consists of connective tissue, elastic fibres, and scanty muscular elements. It requires the same methods of investigation as the equivalent structures of the lymphatic glands (comp. p. 394).

For the study of the splenic nerves the fresh, thoroughly washed-out spleen is to be treated with alkalies and acetic acid; organs immersed in pyroligneous or chromic acid are also to be used.

It is known that the spleen frequently participates in the more general processes of disease. In certain infectious diseases, as in intermittent and typhoid, its swellings present characteristic occurrences. Attention has more recently been paid to a surcharging of the blood with colorless cells, induced by enlargement of the spleen and lymphatic glands. This condition, leucæmia, we have already mentioned at the blood (p. 232). These metamorphoses of the organ, as well as its various degenerations and new formations, are known in their coarser relations, but not, however, or only very incompletely, in their finer texture. In a case of this affection of a high degree I once met with a considerable hypertrophy of the pulp and an astonishing development of the capillary system lying in the pulp-tubes.

Several years ago Billroth, an observer who has accomplished very much for the knowledge of the spleen, made an inroad into this domain by the aid of the improved methods.

The finer changes of the spleen in abdominal typhus are still very imperfectly known. The more or less swollen organ does not show, in injected preparations, the remarkable distention of the veins and capillaries which we have mentioned above as occurring under the same conditions in the lymphatic glands and Peyerian follicles (comp. p. 401 and 395); still, there are certainly slight dilatations of the vessels.

Of interest is, on the contrary, the occurrence in abdominal typhus of the large multinuclear cells in the venous spaces, the same as we have formerly mentioned as occurring in the passages of the lymphatic glands. Here, also, in the later periods, the characteristic molecular ruin of these cell-masses takes place, in so far as they are not previously removed from the spleen by the blood-current.

The numerous granules which are met with in our organ in miliary tuberculosis are, as a rule, located in the tissue of the pulp and only rarely in the Malpighian corpuscles. Their contents is the familiar fine granular substance with shrivelled nuclei and cells.

In the so-called hemorrhagic infarctions of the spleen, which, as is known, are not rare occurrences, the microscopic analysis shows in the overloaded venous passages the appearance and the phases of metamorphosis of masses of coagulated blood.

In the ordinary hypertrophy, the reticular tissue of the pulp may present great thickening, so that sometimes it appears similar to that of the Malpighian corpuscles. In conditions of high degree the lymphatic cells of the latter disappear; in their places fine granular substance and yellowish pigment are noticed.

In cases of malignant intermittent those pigmentated flakes and pigment-cells are produced, which, passing out through the vena lienalis, may give rise to embolia of frequently considerable size, first in the liver and then in other organs, such as the kidneys, brain, etc. (comp. p. 233).

We have already, at the liver, mentioned the amyloid degeneration of the tissue of the spleen which occurs so frequently

The organ, which has become more firm, readily permits of hardening in alcohol, whereby (as was casually remarked at the liver) the capability of reacting of the amyloid substance is not lost, and fine sections permit of the recognition of the deposits in a convenient manner. In many cases we notice the Malpighian corpuscles first attacked; in other cases the parietal layer of the venous canals in the pulp has undergone amyloid degeneration.

The first form of deposit, known to the pathological anatomists under the name of the "sago spleen," shows the arterial walls to be the point of origin.

In the other, more rarely occurring variety, the lardaceous spleen, on the contrary, the transverse sections of the venous passages of the pulp are surrounded by a thicker, homogeneous amyloid layer.

Attempts at preserving such preparations of pathologically metamorphosed spleens must be made according to the directions given for the normal tissue.

Section Nineteenth.

RESPIRATORY ORGANS.

The investigation of the respiratory apparatus presents relatively less difficulties to the microscopist than that of the organs described in the previous section.

The larynx, trachea, and bronchi consist of tissues which have already been described by us in previous chapters, so that the methods there given are to be repeated here.

The epithelium of the parts mentioned, coverings of epithelial cells, with the exception of the stratified epithelium on the lower (true) vocal cords, are examined either by scraping them off in the fresh condition or after being hardened in alcohol on thin stained sections. The latter method also serves for the recognition of the texture of the mucous membrane and the racemose glands which occur here. The latter not unfrequently become changed in consequence of catarrhal processes, their vesicles become enlarged, and their cell-contents changed. The cartilages may be examined either fresh or after being hardened. Calcifications and ossifications of the same, which, as is known, are frequent occurrences in after life, are to be examined fresh or after having been decalcified by chromic acid. The distributions of the nerves are to be studied in acetic or pyroligneous acid preparations; lymphatic vessels are to be injected by the puncturing method from the submucous tissue.

The same methods of treatment serve for the larynx, trachea, and bronchi; their smooth muscular elements require the so-frequently mentioned accessory which is used for the demonstration of that tissue.

The investigation of the lungs is, however, quite different. Portions of the fresh tissue when picked readily show the elastic fibres and membranes, especially after the application of acetic acid or alkalies. The epithelial structures of the alveoli and the finest bronchial ramifications may also be recognized. But in general the results are limited to these, and such examinations are not unfrequently considerably impeded by the numerous air-bubbles in the preparation.

Other methods of treatment are therefore necessary.

These consist of drying, or the use of the same hardening solutions which we have already so frequently mentioned. If possible, the blood-vessels should always be previously injected, and for many investigations it is almost indispensable to have the respiratory canals distended.

The whole lung, or portions of the same, when carefully dried, assume a consistence which permits of sections being conveniently made in all directions. These when softened permit of the satisfactory recognition of the greater portion of their details, and the application of staining methods, of acetic acid and alkalies, constitute further advantageous accessories. It is preferable to moderately inflate the lung which is to be dried by the bronchus or the air-passages, and after tying these, to hang it in the sun or near a stove to harden. The injection of the air-passages (from which the air may be previously removed by means of an air-pump) with uncolored (also colored) gelatine is a very good method. Injections of the blood-vessels with transparent colors and a menstruum which solidifies, such as gelatine, permit of the same treatment. Sections made from these and softened present beautiful appearances, especially if the injection fluid was not too watery. With smaller creatures the injection should be made by the arteria and vena pulmonalis; with those which are larger, generally only by single branches of each of these vessels. The injection is in general to be regarded as an easy one, even with small mammalia, if the syringe is only very cautiously managed.

If the finer textural relations are to be examined, alcohol, chromic acid, and chromate of potash are to be used for the immersion of portions of the uninflated lung or the whole organ, whereby it is also well to inject the bronchi with the hardening fluid. The employment of these fluids also forms the chief means for the recognition of pathological structural changes.

Still better, even here, is the preparatory inflation of a whole organ, or the injection of its air-passages with uncolored gelatine, the blood-vessels having been previously filled with cold-flowing transparent mixtures. If such a lung be suspended by the trachea, in a large vessel filled with alcohol, it affords, after several days, excellent views of its entire structure; and if it was fresh when exposed to this preparatory treatment, even the alveolar epithelium, that cellular covering which was so extensively disputed years ago, and which is nevertheless so easy to recognize, may be seen.

The ultimate terminal ramifications of the bronchial passages pass over into a system of acute-angled ramified canals, which present thin, sinuous walls. These (fig. 265 *c*) are beset laterally as well as terminally by groups of the alveoli or lung-vesicles, the so-called infundibula (*a*), while other alveoli (*b*) constitute the sinuosities mentioned on the walls of these passages (F. E. Schulze). The infundibulum corresponds to the primary lobule of a racemose gland, and may be seen in sections of lungs which are simply dried, or in those of which the air-passages have been filled with transparent materials. Another way of obtaining a view of them is by the corrosion method. The air-passages are to be injected with a colored resinous mass, and then the lung-tissue destroyed by the long-continued action of concentrated muriatic acid. The relation of the pulmonary lobule to its bronchial tube is, however, not easy to recognize.

Fig. 265. Two so-called infundibula of the lungs (*a*), with the finest bronchial twigs (*c*), and the pulmonary vesicles (*b*).

For the closer examination of the air-vesicles and their more minute construction, fine sections of the tissue which has been hardened in fluids are used.

For this purpose an entirely fresh lung, which has been carefully isolated and injected, is selected; it should be immersed in alcohol to harden, and the sections when made are to be carefully colored in the familiar mixture of equal parts of the ammoniacal solution of carmine and glycerine (p. 152), and finally washed out in water containing a little acetic acid. To proceed with more certainty, the sections may be taken from the surface of the organ, as recommended by Eberth. In this way one obtains a great number of surface and profile views of the alveoli, and is protected from mistaking them for transverse sections of the finer bronchial branches.

The walls of the air-vesicles (fig. 266, *b*) are rather thin and consist of elastic fibres (*a*). Between the latter there is a

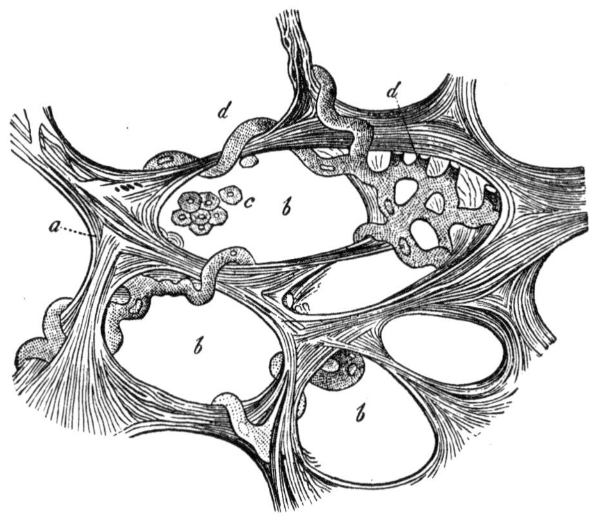

Fig. 266. Section through the lungs of a child 9 months old. Elastic trabeculæ (*a*) between the alveoli, *b*; *d*, capillary vessels curved in a tendril-like manner; *c*, remains of the simple pavement epithelium of the alveoli.

homogeneous connective substance, which is also to be recognized as a limiting layer towards the cavity. These walls are without muscular elements, as any one will be convinced by treating them with chloride of palladium.

We meet with a wonderfully rich network of capillaries, the meshes of which are small, but vary in size with the degree of distention of the alveoli (figs. 267 *c*, 266 *d*). There are one, two, or three small, very transparent, rounded and polygonal nucleated cells (fig. 267 *c*,) lying in each of the meshes according to its size. In transverse sections of the pulmonary vesicles, the epithelial cells are seen to spring forward in a slightly convex manner into their cavities. Very dilute acetic acid may also be used to demonstrate their nuclei; it should not be too concentrated, as a free nuclear formation remains which has been erroneously taken for nuclei of the alveolar tissue. Frequent use of the silver impregnation has also been made of late, and the occurrence of large, non-nucleated cells over the capillary walls (fig. 268) has been recognized with its assistance, so that the appearance of a connected epithelium is presented (Œlenz, F. E. Schulze, and others).

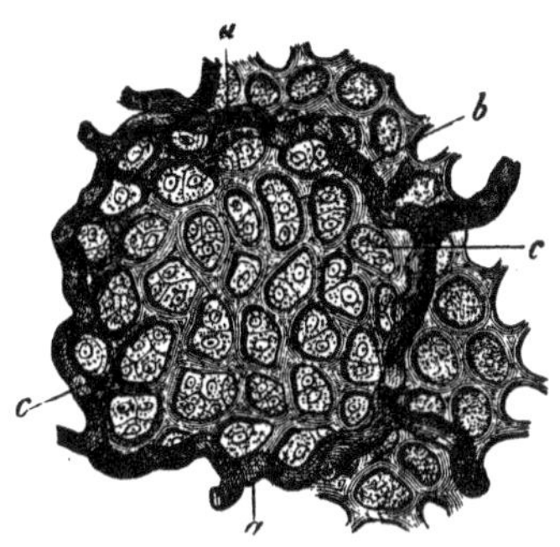

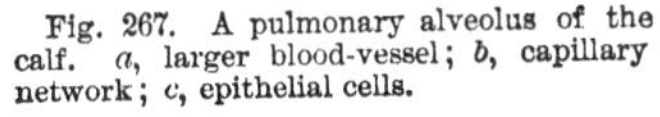
Fig. 267. A pulmonary alveolus of the calf. *a*, larger blood-vessel; *b*, capillary network; *c*, epithelial cells.

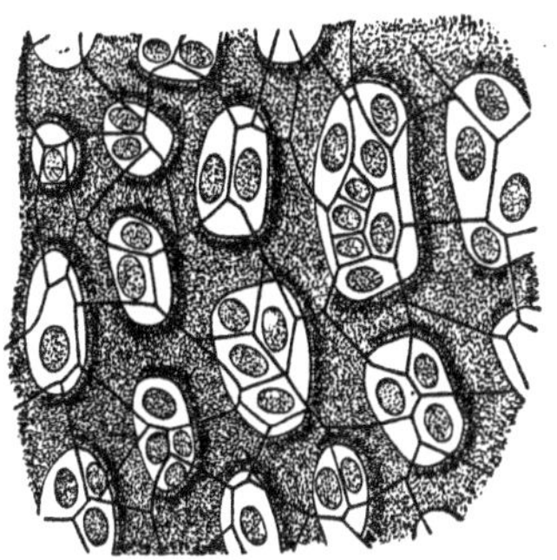
Fig. 268. Pulmonary epithelium of an adult cat, after Elenz.

If, therefore, a simple tessellated epithelium is certainly present in the lungs, we must assume that, though it is interrupted, it nevertheless occurs in a modified form over the tubes of the capillary network, and may, therefore, be continuous.

But we must again have recourse to the injected preparation (figs. 266, 267). If a portion of the capillary network be regarded from the surface, the wavy undulations and loop-like curves of the vessels are seen. If viewed from the side, the tendril-shaped curves are seen to pass more or less beyond the walls of the alveoli, according to the degree of distention of the latter, so that they often project into the air-vesicles as loops of considerable size, projections which, in pathological conditions, may be met with in a much higher degree (Buhl).

The numerous lymphatics of the lungs are to be injected by the puncturing method. A single-layered network with large meshes is found beneath the pleura; this communicates with the deeper lymphatics accompanying the bronchi by means of branches which pass between the lobules toward the interior of the organ. The commencement of the lymphatics appears in the walls of the pulmonary vesicles in the horse, in the form of lacuna-like dilatations (Wywodzoff).

The nerves of the lungs pursue the same course as the bronchi and vessels (especially the pulmonary artery), and may be followed deep into the interior of the organ. Microscopic ganglia occur at the points where their branches are given off. The treatment with chromic acid or diluted pyroligneous acid serves for their primary recognition; osmic acid may be recommended for more accurate studies.

The gland-like structure of the whole organ may be recognized in a beautiful manner in fœtal lungs, especially those of embryos from the first half of intra-uterine life. After being hardened in alcohol, thin sections are to be made and carefully stained; in these the covering of cylindrical epithelium of the glandular canals and the connective-tissue framework (Darmfaserblatt of Remak) are easily seen.

Numerous structural changes of the respiratory organs, especially of the lungs, come under the observation of the physician. The methods of investigation are either the same as, or quite similar to, those for the normal organ. Some of these conditions, which present considerable microscopical interest, may here be briefly mentioned.

Pigmentations, that is, collections of fine melanine granules, which give the organ a spotted appearance, are met with in all human lungs after a certain period of life, so that they are to be denoted as normal occurrences. They sometimes lie in the inter-alveolar elastic tissue, sometimes in the connective-tissue interstitial substance of the pulmonary lobules. The cells of the alveolar epithelium may also undergo this pigmentation, and, evacuated by coughing, occur in the sputum (p. 498), as in other cases they are found to have undergone fatty degeneration.

Now where do these black molecules originate? They are—and we may at the present time assert it with confidence—of a double origin. They may consist of the ordinary dark pigment of the organism, of melanine. Here, as in the bronchial glands (p. 399), the cause may be small apoplectic effusions from the pulmonary capillaries, which are so readily surcharged with blood; likewise transudations of dissolved hæmatine into the tissue. Then also, in civilized life, the individual surrounded by smoke and soot inspires the finest particles of coal. They penetrate into the cell-bodies of the alveolar epithelium, then into the lung-tissue, and from here (probably with the aid of migratorially inclined lymphoid cells) into the bronchial glands. This condition, anthracosis, may be artificially induced in mammalian animals by shutting them up in a sooty room (Knauff). Coal-workers show the highest degree of the disease. Another observation of Zenker's is very interesting. Operatives in factories where they have much to do with oxide of iron, present exactly the same condition of the lungs, only everything is red instead of black.

A senile alteration of the pulmonary tissue and of the alveoli, which accompanies the atrophy of the capillaries, consists of the disappearance of the walls of some of the air-vesicles, and the union of several of these into larger cavities. To prepare such lungs for examination they should be inflated and dried; in certain cases the blood and air passages may be previously injected.

Pathological new formations of the lungs still give rise to many difficulties for the microscopist, especially in recognizing the normal cellular elements of the organ from which the former take their origin.

The lymphoid cells which have emigrated from the blood-vessels are represented here also, at least in part, by the pus-corpuscles. Such an extravasation of these cells appears to be very much facilitated in the pulmonary alveoli, where the so numerous vessels are covered only by a thin layer of epithelium. They may also occur here in the interior of cylindrical or irregularly formed epithelial cells, probably having, however, only penetrated from without, and not having been produced in the latter.

The generally more rapidly progressing inflammation of the pulmonary tissue, the so-called croupous pneumonia, shows at the commencement a considerable distention of the respiratory capillary network, then a filling up of the alveoli and infundibula with coagulated fibrin, as well as with extravasated red and colorless blood-cells. Later the lung-tissue proper also becomes infiltrated with cells. The softened mass is at last met with under the form of pus. The rôle which the alveolar epithelium plays in this disease still remains to be investigated.

The primary microscopical appearances of the above-mentioned contents of the air-passages in a case of pneumonia may be obtained by scraping the cut surfaces. For closer examination the tissue is to be carefully hardened in solutions of chromic acid of increasing concentration, in Müller's fluid, or absolute alcohol. Injections of the vessels of inflamed lungs do not readily succeed, in consequence of the distention of the alveoli and the numerous lacerations of the capillaries.

Tuberculosis of the lungs is of extremely frequent occurrence, partly in the form of so-called tuberculous infiltration, partly in the form of scattered aggregations and numberless small nodules. Very many investigations have been made of this matter, and much has been written about it, but our knowledge of the subject still leaves much to be desired. Although it is well established that the tubercular substance is formed of shrunken nuclei and cells, and from the fragments of these structures and a fine granular matter, and that the neighboring vessels which lie between them also become atrophied, the point of ori in still remains uncertain. The alveolar epithelium is certainly frequently concerned in this process, and hence the position of the tubercular mass within the alveoli is readily accounted for. On the other side, however, the tissue of the lung itself gives rise to such masses. In consequence of the absence of connective-tissue corpuscles from the walls of the alveoli, and the sparseness of this tissue between the primary lobules, attention should be directed firstly to the emigration of the lymphoid cells, and secondly to the cells of the capillary vessels and the adventitia of the finer blood-vessels ; and, in fact, recent investigations have discovered such a point of origin of the miliary tubercles.

The proliferations of the nuclei of vessels which have been noticed by several observers to take place in such cases are rendered all the more probable from the fact that a quite similar process occurs in the adventitia of similar vessels of the brain, and also leads to the formation of miliary tubercle. Further investigations appear to be necessary to determine whether the nuclei of the true primary capillary membrane are also capable of undergoing such a metamorphosis. How important for all such investigations the previous injection of the blood-vessels with transparent masses is, it is unnecessary to mention. For hardening, chromic acid is to be used, commencing with weaker solutions (0.1–0.2 per cent.), and then passing to those which are stronger (0.5–1 per cent.) ; Müller's fluid or absolute alcohol ; the organ should, naturally, be immersed in small pieces.

We must leave the further consideration of these masses of tubercle to the text-books on pathological anatomy. The substance we have described usually becomes softened and leads to the formation of cavities, in consequence of the destruction of the tissue of the lung. If we examine the contents of such a cavern we find softened tuberculous matter, pus-cells, blood-corpuscles, blood coagula, and elastic fibres. The latter may be coughed up and appear in the sputum and thus confirm the diagnosis. We shall return to this point. The walls of these caverns are seen to be formed of compressed lung-tissue.

The examination of the fresh tissue of the pleura may be made by scraping the epithelium and tearing the connective tissue of the serous membrane, with the assistance of the usual reagents. Thin sections may also be made from hardened preparations. These methods, besides being used for the other serous sacs of the body, as the pericardium and peritonæum, may also be employed for the examination of pathological conditions.

Effusions of a watery or purulent nature are to be treated in the same way as other fluids which contain cells ; more solid masses of exudation, which show rounded cells enclosed in coagulated fibrine, are to be examined partly fresh, partly in sections from the hardened preparation. The new formations of connective tissue in the form of looser or firmer bands, which unite both walls of the pleura, require no further mention, as they are to be investigated in the same manner as connective tissue.

The masses expelled by hawking or coughing are called sputa. They do not originate exclusively from the respiratory organ, however, as matters from the mouth as well as from the posterior nares may become mixed with the products furnished by the respiratory apparatus. In the examination of the sputa, therefore, we must always expect to find not only the elements of the respiratory apparatus, but also the epithelium of the two systems of cavities mentioned, fragments of food which have remained in the mouth, grains of starch, for example, the Leptothrix buccalis, etc.

The microscopical treatment is, on the whole, very easy. The specimen for examination is to be obtained, according to its consistence, either with a glass rod, or, if pretty tough, by means of the forceps and scissors; it is then to be examined floating in its natural fluid with a power of medium or greater strength. Reagents are to be employed as circumstances may require, though their action may be impeded, it is true, by the mucus of the fluid.

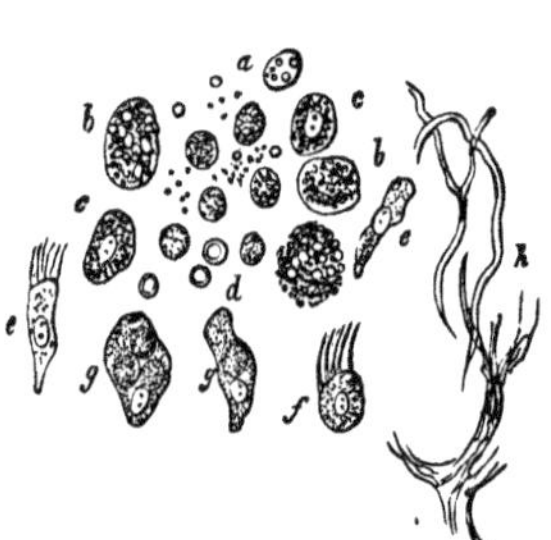

Fig. 269. Elements of the sputum. *a*, mucous and pus corpuscles; *b*, so-called granule cells; *c*, with black pigment (alveolar epithelium); *d*, blood-cells; *e*, ciliated cells after the loss of the cilia and such a cell with cilia; *f*, spherical ciliated cell in catarrh of the respiratory passages; *g*, ciliated cells which have pus corpuscles in their interior; *h*, pulmonary fibres.

It is relatively difficult, however, to preserve such objects as permanent specimens. The attempt may be made to preserve them in camphor-water, very dilute solutions of chromic acid, the Pacinian or some similar fluid (pp. 213 and 214).

The constituents of the sputum (fig. 269) are, together with entangled air-bubbles, epithelium, the cellular elements of glands, mucous and pus cells, blood-corpuscles, pigmentated cells, those in a condition of fatty degeneration, and fragments of pulmonary tissue. Crystals rarely occur and are of minor importance. We find the organized constituents either unchanged or more or less altered by the action of endosmosis and of maceration.

The pavement epithelium is derived from the mucous membrane of the cavity of the mouth, but a few such cells may also come from the larynx, where they cover the lower vocal cords. Smaller pavement or rounded cells come in part from the glands of the mucous membrane, and in part, doubtless, from the alveoli of the lungs, although it is scarcely possible to recognize the latter with certainty in sputum. The quantity of this pavement epithelium from the mucous membrane is naturally quite variable. The tough masses which many persons are accustomed to hawk up in the morning are as a rule rich in them; their number also increases in the sputum when the digestive organs are in an irritated condition. Ciliated cells, which occur, however, by no means frequently in the sputum, originate in part from the posterior portion of the olfactory organs, in part, and chiefly, from the respiratory passages. They may be met with entirely unchanged in form (*e*) or, which is more frequently the case, after their ciliæ have fallen off (*e g*). At the commencement of catarrhal affections of the air-passages one may see, here and there, cells coughed up which still vibrate, partly in their normal shape (*e* below), and partly changed to spherical shapes (*f*). The nuclei appear either single or we notice a few granulated structures (*g*), which are probably mucous and pus corpuscles within the cylindrical cells, so that migratory conditions of these cells, similar to those we have formerly mentioned, are probably also repeated here. Then the granulated elements designated as mucous corpuscles (*a*) are also found in all sputa. Their number, and with them the appearance of the sputum, is subject to very great variations. If the latter be yellow and thickened, the number of the former structures is enormous, and one then speaks of pus-corpuscles. It is evident that this most extensively disseminated element of the sputum is to be met with changed in many ways, which is due in part to endosmotic influences and to maceration, as well as to the various stages of life of the cell. Dark, granulated cells, overloaded with molecules of fat, are considered as older forms, and this view is certainly correct. Larger structures, with similar fat-like contents, are partly due to pus-corpuscles, but partly also to changes of the alveolar epithelium. The name of granule cells or inflammatory globules (*b*) was formerly given them. Their physiological prototypes are represented by many gland-cells overloaded with fat (sebum cutaneum, colostrum).

Similar spherical cells may contain a brown, still somewhat soluble pigment, though they are of rare occurrence. The same cells with black pigment granules (*c*) form more frequent constituents. They are observed in more severe diseases of the lung-tissue, but also in simple catarrhal irritations. They are degenerated alveolar epithelium (p. 492).

On a previous page we mentioned the quite superficial position of the pulmonary capillaries. We can easily understand that the red blood-corpuscles readily pass out through the uninjured capillary walls, and that the latter also frequently become ruptured in consequence of their over-distention with blood, and hence the frequent occurrence of blood-corpuscles in the sputum (*d*). According to the quantity of the former, the latter appears to the naked eye either as blood or as spotted and striped with blood, or, if the mixture be more intimate, more of a yellow, reddish, or rust color. Very small quantities of blood-cells are only to be found with the aid of the microscope. The blood is either still fluid or coagulated, and then the cells, together with other structures, are concealed in the fibrous coagulum. They sometimes appear entirely unchanged, with the familiar depression in their centres (p. 230), sometimes shrunken and in a crenated form, or, finally, swollen into a globular shape, and then, not unfrequently, they are in various stages of discoloration. One sees in part single cells, in part lumpy aggregations, and in part the familiar groups resembling rolls of coin (with which fig. 89, p. 235 is to be compared). The most frequent arrangement of the blood-corpuscles in the sputum is such, however, that the borders of the cells touch each other. The tough mucus may, finally,—and this change of their form is frequently met with,—tear the soft blood-corpuscles considerably.

The presence of elastic fibres and shreds of elastic membrane in a sputum is, finally, of greater importance to the practical physician for the purposes of diagnosis. If these are not fragments of food, which is sometimes the case, they indicate a destruction of the pulmonary tissue in consequence of softened tubercle or gangrene. Still, they are by no means of frequent occurrence in the former widely disseminated disease, so that their absence from the expectoration does not possess any negative importance. One finds in part single fibres, in part several lying near each other, or also hanging together like a network (fig. 269 *h*). The difficult solubility of these structures and their entire optical relations secure those who are somewhat practised from any danger of mistaking them. The beginner might, accidentally, mistake threads of lint and the like for them, and it would therefore be well for him to consult an experienced observer. A highly meritorious observer, Remak, has long since given us a good method for finding the lung-fibres. Each sputum that the patient coughs up should be placed by itself on a dish or, where the whole expectorated mass is obtained for examination, it is to be placed in a glass cylinder filled with water, and smartly shaken. The masses thus separated will, after a short time, form a sediment, and in this the fibres in question are to be sought for.

Crystals of the ammonio-phosphate of magnesia as well as needle-shaped concretions of fatty substances may be met with in decomposed masses of sputum. Tablets of cholesterine are rare.

We cannot leave the respiratory apparatus before having made mention of two organs lying in its neighborhood, the thyroid and thymus glands.

The thyroid gland, an organ which is entirely enigmatical in its physiological regard, belongs to a naturally related series of gland-like, ductless structures, to which, in the human body, we also reckon the supra-renal capsules and the hypophysis cerebri. Although it does not share with these organs the near relationship with the nervous system, it nevertheless agrees in this, especially with the supra-renal capsules, that it is also subjected to an earlier senescence, and, like the latter, is met with in the adult body in a condition of retrograde metamorphosis. While, however, the supra-renal capsule undergoes fatty infiltration, the thyroid gland presents another, namely, the colloid metamorphosis, the commencement of which may, indeed, begin even at the end of embryonic life.

The framework of the thyroid gland (fig. 270 *a*) consists of

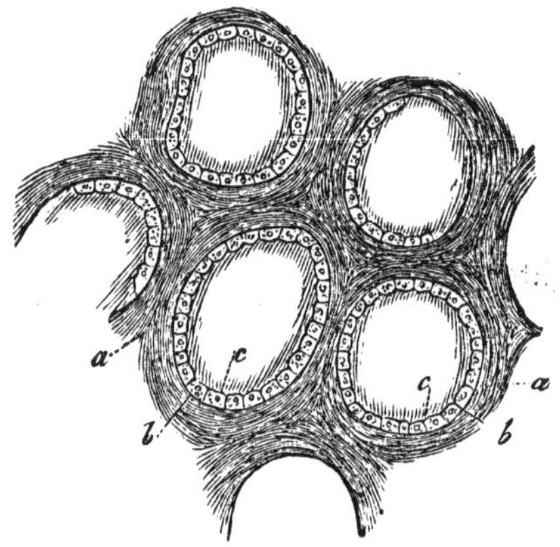

Fig. 270. Portion of the thyroid gland of a child. *a*, the connective-tissue framework; *b*, the rounded cells of the inner surface, lined with epithelium (*c*).

an ordinary fibrillated connective tissue intermingled with elastic fibres, which is permeated by numerous vessels and a not inconsiderable number of lymphatic canals. It encloses groups of rounded cavities (*b*) in which an especial membrana propria is wanting (p. 408). From these groups are formed the lobules, and from the latter the larger lobes.

A fœtal thyroid gland, or one which is not as yet altered, shows the cavity lined by a layer of nucleated cylindrical cells (*c*) which are shorter and more flattened against each other, and within the same a homogeneous viscous fluid. The cavity is surrounded by a close capillary network which is easily injected from the artery. In the connective tissue of a group of cavities run fine canals, originating in the numerous superficial lymphatic vessels with valves, forming sometimes closed, sometimes irregular circular-shaped, sometimes only arched columns. Still finer lymphatic canals not unfrequently pass between a few cavities. Their injection may also be readily accomplished by means of the customary puncturing method in the new-born and the child, in the dog and the rabbit.

Hardening in chromic acid or alcohol serves for the preparatory treatment. Thin sections show many things more beautifully after tingeing than in an uncolored condition. The framework is easy to isolate by brushing. The thyroid gland of the calf, macerated in pyroligneous acid, is to be recommended for the recognition of the nerves (Peremeschko). Osmic acid has not afforded us here any results worthy of mention.

In the place of the viscous fluid mass there occurs (and it is indeed frequently noticed even in the bodies of new-born children), together with dilatations of the glandular cavities, another homogeneous, more solid contained matter, the colloid, a modified albuminous substance (fig. 271). It is formed by the metamorphosis of the cell-contents of the epithelium, whereby the cells are destroyed. We have already mentioned the colloid degeneration, which, though not constituting a very frequent occurrence, appears in the similarly formed hypophysis cerebi, also attacks the cells of carcinomatous neoplasms, and may give rise to colloid carinoma (p. 285). In the slighter degrees, the dilatation of the cavities, and the compression of the interstitial connective tissue coincident therewith, is moderate, so that although narrowed, and here and there atrophied, the lymphatic passages may be rendered apparent by injection. The capillary network retains the old permeability, and the ephithelial cells still appear preserved.

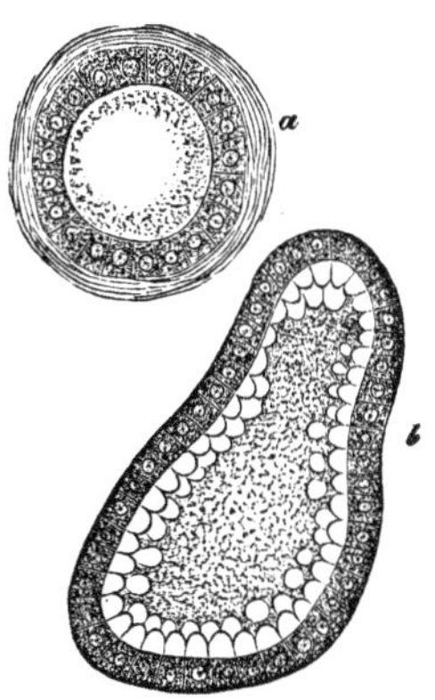

Fig. 271. Colloid metamorphosis. *a*, Gland vesicle of the rabbit; *b*, commencing colloid metamorphosis of the calf.

Higher degrees of this colloid metamorphosis show, together with an increase of volume, that the whole organ is permeated by transparent, sometimes smaller, sometimes larger colloid lumps. The epithelium of the distended cavities has disappeared, and the compression of the connective tissue has become such that, though the blood still passes, an impermeability for the lymph has taken place. All attempts at injection remain unsuccessful, and, from the nature of the colloid matter, a resorption through the capillary walls is no longer to be thought of. Thus arises the goitre, that disease which is still so obscure in its etiological relations.

With further accumulations of the colloid masses the connective-tissue interstices disappear, and as the excavations unite, these masses become joined together. In this way larger and larger spaces become filled with such masses, and the connective-tissue stroma lying between them appears as if macerated. An entire lobe may finally present a single collection of colloid.

Sections of the injected organ, deprived of their water by means of absolute alcohol, may be mounted in Canada balsam; the remaining preparations are to be mounted moist in dilute glycerine.

Not less obscure in its function and in its structure, the thymus appears, at the present time, to be not entirely comprehensible. It also undergoes, although later, a transformation, that is, a metamorphosis into fat-tissue.

The elements which constitute the lobes of our organ have been described by authors as granules or acini. They remind one in their texture of a lymphatic follicle, and show the same connective-tissue reticular framework, with nuclei at the nodal points and permeated with capillaries; likewise the same filling up of all the intervening spaces by an innumerable quantity of lymphatic cells. Nevertheless, a more thorough investigation also reveals many deviations. In fine, transverse sections of hardened organs the thymus follicle contains in its centre a cavity filled with a cloudy fluid, the further explanation of which is obtained by side views. In such, cul-de-sac-like ducts appear to come from the follicle, and these canals from each lobe become conjoined below. Herein lies, according to my view, the rudiment of the further diverticulated, fœtal thymus-gland tube, and not a lymphatic canal-work, as His, in a beautiful work, has declared the same to be. Firstly, notwithstanding numerous attempts, it has not been possible for us to accomplish a lymph injection of the organ and of these passages; then—and upon this point greater weight may be laid—the more recent investigations have demonstrated entirely different arrangements of the lymphatic channels in the lymphoid follicles. A

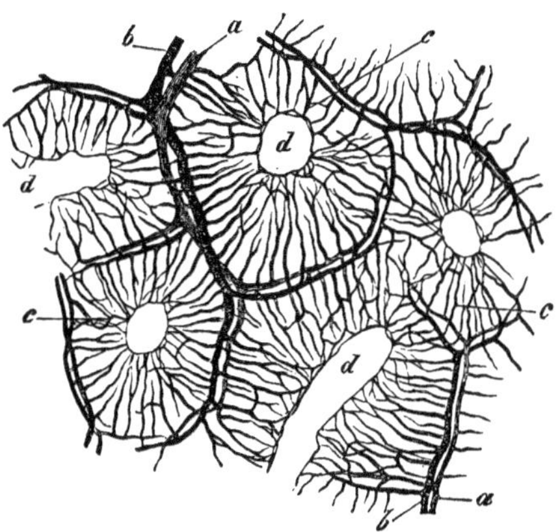

Fig. 272. Portion of the calf's thymus, after His. The rings of the arterial branches (*a*) and venous branches (*b*), with the capillary network (*c*) and the cavities of the acini (*d*).

delicate vascular network (but also not entirely corresponding to the ordinary arrangement of that of the lymphatic follicles) permeates the follicles of the thymus. In the calf (fig. 272) the peripheral portion of the latter is surrounded in a circular manner by arterial (*a*) and venous (*b*) branches, and the capillary network (*c*) resembles that of a Peyerian follicle, but naturally bends with all its tubes around the central canal (*d*) in a loop-like manner (His). In the human thymus, on the contrary, the arterial branches run in the interior of the lobules and follicles. The venous ring of the latter remains, however, the same as in the calf.

Some time after birth (pretty early in well-nourished calves, probably much later in man) commences an extensive transformation of the stellate cells of the thymus stroma into globular fat-cells, and of the neighboring reticular-fibres into a more homogeneous matter which envelops the latter. The microscopic examination shows interesting transformations of the capillary network and a gradual disappearance, frequently united with fatty degeneration, of the lymph-cells of such metamorphosed localities. A quite similar process may also, as I have shown, attack the follicles of the lymphatic glands.

Peculiar structures of the contents of the thymus are presented by the so-called concentric corpuscles. Their stratified investment consists, according to Paulitzky, of pavement-shaped epithelial cells (comp. p. 265).

The methods for examining the thymus gland are various. For hardening, use at first very watery, later somewhat stronger solutions (chromic acid from 0.1–0.2, then from 0.5 per cent., chromate of potash in corresponding strength, strongly diluted alcohol). Only in this way can the reticular framework be brushed out over larger distances. Increased hardening leads to the recognition of the passages described, and of the walls of the blood-vessels.

Kölliker recommends boiling in ordinary water to render the canal-work of the thymus visible. Subsequently hardened in alcohol, such organs are said to permit of good sections being made; that observer also recommends the boiling of this organ in vinegar.

The blood-vessels are not very easy to inject, as it is always necessary to ligate a number of them, or to compress them with the sliding forceps. An opaque mass, chrome-yellow, for example, is very handsome for review preparations (which may be mounted dry); for histological purposes select carmine and Prussian blue. Watery glycerine serves for their preservation.

It has already been remarked above that, thus far, attempts at injection have not shown any lymphatics in the interior. May another be more fortunate, and thus elucidate in this structural relation that organ which, as the at present last of its species, must awake the interest of histologists.

Section Twentieth.

URINARY ORGANS

The investigation of the urinary apparatus, and especially of the secretion produced by it, lays claim in a high degree to the interest of the medical world; indeed, the signification of the urine at the sick-bed has been valued for thousands of years, and often also overestimated in the most ridiculous manner.

The kidney, as is known, constitutes the most important organ of the urinary apparatus.

An external brown-red mass, the cortical substance, envelops in the mammalia and man an internal, paler mass, the medullary substance, which presents even to the naked eye a radiated fibrous appearance. The latter, in most mammalial animals, enters the pelvis of the kidney with a single ridge-shaped point; but is, on the contrary, in man (and also the pig) separated into a number of larger conical-shaped divisions which turn their points towards the hilus. These are the so-called Malpighian or medullary pyramids. The cortical tissue extends down between the lateral surfaces of the same like a septum (columnæ Bertini). A connective-tissue supporting substance permeates both substances, and consequently the whole organ.

The essential conditions of the finer structure of the kidney also seemed for a long time to be firmly settled.

The radiated fibrous medullary substance was regarded by the anatomists and physiologists as consisting of the uriniferous canalicules which opened free at the pyramidal points, and which, from here, with numerous acute-angled divisions and diminutions in size induced thereby, passed towards the cortical substance. In passing over into the latter, they were said to lose their previous rectilinear direction, to assume an extremely complicated tortuous course, and finally, enlarged in a spherical manner, to terminate as capsules of the Malpighian vascular coils (fig. 273.)

Especially after Bowman, in the year 1842, had discovered the manner of termination (or origin) of the uriniferous canalicules just mentioned, the structure of the mammalian kidney was regarded as assured and near to a conclusion.

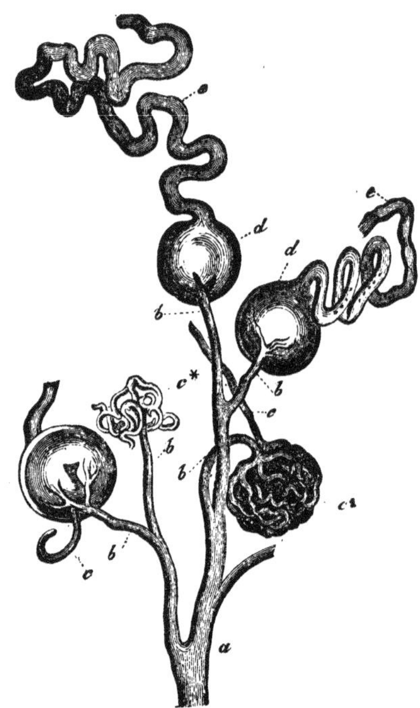

Fig. 273. From the cortical substance of the human kidney. *a*, Arterial trunk giving off the afferent vessels *b*, of the glomerulus *c**, *c′*; *c*, efferent vessels of the latter; *d*, the capsule of Bowman, with its continuation into the convoluted uriniferous canalicule of the cortex (*e*).

The merit is due to Henle of having brought a new element of agitation into this subject. He discovered, a number of years ago, in the medullary substance of the organ, together with the long-known open uriniferous canals, a system of finer, loop-shaped passages (which turn their convexities towards the points of the papillæ). He also succeeded, in several mammalial animals, in injecting, from the ureter, the straight canals of the medullary substance, as well as their rectilinear continuations through the cortex to close beneath the renal capsules. As, however, all attempts to inject, from these passages, the loop-shaped canalicules of the medulla as well as the convoluted ones of the cortical substance failed, this savant took—as we now know, erroneously—the loop-shaped passages for a system of closed canals not connected with the former, and maintained that each of the two sides of the loop terminated, finally, in a convoluted tube ending in a Bowman's capsule in the cortical layer.

Henle came, hereby, in conflict with several older reports of injections, which told of successful injections of the entire canal-work as far as the capsule of the glomerulus, in mammalial animals and in man (Gerlach, Isaacs). Neither could the (occasionally easy) injection of the entire canal-work of the kidney from the ureter, which may be accomplished in the lower vertebrates, be made to coincide with this (Hyrtl, Frey).

In consequence of a large series of new investigations (among which we would designate the work of Ludwig and Zawarykin, as well as that of Schweigger-Seidel as the most important), Henle's statements have been modified and our knowledge of the mammalial kidney not inconsiderably enlarged, although even now there are still many points in the structure of the kidney remaining to be investigated.

The primary fundamental view of the structure of the kidney may be obtained with any mammalial animal; the most conveniently and summarily, it is true, in the organs of very small creatures (Guinea-pig, marmots, moles, quite especially, however, the bat and the mouse).

A fine longitudinal section from the medullary substance of the fresh organ shows the open, uriniferous canalicules, covered by a transparent, low, cylindrical epithelium, and a distinct lumen. Their ramifications may be represented by fig. 274 (a preparation which, however, was obtained by another method). These may be readily distinguished from the blood-vessels, if the latter have been previously injected with cold-flowing Prussian blue. A cautious picking apart with the preparing needle will also isolate a few of these uriniferous canalicules, and lead to the recognition of the acute-angled divisions. With a sharp razor one may also succeed in obtaining sufficiently thin, transverse sections of the cortical substance to show the meandering convolutions of their uriniferous canalicules, the darker, more granular, thick epithelium of the latter, the Bowman's capsules, and (if a moderately large amount of blood has remained) the reddish-yellow Malpighian vascular coils. The latter appear most beautifully and sharply with all artificial injections.

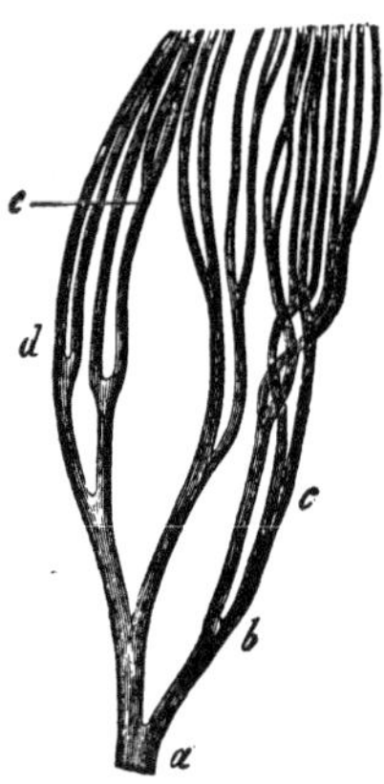

Fig. 274. The ramifications of a uriniferous canal from the medullary substance of the new-born cat (muriatic acid preparation). *a–e*, divisions of the first to the fifth order. (Original sketch by Schweigger-Seidel.)

Even here, a diligent picking enables the observer to recognize isolated continuations at least of the uriniferous canals into the widened capsules (fig. 273 *e d*), although this communication can only be recognized with difficulty in this way. The most favorable for the recognition of the latter are the kidneys of the lower vertebrates, such, for example, as the frog, the triton, and salamander (although their structure is not the same); among the mammalial animals I would most recommend the organs of the mole. The gland-cells become transparent by the addition of alkalies, and their structural condition is not unfrequently rendered more distinct.

The earlier knowledge of the kidney was obtained in this way, and, towards the close of the first half of this century, our information concerning the same remained stationary at about this stage.

The more recent times have made us acquainted with several other very important methods of investigation. Let us mention first the section through the artificially hardened organ. Most (and especially almost all pathologico-histological) examinations are at present made in this manner. Here also the freezing method proves to be extraordinarily conservative. Recourse may, furthermore, be had to chromic acid, its potash-salt, or—which is best—to absolute alcohol. We obtain in this manner, without trouble, very fine and instructive longitudinal views and—what is of the greatest importance for many textural conditions—good representations from transverse sections of the kidneys.

We would also recommend here the previous injection of the vessels, in small kidneys, with cold flowing, in more voluminous organs, with solidifying, transparent masses. The slight trouble will be richly repaid in the subsequent examination. Staining methods are of the highest value for the recognition of the renal tissue in the healthy and pathologically altered condition.

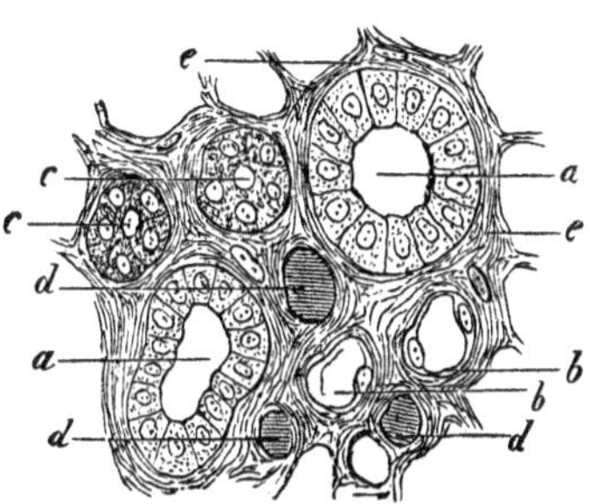

Fig. 275. Transverse section through a renal pyramid of the new-born child. *a*, collective tubes with cylindrical epithelium; *b*, descending side of the looped canal with flat cells; *c*, returning side of the loop with granular cells; *d*, transverse sections of vessels; *e*, connective-tissue framework substance.

In the medullary substance we again recognize, in vertical sections, the relations of the fresh preparation; in transverse sections (fig. 275), on the contrary, the lumina of the uriniferous canals, the straight ones with their cylindrical epithelium (*a*) as well as the loop-shaped ones with, for the most part, quite flat cells (*b*), reminding one of vascular epithelium, and also the connective-tissue stroma of the medullary substance (*e*).

Fine longitudinal sections of the cortical substance (fig. 276) show, on the contrary, how this, the stratum of the convoluted uriniferous canals (*B*), is permeated at rapidly following intervals by thin bundles of uriniferous tubes, having a straight course (*A*), which diminish in size as they pass outwards, and only become lost in convolutions (*d*) just beneath the surface of the kidney. These groups of straight passages, whose calibre is also variable (*a b*), penetrate the layer of the convoluted canals in the same manner, we might say, that a board is perforated by numerous pegs driven closely together into it.

These bundles of straight canals which were previously seen, and which are continuations of the familiar straight passages of the medullary substance, have been called pyramidal processes (Henle) or medullary rays (Ludwig). We shall soon return to the consideration of their signification. The tissue of the convoluted uriniferous canals lying between them may be considered, although indeed only factitiously, as consisting of

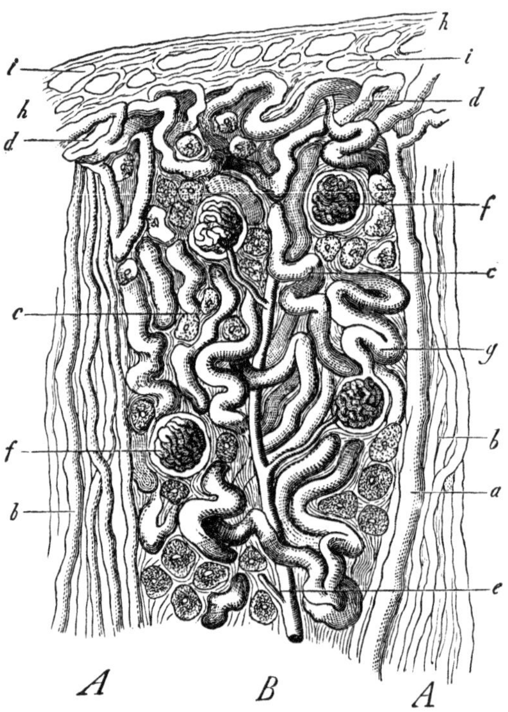

Fig. 276. Vertical section through the renal cortex of the new-born child (semi-diagrammatic). *AA*, medullary rays; *B*, cortical substance proper; *a*, collective tube of the medullary ray; *b*, finer uriniferous canalicules of the latter; *c*, convoluted canalicules of the cortical substance; *d*, their peripheral stratum; *e*, arterial branch; *f*, glomeruli; *g*, continuation of a uriniferous canal into the Bowman's capsule; *h*, the renal tunic with its lymph-spaces, *i*.

individual pyramidal pieces which have their bases turned towards the renal capsules. These are the cortical pyramids of Henle.

Transverse sections of the cortex (fig. 277) show both varieties of the uriniferous canals; those of the medullary rays cut transversely (*a*), those of the ordinary cortical substance (*b*), in all possible forms. The connective-tissue stroma may also be readily recognized at the same time.

If the study of the epithelium be renounced, I would here recommend still another method which became known to me through Billroth. If a portion of kidney be treated for a very short time with boiling vinegar it will, after having been dried, or hardened by means of chromic acid or alcohol, afford very

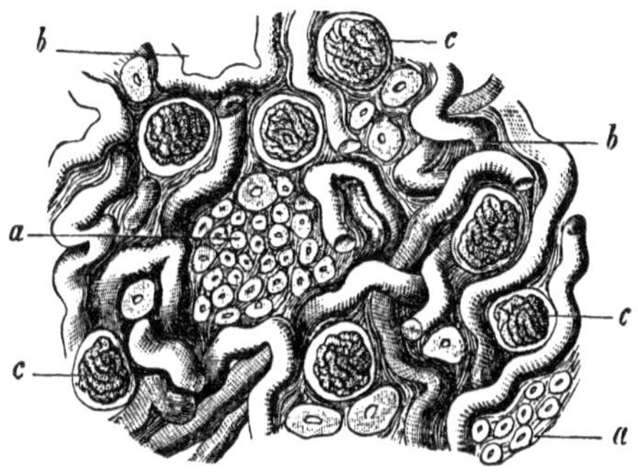

Fig. 277. Surface section through the cortical substance of the kidney of the new-born child (semi-diagrammatic). *a*, Transverse section through the uriniferous canalicules of the medullary ray ; *b*, convoluted canals of the cortical substance proper ; *c*, glomeruli and capsules of Bowman.

handsome views of the glandular passages in the medulla and cortex.

The chemical method of isolation has more recently assumed the greatest importance in the investigation of the kidney. The fresh tissue (or also that which has been hardened in alcohol), treated with strong muriatic acid (p. 129), suffers, after a series of hours, an almost complete destruction of the connective-tissue interstitial substance, while the blood-vessels, and especially the uriniferous canals, remain completely intact; and not unfrequently even their epithelium is almost entirely preserved. These canals may then be isolated by very slightly shaking or a gentle manipulation with the needles, or, already floating in the fluid, they may be fished out with a curved glass rod. The whole has, it is true, become very frail and readily destructible; nevertheless, we may frequently succeed in tingeing them slightly with carmine, and mounting them very handsomely in watery glycerine.

The muriatic acid may be used in various ways for this purpose.

The ordinary commercial muriatic acid has frequently been diluted with water until it ceased to fume, and the object immersed in it from twelve to twenty-four hours. Schweigger-Seidel employed the officinal, pure muriatic acid of the Prussian pharmacopœia (of 1·120 sp. wt.), and allowed the pieces taken from an animal killed about a day previously to macerate in it from fifteen to twenty hours. Stronger acid acts rapidly, but attacks the gland-cells energetically; that which is weaker requires more time. The pieces must afterwards be carefully washed with distilled water, and a subsequent maceration of the same for one or more days in water will, for the most part, essentially further the isolation. Boiling with this acid or with alcohol containing muriatic acid, has also been recommended.

If the chemical isolation has been successfully accomplished (which is not always the case, however), such objects (fig. 274, figs. 278–281) afford the circumspect observer extremely important information.

It is naturally impossible, even with the most conservative manipulation, to isolate a uriniferous canal throughout its entire course; it is here, therefore, only a question of obtaining the longest possible fragments and of their combination. The expert will obtain these in lengths of from 1–2‴, at least here and there. In consequence of the enormous length of the canal-work in question in the kidneys of the larger creatures, a successful result becomes much more difficult in them than in the organs of the smallest mammalia. The kidneys of the mole, the bat, the marmot, the mouse, rat, and Guinea-pig deserve to be chiefly recommended. As Prussian blue is preserved in the acid macerating fluid the blood-vessels are to be previously injected, a precautionary measure which is extremely important for the study of the medullary loops.

If the examination be commenced with the medullary substance from the apex of its pyramids one may recognize how the open canals, with their characteristic epithelial covering, form a number of rapidly following forked divisions (fig. 274 *a–e*, 278 *a b*), and then, the branches having become narrower, assume a straight course and pass unchanged for long distances through the medullary substance (fig. 278 *c*) until they arrive at the external portion of the medulla, which is characterized by tuft-shaped blood-vessels (boundary layer of Henle). Between them appear, throughout all the layers of the pyramids, the much narrower loop-shaped canaliculi (*d*), which are lined with flat, transparent cells. Their returning sides, that is, the ones which again tend towards the cortex, may show themselves to be enlarged and filled with darker granular gland-cells.

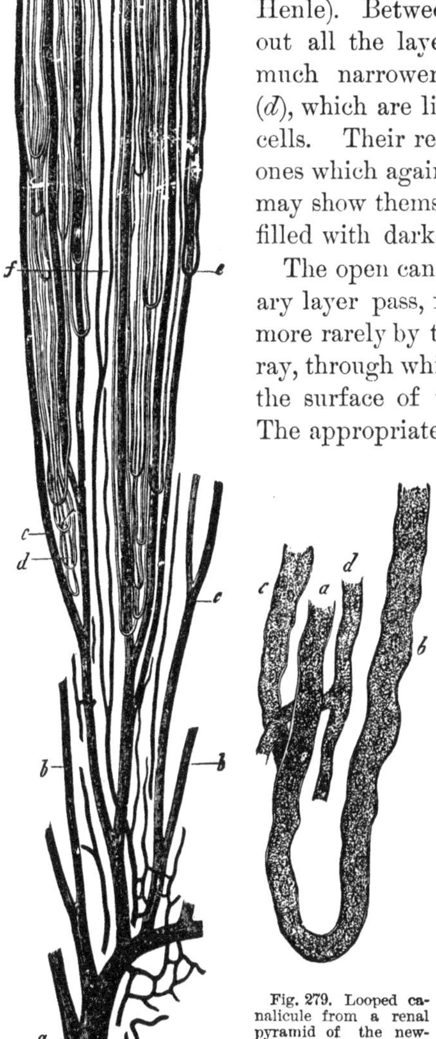

Fig. 278. Vertical section through the medullary pyramids of the pig's kindey (semi diagrammatic). *a*, The trunk of a uriniferous canal opening at the apex of the pyramid; *b* and *c*, its system of branches; *d*, the loop-shaped uriniferous canals; *e*, vascular loops, and *f*, divisions of the vasa recta.

The open canals in leaving the boundary layer pass, for the most part single, more rarely by twos, into each medullary ray, through which they continue towards the surface of the kidney (fig. 276, *A*). The appropriate name of collective tubes has been given them (*a*). The above-indicated differences in the epithelium here become less distinct. The remaining considerably narrower canals of the medullary ray consist of the descending (that is, turned towards the hilus) and the returning sides of the loop-shaped canals (*b*).

Fig. 279. Looped canalicule from a renal pyramid of the new-born child. *a b*, The two sides; *c*, another canalicule; *d*, capillary vessel.

The collective tube, having arrived nearer the surface of the kidney, gives off more numerous branches (figs. 280 *c*, 281 *c*) and terminates above in arched ramifications (figs. 280 *d*, 281 *d*), which, especially in the smaller animals, may present a zigzag appearance ("intercalary portions" or "connecting canals"). From them, but also deeper from the stem of the collective tube, rapidly narrowing canals of various forms arise, the descending sides of the loops (*e*), whose entrance here from the medullary substance has been shown in other macerated preparations.

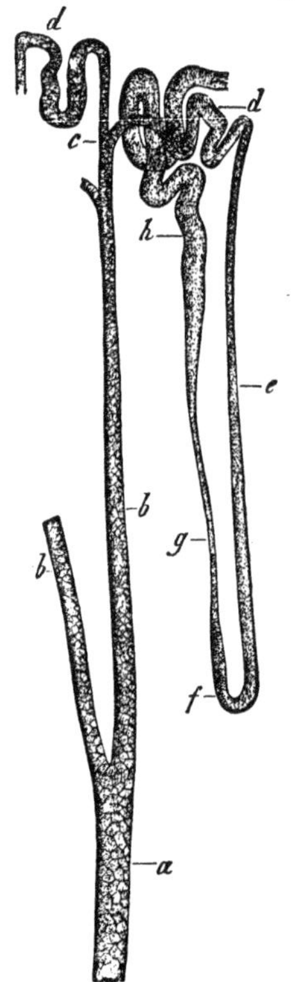

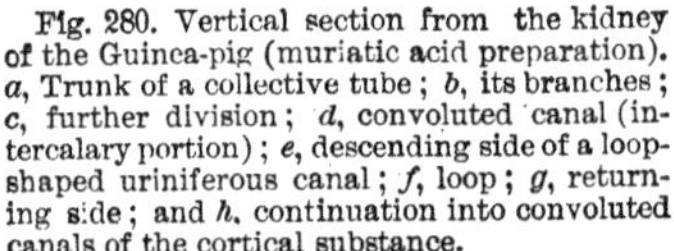

Fig. 280. Vertical section from the kidney of the Guinea-pig (muriatic acid preparation). *a*, Trunk of a collective tube; *b*, its branches; *c*, further division; *d*, convoluted canal (intercalary portion); *e*, descending side of a loop-shaped uriniferous canal; *f*, loop; *g*, returning side; and *h*, continuation into convoluted canals of the cortical substance.

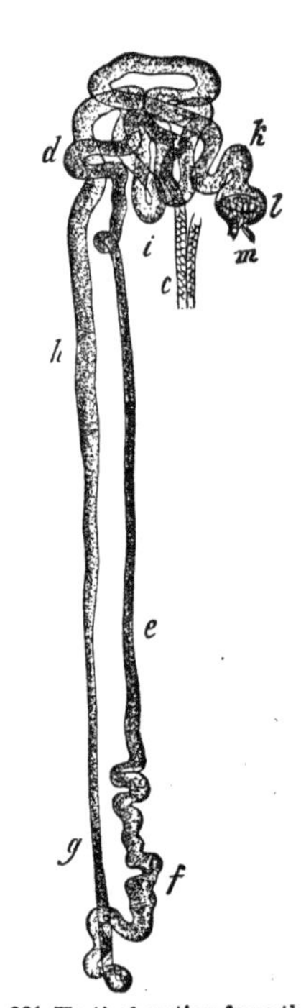

Fig. 281. Vertical section from the kidney of the mole (muriatic acid preparation). *c*, Terminal branch of the collective tube; *d*, convoluted portion of canal; *e*, descending side of the looped canal; *f*, loop; *g h*, returning side and continuation into the convoluted canalicules *i*; *k*, neck-piece of the latter; *l*, Bowman's capsule; *m*, glomerulus.

Having thus become acquainted with the one side as an offshoot or a terminal branch of the open uriniferous canals, the question still arises, what becomes of the other, the returning side (figs. 280 and 281 *g g*).

This, accompanied by the canalicules of the medullary ray, bends deeper or higher from the group in a lateral direction (figs. 280 *h*, 281 *h*), assumes another convoluted course, gains at the same time an increased diameter and a darker granular epithelium, and becomes an ordinary convoluted uriniferous canal of the cortical substance proper, and, finally, with numerous windings and curves, terminates as the Bowman's capsule of the glomerulus (fig. 281 *k l*). We must here pass over in silence many peculiarities of a subordinate nature.

We will only mention one condition here, namely, that of the epithelial lining of the Bowman's capsule (fig. 282). Its inner surface has a layer of pavement-cells (*g*) of considerable size, which may be easily rendered visible by means of nitrate of silver (either by simple immersion or by injecting from the artery). More difficult of recognition, and in a singular manner not permitting of the impregnation with silver, is a layer of smaller and higher cells, which cover the surface of the glomerulus (*f*). They are to be seen in the frozen organ (Chrzonszczewsky).

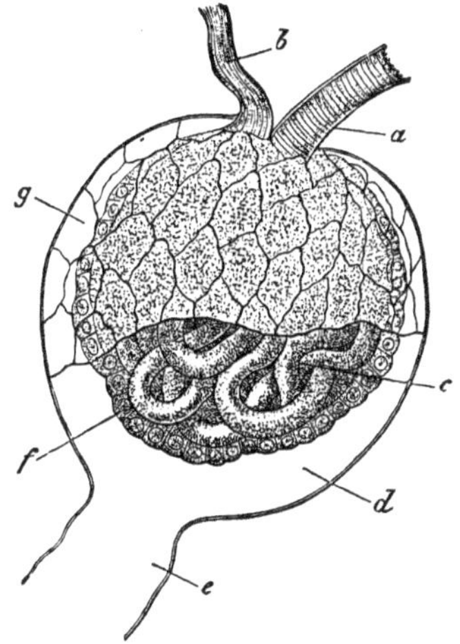

Fig. 282. Glomerulus of the rabbit (diagrammatic). *a*, Vas afferens; *b*, vas efferens; *c*, glomerulus; *d*, under portion of the capsule (without epithelium); *e*, neck; *f*, epithelium of the glomerulus; and *g*, that of the inner surface of the capsule after treatment with silver.

Not less important for the investigation of the structure of the kidneys is the injection of their glandular canals from the ureter. Cold-flowing mixtures are to be used for this purpose. The addition of alcohol is not suitable for such investigations, although also not an absolute hindrance, as has been here and there asserted. It is most suitable to select a watery Prussian blue or carmine, to which glycerine or gum-arabic may be added (see p. 186).

The varying pressure of the injecting syringe is less suitable than the constant pressure of a column of fluid or quicksilver (comp. p. 189), which may be gradually increased. Such injections then require many hours, and, notwithstanding every precaution, not unfrequently remain without the desired result. While the simply-constructed kidneys of a frog or of a coluber natrix may be injected with facility, the attempts miscarry in small mammalial animals from speedy extravasations into the venous system. Only embryonic kidneys, in consequence of the slightly developed medullary substance, occasionally afford the cautious experimenter a successful result. As a rule, the organs of the dog, the sheep, the calf, and the pig are to be used, and in as fresh a condition as possible. The pig's kidney may be injected under the pressure of a quicksilver column of 50–100 millimetres and more.

One succeeds with comparative facility, after filling the open canals of the medulla (fig. 278), in forcing the injection mass to the end of the medullary rays and their systems of branches. The descending sides of the looped canals, which are turned towards the hilus, may also be filled with relative facility, and are characterized by their smaller diameter. The colored fluid passes with greater difficulty through the loop itself, and into the returning side. The most rarely—and it is appreciable from the nature of the contents and the convolutions—does one succeed in forcing the injection mass through the convoluted canalicules of the cortex into the capsules of Bowman. Numerous successful results have, however, been more recently obtained, and thus the inferences presented by maceration confirmed (Ludwig-Zawarykin, Kollmann, Chrzonszczewsky, Hertz, Frey, and others.

Our diagram, fig. 283 (which contains the sketch of two such courses of the injection from the medullary canal (*i*) to the right and left Bowman's capsules), may represent to the reader the results of the injection, of which only the chief points have been described.

To recapitulate, let us again follow the course which the secretion must take from the glomerulus. Surrounded by the capsule of Bowman (*a*), it passes over into the convoluted uriniferous canalicule (*b*), which, after its convolutions, assumes a straight course towards the apex of the papilla (*c*). Changing its epithelium, it passes, more or less downwards, through the papilla, bends around in a loop-like manner, and again returns with the other side to the cortex (*d*). This side, later or earlier, alters its character, becomes broader and more convoluted (*e*), and enters, sooner or later, in connection with other similarly-constituted passages, into the collective tube (*f*), which, uniting with others at an acute angle (*g h*), finally pours out the urine at the apex of the papilla (*i*).

The new method, the self-injection of the living animal, with which Chrzonszczewsky has made us acquainted, has already been mentioned in a previous section of this book (p. 187). Although the appearances thus obtained are variable and not always comprehensible, repetitions of the experiment with injections of a solution of carmine into the jugular of the rabbit have nevertheless afforded me good results.

We have still to mention the connective-tissue stroma, as well as the blood and lymph passages of our organ.

The course of the vessels in the kidney has been so fre-

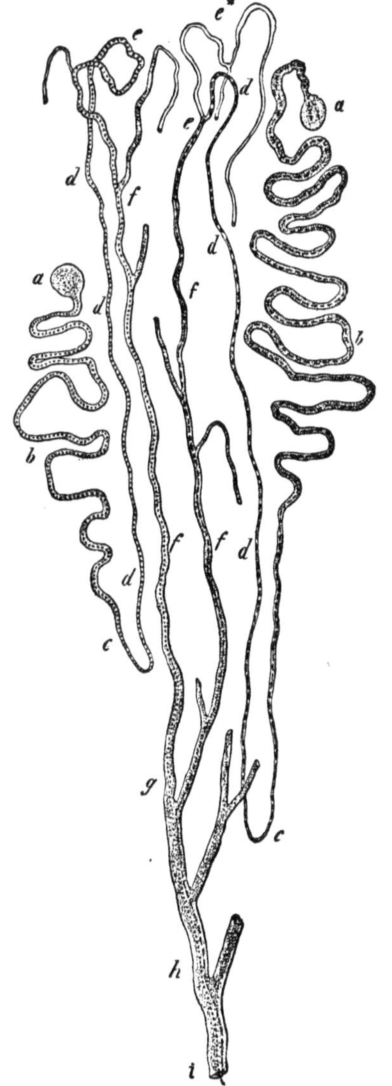

Fig. 283. Diagrammatic representation of the arrangement of the uriniferous canals (using the pig's kidney). *a*, Bowman's capsules; *b*, convoluted uriniferous canal and returning side of the loop, *c*; *d*, descending side; *e*, convoluted tube; *f*, collective tubes, uniting into a larger, open uriniferous canal (*g*), which is joined by others to form the canal *h*; *i*, trunk which opens at the apex of the papilla.

quently described (in an especially excellent manner by Hyrtl), that we may here limit ourselves to the most necessary statements. The branches formed by the division of the renal artery and vein pass through the medullary substance between the several Malpighian pyramids. At the basis of the latter, bow-like arrangements of both varieties of vessels are noticed. From the arterial arches arise then, in the form of branches, the coil-supporting arteries of the cortical substance, which keep in the axial part of a portion of cortex (cortical pyramid) bounded by two medullary rays, and towards the periphery give off the afferent vessels of the glomerulus (fig. 276 *e f*; fig. 283 *b*).

This, the vas afferens, undergoes, in man, further acute-angled divisions within the coil-shaped convolutions (fig. 273 *b*), and

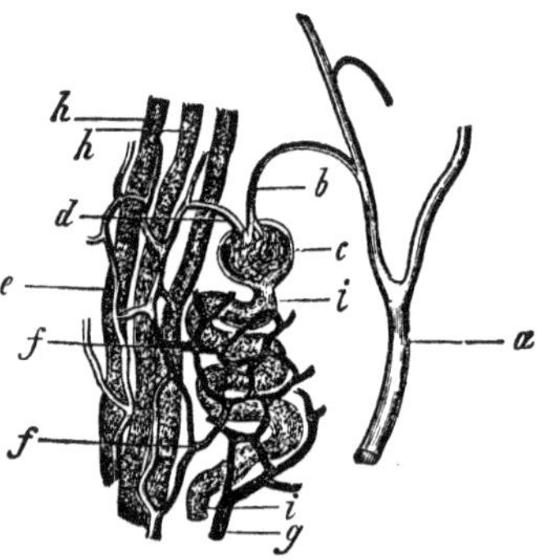

Fig. 284. From the kidney of the pig (semi-diagrammatic). *a*, arterial branch; *b*, afferent vessel of the glomerulus, *c*; *d*, vas efferens; *e*, breaking up of the same into the straight capillary plexus of the medullary ray; *f*, rounded plexus of the convoluted canals, *i*; *g*, commencement of the venous branch.

forms, after the convolutions, by the reunion of the latter branches, the efferent vessel, the vas efferens (fig. 273 *c*; 284 *d*). The latter then disappear in a capillary network with elongated meshes which encircles the straight uriniferous canals (fig. 284 *e*). From the periphery of the latter are first formed those capillary tubes (*f*) which encompass with rounded meshes the convoluted uriniferous canalicules (*i*) of the cortical substance proper.

The most superficial stratum of the cortical substance, which is free from vascular coils, receives its capillaries substantially from the efferent vessels of the superficial glomeruli; much more sparsely (and certainly not in all mammalial animals) from several of the terminal branches of the glomerular arteries, which pass directly and immediately forward to this peripheral layer.

Venous roots appear close beneath the capsule in the form of star-shaped figures; other venous roots arise deeper in the connective tissue. Generally coalescing into larger trunks, both varieties of venous branches empty at the boundary of the cortex and medulla into the arched vessels.

The long, rectilinear vascular tufts, which appear between the uriniferous canicules in the medullary substance (its boundary layer), then extend downward and either pass over into each other in a loop-like manner, or form a delicate network around the apertures of the uriniferous canals at the apex of the pyramid, and are called vasa recta (fig. 276 *e f*). Between these there also appears a capillary network of finer tubes.

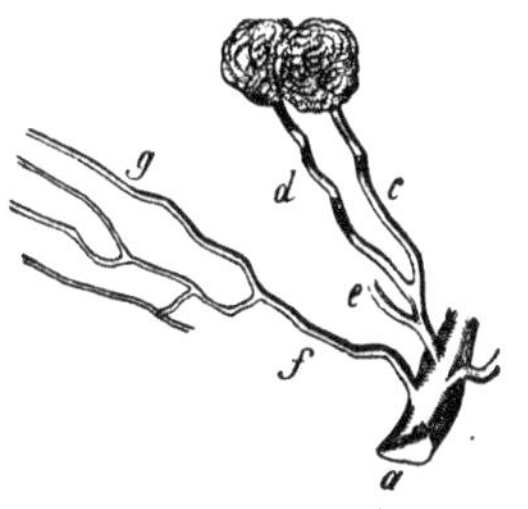

Fig. 285. From the boundary layer of the human kidney. *a*, Arterial trunk; *b*, one branch, and *e* another which furnishes the vasa afferentia of two glomeruli at *c* and *d*; *f*, a third branch (arteriola recta), with its division into rectilinear capillaries of the medullary substance, *g*.

A great diversity of opinion prevails concerning the origin of these vasa recta.

According to our observation, they have essentially, if not also exclusively, a venous character, as they are formed from continuations from the capillary network of the medullary rays. The vasa efferentia of deeply situated glomeruli are joined to them as arterial supplies. Quite unimportant, finally, are the arterial branches (arteriolæ rectæ), which, even before the giving off of the glomeruli, have left the coil-bearing artery, and become buried in the rectilinear vascular district (fig. 285 *f*).

As we have above remarked, the larger trunks frequently divide, in a tuft or tassel-like manner, into these vasa recta.

Exactly the same appearance is, in general, also presented by the meeting of the returning straight vessels. Their insertion takes place in the arched veins, which we have become acquainted with above, as occurring at the margin of the cortex and medulla.

The ascertaining of such extremely complicated conditions premises, naturally, extensive studies of injections and very careful examinations of the preparations.

Notwithstanding the facility with which the kidney may be injected by the arteria renalis (so that it constitutes a good exercise for beginners), and however little it can be called a display of skill to accomplish an extensive injection of the medullary substance, the question as to the finer vessels of the organ requires entire series of other injections. We would advise, first, the very early discontinuance (at various stages) of the injection by the artery, as soon as some coloring matter has reached the cortex. We would then recommend other, somewhat longer continued arterial injections, with which the medullary rays, but not the capillaries of the portions of cortex lying between them, are injected.

Other instructive preparations are afforded by the injection by the vein, which is likewise to be discontinued at various stages. Even an extensive venous injection usually clogs at the glomerulus. Thin fluid masses may be injected into them, however, even by the vein.

The double injection is, finally, very instructive. It should be commenced by the vein, and be continued sometimes more, sometimes less by the arterial or the venous side. Greater practice is here necessary. If a gelatine mass has been selected for the complete filling of the veins, it is advisable, for the recognition of the boundary district of both varieties of vessels, to undertake the subsequent injection of the arteries with cold-flowing masses.

We would chiefly recommend the kidneys of dogs, cats, and rabbits. Of larger animals, those of the pig and sheep are to be used. If the system of uriniferous canals has been filled with Prussian blue, the carmine mass and the transparent yellow of Thiersch (p. 182) is to be selected for the injection of the blood-vessels. Human kidneys, even from bodies which are no longer fresh, frequently afford good results. Injections of the organ in Bright's disease are also usually more or less successful.

A connective-tissue stroma is met with as the framework of the kidney. It consists, in the cortical substance, of a but very little developed connected septal system of connective-tissue cells and homogeneous or striated interstitial substance. It appears somewhat thicker on the adventitia of the larger vessels and the capsules of Bowman, and, metamorphosed at the surface of the organ into a connective tissue with numerous spaces, is continued into the renal capsules. This connective-tissue stroma becomes somewhat firmer in the medullary rays; it reaches its greatest, although absolutely slight development in the medullary substance (fig. 275 *e*). Thin sections from the organ which has been hardened in alcohol or chromic acid afford the very best views when brushed or tinged with carmine The star-shaped connective-tissue cells may be beautifully isolated by macerating in muriatic acid (Schweigger-Seidel).

The attempts to inject the lymphatics of the kidney by means of the puncturing method have been, for the most part, unsuccessful. It succeeds best by the swollen vessels of organs (of the dog) which have been rendered œdematous by ligating the ureter. The parenchymatous lymphatics occupy the interstices of the connective tissue (fig. 276 *i*), abounding in spaces, which lies beneath the capsule, and pass inwards from here through the spaces of the connective-tissue stroma, between the uriniferous canals, around the capsules of Bowman and the finer blood-vessels. While the intercommunication of these lymphatics in the cortical tissue is very free, it is only subsequently that the narrower spaces of the medullary rays, and finally the passages of the medullary substance itself, become filled. The whole, moreover, reminds one strongly of the lymphatics of the testicle (see below).

Through the industry of competent investigators we have become more intimately acquainted with the numerous pathological changes of the renal tissues. Here, also, the predominant participation of the connective-tissue frame-work in the morbid textures was for a long time firmly maintained, and the new formations were also said to take their origin from its cells, while, in this regard, the structureless membrane of the glandular passages was considered as playing a subordinate rôle. The gland-cells themselves were capable, indeed, of swelling, the production of granular contents, of increase, as well as of degeneration (especially the fatty), and of decay (and these occurrences are very frequent), but corresponding to their epithelial nature, did not pass over into other tissue elements,—all acceptations which properly meet with renewed doubts at the present day.

Increase of the connective-tissue framework substance, partly local, partly diffused, is frequently met with in the kidney. After the application of the methods already mentioned, the connective tissue appears sometimes homogeneous and compact, sometimes split up into fibrillæ with its cells usually more distinct. The related substance of the membrana propria, especially in the capsule of Bowman, also undergoes thickening, now and then with a stratified appearance. Whether the cell-like bodies which may be rendered visible under such conditions are actually connective-tissue cells belonging to the capsular membrane, we will leave undecided. From these connective-tissue cells commence additional processes of multiplication, which may lead, in part, to the formation of new connective-tissue corpuscles, in part to the production of spherical cells, similar to the elements of the lymph and pus, whereby, however, the emigration of the coloress blood-corpuscles will also play its part. The pus of the renal tissue also consists of such cells. A similar process of proliferation, but accompanied by shrivelling and fatty degeneration, is produced by tuberculosis of the kidney, while miliary tubercle here also frequently takes its origin from the sheaths of the arteries. Other, and especially carcinomatous new formations are also said to take their origin from this connective tissue. This has recently been denied, however, so far as the cells of the former are concerned, as they are thought to originate from the glandular epithelium (Waldeyer).

Brief mention may here be made of the deposits of fatty and pigment molecules as well as of the amyloid degeneration. Even in the normal kidney one may meet with a few molecules of fat in the fine granular contents of the glandular epithelium; the number of the same is occasionally not inconsiderable. Large collections of them, which may produce a fatty degeneration of these cells leading to their destruction, are, in pathological conditions, of extraordinarily frequent occurrence. These fat-granules also appear within the trabeculæ and the connective-tissue corpuscles of the connective-tissue framework. Gradually flowing together in the connective-tissue cells, they may lead to the formation of globular fat-cells.

Remarkable pigmentations of the kidneys (preponderating, however, in the gland-cells) may be met with in persons who have been destroyed by an obstruction of the biliary ducts. We have already (p. 470) mentioned the changes which occur in the liver-cells with such a retention of the bile. Such kidneys present an olive-green color. Variously tinged epithelium is seen in the uriniferous canals of the medullary substance, likewise epithelium with variously colored masses of pigment in the cell-bodies. In cases of high degree the uriniferous canals are observed to be filled with lumps of hard, brittle black substances. A similar but blacker pigmentation is also met with in the convoluted uriniferous canals of the cortex, as well as in the capsules of Bowman, that is, in the epithelium of the glomerulus.

Melanæmia, the passage of pigmentated cells and flakes from the spleen in malignant intermittents (p. 485), causes embolia in the renal vessels by means of the structures mentioned. The masses of pigment are found in the vessels of the glomerulus, the capillaries of the cortex, more rarely in those of the medulla. A few such pigment aggregations may be met with even in the uriniferous canals.

The participation of the gland-cells is probably somewhat greater in the not unfrequent amyloid degenerations of the kidneys. They become transformed into the characteristic flake-like bodies, similar to those which we have mentioned above (p. 475) at the equivalent degeneration of the liver. The seat of the degeneration is predominantly in the vascular walls, especially those of the glomerulus (vas afferens, convoluted canals, and efferent vessels). The membrana propria may also undergo the process of degeneration.

An interesting succession of the just-mentioned metamorphoses of these several elements—the glandular and the connective-tissue elements, together with the vascular—is shown by the process called "Bright's disease." This commences with an increased inflammatory congestion and swollen, granular gland-cells, leading to an extensive destruction of the gland-cells of the organ as well as of its blood-vessels, with which is also associated a considerable increase of the connective-tissue frame-work substance, and a further metamorphosis of the glandular tissue.

At the commencement periods, especially of severe and rapidly progressing cases, one notices in the cortical substance, where these pathological processes first take place, a considerable repletion of the finer vessels with blood, and somewhat cloudy, granular gland-cells. The vascular coils appear more distinct, small extravasations from ruptured vessels are frequent, and glassy cylindrical masses of albuminous substances begin to appear in the straight uriniferous canals. These "fibrin cylinders" (which are to be distinctly recognized in hardened kidneys as masses filling the glandular canals) sometimes present more of an appearance of pure fibrin, sometimes they appear to be more impregnated with a few blood-corpuscles and separated gland-cells. At a later period the quantity of blood contained in the cortex of the kidney diminishes; injections of the organ, which is frequently increased in volume, now succeed with greater difficulty. A fatty degeneration takes place extensively through the gland-cells, and even the fibrin cylinders frequently contain such fragments of cells and free granules of fat. Other gland-cells shrivel, without presenting these fat-molecules. In well-hardened preparations the connective-tissue framework substance is, for the most part, found to be increasing by proliferation. If these cylinders are not washed away by the current of the urine (in which case they appear as urinary constituents), the obstructed uriniferous canals become widened and sinuous, and in this way may give origin to the formation of cysts. If the process continues, the glandular canals are found to be deprived of their epithelium, filled with detritus, and, in part, collapsed and gradually disappearing in the increasing connective tissue. Concentric depositions of connective tissue also occur around the shrinking capsules of Bowman. In this manner are formed, here and there, these metamorphosed connective-tissue places in kidneys which are decreasing in volume. Between them remain portions of glandular tissue, widened canals filled with granular masses, etc. These are the so-called "granulations" of pathological anatomy.

The structural changes in question can only be inadequately and unsatisfactorily followed in the fresh organ, although such examinations should always be made, especially on account of the metamorphoses of the cells. Hardened kidneys must serve for further investigations. This procedure may present some difficulty if the softness be extreme. The object will be obtained after a time, however, especially if the pieces immersed are not too large and a certain accuracy be observed. The injection should, so far as possible, always precede the immersion; in many processes, such as the formation of tubercles, amyloid degeneration, and Bright's disease, the microscopical preparations frequently gain thereby a surprising intelligibility. Tingeing with carmine and with aniline blue also deserves to be urgently recommended to the physician for these cases. Where more considerable connective-tissue new formations are concerned, the preparation is to be boiled in vinegar, and then immersed either in alcohol or in chromic acid. With the latter treatment, especially, many things become very handsome.

Several extensive deposits in the renal canaliculi, originating in the urinary constituents, are also to be mentioned here. The so-called uric acid infarction, which occurs in infants in the first days after birth, is a common occurrence. A yellowish, somewhat red substance fills, in streaks, the open uriniferous canals of the pyramids, and may be readily pressed from their openings with the fingers. The microscope shows, mingled with glandular epithelium, a sometimes homogeneous, sometimes coarse granular mass of uric acid salts, from which the characteristic uric acid crystals may be separated by means of a drop of acetic acid. The altered assimilation which the pulmonary respiration induces in the body of the new-born is probably the cause of this condition, which is not of itself important. Not unfrequently such masses also appear in older persons, and may become united to concretions of uric acid salts. They are met with, for example, in Bright's disease.

Molecules of the carbonate of lime, as dark granular masses, may, especially in advanced age, obstruct the loop-shaped uriniferous canals (lime infarction). They dissolve with effervescence by the addition of acetic acid under the microscope.

Kidneys injected with transparent masses tinged with carmine, and deprived of their water by means of absolute alcohol, afford handsome preparations for a collection when mounted in Canada balsam. Other preparations are to be preserved in the customary manner with glycerine. Concerning the methods of examining the efferent portion of the urinary apparatus, the ureters, bladder, urethra, etc., a few remarks may suffice.

The calices and pelvis of the kidney, the ureters, and the bladder scarcely require discussion, as the methods of investigating their constituent layers are sufficiently well known to the reader. The stratified epithelium of these parts presents numerous and peculiar forms, with which it is necessary to be acquainted in order to avoid embarrassment in examining the urine. The most superficial layer of the epithelium of the bladder (fig. 286 *c*) shows large, more flattened cells with depressions on the under surface, the one turned towards the next following layer of cells. The arched ends of the cylindrical cells of the following layer fit into these cavities; nevertheless, the cells of the deepest of these two layers are extremely irregular in form. A similar condition is also shown by the ureters and the pelvis of the kidney. The cells of the deepest layer appear more rounded.

Of greater importance for the practical physician is the chemical and microscopical examination of the urine, of which, however, we can here only consider the latter.

Fresh normal urine presents a clear fluid which holds its numerous organic and inorganic matters in watery solution, and contains but few elementary constituents from the mucous membrane of the urinary passages. The latter, pavement epithelium and mucous corpuscles, usually sink to the bottom of the vessel as a light cloud. A more abundant admixture of tissue elements may appear in the urine as a result of pathological conditions of the urinary apparatus, as well as of the efferent passages. They may cause cloudiness and changes of color in the fluid just evacuated, which, after standing, deposits sediments. Among these are to be enumerated the pavement-shaped epithelial cells of the bladder, ureters, and pelvis of the kidney, pus and mucous corpuscles, blood-cells, gland-cells of the uriniferous canals, and so-called exudation cylinders of the latter (fig. 286). To these may also be added parasitical structures.

Nearly all these elements have been already mentioned. Pus and mucous cells (*a*) usually occur in considerable numbers in the urine in catarrh of the bladder; at the later periods only with a very scanty admixture of pavement epithelium (*c*). At the commencement, the latter cells are more abundant, and just in the first periods larger metamorphosed epithelial cells are met with, which present, together with their nucleus, a number of these pus corpuscles in the cell-body. So that even here the epithelial origin of these structures has been accepted, as has already been stated concerning other mucous membranes, under similar processes. Blood-corpuscles (*d*) appear spherically distended in the thin fluid medium of the urine; gland-cells (*b*), washed out of the uriniferous canals, present various appearances.

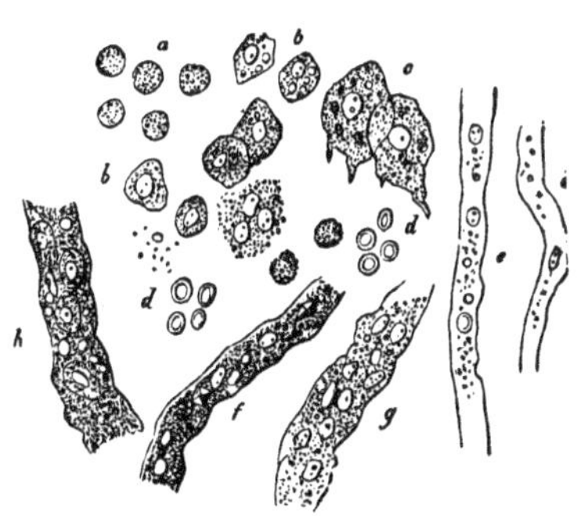

Fig. 286. Organized constituents of the urine. *a*, mucous- and pus corpuscles; *b*, gland-cells of the uriniferous canals, partly filled with fat, partly breaking down; *c*, pavement epithelium of the bladder; *d*, blood cells; *e*, *f*, *g*, *h*, *i*, various forms presented by the fibrin cylinders.

We have already, in sketching the Morbus Brightii, mentioned the fibrin or exudation cylinders (*e–i*) characteristic of this disease. In the rapidly progressing form of the disease the urine is usually at first bloody. It deposits a sediment in which appears, together with swollen blood-cells, mucous and pus corpuscles as well as epithelium from the pelvis of the kidney, the ureters, and bladder (*c*), homogeneous fibrin cylinders enclosing (sometimes numerous, sometimes scanty) blood-cells (*e*). These occasionally contain crystals of uric acid or oxalate of lime (*f*). At a later period these exudation cylinders no longer contain any blood-cells, but, on the contrary, gland-cells of the uriniferous canalicules or their remains (*h g*). If the epithelium of the passages has been entirely destroyed, one may meet with completely hyaline, homogeneous exudation cylinders (*i*). In the slowly progressing form of the disease with which we are occupied, this admixture of blood-corpuscles is wanting. Mucous corpuscles and gland-cells of the uriniferous canals (*b*) appear, and likewise, varying exceedingly in their characteristics, the fibrinous coagula. They are at first covered by the gland-cells, if the exudation has taken place in uriniferous canals which are still uninjured. Such a covering of cells may also be met with on these exudation cylinders, even in later stages, if the latter have been formed in passages which, up to that time, have remained intact. If, on the contrary, the fibrinous effusion has taken place in canals which have already lost their epithelium, pure fibrin cylinders, or those containing only a few fat-granules may appear. If rapidly expelled, they are pale; after remaining for a longer time in the uriniferous canals they become darker contoured and more yellow, and do not rapidly become pale on the application of acetic acid. If a more considerable fatty degeneration of the gland-cells has taken place, such cells, their remains, or fat-molecules may occur on or in the cylinder (*f g h*). Shrivelled cells may also show themselves in the fibrinous coagulum, and even one and the same exudation cylinder may appear different in its various portions.

The quantity of the fibrinous coagula, affording a measure of the extension of the process in the kidney, is extremely variable. These exudation cylinders in the urine generally form an expression of the renal changes; but not an accurate one, as the degeneration may be met with in different stages in various portions of one and the same kidney; furthermore relapses, that is, a local recommencement of the process, may occur (Frerichs). Further remarks concerning the methods of investigation are unnecessary.

Among the vegetable parasites which occur in the freshly-evacuated urine may be mentioned the sarcina, with which we have become acquainted from the contents of the stomach (p. 433). The urine may receive casual admixtures from the semen, as well as from other secretory productions of the male and female genital mucous membranes.

Our fluid much more frequently forms sediments from amorphous and crystalline precipitates of the organic and inorganic constituents dissolved in it. Among these are to be enumerated in the first line, as the most extensive, the precipitates of uric acid, uric acid salts, oxalate of lime, and the ammonio-phosphate of magnesia. With these are associated other rarer forms.

These precipitates, to which allusion is here made only in so far as their forms are concerned, are caused in part by the phenomena of decomposition taking place in the evacuated urine, the acid and alkaline fermentation, and are therefore constant occurrences, in part depending on increased concentration and altered composition, and are therefore isolated, and frequently pathological occurrences.

All strongly-concentrated human urine deposits, on cooling, a finely granular yellow or brick-colored sediment, which shows, by microscopical analysis, small, dark-contoured yellowish molecules which appear in irregular groups and aggregations, and in part united in dendritic figures (fig. 287). This is the urate of soda, which is soluble on warming. It was formerly erroneously regarded as a combination of uric acid with ammonia. The figure mentioned shows in its lower portion such precipitates of the uric acid salt in question. In the upper portion we perceive developed crystals which come from urine which had been evacuated for a longer time, and in which the acid fermentation had passed off and the alkaline had commenced. Several crystals of the oxalate of lime appear among the molecular sediments.

The uric acid soda salt also occurs in gouty concretions.

Urine which has been exposed to the atmosphere for some time after its evacuation undergoes at first, for several days (occasionally weeks), an acid fermentation whereby lactic and acetic acids are formed and the acid reaction increases. This fermentative process usually commences rapidly in febrile diseases. In consequence of the same, the uric acid salt (urate of soda) becomes decomposed, and the uric acid, which is difficult of solution, is separated and forms a reddish sediment.

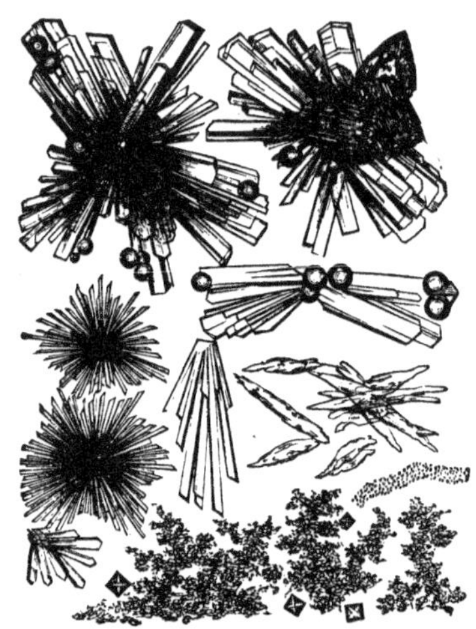

Fig. 287. Crystalline and amorphous precipitate of the urate of soda.

Our fig. 288 shows the crystals of the same which are thereby formed. One usually recognizes rhomboidal tablets with rounded, obtuse angles, and colored by the urinary pigment, as represented below and to the right of the figure. They have received the name of the "whet-stone form" ("Wetzsteinform"). By their union are formed the druses which are shown at the upper half of the right side. Viewed from the side, these "whet-stones" frequently present cask-shaped figures. When slowly precipitated, uric acid (fig. 288 to the left) may form druses of four-sided prisms with straight terminal surfaces which remind one of those of the urate of soda.

Fig. 288. Crystals of uric acid from the acid fermentation of the urine.

That these, however, are not the only crystalline forms of uric acid, that the latter much more presents the greatest changes, is well known.

If the acid with which we are occupied be precipitated from

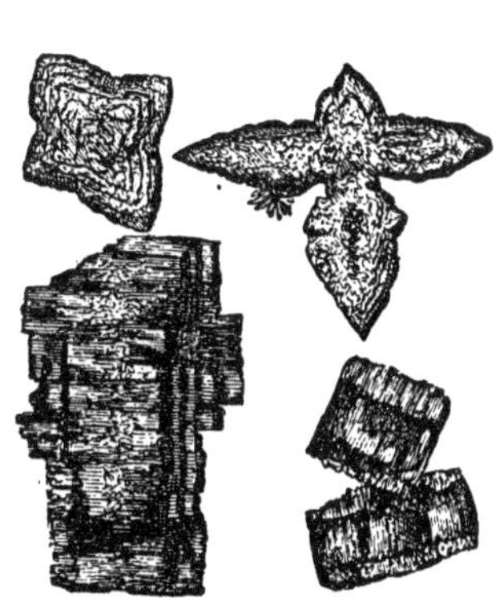

Fig. 289. Crystals of uric acid artificially precipitated.

Fig. 290. Uric acid in its various crystalline forms. At *a a a* crystals such as are obtained by the decomposition of uric acid salts; at *b*, crystallizations of uric acid from the human urine; at *c*, so-called dumb-bells.

fresh urine by the addition of several drops of muriatic acid, large, tinged, often peculiar forms of crystals are produced, a few of which are represented by our fig. 289. Other forms are obtained by precipitating the pure uric acid (it is to be dissolved in a solution of potassa and the potash salt reduced by means of muriatic acid). The forms *a* of our fig. 290 are then produced.

Abortive forms of the uric acid crystals constitute those peculiar masses of the figure *c*. They have been called "dumb-bells." Their form is partly that of a drum-stick, partly that of the dumb-bells used by gymnasts. They occur at times naturally in the urine, at times artificially, from the decomposition of urate of potash.

The surprising variety of forms in which we meet with the uric acid crystals sometimes make it very desirable for the microscopist to test them chemically under his instrument. This may be done with great facility. By the addition of a few drops of a potash solution, the crystals in question may be dissolved, to be then reprecipitated in the ordinary crystalline forms (fig. 290, *a*) by the aid of muriatic acid.

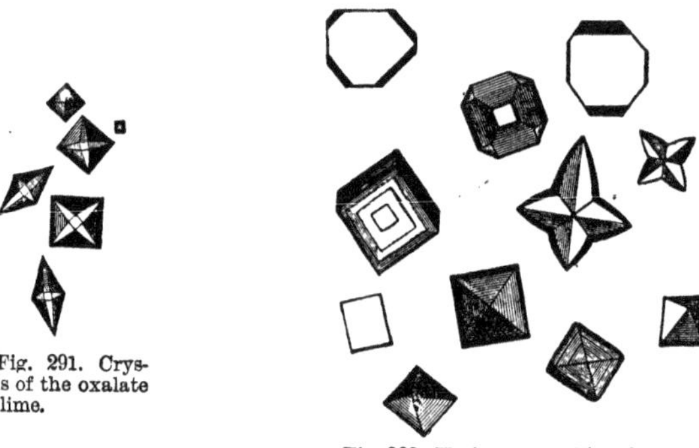

Fig. 291. Crystals of the oxalate of lime.

Fig. 292. Various crystalline forms of chloride of sodium, mostly from animal fluids.

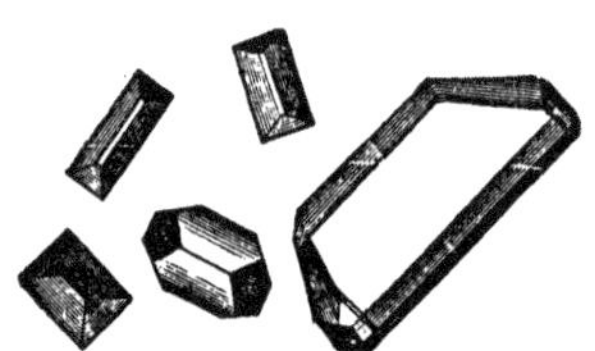

Fig. 293. Crystals of ammonio-phosphate of magnesia.

The acid fermentation not unfrequently leads to the precipitation of crystals of the oxalate of lime, the familiar octohedra which are shown by our fig. 291. Under what conditions this combination here takes place has not yet been determined. They may also occur in neutral and alkaline urine, and may likewise form constituents of pathological sediments. Chloride of sodium (fig. 292) also assumes the form of octohedra from the presence of urea. In consequence of its ready solubility, however, it never crystallizes from fluid urine. To demonstrate it, the drop of fluid must be allowed to evaporate.

Numerous small fermentative fungi make their appearance in the urine as a sign of the acid fermentation. They quite remind one of the yeast fungus (Cryptococcus cerevisiæ), but are smaller. Comp. fig. 295 (to the right and below.

If the urine remains standing for a longer time after its evacuation it becomes decomposed, and the fluid assumes a neutral and subsequently an alkaline condition, produced by the decomposition of the urea into carbonate of ammonia. The urine hereby grows somewhat paler; the former sediments disappear, it becomes more and more offensive to the smell, is

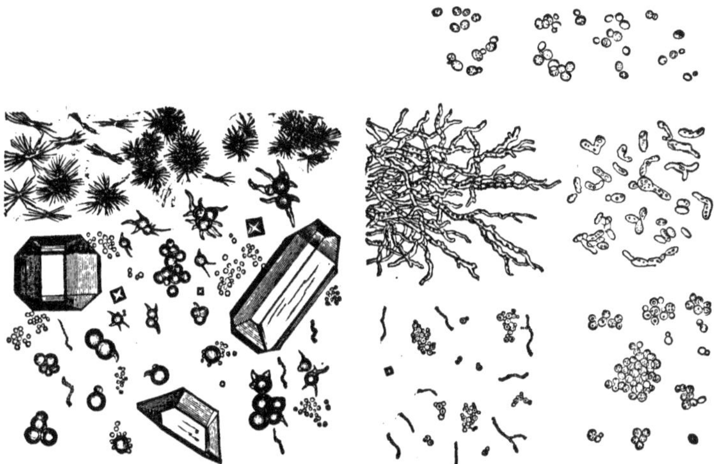

Fig. 294. Precipitates of the urate of ammonia from alkaline urine, together with crystals of the oxalate of lime and ammonio-phosphate of magnesia.

Fig. 295. Fermentation—mould and vibrionic formations in the urine.

cloudy, a whitish pellicle is formed on its surface, and a similarly colored sediment is deposited at the bottom of the vessel. This consists of the familiar crystals of the ammonio-phosphate of magnesia (fig. 293). The precipitates of the urate of ammonia also show themselves. This consists of strongly contoured, often quite dark globules, which are frequently covered with fine points, and thus remind one of morning stars, or they may also have club-like, corrugated processes on them, and thereby assume an appearance similar to bone-cells. Fine needle-shaped masses may also be met with. Fig. 294 represents these conditions, together with crystals of the oxalate of lime and the ammonio-phosphate of magnesia.

The fermentative fungus also disappears from the acid urine, and in its place appear the elements of mould and numerous confervoid growths. Numerous fine granular masses, vibrionæ, likewise make their appearance. Our fig. 295 may, in its middle portion, represent such mould formations, while to the left and below vibrionæ are delineated. The upper portion is occupied by the fungus of the cryptococcus cerevisiæ from yeast, the right lower corner by the fermentative fungus of diabetic urine.

Alkaline fermentation of the urine may take place, in an abnormal manner, very soon after its evacuation. In consequence of the fermentative action of the mucus and pus of the bladder, urine which has been retained there is decomposed into carbonate of ammonia, and may thus be evacuated in an alkaline condition. The needle-shaped groups of urate of ammonia represented in the upper part of fig. 294 came from such urine in a case of paralysis of the bladder.

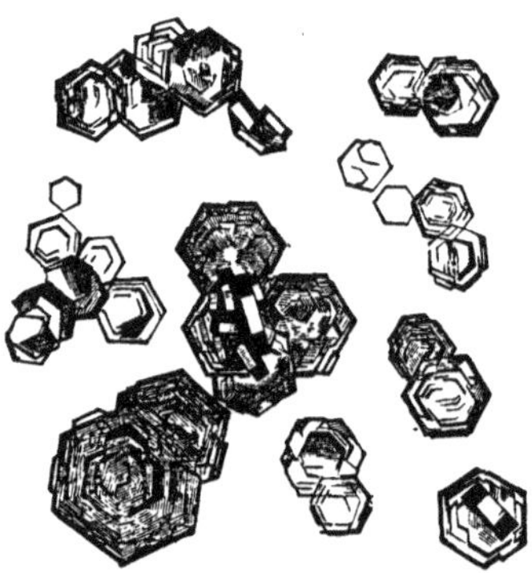

Fig. 296. Crystals of cystine.

Spontaneous deposits of other matters are rare phenomena. In a few cases only have crystals of cystine—those readily recognizable, delicate six-sided tables, such as are represented by fig. 296—been found in human urine.

The remarkable and rapid destruction of the liver-cells, which has received the name of yellow atrophy of the liver, has been mentioned on earlier pages of this book (p. 471), and it was remarked that this degeneration produced large quantities of leucine and tyrosine. These, excreted by the kidneys, appear in the urine of such patients. Brownish spherical druses of tyrosine have been noticed in the urinary sediment deposited. A drop evaporated on the microscopic glass slide shows yellowish tyrosine druses embedded between membraniform and globular deposits of leucine (Frerichs).

Among the remaining precipitates of urinary constituents which are only to be obtained as a result of further chemical procedures, let us here mention only the crystalline forms of combinations of urea with nitric and oxalic acids (fig. 297). Their production, as also the occurrence of other matters, such as sarcocine, xanthine, etc., we must leave to the text-books on physiological chemistry.

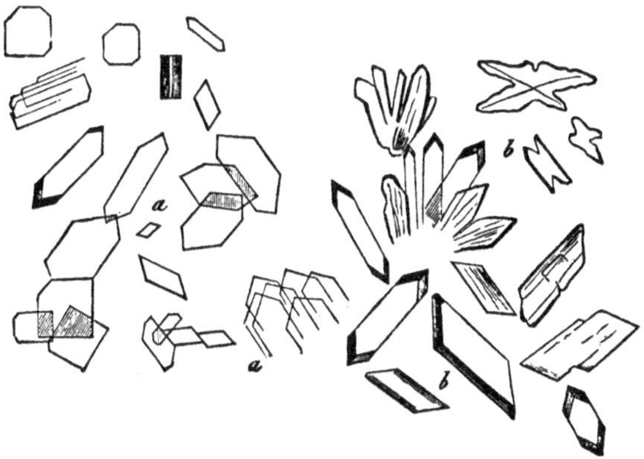

Fig. 297. Crystals of the combinations of urea with nitric and oxalic acid. *a a*, nitrate, *b b*, oxalate of urea.

The anatomical methods of examining the various sediments of the urine are of a very simple nature. After standing for some time, the clear fluid is poured from the vessel and the remainder is placed in a watch-glass, glass box, or glass beaker, from which a drop is to be removed by means of a glass rod or a pipette and placed on the microscopic glass slide. It is convenient to have a small burette with a caoutchouc tube and clamp, after the manner of the larger one used fortitrition (fig. 74, 1, p. 145), with a fine glass tube for the exit. The burette is to be filled with the sediment, or the still clear urine which is to form a sediment, and the drops are allowed to flow on to the slide by opening the clamp.

With regard to the preservation of urinary sediments in the form of objects for a collection, it may be stated that those which consist of tissue elements are not capable of being kept permanently. Crystalline sediments, on the contrary, are to be allowed to dry on a glass slide and then mounted with Canada balsam.

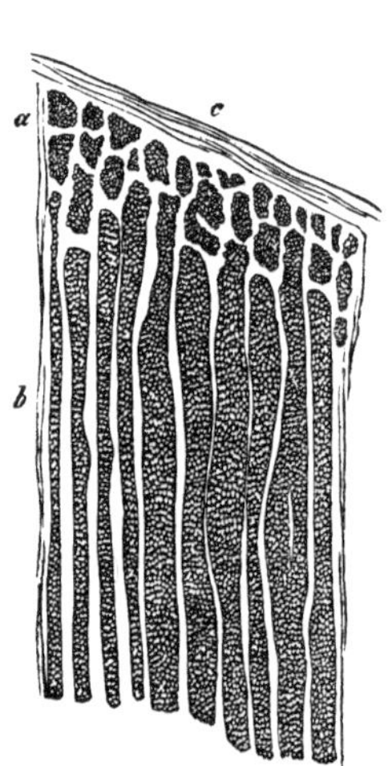

Fig. 298. Cortex of the human supra-renal gland in vertical section. *a*, smaller, *b*, larger gland-cylinders; *c*, capsule.

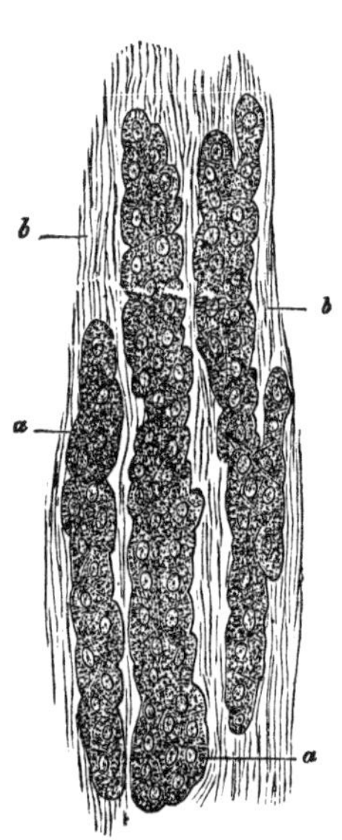

Fig. 299. Cortex of the human supra-renal gland more strongly magnified. *a*, gland-cylinder; *b*, interstitial connective tissue.

A few words may be devoted, at the close of this section, to the supra-renal glands. These organs, which are remarkably developed in the earlier fœtal period, occur in a less vigorous condition in the adult, and are then frequently very fatty, almost fatty degenerated. They show, as is known, a firmer, reddish-yellow cortex (fig. 298) which, in human beings, also permits of the recognition of a narrow, darker, and, after death, not unfrequently liquefying inner zone, and a softer, grayish-red medullary substance. The former (fig. 299) consists of the same connective-tissue stroma (*b*) which we have already described for the hypophysis cerebri and thyroid gland, and which is prolonged inwards from the capsule in radiated lamellæ. Numerous cavities are found in it; they grow smaller and smaller in an outward direction (fig. 298 *a*); in the middle they are oblong and cylindrical (*b*). Their contents are a varying number of granular cells. In the medullary substance there is a much finer connective-tissue stroma which contains transversely oval cavities filled with variable cells, but which contain but little fat. The latter, but not the cellular elements of the cortex, become brown in a remarkable manner, as Henle found, from the action of bichromate of potash. In certain mammalial animals the medullary substance is very rich in nervous plexuses containing ganglion-cells, and, indeed, a connection of our organ with the embryonic sympathetic can scarcely be denied. The number of the blood-vessels is also quite considerable. Delicate fine capillaries, formed from the numerous smaller arterial branches from the capsule, encircle the cavities of the cortex and pass over into a highly developed venous network. The vessels of this network are, however, increased in diameter; they pass through the connective tissue of the medulla and lead into the large single or double veins which lie in the interior of the organ. The lymphatics require more accurate investigation; the puncturing method has thus far yielded me no results, while the blood-vessels, for example, those of the calf, may be readily injected by the artery as well as by the vein. Very handsome injections may be obtained with the Guinea-pig, as well as the rat, from the aorta and lower vena cava.

The supra-renal bodies of the new-born and very young animals, and also those of embryos, from the later periods of embryonic life, are to be selected for examination.

Some things may be recognized, even in sections from fresh organs, with the aid of acids and alkalies. Far better views are afforded by supra-renal glands, which have been hardened in chromic acid, Müller's fluid, or absolute alcohol, with the assistance of brushing and tingeing. The nerves may be studied on the fresh organ with the addition of alkalies, or on preparations which have been immersed in dilute acetic or pyroligneous acid, as well as in weak chromic acid. Sections deprived of their water by means of absolute alcohol are to be mounted in Canada balsam, or, when moist, in glycerine.

Section Twenty-first.

SEXUAL ORGANS.

Among the female generative organs the uterus and the lacteal glands are the most important.

The ovary (fig. 300) shows, as is known, the rounded, closed glandular capsule (*b c*) which contains the primitive ovum embedded in a firm connective-tissue framework or stroma. These

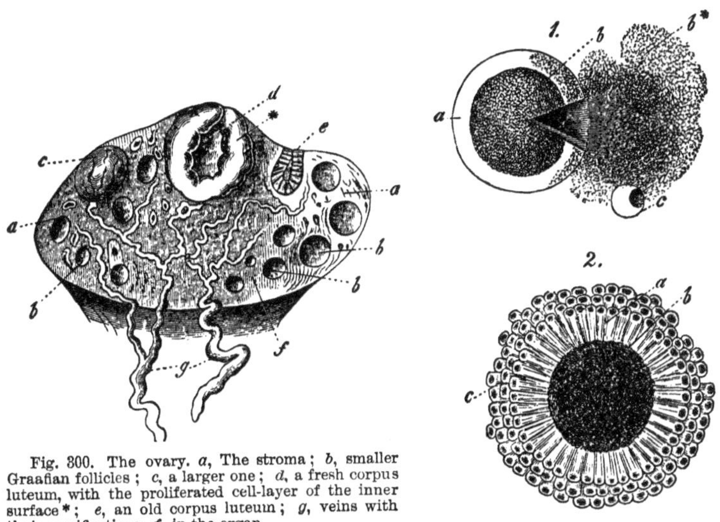

Fig. 300. The ovary. *a*, The stroma; *b*, smaller Graafian follicles; *c*, a larger one; *d*, a fresh corpus luteum, with the proliferated cell-layer of the inner surface*; *e*, an old corpus luteum; *g*, veins with their ramifications, *f*, in the organ.

Fig. 301. The mammalial ovum. 1. One which, by a rupture of the envelope, *a*, of the ovum, permits of the partial escape, *b* *, of the yolk, *b*; *c*, the expelled germinal vesicle with the germinal spot, *d*. 2. A ripe ovum, covered by epithelial cells, *c*, arranged in a radiated manner, with the chorion, *a*, and the yolk, *b*.

ova are set free by the rupture of this capsule, or the Graafian follicle; this takes place in the human female at periods of four weeks, corresponding to those of menstruation; in the mammalia it occurs at the period of heat. The follicle itself cicatrizes by a connective-tissue formation, and disappears. By this metamorphosis it produces the so-called corpus luteum (*d e*).

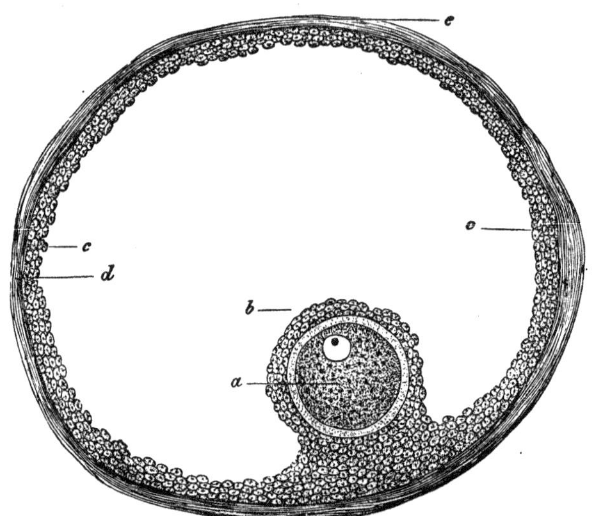

Fig. 302. Mature follicle. *a*, Ovum: epithelial stratum covering the same, *b*, and lining the cavity, *c*; *d*, connective-tissue wall; *e*, outer surface of the follicle.

If one desires to obtain a primary view of the ovum (figs. 301 and 302 *a*), this most beautiful cell-formation of the body, the ovarium of a mammalial animal just killed is to be used. The larger Graafian follicles (fig. 302) may be readily cut out from the stroma with a curved scissors and opened on the microscopic glass-slide. The ovum may be perceived as a small white point in the exuding, slightly cloudy contents, by a sharp eye, even without any further accessories; while a less perfect organ of vision requires a loupe, or the very weak magnifying power of a microscope, for its discovery. The adherent, frequently thick covering of follicular epithelium (fig. 301, 2 *c*; fig. 302 *b*) is to be removed by means of a cataract needle; a very thin and light covering glass is to be used, with the interposition of a piece of human hair. The cell capsule, zona pellucida (fig. 301, 1 *a*, 2 *a*), the contents, the yolk (1 *b*, 2 *b*) and the nucleolus, the so-called germinal spot (*d*), may be readily seen; the recognition of the fine contours of the germinal vesicle of the nucleus (1 *c*) will, on the contrary, cause some difficulty.

A 3–400-fold enlargement is to be employed for this purpose. A cautious pressure made on the covering glass with the point of a needle, while the observer looks through the instrument, will then cause the rupture of the thick envelope of the ovum (1 *a*) and permit of the recognition of the nature of the exuding yolk substance (*b b**), as well as the germinal vesicle with the germinal spot.

With human females, the freshest possible ovaries of youthful individuals are to be selected, preferably those who have died suddenly. Persons who have been lying sick for a long time, and those of a more advanced age, frequently cease to show the ova with any distinctness.

If the trouble of separating the Graafian follicles from their attachments, especially the very diminutive ones of our smaller mammalial animals, be feared, the ovulum may also be obtained by scraping the cut surface of the ovary. An indifferent fluid medium will here be necessary.

Young, smallest possible follicles, carefully separated from the stroma, may be reviewed in their totality with low magnifying powers; in this way they will show the ovulum, the epithelium, and the parietes of the glandular capsule.

The examination of the fresh ovaries may also suffice to establish the main points concerning the nature of the framework substance, as well as the cell changes which take place in the corpus luteum.

If, on the contrary, a more accurate analysis of the ovary is to be made, the fresh organ must be hardened. If the freezing method be omitted, the ordinary fluids are to be employed, among these I would give the first place to absolute alcohol and chromate of potash. The blood-vessels should also, when possible, be previously injected. Tingeing with carmine, combined generally with washing in acetic acid water, constitutes an additional excellent accessory.

The connective-tissue framework forms towards the centre a nucleus of the organ, and is extremely rich in blood and lymph vessels; externally it is a non-vascular framework, in the smaller and larger spaces of which are contained the ova. The youngest of the latter appear in extraordinary numbers in its periph-

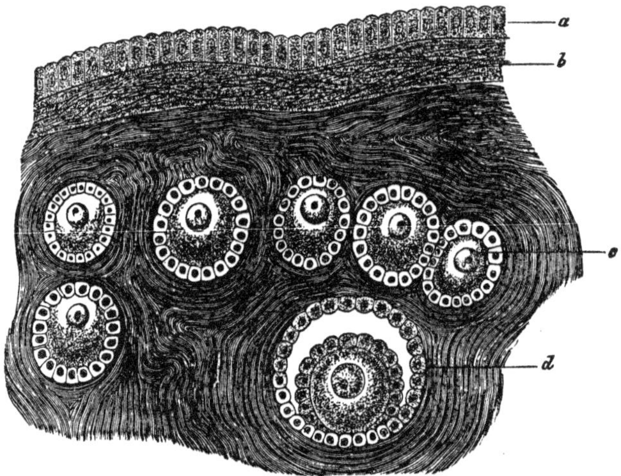

Fig. 303. Ovary of the rabbit. *a*, epithelium (serosa); *b*, cortical or external fibrous layer; *c*, youngest follicles; *d*, a somewhat more developed older one.

eral portion (fig. 303 *c d*), the "zone of the primordial follicles." Here lie the most unripe ova, beautiful cells without envelopes, surrounded by a mantle of small, epithelium-like elements (fig. 304, 1). The developing ovulum (2) soon shows the latter cells as a double layer, and on it itself a capsule, the zona pellucida (2 *a*) is perceived. A cavity (fig. 303 *d*) is afterwards formed by the separation of the follicular epithelium, and finally we receive the appearance of the ripe follicle, represented in fig. 302. The latter occur only in limited numbers in the ovarium. They have a connective-tissue wall. An inner stratum (*d*) shows an extraordinarily rich network of capillaries, while larger vessels are contained in a more peripheral layer. If we add the layer *d* of our fig. 303 as a connective-tissue boundary layer to the organ, and remark, finally, that a stratum of

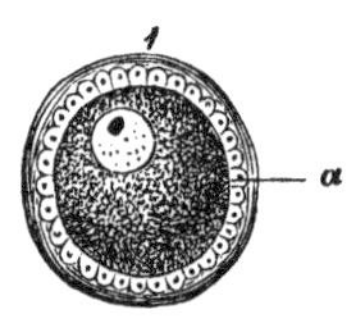

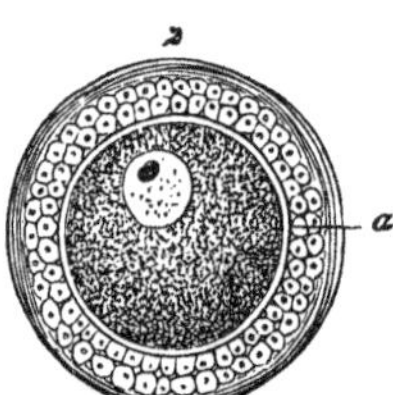

Fig. 304. Youngest follicles from the ovary of the rabbit. At 1 the ovulum *a* is still without a zona pellucida; at 2 this commences to surround the ovum *a*.

cylindrical cells, the ovarian or germinal epithelium (*a*), covers the surface, then we have presented in a few sentences an outline of the structure of the ovary.

Fine transverse sections of the hardened ovary show these relations without difficulty. If the plane of the section be favorable, the ovulum may also be perceived in large follicles, embedded in the epithelial layers of the latter (very frequently lying to the inner side). Occasionally, with strongly hardened ovaries, such fine sections of the smaller follicles may be obtained with a sharp knife, that the ovulum may likewise be seen in the section; sometimes only the zona, after the loss of the yolk and nucleus.

We have more recently received very important information, through Pflüger, concerning the formation of the Graafian follicles, which established the truth of the older but no longer regarded observations of Valentine and Billroth, and afforded an interesting parallel between the testicle and the ovary. Other observers afterwards coincided, and Waldeyer has produced a beautiful monograph. According to this, the ovary consists originally of ordinary oblong, occasionally, however, also of quite irregularly formed cell-aggregations, the follicular chains or ovular strands (fig. 305). In these primordial follicular rudiments are formed the ova; the follicles become separated from them by constriction, but may still remain connected with each other in rows, and may continue to increase in size (Pflüger's "follicular chains"). The whole formation, although possibly it may also repeat itself in after-life, is, however, very transitory, and was

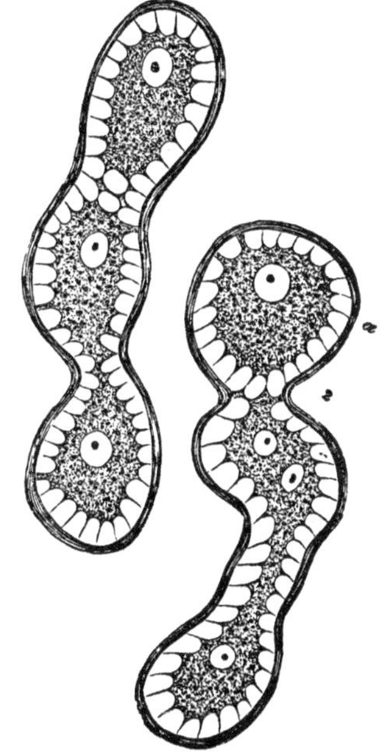

Fig. 305. Follicular chains from the ovary of the calf. 1. With ova forming. 2. At *a*, showing the constriction into a Graafian vesicle.

consequently overlooked for so long a time. Young kittens in the first weeks of their life are to be recommended here; as fluids, weaker solutions of chromate of potash, or the Müller's eye-fluid. A stratum of free ovum cells, which is said to have been observed close under the surface of the ovary (Schrön, Grohe), does not exist, as the small cells of the so-called formatio granulosa surrounding these ovula were destroyed by the action of the reagents (alcohol and strong chromic acid).

Now whence came the ovular strands and the ovules which they contain?

The peculiar ovarian epithelium, which we have previously mentioned, sends conical proliferations into the peripheral stratum of the organ. Some of the cells of the cone become ova, and by separation from the epithelial matrix the ovular strand is formed.

The Graafian follicle constantly approaches the surface of the organ in proportion as it approaches its maturity, so that, finally, entirely matured, it is only covered by a thin fibrous layer of the albuginea.

It is well known that as a result of increased congestion of the walls of the follicle, the collection of fluid in a Graafian vesicle becomes greater and greater, and that thus a rupture of the latter may take place, naturally at the point of the slightest resistance, that is, at the surface of the ovary.

This bursting of the follicular walls, which frees the ovulum and renders its further development possible, is also promoted by still another phenomenon, a cell-proliferation at the base and on the lateral walls of the follicle.

A recently ruptured follicle of the human female occasionally presents us a lump of coagulated blood (coming from the lacerated parietal vessels), but always, however, the layer of plicated substance which has a yellowish appearance from the fat it contains. Our fig. 300 at *d** shows this proliferating stratum, which probably consists of derivatives from the capsular epithelium, but principally, however, of the cells of the inner parietal layer, and which also, at this time, contains numerous emigrated lymphoid cells (Waldeyer). While a portion of these cells are destroyed by fatty degeneration, an active formative process is maintained in others, in consequence of which a vascular young connective tissue is formed, which diminishes the interior space more and more, and not only completely fills the cavity of the follicle, but also causes a considerable hypertrophy. The presence, at this time, of a rich delicate vascular network in the corpus luteum is proved by injection. We recommend for this readily practicable experiment the ovarium of the sow, in which the oviducts and uterus also afford very handsome objects.

The progressive metamorphosis has herewith reached its height. The young connective-tissue contained substance shrivels more and more (probably with a simultaneous atrophy of the vessels), the tissue becomes firmer, more like a cicatrix. Such remains of the corpus luteum may still be seen for a longer period. The whole process, however, proceeds much more rapidly in a corpus luteum caused by an ordinary menstruation than in one where the escaped ovum has been impregnated. It has been asserted, in accordance herewith, that there are two forms of the corpora lutea.

In the portion of the blood-clot which remains behind, the crystallization of the hæmatoidine (fig. 306), which we have already mentioned, takes place.

Fig. 306. Crystals of hæmatoidine.

Among the pathological occurrences, the formations of cysts are, as the practical physician knows, of extraordinary frequency in the human ovaries. A portion of these--and, indeed, the greater—certainly correspond to hydropically distended Graafian vesicles. Others of these formations originate, on the contrary, from a proliferation of the stroma of the ovarium. The walls are formed of connective-tissue substance, and the mucilaginous contents of colloid degenerated confluent cells. Innumerable quantities of such structures, with very slight dimensions, may be found in an ovary. A number of larger ones mav be met with. or one grown to a gigantic size may be found. The most remarkable form of ovarian cyst is that, however, where a part of the parietes has assumed the structure of the corium, with papillæ, hair-follicles, sebaceous and sudoriparous glands, and where hairs, occasionally united into long bundles, are met with (dermoid cysts). Even teeth, pieces of bone, and hyaline cartilage may be found in such cysts. The remaining contents are formed by a pap-like mass, consisting of desquamated epithelium, fat molecules, and crystals of cholesterine. (Similar capsules, with such strange contents, have also been found in other organs; for instance, in the lungs.) An explanation of the remarkable production is, at the present time, impossible.

The contents are to be examined in the fresh condition, bones and teeth after the manner of the normal structures, and the walls on objects hardened by means of alcohol.

Preparations of the ovaries, tinged and deprived of their water by means of alcohol, may be very suitably mounted in Canada balsam; otherwise, dilute glycerine is to be selected.

Concerning the efferent passages, the oviducts, it may be said that their mucous membranes, muscular and serous layers, are to be examined in the same manner as those of other large glandular canals. The ciliated epithelium requires quite fresh objects; a previous hardening is most suitable for the remainder.

The womb or uterus also possesses an epithelial layer formed of ciliated cells, and a tubular-shaped mucous membrane containing glands. These tubular follicles, lined with cylindrical cells, are to be observed in fresh female mammalial animals, partly immediately, partly after hardening, by means of vertical and horizontal sections. In the human female the uterine follicles appear particularly fine during menstruation or in the first months of pregnancy.

The increase in volume of the womb during pregnancy is exhibited principally by its muscular portion, which consists of contractile fibre-cells. We first see an increased growth of these elements, in part into structures of gigantic length. Even on this account the massiveness of the muscular portion must be considerably increased. Besides this, a new formation of such muscular cells also takes place (although in its details not yet explained), especially in the first half of pregnancy. The mucous membrane also increases considerably, becomes loosened in its connection with the muscular layer, and forms the decidua of the ovum.

After the birth the contractile fibre-cells return to a shorter length; a part of them are, however, without doubt, destroyed by fatty degeneration. Numerous depositions of small fat molecules in the fibre-cells are, besides, a quite extended phenomenon.

The remains of the mucous membrane are then removed in child-bed by the lochial secretion. The manner in which the new mucous membrane of the uterus is formed requires more accurate investigation.

The energetic proliferating vegetation, the active change of its elements which the uterus presents under abnormal conditions, asserts itself in the sphere of pathology, and thus produces the so frequent new formations, among which the so-called fibroid, hard fibrous tumors, are the most widely diffused. These consist sometimes exclusively of fibrillated connective tissue permeated by blood-vessels, sometimes of the above, mingled with contractile fibre-cells, occasionally, however, also, almost entirely of smooth muscular tissue. They supplant the normal tissue in proportion to their growth. If connected with the wall of the organ by means of a pedicle, they are called uterine polypi. Their examination is to be made in accordance with the directions which have been given for the developed connective tissue.

Cancerous tumors of the womb are also of frequent occurrence. They frequently appear in the form of epithelial cancer.

We may pass over the methods of investigating the uterus, the vagina, and the external genitals. The accessories are in part the same as those which we have already mentioned above for mucous membranes and smooth muscles, in part those of the skin, which are to be discussed in the following section. Let it here be mentioned, however, that it has been recommended for the uterine muscular tissue to boil the uterus for a few minutes, and to join with this an immersion in carbonate of potash. Furthermore, the maceration in pyroligneous acid, and the employment of alcohol, with subsequent drying, after which thin sections are to be exposed to the action of the 20 per cent. nitric acid.

Cabinet preparations of the tissue of the womb and of the textures of the external genitals are to be prepared by the methods now employed for the mucous membrane, the skin, and the muscular tissue.

The mucous secretion of the female genitals originates first and principally from the cervix uteri, the mucous membrane of which contains numerous fossæ or mucous follicles; and then from the glandless vaginal mucous membrane. The former has an alkaline reaction, appears hyaline, tough, and tenacious, and contains numerous mucous corpuscles together with scanty pavement epithelial cells. In contact with the acid vaginal mucus it becomes cloudy. The latter, an almost limpid fluid substance, is very scanty in young maiden bodies, except during the menstrual periods; in blennorrhœas of the genital mucous membrane, as well as with women far advanced in pregnancy, it increases in quantity, and the vaginal mucus becomes cloudy, milk- or pus-like. The elements of the vaginal secretion, which are shown by the microscope to increase in quantity in proportion to the increase of consistence and opacity, are again mucous corpuscles and pavement epithelium.

In the vaginal mucus of non-pregnant persons, but especially, however, in the pregnant, there occurs, together with several vegetable parasites, an interesting animal parasite, the Trichomonas vaginalis, discovered by Donné. This is an infusorium which is provided with a flagellum and vibratile ciliæ; it moves actively in pure mucus, but very sluggishly, on the contrary, in that diluted with water. This infusorium appears to be entirely absent from the normal secretion of the vagina of unimpregnated females (Kölliker and Scanzoni).

A speculum is to be employed in order to obtain the secretion in question. The vaginal mucus may be obtained by scraping with a spatula. It is difficult to obtain the mucus from the cervix unmixed with vaginal secretion. The addition of water is naturally to be avoided in the microscopical examination of the Trichomonas.

The menstrual blood from the ruptured capillaries of the uterine mucous membrane has usually (perhaps from the admixture of secretions from the mucous membrane) lost its coagulability. It shows, together with the blood-cells, numerous granulated structures, mucous corpuscles, also separated ciliated cells which have lost their vibratory ciliæ. Shreds or connected masses of the uterine mucous membrane may be caused by more considerable separations, so that a formal decidua spuria has been spoken of.

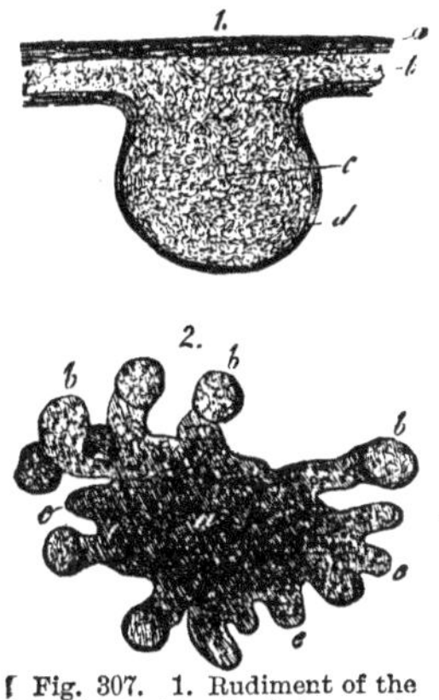

Fig. 307. 1. Rudiment of the lacteal gland in the fœtus. *a b*, Epidermis; *c*, cell aggregation; *d*, fibrous layer. 2. From a seven-months' fœtus. *a*, central substance; *b*, larger, *c*, smaller outgrowths.

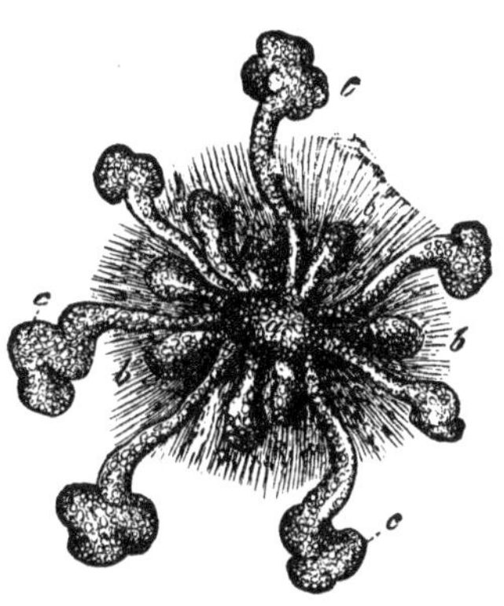

Fig. 308. The milk-gland of another embryo. *a*, The central club-like mass, with smaller internal, *b*, and larger external, *c*, outgrowths.

The lochial secretion consists at first almost entirely of blood which comes from the torn vessels of the contracting uterus. In the first days after the birth, when a brown-red mucous fluid, with a few flakes and shreds, usually come away, the microscope shows as elements, together with sometimes unaltered, sometimes swollen or indented blood-corpuscles, pavement-shaped cells, granulated structures (mucous and pus corpuscles), effete cells as well as their ruins, fat-molecules, and likewise, here and there, cholesterine tables. In the latter periods, where the blood-corpuscles decrease more and more in numbers, and finally disappear altogether, the number of the granulated cells usually increases in an inverse ratio. Towards the end the lochial secretion gradually assumes the character of a mucus rich in cells. The examination presents no kind of difficulty. For its reception, flat oval plates may be used (Werthheimer).

The lacteal glands are formed in the fourth and fifth month of human embryonic life, after the manner of other cutaneous glands, by solid papillary projections of the fœtal epidermoidal cells, covered by a fibrous layer of the corium (fig. 307, 1, *d*). Several weeks later (fig. 307, 2, and 308) such a club-like papilla (*a*) has, by division of its cells, sent down new papillæ, from which the chief excretory passages are afterwards formed, which, by further proliferations of this kind, produce the first rudiments of the gland-body. Even at the hour of birth, however, a rudimentary formation of gland vesicles has not taken place, and, while the ducts become hollow, their outgrowths remain at the stage of solid cell-aggregations. The larger gland divisions keep to the periphery, the smaller ones to the inner portions of the whole organ.

The lacteal glands preserve this undeveloped, more fœtal character, even during the period of childhood, in the male as well as in the female sex.

While it is true that even here the lacteal gland of the female has advanced more rapidly than that of the male, it is only at the entrance of puberty that a further development of the former becomes energetic. Numerous gland-vesicles are the result. Nevertheless, the organ still remains far behind its complete development, for which the first pregnancy is necessary. After the delivery it generally receives that organization. Atrophy is first introduced by the period of involution; fat-tissue takes the place of the gland-body.

The lacteal gland of the male, on the contrary, remains at the lower stage throughout the whole life.

The developed gland of the sexually mature female contains, in a condition of rest, an epithelial lining of ordinary rounded, polygonal gland-cells.

In the gland-vesicles of the cow, the same finest reticular canal-work which we mentioned formerly (p. 459) at the pancreas has been injected (Gianuzzi and Falaschi).

The mammary gland, as is known, presents manifold pathological new formations. In many of them the development takes place from the proper gland-bodies. Thus, for example, at the period of involution small cysts, filled with a mucilaginous fluid, are of frequent occurrence. They arise from a metamorphosis of the gland-lobules, the distended vesicles of which become united with each other. New formations of gland-substance, under pathological conditions, have also been ascribed to the organs under consideration, and have been described as "adenoid" tumors. Soft, but especially hard cancers, cysto-sarcoma and simple sarcomatous tumors also occur. Concerning the point of origin, the same uncertainty prevails here as elsewhere.

The examination of the lacteal glands (normal as well as diseased) is, for the most part, to be made (in the usual manner) on hardened organs, by means of fine sections. A preparatory immersion in very dilute acetic acid, in diluted pyroligneous acid, or a brief boiling, is useful. For the first rudimentary appearances, human embryos at about the fifth month are to be selected; for the later appearances, the bodies of children. The material necessary for the recognition of the completely developed gland can only be obtained from a female who has borne children. The active organ may be obtained from the bodies of the lying-in. Injections of the larger glandular passages from the lacteal sinuses succeed with tolerable facility; the constant pressure is to be employed for the injection of the finest passages.

The milk of the human female and of the mammalia is formed by the liberation of the fat produced in the cells of the lacteal glands, which thus becomes suspended in the glandular fluid, rich in albumen and sugar.

The secretion under consideration presents a near relationship, in this regard, with the less fluid substance secreted by the sebaceous follicles of the external integument, and, in fact, we are in a position to vindicate, by the aid of facts in embryology, the same manner of origin of both varieties of glands.

Fig. 309. Elementary forms in milk. *a*, Ordinary milk-globules; *b*, so-called colostrum corpuscles.

Ordinary milk shows, in a clear fluid, an innumerable quantity of globular fat-drops, the so-called milk globules (309 *a*).

These, which may be examined even with medium powers, never flow together after the manner of free fat, but rather possess a delicate investment of coagulated casein. It is only when this is dissolved by means of acetic acid or alkalies, that the union of the free drops of fat is noticed under the microscope.

If the secretion of the milk takes place less energetically, as is the case with the so-called colostrum, and the secretion which occurs during the latter period of pregnancy, as well as in the first days after the delivery, this rapid destruction of the gland-cells is absent, and we still meet with them in part (overloaded with fat to a high degree, it is true) as constituents of the evacuated fluid. We also meet with fragments of these cells and membraneless conglomerations of fat. These are the so-called colostrum corpuscles of the authors (fig. 309, *b*); they do not appear to be wholly without vital contractility (Stricker, Schwarz). A few remain for a long time in the milk of women. A larger number of such structures in the milk, months after the delivery, must, on the contrary, be denoted as abnormal.

Abnormal constituents of the milk are of secondary importance. Blood-cells, and also lymphoid corpuscles, may be met with in it; their recognition is not difficult.

Remarkable colorations may occur in milk which has stood for some time; thus blue and yellow colors have been observed. In such cases the microscope has shown vibrionic, and also protococcus formations.

A drop of milk spread out in a thin layer will, without any further manipulation, present its elements to view in the same manner as other cell-containing fluids, as, for instance, the blood. Strong magnifying powers are unnecessary; permanent preservation will hardly be undertaken.

Among the parts of the male generative organs we will first speak of the testicles, taking it for granted that the coarser structural conditions are already known.

Numerous, but not complete connective-tissue septa, arising from the fibrous tunic (the tunica albuginea) converge to unite, in the upper part of the testicle, in a firmly woven, connective-tissue, wedge-shaped mass, the so-called corpus Highmori. The gland substance, consisting of reticularly united and convoluted canals, the so-called seminal canaliculi, is thereby divided into conical, lobular convolutions.

The appearance of such a seminal canalicule may be represented by the adjoining fig. 310. The inner surface of the membrana propria is lined by a simple layer of rounded polygonal cells (*b*). A fibrous connective-tissue tunic, with longitudinally arranged connective-tissue corpuscles (*a*), is spread out externally around the structureless gland-membrane.

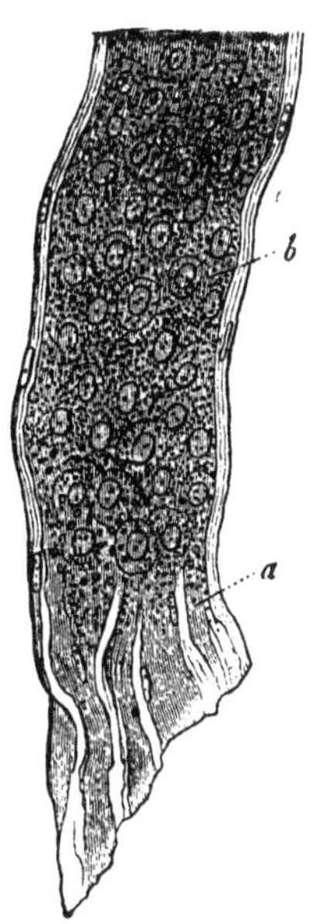

Fig. 310. Human seminal canaliculi, with the gland-cells *b* and the connective-tissue tunic *a*.

The cell-body of the glandular epithelium consists, in youthful individuals, of a finely granular substance, which, in later years, becomes richer in fat.

A soft, loose connective tissue is met with between the seminal canaliculi. In smaller mammalial animals this is very scanty and soft.

The seminal canals then coalesce and finally form a single vessel, a tolerably large canal, which, with innumerable convolutions, forms the so-called body and tail of the epididymis; it afterwards straightens and becomes the vas deferens. The epididymis shows, in addition, a ciliated lining of its convoluted seminiferous canals (Becker), in places with gigantic cells and vibratile ciliæ.

The blood-vessels, which are very easy to inject, pass from without and from the corpus Highmori into the organ, permeate the septula, and finally encircle the seminiferous canals with a wide-meshed (but not particularly abundant) network of capillaries.

The first accurate information concerning the lymphatics was furnished by Ludwig and Tomsa. In fact, nothing is easier than the injection of the organ by means of a puncture. A surprising picture of numerous vessels (Ludwig and Tomsa)

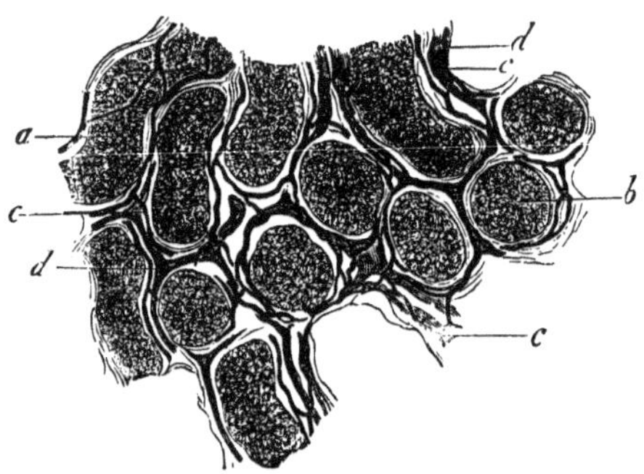

Fig. 311. From the testicle of the calf. *a*, seminiferous canals seen in more oblique, *b*, in more transverse sections; *c*, blood-vessels; *d*, lymphatics.

unfolds itself here, and, as it appears, in exactly the same manner in all mammalial animals. A very extensive network of lymphatics, with valves, lies under the serous covering, permeates with its branches the albuginea, and spreads itself out beneath the same to a likewise very compact network of passages enclosed by connective tissue. Some of these passages pass at once between the seminal canals; the greater portion, however, first pass through the familiar septula, and finally likewise enter the loose connective tissue lying between the glandular passages (fig. 311, *a b*). The spaces of this connective tissue, in so far as they are not occupied by seminiferous canals and blood-vessels (*c*), are filled with lymphatic fluid (*d*). We meet with this in a remarkable manner, in small mammalial animals especially, the seminal canals of which, having but very little interstitial connective tissue, are regularly bathed in lymph. For the recognition of the vascular cells of our lymphatics, either the injection of a solution of nitrate of silver or the immersion of the sections in such a solution of 4 per cent. may be employed (Tommasi).

The most frequent pathological new formations of the testicle are soft tumors, appearing in the form of medullary carcinoma and sarcoma. In the so-called cysto-sarcoma we meet with larger or smaller vesicles partly filled with water, partly with colloid substance, which proceed from transformations of the seminal canals.

For the examination of the testicles, human bodies or those of larger mammalial animals may be selected. The canals of the fresh organ may be readily isolated with the preparing-needles and their cell-contents recognized. Acetic acid and alkalies may thereby be suitably employed. To obtain the entire arrangement of the seminal canals, they are to be injected with transparent cold-flowing masses or with gelatine.

Gerlach gives us the following directions for the injection of these passages with gelatine: The testicle is to be placed in a weak solution of potash for 4–6 hours, to dissolve out the cells and the entire contents as much as possible. An attempt is then to be made to remove the mass by cautious pressure, after which the organ is to be washed off in water. The air contained in the glandular canals is to be drawn out as thoroughly as possible, and the injection fluid (colored with carmine or chromate of lead) is to be forced in very slowly. During this procedure the organ should be kept in warm water.

I would here also recommend the organ hardened in pieces in absolute alcohol, that for instance of the calf, in which the blood and lymphatic vessels have been previously injected with blue and red transparent substances. The vas deferens must be studied hardened in fluids. A mammalial animal just killed is to be used for the ciliated epithelium of the epididymis.

Concerning the deeper, efferent, and copulating organs of the male generative apparatus, it may be remarked that the ductus ejaculatorii and seminal vesicles have the same structure as the vas deferens, and are to be examined in a similar manner. In the latter is found, together with spermatic filaments, a transparent albuminous substance, which undergoes a gelatinous coagulation, to afterwards resume a fluid condition. It is the same substance which the evacuated sperm contains.

The prostate, a racemose glandular aggregation, is very rich in smooth muscular tissue. The latter elements may be examined in the fresh organ with the reagents usually employed for this tissue, such as the potash solution or a 20 per cent. nitric acid. For the investigation of its further structure, the organ is to be immersed in pyroligneous acid or hardened in alcohol.

The prostatic secretion appears to contain the same albuminous body which we have just mentioned at the vesiculæ seminales. The prostatic stones, as they are termed, concen-

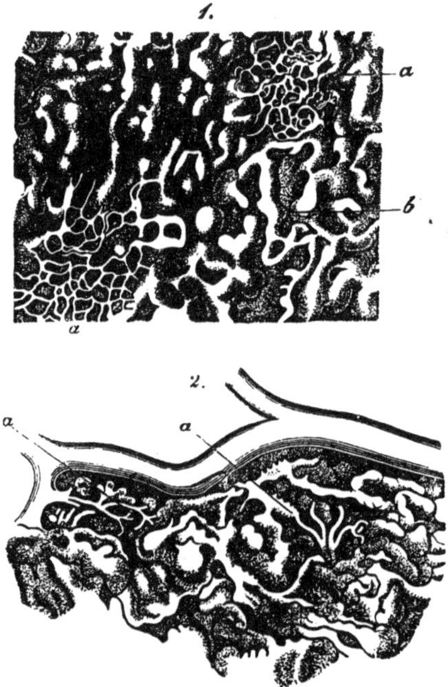

Fig. 312. From the peripheral portion of the corpus cavernosum penis, by a low magnifying power. 1. *a*, so-called superficial, and *b*, deeper cortical network. 2. Insertion of arterial branches (*a*) in the passages of the deeper cortical network (copied after Langer.)

tric structures, occasionally of considerable size, consist of this substance (Virchow).

The glands of Cowper are to be examined in the same way as other racemose glands.

The tissue of the cavernous organs consists of elastic and connective-tissue fibres, intermixed with smooth muscles. The latter are to be studied in a fresh condition, the remainder on alcoholic preparations, where we would recommend the previous injection with uncolored gelatine. These also afford opportunity, especially in transverse sections, for studying the male urethra. To follow the vascular arrangement of the corpora cavernosa (fig. 312), injections are to be made with transparent blue or red gelatine fluid, and the preparation somewhat strongly hardened. The lymphatics of the glans penis are to be injected by the puncturing method (Belajeff).

We have finally to consider the semen (sperm). A drop of evacuated human seminal fluid spread out in a thin layer, without any further addition, on the microscopic slide, shows, with a magnifying power of about 400 diameters, a number of peculiar structures, the so-called spermatic filaments, spermatic animalcula (spermatozoa, zoosperms). These (fig. 313) permit of the recognition of an anterior broader flattened portion, the head (*a*); and a more posterior long filament, with a relatively thick commencing portion, the so-called middle-piece (*b*); and a terminal filament (*c*) of extreme fineness.

Fig. 313. Spermatozoa of the sheep, after Schweigger-Seidel. *a*, head; *b*, middle-piece; *c*, tail.

The remarkable movements which these structures present in evacuated living semen have at all times awakened the astonishment and interest of the observer; and, in fact, a wonderful appearance is presented on looking down into this confused mass and observing the spermatic filaments darting wildly about. A closer investigation of this busy multitude shows that the individual spermatic element makes undulating and whip-like movements of the filament, and is thereby pushed from the place.

An independent change of position, directed towards a definite object, for which earlier observers mistook the phenomenon (and in harmony therewith declared the spermatic elements to be animal beings) is, however in no wise the case. If the phenomenon be followed throughout a longer period, it will be seen how, after the manner of the nearly related ciliary motions, the movements gradually become extinct; how the energy of the filamentary movements decreases more and more, and thereby the change of place ceases; how weaker and weaker contortions of the filaments are then to be noticed in the spermatozoa, which can no longer move from the place, until finally the whole becomes quiet. We would, in addition, repeat here a remark which we have already made; namely, as each excursion appears very much exaggerated by the strong objective (p. 101), the irregular advance of the spermatozoa should not be overestimated. It is in reality but very slow.

More indifferent fluids, blood-serum, lymph, white of egg, iodine-serum, solutions of sugar (1060–1030 sp. wt.), urea (10–5 per cent.), neutral salts of the alkalies (chloride of sodium, one per cent.), and earths may be employed as media. Pure water increases the energy of the movement of mammalial spermatozoa, at most for a very short time, to lead to a rapid cessation, whereby the filamentous extremity bends itself into the form of a loop. Everything, on the contrary, which acts chemically, generally arrests the movement once for all. Spermatozoa which have become quiet from too watery media may frequently be temporarily restored to life by means of a concentrated solution (of sugar, chloride of sodium or albumen), and inversely. As on the motus vibratorius, so also on the movements of the spermatozoa a peculiarly stimulating effect is exerted by dilute solutions of the alkalies and caustic potash of 1–5 per cent. (Kölliker). The alkaline fluids of the body therefore also maintain the vitality of the spermatic filaments for a long time. In a similar manner a suitable solution of sugar with 0.1–0.05 per cent. of caustic potash also acts excellently. Moreover, according to Montegazza, human spermatozoa preserve their vitality and capability of motion within the broad thermometric limits of from -15 to $+47$ degrees of the centigrade thermometer.

The origin of the spermatozoa, from cells which are formed by the metamorphosis of the ordinary glandular epithelium of the seminiferous canals, removes all doubt—if, indeed, such were still possible—concerning the nature of these structures as tissue-elements.

For the investigation of this method of origin (the details of which are, however, still matters of controversy), a sexually mature mammalial animal, such as a male dog, rabbit, or Guinea-pig, is to be selected, and the contents of the testicle examined immediately; and, in consequence of the extreme alterability of the spermatic cells, with indifferent fluids. Strong magnifying powers will be necessary.

The relatively resistant substance of which the spermatic filaments consist readily permits of their being preserved dry as cabinet preparations, and also of being softened with water from dried seminal stains, and thus recognized with the microscope.

From the great importance which the recognition of the latter is for the medical jurist (and it may still be accomplished after years), the simple procedure may here find its place.

The suspicious portions are to be cut from the body or bed linen, and, having been reduced to small pieces, placed in a watch-glass or glass box, with the addition of a small quantity of water. After a time, a quarter or half hour, during which the pieces of linen have been several times stirred about in the water with a glass rod, this fluid is to be examined, and then the fluid pressed drop by drop from the fragments on to the microscopic slide. Any spermatozoa which may be present will thus be discovered with certainty, and there is scarcely any possibility of mistaking them.

Section Twenty-Second.

ORGANS OF SENSE.

1. THE human skin consists of the epidermis, the corium, and the subcutaneous cellular tissue; the latter being rich in fat. Numerous nerves, blood-vessels, and lymphatics permeate it; innumerable glands lie embedded in it; the hairs and nails, finally, constitute special organs. All these have already been individually mentioned in previous sections, so that we can here only present a short recapitulation of the whole.

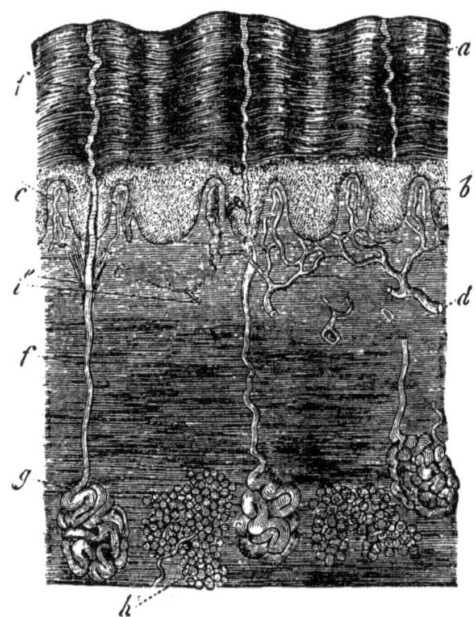

Fig. 314. Human skin in transverse section. *a*, superficial layers of the epidermis; *b*, Malpighian rete mucosum. Beneath the latter is the corium, forming the papillæ above at *c*, and terminating below in the subcutaneous connective tissue, in which at *h* aggregations of fat-cells appear; *g*, sudoriparous glands, with their excretory ducts *e* and *f*; *d*, vessels; *i*, nerves with tactile bodies.

The structure of the skin may be represented by fig. 314, a

vertical section of the same from the point of the finger. The cornified epidermis appears at *a* with its numerous layers of flattened cells; the (punctated) layer beneath it (*b*) represents the so-called Malpighian rete mucosum. The papillæ of the cutis appear at *c*, and beneath them commences the superficial extension of the corium, which is sometimes thinner, sometimes thicker, and passes into the subcutaneous cellular tissue without any sharp line of demarcation. Among the constituents of the latter we perceive the coil-shaped bodies of the so-called sudoriparous glands, *g*, the ascending ducts of which may be recognized at *f*, as well as the aggregations of fat-cells *h*. A similar section through a hairy portion of the integument would present us, in addition, the hairs, with their follicles and the sebaceous follicles.

Such preparations may be obtained in various ways from the freshest possible corpses. We can, although with some trouble, still without any further addition, prepare pretty thin sections, and render them transparent by means of weak alkaline solutions. If the fluid medium is of a proper degree of concentration, we then obtain a satisfactory, although very perishable preparation. It is preferable, however, even here, to previously give the object a greater degree of firmness by means of artificial hardening. The drying method as well as (which I prefer) the immersion in absolute alcohol accomplishes this purpose. Many views will be obtained in a better form if boiling in vinegar precedes the drying, or a precursory immersion in pyroligneous acid, after which the object is placed in alcohol. A prolonged maceration with pyroligneous acid is also not bad. Chromic acid has not presented me any advantage over alcohol for the skin.

Tingeing with aniline blue, and especially with carmine and subsequent washing in very dilute acetic acid, is excellent. Such a section from an injected portion of the skin, tinged with carmine and deprived of its water by means of alcohol, affords a very handsome review preparation when mounted in Canada balsam.

To enter here into the methods of examining the epidermis would be superfluous, as the essentials have already been mentioned at pages 261 and 262, and the peculiar surfaces of the younger cells discussed. The nails (p. 264) and hairs (p. 265) have likewise been treated of in the same section.

Fine sections of dried preparations, or those hardened in alcohol, with the addition of acetic acid, serve for the discovery of the elastic fibres, as well as the connective-tissue corpuscles of the corium. A longer immersion in undiluted glycerine, by rendering the connective-tissue bundles extremely transparent, permits us to see the elastic elements.

The same methods are also used for the recognition of the sudoriparous glands and sebaceous follicles. It is well, however, not to select too thin sections. The former gland-formation, which attains to gigantic dimensions beneath the skin of the axilla, may be readily isolated in this place in a fresh condition, and, with the employment of the familiar methods, the structure of the walls and the nature of the cells may be examined.

Surface sections are of relatively less frequent necessity. Made through the upper portion of the epidermis, however, they are of importance for the ducts of the sudoriparous glands, and laid deeper on the border of the former towards the corium, for the study of the papillæ.

The primary examination of the sebaceous follicles may be made on the labia majora, likewise on the skin of the scrotum, by picking fine sections with the needles and employing acetic acid. Good views may likewise be obtained, where the gland-cells are not to be preserved, by means of dilute alkaline solutions. The dried integument of other portions of the body, by similar treatment, also shows the organs in the vicinity of the hair-follicles. Preparatory boiling in vinegar affords a good accessory for such skin. If it is proposed to examine the cells and the remaining contents of the sebaceous follicles, the skin is, according to Kölliker, to be previously softened; then, with the epidermis, the hairs, with their root-sheaths and the cell

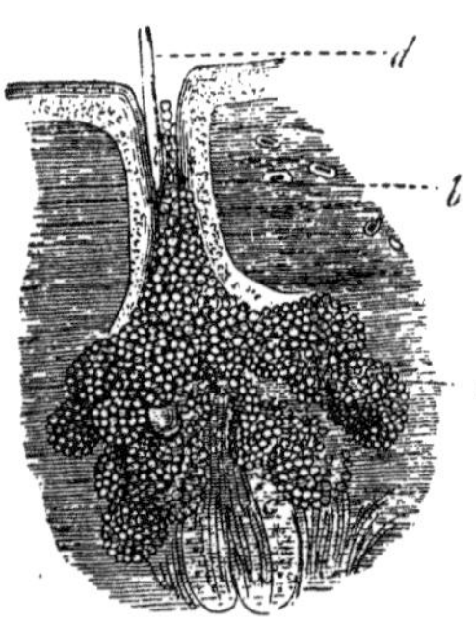

Fig. 315. A sebaceous follicle. *a*, the gland-vesicles; *b*, the excretory duct; *c*, the sac of a lanugo-hair; *d*, the shaft of the latter.

masses of the sebaceous follicles, are often rendered very finely prominent.

The blood-vessels of the skin are to be examined on fine vertical and horizontal sections of transparently injected organs. In the papillæ, at the point of the finger, an extensive natural injection is frequently met with, so that the transverse section of the dried skin, by the addition of a 30 to 40 per cent. solution of potash, affords very handsome views of the vascular loops.

The puncturing method serves for injecting the lymphatic passages which are likewise only enclosed by connective tissue. These consist, for the most part, of a double horizontal reticulum; a deeper one with wider passages, and a superficial one with narrower canals. Central canals, with cæcal terminations, pass from the latter into the papillæ (Teichmann).

For the study of the muscular tissue of the skin, the tingeing with carmine and subsequent treatment with acetic acid, the already frequently mentioned double tingeing of Schwarz, and the application of the chloride of palladium (I. Neumann) is to be recommended.

Concerning the cutaneous nerves, we would first refer to what was mentioned at page 369 about the tactile bodies. For the study of these elements in other localities, the same methods, the treatment of thin sections from the fresh or dried skin, with acetic acid and alkalies, and furthermore with Müller's fluid or chloride of gold, are to be taken into consideration.

The Pacinian corpuscles, which, together with terminal knobs, also occur on the external genitals of both sexes (Krause, Schweigger-Seidel), are to be examined by the methods customary for these structures.

Peculiar organs nearly related to the terminal knobs were met with by Krause on the sensible nerves of the penis and clitoris. These, the "genital nerve-corpuscles," are embedded in the corium and mucous membrane proper (not in the papillæ), and differ from the ordinary terminal knobs by their more considerable dimensions and more irregular forms. For their examination, the discoverer recommends: firstly, quite fresh, when possible still warm, preparations without any media; then injections, and an immersion in 3 per cent. acetic acid.

More recently, in other portions of the integument, fine, non-medullated nerve-fibres are said to have been seen to terminate with button-shaped swellings between the cells of the Malpighian rete mucosum (Langerhans). The treatment of the freshest possible thin sections with chloride of gold has been recommended for their recognition.

The embryos of man and the mammalial animals, hardened in chromic acid, are used for the fœtal skin. On small fœtuses the latter generally separates very readily, and is to be examined on surface sections with glycerine, whereby a conservative tingeing with carmine renders good service. With older embryos fine vertical sections are to be made with a razor. It is relatively easy, on such, to see the first rudiments of the sudoriparous glands and hairs, and also on the latter the sebaceous follicles, and to follow their further development.

The pathological changes of a part so complicated in its structure as the human integument are of a very manifold nature. A few, such as are connected with the epidermis, have been already mentioned (p. 265). Inflammatory conditions show sometimes an implication of the whole skin, sometimes of only the superficial portions. Extensive emigrations of colorless blood-corpuscles occur there (Volkmann and Steudener). Desquamations of entire layers of epidermis (scarlatina), local separations of the horny layer from the Malpighian stratum, by collections of a fluid containing pus-cells, occur as a result of this vascular congestion.

The numerous diseases of the skin affect sometimes the epithelial, sometimes the connective-tissue portion, sometimes both at once.

Elephantiasis shows a more extended and enormous hypertrophy of the corium and of the subcutaneous cellular tissue. Local proliferations of the tactile papillæ of the skin are presented by the warts and condylomata, whereby dilatations and enlargements of the capillaries are met with. The vascular nævi and telangiectasia in general form more extended superficial occurrences of the latter kind. Follicular tumors and cysts, of frequent occurrence in the skin, proceed undoubtedly in many cases from dilatations and degenerations of the hair-sacs and their sebaceous follicles. These frequently present the so-called atheromata; that is, the follicle, lined with a pavement epithelium, contains a grit-like pulpaceous mass, in which the microscope enables us to recognize exfoliated epithelial cells, fat-molecules, and crystals of cholesterine.

The maggot-pimples or comedones present slighter metamorphoses of the sebaceous follicles and hair-sacs, produced by accumulated secretions. If this accumulation is confined to the sebaceous follicles (to be explained by impeded evacuation), hordeolum or milium is produced. The degeneration of the hair-sacs is accompanied by that of the respective sebaceous follicles.*

The number of vegetable and animal parasites found in and on the human skin is a considerable one. Many of these constitute quite indifferent phenomena; others cause appreciable effects, and become causes of diseases, the appreciation of which dates from the discovery of these structures by means of the microscope, and which appear in part on and in the hairs, in part on the horny layer of the epidermis, partly also in the nails (though the nail fungi require further investigation).

Among the epiphytes or vegetable parasites we will next mention the Tricophyton tonsurans of Malmsten. It leads to the destruction of the hairs of the head in the form of rounded patches (herpes tonsurans). One finds only spores of about 0.0022‴, or also rows of the same. These first develop in the root of the hair, then in the shaft, which it splits extensively, so that in consequence the hair breaks off at about a line beyond its exit; the root and shaft of the hair are likewise destroyed. The systematic position of the Tricophyton tonsurans is still a matter of controversy.

Another vegetable parasite of the hairs of the human head, the Microsporon Audouini of Gruby, which causes the porrigo decalvans, behaves in a similar manner. It consists of rounded and oval spores (of 0.0004–0.0022‴) and a network of curved undulating filaments. These develop externally on the shaft

* For more particular information concerning the pathological changes of our organ, we would refer to I. Neumann's beautiful text-book on skin diseases.

of the hair, and occur in such numbers around the portion of hair which projects from the skin that it becomes destroyed, and remnants ½ to 1 line long stand out from the skin.

Another fungus of the same name, the Microsporon mentagrophytes of Robin, grows by preference in the follicles of the beard-hairs, and causes an inflammation and the formation of pus around the hair-follicle—the so-called mentagra. The microscope shows us between the hair sac and shaft larger spores and filaments than in the previous variety. The last two varieties are uninvestigated botanically. Finally, in the Microsporon furfur of Robin groups of double-contoured spores of 0.002‴, longitudinally arranged cells and branched filaments of 0.0004–0.0002‴ in diameter may be recognized. The basis for the development of this epiphyte is, however, different; namely, the horny layer of the epidermis, where it causes yellowish spots and a furfuraceous exfoliation (pityriasis versicolor).

The favus-fungus, Achorion Schœnleinii of Remak, occurs principally on the hairy portions of the skin of the head, and is the cause of the scall, porrigo favosa—an eruption occurring mostly during childhood. It becomes developed first in the hair-sac, where it surrounds the hair and grows into it; then, and indeed principally, on the epidermis. There may be distinguished, according to Robin, the 0.0013‴ broad inarticulate filaments of the mycelium, the somewhat broader unbranched, but articulated receptacula, in the interior of which series of round and oval spores, 0.0013–0.0026‴ in size, develop.

The nature of this parasite is still doubtful; possibly it is a mould-fungus. The favus-scab shows under the microscope a fine granular mass, which surrounds the fungus-substance proper. Externally this consists principally of the mycelium, more internally of the receptacula, and quite internally of the spores.

The examination of all these epiphytes requires in general strong, 4–600-fold, magnifying powers. For the study of the hair fungi, the stumps are to be pulled out with forceps and rendered transparent by means of pure glycerine or oil of turpentine. With the fungi growing on the epidermoidal scales, the addition of alkalies, the dilute solutions of potash and soda are to be employed.

Among the animal parasites, epizoa, of the human skin, two may here be mentioned, both mites of lower organization; the hair-sac mite, Demodex folliculorum, Owen; and the itch-mite, Sarcoptes hominis. Both live in the skin and produce quite different effects. While the first animal constitutes a quite indifferent parasite, the Sarcoptes scabii causes the familiar complication of symptoms known under the name of the itch (scabies).

Fig. 316. Hair-sac mite, Demodex folliculorum.

The Demodex folliculorum (fig. 316) shows a sometimes more, sometimes less elongated body without bristles or hairs. On the fore part of the body in the young creature there are three, on the mature animal four pairs of stump-like legs. The length of this small parasite is from $\frac{1}{2}$–$\frac{1}{8}$‴. It lives generally in scanty numbers in the excretory ducts of the sebaceous follicles and hair-sacs, that is, the space between the hair-shaft and the root-sheath, and deposits its eggs in its dwelling-place. It occurs in the sebaceous follicles of the face, and with especial frequency in those of the nose. If the respective glands of the latter locality are strongly developed, the smegma may be squeezed from the opening by pressure, the mass spread out in water, and the mites examined. With dead bodies vertical sections of the skin are to be prepared.

Not to be confounded with the hair-sac mite is the larger itch-mite. This, which is represented very much enlarged in our fig. 317, has a rather broad, oblong body, from $\frac{1}{5}$–$\frac{1}{4}$‴ in length, covered with hairs and bristles. The first two pairs of legs are placed very far in front; they are short, and with a stalked sucker. After a considerable interval follow the last two pairs of stump-like legs, which terminate in long bristles. The ovum, which is not unfrequently found in the body of the female animal, is (as also in the Demodex) of considerable size, and the young animal is likewise six-footed.

The itch-mite most frequently infests the human skin between the fingers and on their inner surfaces; they may, however, occur on all parts of the body. It penetrates beneath the epidermis, and forms beneath the same a serpentine passage, which appears brown from the fæces of the animal. The animal is met with as a small white point at one end of this passage.

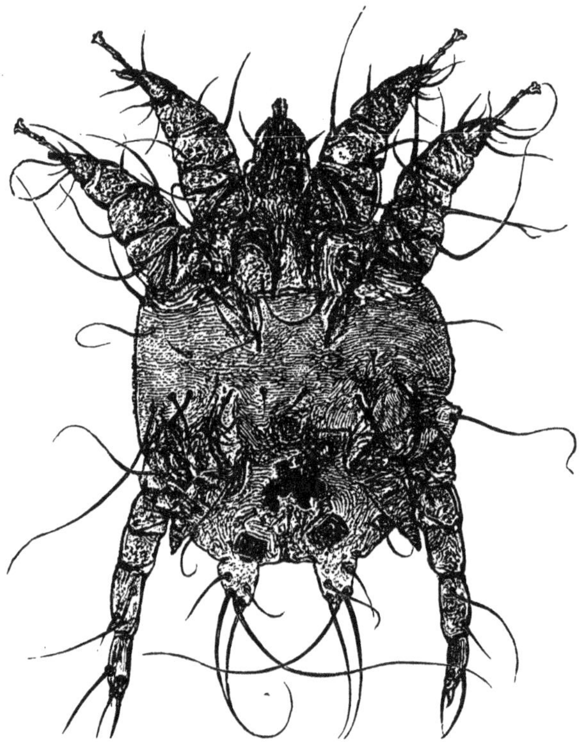

Fig. 317. The itch-mite, Sarcoptes hominis, from a photograph.

To obtain the mite for microscopic examination (which does not require a strong magnifying power), the passage is to be slit up with a cataract-needle, and the white point lifted out on the point of the needle. For a more accurate study, the portion of skin which contains the acarus is to be made into a fold, and this—epidermis and upper portion of cutis—removed with a curved scissors. Spread out on the microscopic slide, the preparation is to be allowed to dry gradually, and then rendered transparent by means of oil of turpentine or Canada balsam. A longer immersion of the moist piece of skin in concentrated glycerine also affords the necessary degree of transparency.

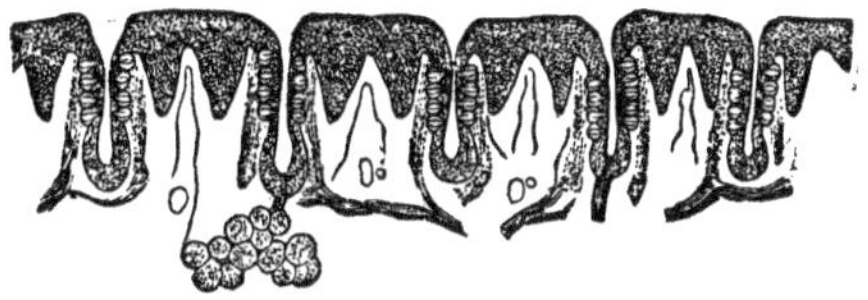

Fig. 318. From the lateral gustatory organ of the rabbit. The gustatory ledges in vertical section, after Engelmann.

2. The organ of taste has already been discussed in describing the organs of digestion (p. 419); so that we may refer to the same, and here treat only of the manner of termination of the nerves of sense.

Investigations which were earlier instituted on the human and mammalial tongues, on their papillæ fungiformes and circumvallatæ, in regard to the distribution of the nerves, yielded an unsatisfactory result, and showed only the nerve-trunks with divisions and plexiform communications, terminating finally in pale, non-medullated fibres. Terminal knobs were described here years ago by Krause. In the circumvallated papillæ of man and the mammalia, Lovén and Schwalbe recently discovered peculiar terminal apparatuses, denoted by the name of the "gustatory buds." They occur especially in the lateral walls of these papillæ, but not unfrequently also at the inner surfaces of the surrounding ledges of mucous membrane.

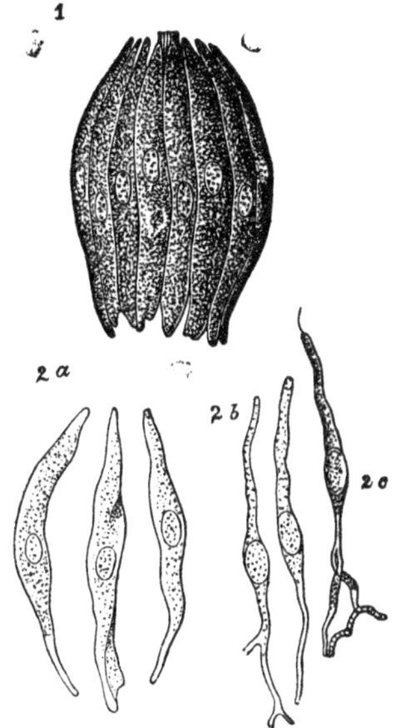

Fig. 319. 1. Gustatory bud of the rabbit. 2 *a*, cover-cells; 2 *b*, rod-cells; 2 *c*, a rod-cell with a fine terminal filament.

Our fig. 318, a transverse section through the lateral gustatory organ at the root of the tongue of the rabbit, discovered by Engelmann and Wyss, may afford us a primary representation of this terminal apparatus of the gustatory nerves embedded in the epithelium. We may obtain a more accurate appreciation from fig. 319.

The parietes of the gustatory bud (1) consist of flattened, lancet-shaped cells (2 *a*), which stand perpendicularly by the side of each other, comparable to the staves of a barrel or the sepal leaves of a flower-bud. These are the "cover-cells."

The pointed portion of our organ perforates the epithelial covering. Small roundish spaces occur here. They are formed in part by several epidermis cells, in part only by twos, or finally even by a single one.

In the interior of the organ appears a second cell-formation, the "gustatory cell" (2 *b*). A spindle-shaped body, as we see, extends above into a rod, and is prolonged below to a thin branched filament. It penetrates into the tissue of the mucous membrane. At the summit of the rod, finally, a short fine cilium shows itself.

A plexus of medullated and non-medullated nerve-fibres has been met with beneath the gustatory bud. The communication of these with the lower filamentous termination of the gustatory cells still remains to be proved.

Even earlier they succeeded in recognizing, with some probability, the termination and the terminal structures in the frog's tongue (Shultze, Key).

Fungiform papillæ stand separated over the tongue of the frog. The lateral portions of these projections and the edges of the surfaces are covered by ordinary cylindrical epithelium. The plateau of the papillæ shows, on the contrary, surrounded by ciliated cylinders, another covering of non-ciliated cells, which, on suitable chromic acid preparations, after brushing off the ordinary epithelium, may sometimes be brought to view sitting like a crown on the gustatory papillæ. Between these non-ciliated cells lie other structures, spindle-shaped cell-bodies, which pass upward into a fine rod, terminating at the surface of the epithelial crown. Downwards, on the contrary, it runs out into a very fine filament, which, with certain reagents, appears varicose, and which must be regarded as the terminal branch of an axis-cylinder which has been split up in a tuft-like manner. The nerve-fibres, therefore, split up into fine fibrillæ, pass over into these rod-bearing gustatory cells. These statements of Schultze and Key have, however, been more recently modified by Engelmann. He denies this connection of the nerve-fibres with the "gustatory cells," and describes a third (hitherto confounded with the gustatory cells) structure, the "forked cell," with a small ellipsoid body, which is prolonged above and below into forked processes. The central processes of the forked cells, with great probability, pass over, under further division, into the axis-cylinders of the nerve-fibres. Possibly, however, both varieties of cells are of a nervous nature.

Engelmann has recently collected the methods of investigation.

For the primary examination, the dried mammalial tongue (appropriately that of the rabbit) may be employed. The sections are to be softened in dilute acetic acid and glycerine. The organ may also be hardened for a day in osmic acid (0.5–1.5 per cent.). The freezing method also affords good results.

For the study of the finer structure, the maceration in iodine-serum and the immersion for several days in chromic acid (1–2 per cent.) to which an equal volume of glycerine may be added, are to be recommended. Such preparations must then be subjected to a very careful picking under the simple microscope. According to Engelmann's experience, extremely fine-pointed glass rods are superior to the finest steel needles. Wyss gives the preference among all methods to an immersion for about three weeks in Müller's fluid.

Dried tongues, or those which have been frozen in a suitable manner, serve for following the nerves. Sections obtained by the freezing method may be subsequently treated with chloride of gold (0.1–0.5 per cent.) or osmic acid (0.25–2 per cent.). The nerve expansions under the gustatory nerve became distinct for Schwalbe after a maceration for several days in chromic acid (0.02 per cent.) or bichromate of potash (0.5–1 per cent.). Wyss made use of the gold method.

3. Further advanced, especially by the admirable investiga-

tions of M. Schultze, is, on the contrary, our knowledge of the organ of smell, that is, of the manner of termination of the olfactory nerve. Before we consider the remarkable structural conditions of this locality, however, let us mention the other portions of the organ of sense.

No portion of either of the chief cavities, except the uppermost parts, participates immediately in the perception of smell, and does not contain any fibres of the specific nerves, but only those from the trigeminus, the terminations of which are at present still unknown.

Disregarding the entrance to the nose, a ciliated epithelium is found as a covering to the proper nasal fossæ and the accessory sinuses. The mucous membrane, thinner in the accessory sinuses, is, in its submucous connective tissue, firmly united with the bones, so that it constitutes at the same time a periosteum. In the proper nasal fossæ it becomes thicker, and very rich in blood-vessels and racemose mucous glands.

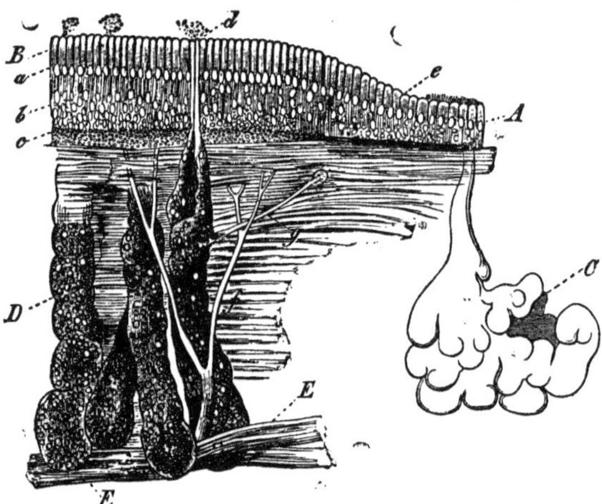

Fig. 320. The regio olfactoria of the fox in vertical section. *B*, The cylindrical epithelium of the same. *a*, Stratum of the nuclei; *b*, of the olfactory cells; *c*, of the pigment. *A*, The adjacent ordinary ciliated epithelium. *e*, The boundary between them. *C*, Ordinary racemose mucous glands. *D*, Bowman's glands, with the duct *d*. *E*, Branch of the olfactory nerve. *f*, Ascending branch with further divisions.

The particular manner in which these parts, as well as the cartilage and bones of the parietal system, are to be investigated does not require further discussion.

In catarrhal conditions of the nasal mucous membrane we

see at first an extensive exfoliation of the ciliated cells, which are met with in the mucus coming under examination, partly still ciliated (and even in motion), partly without ciliæ. Together with the regularly shaped cylindrical cells, others of a more irregular and more rounded form are met with. Large cellular structures, which have arisen from the metamorphosis of the normal epithelial formation, contain, at this commencing period of the nasal catarrh, in addition to their nucleus, granulated lymphoid cells (mucous- or pus-corpuscles) which have penetrated from without. Nearly all these elements soon disappear, with the exception of the last-mentioned structures, which are met with in enormous quantities in the thick yellowish secretion of the later periods. Phenomena which we have previously alluded to under similar irritated conditions of the respiratory organs (p. 497) and of the urinary bladder (p. 527) also repeat themselves here.

As was said, the greater portion of the olfactory region does not participate immediately in the perception of smell, as the termination of the specific nerve of sense is met with only in a limited portion. Such places, called regiones olfactoriæ (fig. 320), occur in all vertebrate animals, but present numerous differences. While the walls of the remaining portion of the olfactory cavity are covered with ordinary ciliated cells (*A*), there appears, as a covering for the regio olfactoria, a likewise unstratified, but non-ciliated cylindrical epithelium of a peculiar kind (*B*), mingled with cells terminating in rod-like processes, similar to those we have just become acquainted with in the frog's tongue. The signification of nervous terminal cells cannot be denied to these structures, although the continual transition of the lower varicose terminal fibre into the fibrillæ of the olfactory nerve, cannot yet be demonstrated with certainty (either by Schultze or others, as for instance C. K. Hoffmann). The extraordinary delicacy and decomposability of the tissue-elements under consideration (which can only be managed by macerating and preserving fluids of a definite composition), render it appreciable that during a long period the microscopists either did not recognize the complicated structure at all, or interpreted it erroneously.

In the mammalia and in man the regio olfactoria distinguishes itself from the remaining nasal mucous membrane by a peculiar color, by a yellow or yellow-brown tinge. This proceeds from fine pigment molecules, which are embedded partly in the bodies of the non-ciliated cylindrical epithelial cells, partly in the cells of an especial gland-formation occurring here. Vertical sections of the part which has been hardened in strong chromic acid serve for the primary survey. These nucleated cylindrical cells (fig. 321, 1 *a*, 2 *a*) may be recognized from suitable side views. They send filamentous processes in a downward direction, which pass into communication with each other by means of branches, and, having arrived at the margin of the mucous membrane, they undergo a further more profuse division, so that they, at least in places, pass over into a reticulum which is very delicate and difficult to understand, and which often spreads out into a kind of homogeneous lamella (similar to the membrana limitans of the retina). Between these cylindrical cells are noticed in considerable numbers the so-called olfactory cells (fig. 321, 1 *b*, and 2 *b*), structures which are analogous to the gustatory cells. At very different elevations between the epithelial cells lies a spindle-shaped nucleated cell-body (1, 2, *b*), which extends upwards into a fine rod (*c*), and downwards into an extremely fine varicose filament (*d*).

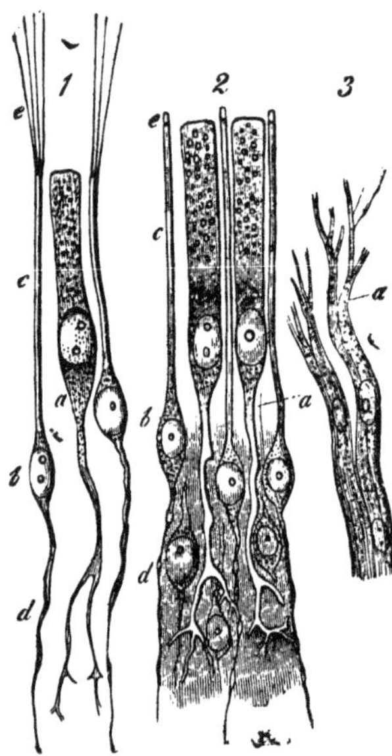

Fig. 321. 1. Cells of the regio olfactoria of the frog. *a*, An epithelial cell, terminating below in a ramified process; *b*, olfactory cells with the descending filament *d*, the peripheral rod *c*, and the long vibratile ciliæ *e*. 2. Cells from the same region of man. The references the same, only short projections, *e*, occur (as artefacts) on the rods. 3. Fibres of the olfactory nerve from the dog; at *a* dividing into fine fibrillæ.

The end of the rod which has reached the surface appears to terminate quite naked in all mammalial animals; small and quite short styliform projections (2 *e*), which may be noticed on it, are the contents which have swollen out from the effects of reagents. This appendix is also absent in fishes which smell in water. In birds and amphibia, which smell in the air, on the contrary, quite large in part, extremely long ciliæ, sometimes slightly, sometimes not at all movable, occasionally single, occasionally in numbers, appear on the free extremity of the rod, so that the surface of the regio olfactoria is surmounted by a regular hair-forest (fig. 321, 1 *e*).

Seen from the surface, one may recognize how the pigmented cylindrical cells are surrounded in a circular manner by these rods; while by the side view the rods are to be perceived between the cylinders, as well as stratified in a deeper position, the spindle-shaped cell-bodies of the structures with which we are at present occupied.

Very fresh cadavers are necessary to obtain the same structures, cylinder- and olfactory-cells, in man. Particularly worthy of recommendation for this purpose are the bodies of new-born children. In the adult, where the numerous nasal catarrhs have preceded, the sharp difference in color between the regio olfactoria and the remaining portion of the nasal mucous membrane is, for the most part, wanting, and the textural peculiarities are likewise, as a rule, not so accurately demarcated as in the mammalial animal. Otherwise, entire conformity prevails.

Peculiar glands (fig. 320 *D*), intermediate in form between simple cylinders and racemose glands, and called "Bowman's glands" by Kölliker, in honor of their discoverer, are situated in the middle portion of the regio olfactoria, and permeate this remarkable cell-stratum with their narrowed excretory ducts. Their bodies, lying in the connective tissue, have no membrana propria, and consist of those yellow or yellow-brown pigmented gland-cells of which mention has just been made. The adjacent mucous membrane, on the contrary, shows ordinary racemose mucous glands (*C*). Although an ordinary ciliated epithelium is found in places in the human regio olfactoria, yet these true racemose gland-formations follow immediately. Of interest is the circumstance that the Bowman's glands occur in all the higher vertebrate animals, but are absent in fishes which smell in the water.

The olfactory nerve (fig. 320 *E*) presents only non-medullated elements in its branches. These appear at first as pale nucleated fibres, quite similar to those which we meet with in many sympathetic nerves, as, for instance, in those of the spleen. By suitable treatment, however, one succeeds in separating the olfactory nerve-fibre into extremely fine fibrillæ enclosed in homogeneous sheaths; it is therefore a primitive bundle.

The finer branches of the olfactory nerve (fig. 320, *fg*) ascend between the glands of the regio olfactoria, and thus arrive at the margin of the epithelium. Here they divide into the finest filaments, or primitive fibrillæ. These, quite similar to the processes of the olfactory cells, and appearing varicose under the same conditions as those, permeate the fine latticed reticulum formed by the spreading of the processes of the cylinder-cells, to finally, as we must admit and have already remarked, become united with the processes of the olfactory cells (fig. 322).

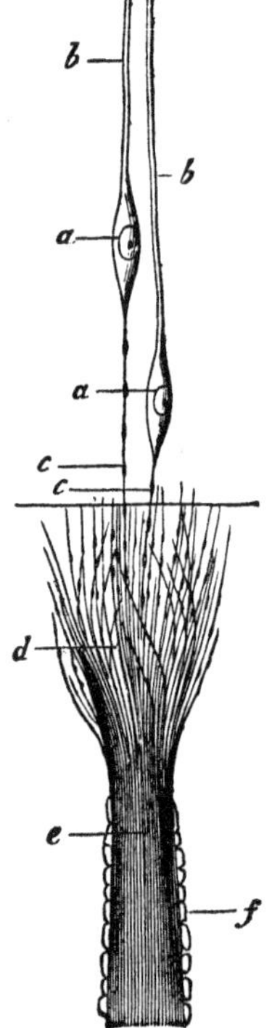

Fig. 322. Probable termination of the olfactory nerve in the pike (after Schultze). *a*, Olfactory cells; *b*, rods; *c*, lower varicose filament; *e*, axis-fibrillæ in the sheath *f*; *d*, spreading out of these; at ——— wanting connection with the same fibrillæ *c*.

In his excellent monograph, Schultze has given us a long series of directions for the demonstration and investigation of these extremely subtile textural relations, and thereby made an extremely important contribution to microscopic technology.

To obtain a primary view of the cells of the regio olfactoria from the body of a mammalial animal just killed, thin sections obtained with the scissors may be placed under the microscope, with the addition of the most indifferent possible fluid media. In these the olfactory cell-rods will be discovered between the non-ciliated epithelial cylinders as hyaline rods. However, even with the employment of vitreous fluid, one will soon see hyaline drops, which proceed from the decomposing olfactory cell-rods, coming out over the margin of the epithelial surface, a decomposition which takes place with even greater rapidity from the addition of water. Schultze found the addition of a not too watery glycerine useful. Fine vertical sections of organs, hardened in stronger chromic acid, or dried and softened in acidulated water, also fulfil this purpose.

To isolate the epithelial structures (and this separation is more difficult to accomplish in warm-blooded vertebrates than in cold-blooded ones), the employment of conservative and macerating fluids is necessary. This effect is rapidly and thoroughly obtained by the use of the 30–40 per cent. solution of potash, or one of soda of 20–25 per cent. If quite fresh pieces of the ethmoid bone, with the adherent mucous membrane, be immersed in this, and, after the lapse of a half or a whole hour, the epithelium be scraped off, the separation may be accomplished on the microscopic slide. With weaker solutions one must wait two or three hours. The well-preserved cylinder-cells and rods, a portion of them still in connection with the spindle-shaped olfactory cells, may then be readily recognized, and in amphibia and birds even the olfactory ciliæ; on the contrary, there is usually nothing preserved of the descending fine filamentous processes of the latter.

To obtain a surface view, the epithelial covering, macerated in a solution of potash or treated with glycerine, is to be used.

Better, though much more slowly, commencing (from two to three days) effects may be obtained by maceration in a very dilute solution of chromic acid (whereby the immersed portion should not be too small, nor the quantity of fluid too great). 0.05–0.03 per cent. solutions are to be recommended for the quite fresh mammalial animal. For the human olfactory organ, when it can be obtained, about twelve hours after death, Schultze employed the action of a chromic acid solution of 0.05 per cent. for from one to three days. Cold-blooded vertebrate animals require somewhat stronger solutions than the mammalia; birds somewhat weaker ones (to 0.01 per cent.). (The division of the bundles of the olfactory nerves into primitive fibrillæ may also be accomplished in this way.)

The extraordinary advantage which such solutions present for the study of the regio olfactoria, consists in rendering visible varicose swellings on the very fine lower filamentous processes of the olfactory cells, as well as the finest terminal fibrillæ of the nerves of sense (a superiority which also belongs to the reagent for analogous textural conditions of the remaining higher nerves of sense). As has already been frequently mentioned, the bichromate of potash may be employed instead of chromic acid; its action takes place slowly. Schultze employed solutions of 0.1–0.5 per cent., and obtained the desired preparations after 1–6 days.

The Müller's fluid, which is, as I found, very suitable for the examination of the cochlea when diluted with water, I also recommended years ago in several degrees of dilution. According to Hoffmann's experience, diluted with an equal part of water, and sometimes acting only one or two days (frog), sometimes for nearly two weeks (mammalia), it in fact constitutes the best of all macerating media.

Schultze has discovered and recommended still other similarly acting fluids.

The concentrated watery solution of oxalic acid preserves (p. 133) the olfactory cells, their rods and varicose filaments (but not the cylinder-cells) quite exquisitely; and one has the great advantage of not being too dependent upon the time, so that the examination may be made even after a few hours, and also after days. The connective tissue swells in it and becomes more transparent, while albuminous tissues retain their sharp contours and become somewhat harder.

Sulphuric acid in a condition of high dilution, in medium of 0.6 per cent. (0.2–1 per cent. and more), also preserves the olfactory cells very well, and still more diluted it renders the filaments varicose. The connective tissue, however, does not swell in it as in the previous acid, but is rendered much more beautiful and sharp. Here also too small pieces should not be taken, and the preparation is to be tried even after a few hours. The immersed pieces keep for days and weeks if they are not ruined by the formation of mould. This acid is less praised by Hoffmann.

To obtain degrees of hardening which are suitable for the preparation of thin sections, and which should present the arrangement of the mucous membrane, the glands of Bowman and the course of the nerves, one may employ, together with the Müller's fluid, higher degrees of concentration of chromic acid and chromate of potash, and subsequently examine with glycerine, acetic acid, etc. Moleschott's so-called strong acetic-acid mixture (p. 142) has also been especially recommended by Balogh.

Strongly hardened objects, as well as the preparations of the remaining portions of the nasal mucous membrane, one should attempt to mount in glycerine. The olfactory cells and the cylindrical epithelium lying betwen them may be still more suitably preserved in Müller's fluid diluted with an equal quantity of water.

4. The organ of vision, from its greater complication, requires a more detailed discussion.

The eyelids, with the cutis accompanying them, their connective-tissue so-called tarsal cartilage, and the embedded Meibomian glands, which in their form remind one of those of Bowman in the olfactory organ, as well as the conjunctiva of their posterior surface and of the eyeball, together with the epithelial covering which clothes it, do not require discussion. Their tissues are to be examined in accordance with previous directions. The Meibomian glands may be recognized with facility, in their coarser relations, in the eyelids of small mammalial animals rendered transparent by means of alkalies or by immersion in vinegar; fine sections from the dried organ serve for the investigation of the finer structure. The lachrymal glands are to be examined in the same manner as other racemose glands.

The conjunctiva of the eye (frequently a lymphoid infiltrated connective tissue) contains throughout the entire line of transition numerous racemose mucous glands, while in the conjunctiva of the eyeball (the portion surrounding the cornea) of the ruminantia, coil-shaped glands, quite similar to those of the skin, have been discovered (Manz). Maceration in dilute acetic acid or pyroligneous acid will make them readily visible. For the recognition of the peculiar nerve terminations in the Krause's knobs (fig. 223), the fresh, still warm eye of one of our slaughter-house animals may be used. The conjunctiva is to be removed with rapidity, but with the utmost caution, and searched with low powers and without the addition of any medium. The necessary details concerning the reagents have already been mentioned at page 369.

We have also previously mentioned (p. 366 and fig. 184) the remarkable termination of very fine nerve fibrillæ in the epithelium of the conjunctiva.

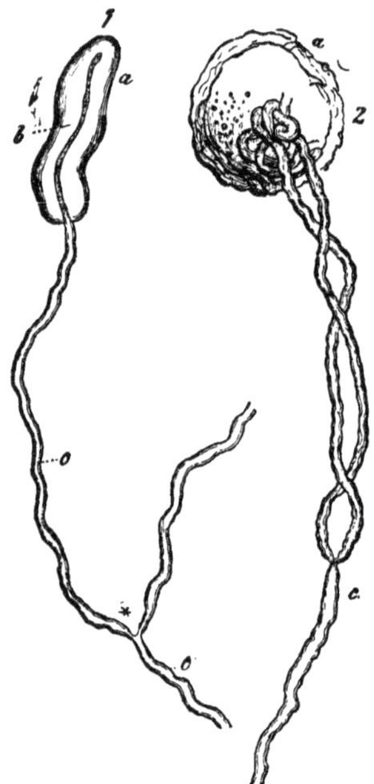

Fig 323. Terminal knobs: 1, from the calf; 2, from man.

The blood-vessels of the conjunctiva present nothing remarkable. The lymphatics of the human conjunctiva form a highly developed network of passages of considerable size over the sclerotic, which also occupies the peripheral portion of the cornea for about one millimetre in breadth, and consisting here of finer canals with bow-like terminations (Teichmann). Among mammalial animals I have succeeded in injecting such canals in the calf.

Interesting phenomena of the conjunctiva are presented by the trachoma glands, as they have been called: lymphoid follicles very similar to those of the intestinal canal, and quite variable in number and arrangement.* The injection in the ox (fig. 324) shows that knotty lymphatics (*a*) of considerable size run towards their lower surface, and, after the loss of their vascular walls, form a highly developed network of lymphatic canals around them, from which finer reticular canals encircle the follicle and spread themselves out in a delicate manner (*c*) in the lymphoid stratum uniting the follicles (*b*). Their most superficial portions, that is, those

* According to the statement of Blumberg, our structures are still entirely absent in the conjunctiva of quite young animals.

turned towards the epithelial layer, run more horizontally, and send off fine terminal canals which present quite superficial cæcal terminations. The whole is enclosed in connective tissue, and the whole arrangement is related in the most intimate manner to that of a Peyerian plaque; only the blood-vessels of the follicles are here less abundant and less regular.

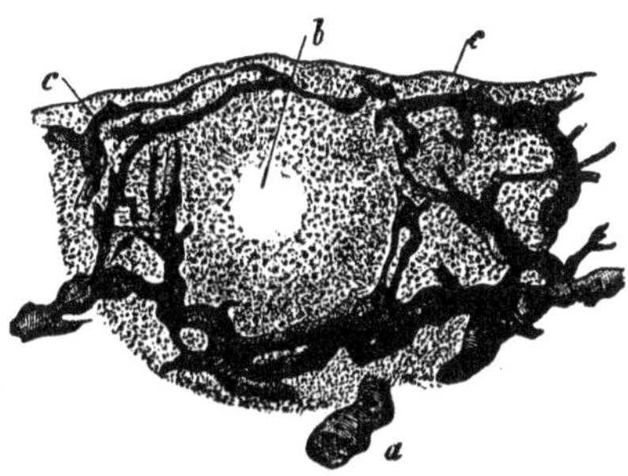

Fig. 324. Trachoma gland of the ox, with injected lymphatics, in vertical section. *a*, submucous lymphatic vessel; *c*, its distribution to the passages of the follicle *b*.

The eyes of younger oxen cr older calves, and cold-flowing mixtures are to be used for injections, which should be made in the so-called Bruch's aggregation of the trachoma glands of the lower eyelid. The other aggregation may, however, also be very readily injected; and furthermore, the injection of the blood-vessels by the smaller arteries also succeeds without great difficulty, while with smaller mammalial animals the entire head must be injected from the aorta with colored gelatine.

The procedure is difficult in man and many other mammalial animals. The preparations are to be hardened in alcohol for examination.

With regard to the eyeball itself, it may be said that its examination is one of the most remunerative labors of the microscopist; but at the same time, however, in one of its elements (the retina) combined with the greatest difficulties. One should always use the quite fresh, still warm eyes of the larger slaughtering animals, especially those of the ox, calf, and sheep, as well as indifferent fluid media, such as the vitreous and aqueous humors which are always at hand. If the eyes have been removed with a little care, the artery may be readily found lying near the optic nerve and employed for injecting (the injection of the human eye is more difficult in consequence of the smallness of the artery). Such injections, when they are to be followed by histological investigations, are always to be made with cold-flowing mixtures, the Prussian blue or carmine. The injection of an eye of one of the larger animals, after the numerous divided vessels have once been ligated, usually succeeds in from two to three minutes. Even after a quarter of an hour one may commence with the dissection and examination. The system of the uvea, especially, shows many things in a much more instructive manner than in the injected organ, and the unpigmented tapetum of such eyes affords a further advantage for many investigations. Where especial reference is to be made to injected preparations, inject with carmine gelatine. The eyes of the smaller mammalial animals are to be injected from the aorta simultaneously with and under the same precautions as the brain (p. 356). White rabbits afford excellent objects. The eyeballs of this animal, divided in halves and mounted in glass cells with Canada balsam, which have been put in circulation by Thiersch, may be recommended as true models of the modern injection *technique*. If it be desired to accomplish a double injection, Prussian blue is to be first thrown into the artery, and a second injection of carmine gelatine is then to be made by the same vessel. Leber has recently instituted excellent studies of this kind, with cold-flowing masses and the use of the constant pressure.

The investigation of such fresh eyes requires in part transverse sections, as of the cornea and sclerotic ; but generally, however, a dissection of the membranous structures. These may be first reviewed unpicked with vitreous fluid, or the addition of reagents, and hereby folds which are artificially made are for the most part very instructive, or they are to be divided with fine needles. Much may be recognized in such a manner concerning the texture of the eyeball, and even of the retina. In this way all the former (and in part adequate) knowledge of the same was obtained ; and even with the employment of other more modern methods, the criterion of the fresh condition can never be dispensed with. Certain elements of the eyeball, on the contrary, are in part so transparent, in part so delicate and soft, that hardening (and darkening) methods of treatment become indispensable. Many structural conditions, the termination of this and that structure, the relations of continuity of the one with the other, etc., can only be ascertained with adequate certainty from such preparations. Those two methods which we have already had to mention in connection with so many organs, the drying and the hardening by means of reagents, are also employed here. For the former purpose the eyeball is to be divided at the equator, and the vitreous body (generally also the lens) removed. Both segments should be spread over the surface of cork, cut into semispheres. For hardening, use either absolute alcohol or—what is far more customary—chromic acid (0.5–0.2 per cent.) and chromate of potash. The bulb may be either divided, only cut open, or even left entirely unopened (in which case a stronger solution of the hardening medium is to be selected). The Müller's eye-fluid (p. 139) is very excellently adapted for hardening the unopened immersed eyeball. It is necessary, it is true, to wait two or three weeks for the adequate effect; but the eye may also be allowed to lie in it for months and even years without any harm, and with this accessory very handsome specimens of most parts of the bulb may be obtained. The mixture has therefore, with propriety, come more and more into use among ophthalmologists. If a weaker action is aimed at, it is to be diluted with water; to obtain stronger hardening, a little chromic acid is to be added. Injected eyes may also be hardened in this manner, though the color suffers somewhat. If it be desired to avoid this, employ cold-flowing baryta masses (p. 185).

Let us next examine the methods for investigating the capsular system, the cornea and sclerotic.

The structure of the cornea (fig. 325), with both its epithelial layers, the stratified of the anterior (*d*), and the simple cell-covering of the posterior surface (*e*), with the two hyaline boundary layers of the so-called lamina elastica anterior (*b*), and the membrane of Descemet (*c*) appearing under them, as well as the ordinary corneal substance (*a*) and their cellular elements, have been so frequently treated of and discussed of late that it would be superfluous to enter further into the textural relations in question. The best descriptions of the cornea are those of His, Kühne, Engelmann, and Schweigger-Seidel.

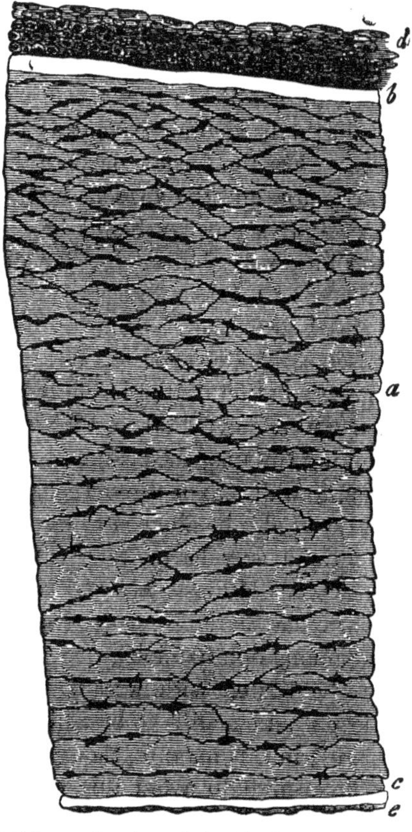

Fig. 325. The cornea of the new-born in vertical section (but considerably shortened). *a*, Corneal tissue; *b*, anterior, *c*, posterior hyaline layer; *d*, stratified pavement epithelium; *e*, simple epithelial layer.

With regard to the methods of examination, it is to be said that the quite fresh transparent cornea of an animal which has just been killed may be employed. It may be cut through from the side, or, although with some trouble, fine vertical and surface sections may be removed from it with the aid of a very sharp razor. Humor aqueus serves for moistening, and a moist chamber (p. 103) for the preservation of the object.

For the recognition of the effete corneal corpuscles and their contents, use very weak acetic acid, or extremely dilute chromic acid solutions of 0.01 per cent. Here, however, as with all the following methods, artificial (and often very considerable) alterations are not to be avoided, as the interstitial substance swells and the cells usually shrink.

Drying is also useful. Very thin sections, either only softened in weakly acidulated water, or first tinged in carmine and then washed out with diluted acetic acid, afford good review specimens (fig. 325).

His, a meritorious investigator of our membrane, recommends first the acetic acid with iodine staining, to cause the corneal cells to appear through the transparent interstitial substance.

According to him, however, the immersion in purified pyroligneous acid, diluted with an equal volume of water, or even more, constitutes a main accessory. The cells, with more cloudy contents, then appear through the somewhat swollen, more transparent interstitial substance. The hardening property which, as is known, is also associated with the distending power of the pyroligneous acid, is also of great worth here for rendering fine transverse sections feasible. These may be vertical, exactly similar to those from the dried object, and then (which is not possible with the latter material) the very much more instructive horizontal sections. Entire corneas may also be preserved for years in pyroligneous acid.

Another mixture, mentioned by Remak, consisting of dilute pyroligneous acid, watery alcohol, and a weak solution of sulphate of copper, causes less swelling, but otherwise yields quite similar appearances. Chromic acid presents no advantage over pyroligneous acid.

Other reagents, such as concentrated chromic acid, Müller's fluid, saturated solutions of sugar, dilute alcohol of 50 per cent., and Merkel's chromic acid and chloride of platinum mixture (p. 140), exert a shrinking effect on the interstitial substance, and have been recommended for demonstrating a fibrillary structure of the corneal tissue.

Frequent use has also been made of the silver method in the

examination of the cornea and a reticulum of star-shaped figures, which appear sometimes bright out of a darker basis substance, sometimes dark with brighter surroundings, pronounced to be the network of corneal cells (His).

Much more certain views are afforded, however, by the cornea treated with chloride of gold. The cellular elements and the nerves are alone colored, and the former preserve all their details. This reagent is here of the first rank, as Cohnheim rightly remarked. Its hardening property permits of the preparation of vertical and surface sections.

Frequent use has more recently been made of the most uninjured possible cornea of the frog (Kühne, Recklinghausen, Engelmann). One should proceed here in accordance with the directions given at page 249. The cornea is to be examined with its posterior surface turned upwards, preferably without any covering glass and with an immersion system.

At first one sees next to nothing in the hyaline transparent tissue; at most, striations of the same and traces of the corneal nerves. After a more close examination one finds small isolated, dull glistening structures of a sometimes rounded, sometimes elongated, occasionally crooked form. The observer convinces himself that these bodies stretch out delicate processes and draw others in; in short, constantly change their form and location. These are the already mentioned (p. 250) wandering cells of Recklinghausen.

If we wait half an hour longer, the corneal corpuscles begin to stand out from the tissue in the form of extremely pale, polygonal appearing dull spots. If about another half-hour be allowed to pass, our corneal corpuscles become more distinct; the dull spots are connected with each other by radiated processes; the reticulum of cells is visible. Nuclei are not yet to be discerned in the latter. Kühne asserts that he has convinced himself of a vital contractility in these stellate cells. Engelmann saw no trace of this in his re-examination. The change of form and location of the wandering cells is, on the other hand, to be observed now as before (and with proper treatment for a long time yet).

We will here mention still another interesting and important observation concerning the latter cell-formation. We have already (p. 102) alluded to the reception of small granules into the interior of such amœboid structures. If a small incision be made at the sclerotic border of the cornea of a living frog, and granules of cinnabar or carmine be rubbed into it, the cornea, isolated after twelve hours or more, will show a number of these cells with the colored molecules in their interior, occasionally considerably removed from the wound, wandering through the tissue.

Concerning the two hyaline limiting layers of the cornea, it may be stated that the membrane of Descemet may be readily isolated by scraping with the firm pressure of a scalpel. An incomplete separation of the membrana elastica anterior from the deeper corneal tissue may be accomplished by maceration in muriatic acid.

Dried sections mounted in Canada balsam are to be employed for recognizing the double refraction of the interstitial substance.

The organ of smaller embryos affords handsome preparations for the cells and interstitial substance. His recommends about two-inch fœtuses of the cow and hog.

We have already discussed the examination of the nerves of the cornea in a previous section of our work (p. 366), so that we must refer to the details there presented.

The blood-vessels occupy only the peripheral portion of the organ in the adult, as we learn from artificial and natural injections.

The magnificent transparency of the so accessible cornea renders it more than any other structure appropriate for artificial inflammatory irritations, and the study of the tissue-metamorphoses which take place thereby. It is therefore frequently employed for such investigations, and the facts obtained interpreted in accordance with the prevailing pathological views. While years ago the profound work of His appeared to lend an important support to Virchow's theory concerning the participation of the connective-tissue corpuscles in the inflammatory process; at the present time the case is entirely reversed, and the cornea has become a favorite object of study for demonstrating the correctness of the Waller-Cohnheim theory of the immigration of lymphoid cells. The doubters have also resorted to this field.

To induce the inflammation of the cornea (keratitis), our structure is to be irritated by the application of tincture of cantharides, a stick of nitrate of silver, or by the insertion of threads and silver wires.

The desired inflammation is obtained in rabbits after 24 hours, in summer frogs after 2–3 days, but only after double this time in hibernating frogs.

If cinnabar, carmine, or, still better, aniline blue suspended in water has been injected with a Pravaz' syringe into one of the lymph spaces of such a frog, no serious disturbance of the health of the animal is caused. The stuffed lymphoid cells now penetrate as pus-corpuscles from the periphery into the cornea, to adhere to the place of irritation. It is only a few of these cells, however, which bear the colored granules in their bodies as a mark of their origin. A considerably larger number of these can only be obtained when this coloring matter has been injected for several consecutive days into the various lymphatic spaces.

Either the fresh organ, incised several times at its margins, or the organ impregnated with gold, may serve for examination.

However, even in a cauterized cornea, if it is only obtained living, lymphoid cells accumulate in such quantities around the point of irritation, that the migratory cells present in it at the moment of separation do not suffice to supply the demand (Hoffmann and Recklinghausen). An origin of these cellular elements from the stellate corneal corpuscles cannot, according to this, be denied.

The network of blood-vessels which may cover the anterior surface of the cornea, as a result of inflammation, requires, after the beautiful statements made by Thiersch concerning the vascularization of wounds (p. 392), a renewed investigation.

Portions which have been removed are regenerated by newly formed corneal tissue. The yellowish margin which the cornea shows in the so-called arcus senilis, consists of a deposition of fat in the corneal cells, and also in their interstitial substance;

it is therefore one of those commencing fatty degenerations, such as likewise make their appearance at a more advanced age in other parts of the body.

Corneal preparations may be tinged, and, after extraction of their water by means of absolute alcohol, mounted in Canada balsam. A moist mounting in watery glycerine is, as a rule, employed.

The tissue of the sclerotica, as is known, is continuous with that of the cornea, but consists, after the manner of the fibrous membranes, of a fibrillated intercellular substance, which is changed by boiling into ordinary gelatine, and not, like the cornea, into chondrin. The flattened bundles of these fibrillæ cross each other nearly at right angles. Fine elastic elements and a reticulum of connective-tissue corpuscles stand out from the hyaline interstitial substance after the application of acetic acid.

The fresh tissue teased out into fine pieces, and objects hardened or dried in the same manner as the cornea, serve for the examination. If the iris and choroid have been retained on them, the immediate transition of these tissues into that of the sclerotic may be finely observed; likewise the transverse section of the canal of Schlemm, the origin of the musculus ciliaris and the prolongation of the membrane of Descemet into the so-called ligamentum iridis pectinatum.

The system of the uvea consists of the choroid and iris, membranes which contain muscular fibres and are rich in pigment and vessels. They are covered on their inner surfaces by a pigmented pavement epithelium (fig. 326), the so-called polyhedral pigment-cells of an earlier epoch. For the demonstration of these cells (which are, however, with much greater propriety to be included with the retina), the fresh eye may be used, or one which has been divided and hardened either by means of chromic acid, chromate of potash, or Müller's fluid. Small portions of the black covering of the exposed inner surface may be removed with the scalpel or the cataract-needle. They are then

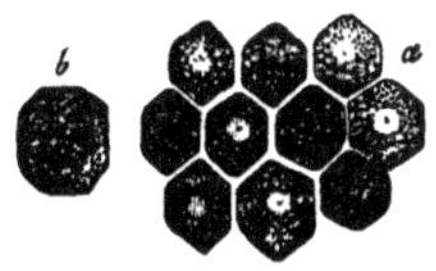

Fig. 326. Pigmented pavement epithelium (so-called polyhedral pigment cells) of the sheep.

to be spread out with needles or the brush, and covered with a right thin covering glass. Strongly hardened eyes permit of transverse sections of the entire uvea, and hence profile views of the epithelial covering.

To convince yourself that we have here, in fact, a pavement epithelium which has assumed a peculiar heterogeneous appearance from the excessive pigmentation, take an Albino eye, that of a white rabbit, or the unpigmented covering on the so-called tapetum of one of our ruminantia. An ordinary mosaic of polyhedral cells will be perceived, and, in the latter locality, appearances will be at the same time obtained which show that on the peripheral portion of the tapetum cells, with a scanty deposition of melanine, form the transition into ordinary pigment-cells.

The choroid proper consists, as is known, of a soft connective tissue, showing stellate cells united in a reticular manner, and which is permeated by an extraordinary quantity of blood-vessels. These cells characterize themselves by a great disposition to develop pigment molecules in their bodies and to thus become stellate pigment-cells (fig. 327).

Fig. 327. Stellate pigment-cells (pigmented connective-tissue corpuscles), from the mammalial eye.

We distinguish several layers of the choroid. An external looser stratum of soft connective tissue, rich in pigmented cells (lamina fusca, supra-choroidea), serves for the connection with the inner surface of the sclerotic. Its structure may be readily recognized in fresh-picked preparations; likewise its relations with the neighboring tissues in sections through the sclerotic and uvea of an eye strongly hardened in chromic acid.

Beneath the lamina fusca follows a middle layer of this connective-tissue substance, presenting the larger arterial and venous ramifications. The fresh tissue, or an eye which has been injected with a transparent mixture, serves for the recognition of this slightly pigmented layer. Finally, as a third layer appears, a more homogeneous unpigmented stratum, the so-called chorio-capillaris, which contains a remarkably rich, very narrow-meshed reticulum of delicate capillary vessels (fig. 328). Here also recourse may be had either to an injected organ (calf, sheep, cat), or a portion of choroid may be taken from a chromic acid preparation and freed as well as possible from the external layers, and, by cautious brushing in glycerine, from the pigmented pavement epithelium which covers the inner surface. A sufficient quantity of blood-corpuscles will generally be found retained in the capillary network.

Fig. 328. Capillary arrangement from the chorio-capillaris of the cat.

The fine hyaline, more independent boundary layer of the chorio-capillaris, towards the pavement epithelium, has been designated as the elastic layer of the choroid. For its primary recognition a fold of the fresh choroid may be used; acids and alkalies serve as media.

These lamellæ undergo interesting senile metamorphoses, thickenings, spherical and druse-like concretions, frequently with deposits of lime-molecules, which may dislodge the pigment epithelium and compress the retina (Müller). Other hyaline membranes of the eye also increase in thickness with age.

The injection preparations mentioned, deprived of their water with alcohol, may be beautifully mounted in cold Canada balsam (diluted with chloroform); the others are to be mounted moist.

For the primary recognition of the ciliary muscle, sections from a dried eye may be used. Here the rows of fibres running in a meridional direction may be perceived, and on good sections also the circularly arranged ones which Müller discovered. For further investigations, use the reagents customary for the connective tissue and the contractile fibre-cells, the 30–40 per cent. potash solution, the chloride of palladium with a subsequent carmine tingeing, the double staining of Schwarz, etc. According to Flemming the contractile fibre-cells may still be isolated, after hardening in chloride of palladium, by means of the potash solution mentioned. A long, 24 hours' action of the latter is then necessary.

The examination of the ciliary body is to be made on fine sections from an eye which has been injected with transparent gelatine fluids, and hardened in chromic acid or alcohol. The delicate rich reticulum of vessels may in this way be most accurately followed. Here, as with the iris, the eye of the white rabbit injected with carmine deserves the preference.

For the primary recognition of the structure of the iris, avoid dark-eyed creatures, as the stellate pigment-cells which occur in their tissue increase the difficulty of the examination very much. The eye of a new-born or of a bright-eyed child deserves to be recommended here. The methods consist, after the removal of any pigmented epithelial layer which may be present (and which may be previously macerated in acetic or oxalic acid) by means of the brush, firstly in tearing, then in the examination of whole pieces with the use of acetic acid for connective tissue, and of the dilute soda solution for the nerves. For the smooth muscular tissue use the reagents now generally employed for that tissue, and which have just been mentioned. One may thus convince one's self of the existence of a dilatator pupillæ, concerning which numerous controversies have lately arisen, and which is, nevertheless, not so very difficult to recognize. The whole or half of the iris of a small white rabbit may be used for the study of the coarser arrangement of the muscles when treated with acetic acid, and also with the addition of soda for following the nerves of the iris with lower magnifying powers. Such objects, tinged with carmine and washed out in weak acetic acid, become very handsome, as do also transparent injections of the blood-vessels.

There is nothing especial to be remarked concerning the methods of preservation.

Concerning the refracting organs, the lens and the vitreous body, the tissue of the latter has already been mentioned in one of the preceding sections of our book (p. 269); the crystalline lens, on the contrary, although in reality an epithelial structure, has not yet been discussed.

The fresh eye of any somewhat large mammalial animal

may be employed for the examination of the lens-capsule (fig. 329 *a*) and of the extremely delicate pavement epithelium (*b*) which occurs on the posterior surface of the anterior segment of the capsule. The capsule, isolated with the lens, is to be liberated by an incision and placed in fragments under the microscope, with the addition of vitreous fluid. Weak powers, with a strongly shaded field, show the borders and folds of the hyaline membrane at once. Stronger objectives show the thoroughly homogeneous structure of the hyaline membrane, and with repeated considerable shading, the pavement-shaped epithelium may be recognized. The addition of aniline red is here very convenient, as the tingeing follows very rapidly

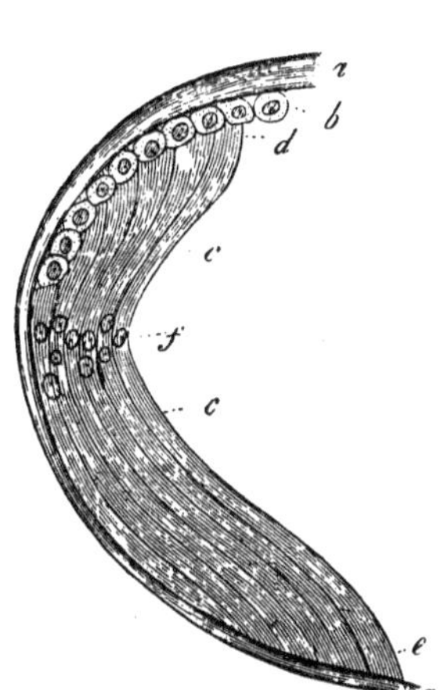

Fig. 329. Diagrammatic representation of the human crystalline lens. *a*, the capsule; *c*, the lens-fibres with widened ends, (*d*) becoming inserted at the anterior layer of the epithelium (*b*), and also inserted posteriorly into the capsule (*e*); *f*, the so-called nuclear zone.

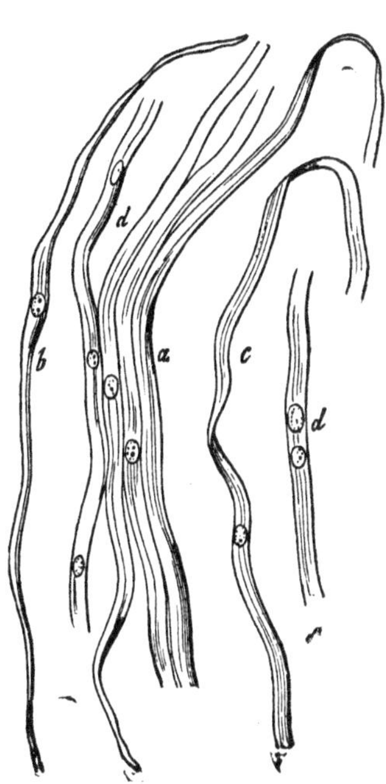

Fig. 330. Lens-fibres of the human embryo of eight months. *a*, Fibres with a nucleus; *b*, one which still presents the cellular character; *c*, the flattened form of the side view; *d*, fibres with two and three nuclei.

and without any alteration of the tissue. Other tingeing methods also accomplish the object.

On a fresh section of a lens, however, even with the use of these two accessories, one would be able to recognize the lens-fibres (fig. 330) only very incompletely.

Various accessories are to be recommended here, as the maceration in highly diluted sulphuric acid (4–5 drops of the acid of 1.839 sp. wt. to one ounce of water), which isolates the lens-fibres (v. Becker); in muriatic acid of 0.1–1 per cent. (Moriggia); further, the preparatory hardening in chromic acid, bichromate of potash, or Müller's fluid (Zernoff). The object may likewise be obtained with alcohol, though not so well; and also by peeling off thin scale-like portions from a dried lens. The organ being of itself quite transparent, with many chromic-acid preparations, the use of media which tend to increase the transparency, such as glycerine, is to be avoided. Such objects may occasionally be very suitably tinged with aniline. Others, which are more opaque, can be again rendered transparent by means of glycerine or acetic acid. An immersion, lasting for 15 minutes, in a nitrate of silver solution of 0.125–0.1 per cent. has also been praised (Robinsky).

The lens-tubes and the nuclei of the equatorial zone readily appear (fig. 329 *f*). To recognize the relations of origin of these fibres to the epithelium of the capsule, one may use a strongly hardened lens which is within its capsule, and the equatorial as well as meridional sections from that region.

Equatorial sections may be obtained from the adequately hardened crystalline lens, and may thus permit of the recognition of the delicate mosaic of the lens-tubes cut at right angles. A lens which has become considerably dried in the air, if employed at the proper moment, has not unfrequently acquired such a degree of consistence as to permit of convenient cutting without splintering. From it very handsome transverse sections may be obtained (fig. 331). A similar appearance may be obtained from ground sections of a strongly hardened lens. The previous saturation with a mixture of gum mucilage and a little glycerine is also to be tried in drying.

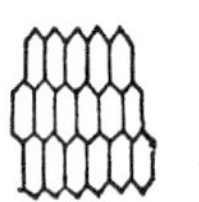

Fig. 331. Transverse section of the lens-fibres from the dried organ.

Opacities of the lens capsules, in part with depositions of elementary granules, likewise embedments of fat-molecules in the epithelial cells and lens-fibres, of granules between the latter, deposits of lime, etc., are not unfrequent occurrences. The methods of examination are the same.

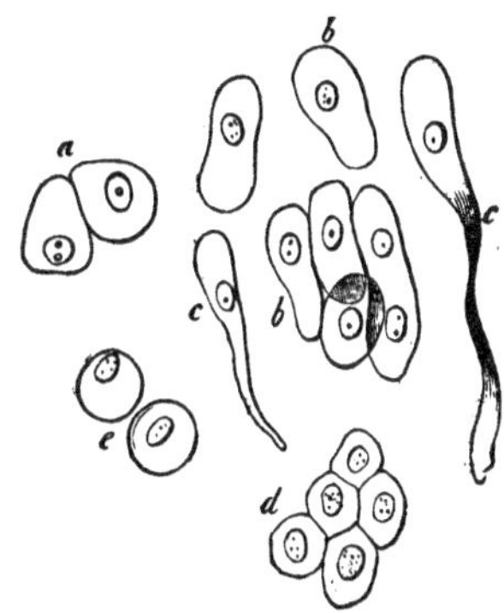

Fig. 332. *a—c*, Lens-cells of a two-inch fœtus of the pig. *a*, Original cells; *b*, oval elongated; *c*, grown longer in the transition into lens-tubes; *d*, epithelium of the lens, from an eight months' human embryo; *e*, cells of the so-called humor Morgagnii.

Human and mammalial embryos, hardened in absolute alcohol or chromic acid, serve for ascertaining the primary origin and the fœtal structural conditions of the lens (fig. 332) and the vitreous body. In embryos of the sheep of 6–7‴, everything is still cellular; in human fœtuses of about 8–9 weeks, the lens also seems to be constituted of only delicate spindle-shaped cells (Kölliker). The condition of a pig's fœtus of two inches is shown in our figure. Fœtuses of this animal of 3½″ have already a fibrous nucleus (Schwann).

Preservation is to be attempted with strongly watered glycerine.

The membrana hyaloidea is readily recognized in the hardened, and also in the fresh organ.

In the latter condition, after brushing off the epithelium, the fibres of the zonula Zinnii may be perceived with some difficulty, but they appear far more beautifully and sharper in the hardened eye.

If we now proceed to the nervous portion of the eyeball, the retina, there lies before us in the so difficult to comprehend, extremely complicated structure of the most delicate and changeable membrane, one of the most troublesome, but likewise also most attractive objects of microscopic investigation. An infinite amount of work has already been done in older and more recent times in the investigation of the so wonderful retina; and, though the knowledge concerning this membrane has made very great progress by the aid of the modern accessories, there remain to the present hour unsolved many textural questions of physiological importance.

To obtain the primary review preparations recourse will nowadays be generally had to an artificially hardened eye. An opened eyeball may be immersed in chromic acid of 0.5–0.2 per cent. (if unopened the concentration may be increased), or the corresponding quantity of the bichromate of potash. At present we can recommend nothing more highly for the unopened bulb than Müller's fluid. It preserves the cones and rods very beautifully, which, with the other solutions, does not succeed or only very incompletely; the examination may be made after 2–3 weeks (but also much later). Alcohol, which was formerly regarded as inappropriate, has more recently been highly recommended (Henle, Ritter); likewise, for the connective-tissue part at least, the mixture (mentioned at p. 140) of chloride of platinum and chromic acid (Merkel).

Thin vertical sections from the fundus of the bulb may also be readily made with a sharp knife with such retinas.

Such a vertical section, examined in the hardening fluid with the addition of a little glycerine (according to circumstances it may very suitably be previously tinged with glycerine-carmine), and cautiously covered with a very thin covering glass, shows at once the numerous layers of the retina which were conquered for science with such difficulty, and of which the adjacent sketch, (fig. 333) may recall the necessary conception to mind. The various localities of the retina, if treated in the same manner, will present their primary structural peculiarities.

Assisted by the advanced knowledge of the connective substances, one may at present recognize in any retina a considerably developed connective-tissue framework substance, the perception of which is, however, considerably impeded by the extraordinary fineness of the elements. The best investigation of this framework substance was made by M. Schultze.

This is permeated by the nervous elements, among which are to be enumerated the layer of optic nerve-fibres (fig 333, 7), and of the ganglion-cells of the inner portion (6); then the cones (and rods) of the outer layer (1), and likewise a part of the elements of the granular layers (2, 4); as also, finally, a system of radially arranged finest nerve-fibres, which have only recently been distinguished from the connective-tissue supporting fibres.

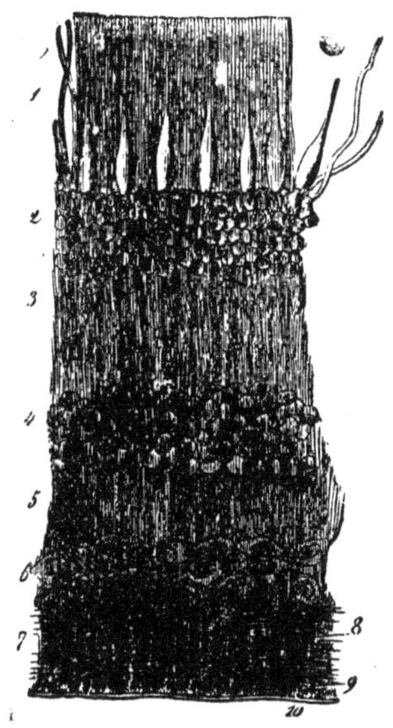

Fig. 333. Vertical transverse section of the human retina (made about half an inch from the point of entrance of the optic nerve). 1. Layer of rods and cones; 2, external granular layer; 3, inter-granular layer; 4, internal granular layer; 5, fine granular layer; 6, layer of ganglion-cells; 7, expansion of the optic nerve-fibres; 8, radial fibres; 9, their insertion into the inner limiting membrane, the membrana limitans interna, 10.

The verification on the fresh eye will be necessary even for what has thus far been described. The eye is to be opened under iodine-serum in a dish, and a portion of the retina removed. A fold having been carefully made, and a piece of thin glass placed near it, to protect it from the pressure of the covering glass, the various layers may be more or less distinctly recognized. Fine vertical sections are, nevertheless, more suitable. Do not believe that immense skill is necessary for their preparation. The piece of retina carefully separated from a fresh ox-eye is to be placed on the microscopic glass slide, or on a cork plate, and a little vitreous fluid or iodine-serum added. The attempt is then to be made with a sharp moistened blade, such as that of a cataract-knife, or the convex one of a small scalpel, by careful pressure and a rocking motion, to obtain as fine as possible sections. Many of these attempts will fail, but a few objects will possess sufficient thinness to permit of a successful examination, if treated with the same precautions as a fold.

For further studies such sections (in which it is true one is not protected from a displacement of the elements, and which therefore require a comparison with other methods) may be further picked. It is also suitable to employ with them a weak chromic acid or the dilute Müller's fluid.

A portion of the above-mentioned connective-tissue framework of the retina may be recognized even by the aid of the previous methods; though even a half-way sufficient view is never obtained. As Schultze has shown, other methods are necessary for this purpose, the same as have already been mentioned at the organs of smell.

Among these are to be enumerated chromic acid in a condition of extreme dilution (p. 131), the very watery sulphuric acid (p. 127), and the concentrated oxalic acid solution (p. 133).

To obtain a primary view of the connective-tissue framework structure of the retina, that investigator says that the eye of a fish is to be used, as the arrangement is easier to understand here than in the mammalial animal. The bulb of a river perch which has just been killed is to be divided through the equator and placed for three days in the familiar highly diluted chromic acid solution, which contains $\frac{1}{4}$, $\frac{1}{5}$, $\frac{1}{6}$ grain of the acid (or $\frac{1}{2}$–2 grs. of bichromate of potash) to the ounce of water. The examination is then to be made, cautiously picking, and using the high magnifying powers of a Hartnack's immersion system. While the connective-tissue framework substance is thus rendered recognizable by the chromic-acid solution, the latter has likewise the already mentioned excellent property of causing varicosities on the finer nerve-fibres, and of rendering it possible to distinguish the two systems of fibres in the retina as in the regio olfactoria.

The concentrated watery solution of oxalic acid is also an excellent medium for this examination and discrimination, as it renders the connective-tissue framework paler and hardens the nervous elements somewhat, and thus renders them more distinct. One is not confined to a definite time by this means, as the examination may be made after a few hours, or even after several days.

Sulphuric acid of 0.6 per cent. preserves the nervous elements very well, but also at the same time those of the connective-tissue framework.

The scientist mentioned afterwards found in osmic acid an important accessory for the investigation of the textural conditions under consideration. We shall return to the same.

By such methods the connective-tissue framework substance has presented the following arrangement (fig. 334, *A*).

The entire retina is permeated by it, with the exception of the bacillar layer. A system of radial or Müllerian supporting

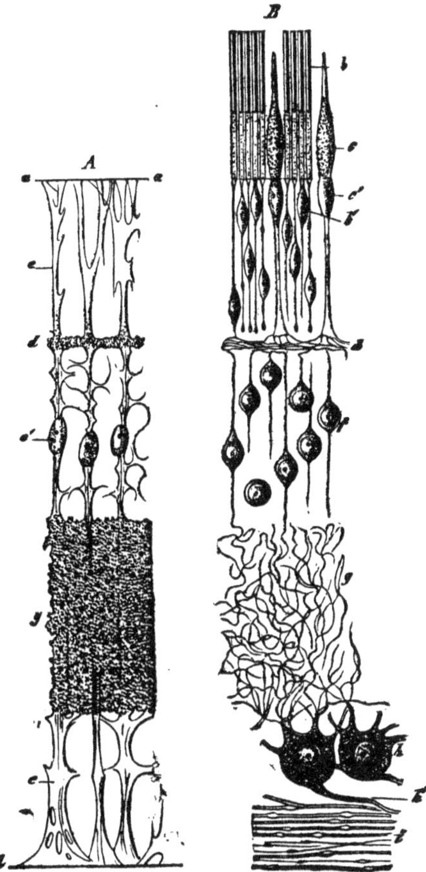

Fig 334. Diagrammatic representation of the retina of man and the vertebrata, after M. Schultze. *A*, connective-tissue framework of the retina. *a*, membrana limitans externa; *e*, radial or Müllerian supporting fibres, with their nuclei *e'*; *l*, limitans interna; *d*, framework substance of the inter-granular, and *g* of the fine granular layer. *B*, nervous elements of the retina. *b*, rods with outer and inner portions, as well as the rod-granule (*b'*); *c*, cone with the rod and granule (*c'*); *d*, expansion and apparent termination of the cone-fibres in the inter granular layer, with the transition into finest fibrillæ; *f*, granules of the inner-granular layer; *g*, confused mass of finest fibrillæ in the fine granular layer; *h*, ganglion-cells; *h'*, their axis-cylinder processes; *i*, layer of nerve-fibres.

fibres (*e*) forms, with its innumerable fine processes, a delicate network which, in two places, namely, in the inter-granular layer (*d*) as well as the fine granular layer (*g*), assumes an extraordinary fineness and compactness, and here becomes changed to a regular spongy tissue, related to that of the gray substance of the brain. Inwards, these supporting fibres, uniting together and spreading out in a peculiar manner, form a hyaline connective-tissue boundary layer, the membrana limitans interna (*l*). Externally, over the so-called external granular layer, a second similar boundary layer, but finer and perforated in a sieve-like manner, may be seen. This is the limitans externa (*a*).

If the finer textural conditions of the nervous elements of the retina, as well as finally the connection of the same, are to be investigated, the methods which have until recently been employed in this difficult domain have already been mentioned in the preceding.

Maceration may be accomplished with the various acids mentioned for the connective-tissue framework, among which the highly diluted solutions of chromic acid have been most employed. Corresponding solutions of the bichromate of potash, as well as Müller's fluid diluted with water, are also to be recommended as suitable. A careful picking naturally follows.

For hardening, to subsequently obtain very fine sections, the stronger solutions of chromic acid and its potash salt, as well as, above all, the Müller's fluid, are employed. A very conservative tingeing with carmine will render many things more distinct, although its value proves to be less here than for many other organs.

We scarcely need to remark that for such infinitely delicate textural conditions the strongest objectives must be used.

The rods and cones usually keep well in weak solutions of chromic acid and chromate of potash; the extremely dilute solutions mentioned by Schultze are unserviceable. The Müller's eye-fluid preserves them well. Schultze found the rods excellently preserved in the concentrated oxalic acid; the above-mentioned sulphuric acid of 0.6 per cent. is also useful for the rods. The recognition of the latter with the external and internal portions is relatively easy in the quite fresh eye, with the addition of humor aqueus and vitreous or iodine-serum, whereby we meet at the same time with a quantity of fragments and in part strangely disfigured specimens. A portion of fresh retina, with the external surface turned upwards and placed under the microscope without a covering glass, forms the best object for the recognition from above of the mosaic of the rods and cones.

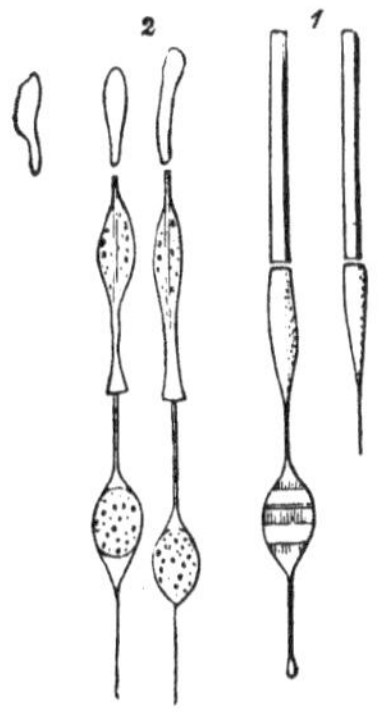

Fig. 335. Structure of the rods (Schultze). 1. Those of the Guinea-pig in a fresh condition, with inner and outer portions to the left, in connection with a transversely striated granule. 2. Those of the Macacus cynomolgus, macerated in iodine-serum, with Ritter's filament.

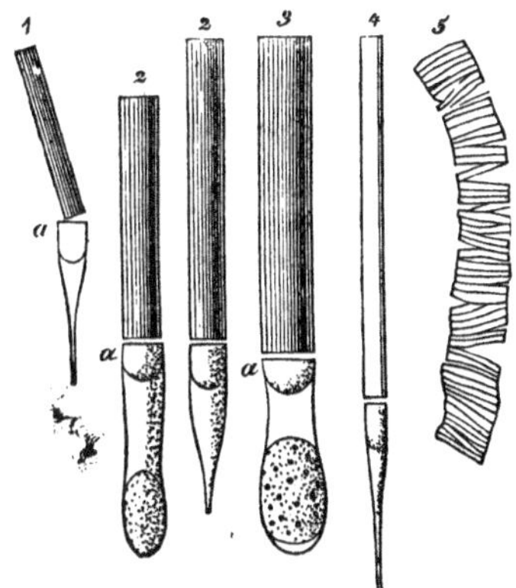

Fig. 336. Structure of the rods (Schultze). 1. from the chicken; 2, from the frog (in a fresh condition); 3, from the salamander (likewise fresh); 4, from the pike (also fresh); 5, the division into plates of a rod from the frog treated with acetic acid; *a*, lenticular body.

The external and internal portions of the rods, the former (fig. 334 *B. b*, fig. 335 and 336) of stronger refractive power, the latter delicately contoured and becoming reddened in the carmine solution (Braun), are to be discovered with tolerable facility; likewise the very fine and perishable filament which arises from the pointed end of the inner member. Schultze succeeded years ago, with his familiar highly diluted chromic acid solutions, in demonstrating varicosities on these, and thereby their nervous nature in contradistinction to the connective-tissue supporting fibres.

The impossibility of following these finest rod-fibres through the entire thickness of the retina was known even at that time, as their course is maintained in a radial direction over limited portions only.

The recognition of the cones (fig. 334 *c*, fig. 327), as well as their rod-shaped terminal portions (cone-rods), also succeeds with the older methods, although the extraordinary changeability of the cone-rod renders the perception of its true structure very difficult.

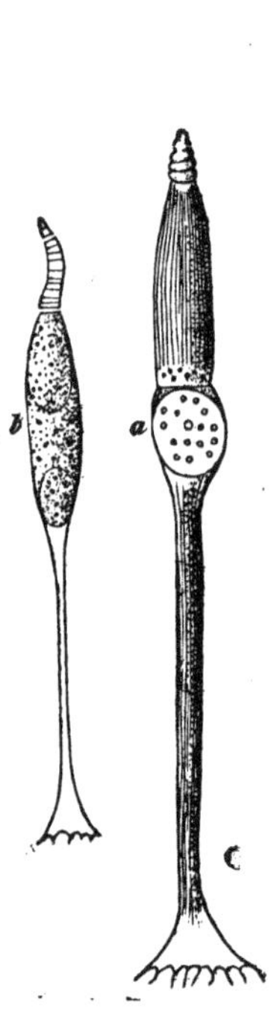

Fig. 337. Cones. *a*, from man, with a decomposed external portion and fibrillary appearing inner portion; *b*, from the Macacus Cynomolgus, after maceration in dilute nitric acid, with the lenticular body (Schultze).

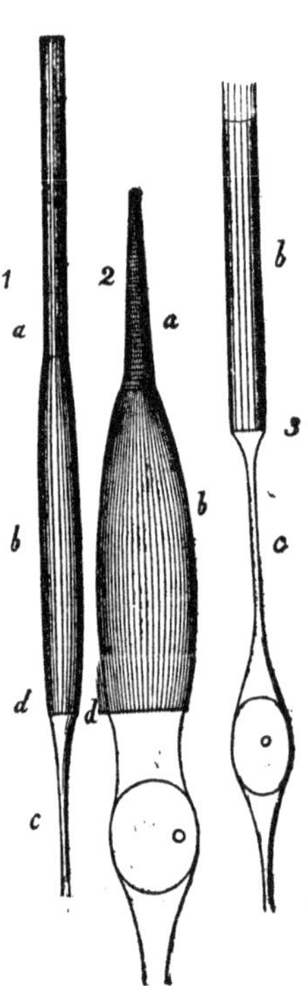

Fig. 338. Fibrillated covering of the rods and cones, after Schultze. 1. Rods, 2, cones of man; *a*, outer, *b*, inner portion; *c*, rod-filament; *d*, limitans externa. 3, Rods of the sheep. The fibrillæ project beyond the inner portion; the outer portion is wanting.

The external granular layer, occurring under the limitans externa, shows with tolerable facility the varicose rod-fibres, as well as the small spindle-shaped and transversely striated cell (rod-granule) embedded in it, with a nucleus and nucleolus.

One likewise perceives, joined to the inner extremity of the cone, the analogous (but not, as in the rod-granule, fig. 335. 1, transversely striated) structure, the cone-granule. Years ago H. Müller correctly recognized differences in these rod and cone-granules. A difference could be recognized between the broader fibres passing from the cones and the finer varicose rod-fibrillæ, and they seemed to terminate in a strange manner with conical, widened, terminal portions at the margin of the inter-granular layer (Müller, Henle), so that, for a time, even Schultze had doubts of their nervous nature. Against this, however, the macula lutea, formed entirely of more slender cones, with their obliquely arranged cone-fibres, formed an important objection.

The inner granular layer (fig. 334 *B f*) likewise shows without difficulty a small cell with a nucleus and nucleolus, analogous to that which appeared to us as the rod-granule in the external granular layer of the retina. From both poles of a number of these so-called granules arise very fine radial filaments, in which, however, no connection with the varicose fibres of the rods can be recognized. Years ago Müller and Schultze succeeded in distinguishing the oval nuclei of the framework of supporting fibres (*A*, *e*) from these granules.

By the aid of the older methods one may recognize with relative facility, especially in large mammalial animals, the layer of the multipolar ganglion-cells (*B h*) and its varying thickness in the various portions of the retina. Neither does the flattened extension of the (as a rule non-medullated) retinal fibres (*i*) in the inner layer of nervous elements present any great difficulties, either as surface views or their vertical sections. An exquisite object is afforded by the retina of the rabbit and the hare, where our nerve-tubes, streaming in by way of exception, as two rows of medullated fibres, may be everywhere readily noticed.

We have here mentioned with the most concise brevity the chief results of earlier investigations. The great differences which the retina presents in the various groups of vertebrate animals also becomes more and more apparent (Müller). The gigantic rods of the frog, the peculiar twin-cones of the osseous fishes, the often delicately colored fat-globules at the base of the cone-rods in birds and squamigerous reptiles, must captivate the interest of the observer. We can at present say that rods and cones are widely diffused among the vertebrate animals, but are by no means everywhere present. Most mammalial animals (ape, ox, horse, dog, etc.) have them similar to those of man. The eye of the bat, the hedgehog, the mole, the mouse, and the Guinea-pig has, however, only rods and no cones. The latter structures are quite scanty and undeveloped in the retina of the rabbit and the cat. Birds have an excess of cones (only in owls these elements recede entirely and colored fat-globules are wanting). Only cones, and no rods, appear in the retinas of lizards and serpents. Rays and sharks, in contradistinction to the osseous fishes, are entirely without cones (Schultze). We cannot here enter further into the important physiological consequences of these remarkable conditions. It suffices for us to have made mention of them for practical purposes, for the selection of materials for investigation.

As was mentioned, a further excellent accessory for the examination of the retina has become known through M. Schultze in osmic acid (p. 162), and this reagent has been employed in superior investigations with the greatest success.*

For its application a 2 or 1 per cent. solution should be kept on hand, so that it may be diluted at pleasure in a measuring glass. One may go down to 0.1 per cent., or even lower. Stronger solutions of 1–0.25 per cent. cause rapid hardening (without inducing coagulations), so that after an immersion of even half an hour portions of the retina may be divided, in the direction of their radial fibres, into lamellæ, and the nervous elements may be recognized, while the connective-tissue supporting apparatus is rendered but slightly prominent. Such preparations may be left for a day in the solution, and, washed out in water (which also serves as a medium for osmium pre-

---

* Chloride of gold and chloride of gold and potassium deserve a more accurate trial for the retina.

parations), it may be preserved for days for further examination likewise in alcohol and acetate of potash.

The blackening, which appears very rapidly, is more regular at the commencement. Later, the nerve-fibres frequently become colored, the fine granular and inter-granular layers more intensely than the other portions. The outer portion of the rod appears, as a rule, darker and sharply contrasted from the inner portion, quite especially and very remarkably so in the frog and in fishes.

Weaker degrees of concentration of the osmic acid of 0.2 per cent. and less, no longer cause hardening alone, but also exert at the same time a macerating effect. The preparation is now less brittle, and permits of picking with needles. It is generally sufficient to allow the action to continue for a half or a whole day. The nerve-fibres may assume varicosities in these watery solutions.

The connective-tissue framework becomes hardened later than the nervous elements.

To thoroughly preserve the rods and cones, take a vitally warm eye, remove the posterior segment of the sclerotic to beyond the equator, and immerse in a solution which contains about two per cent. of the dry acid. The desired effect is obtained in a few hours. The fluid media and preservative fluids have already been mentioned.

Schultze arrived at important results for the external portion of the retina. The rod-fibres arrive, as far as the inter-granular layer (fig. 334 *B*), to here, with slight intumescences, withdraw from observation. The broader cone-fibres are quite similar to an axis-cylinder, permit of the recognition of delicate longitudinal striations (perhaps as an indication of a further composition), and form at the same locality the already mentioned conical expansion (*d*). From the base of the latter arises a new system of extremely fine fibrillæ, which, with numerous divarications, assume another, and, indeed, horizontal course. Varicosities speak for the nervous nature of the latter fibrillæ.

The connection of the nervous elements in the inner layers of the retina still remains quite obscure. Whether the radial fibres of the inner granular layer (*f*), which are united to a granule, are connected with the confused mass of finest fibrillæ which arise from the resolution of the cone-fibres, we are not yet able to say.

The similar mass of fibres, perhaps arising from the radial fibres of the inner granular layer, also passes through the fine granular layer (*g*), to finally pass over into the fine or so-called protoplosma processes of the ganglion-cells (*h*), (comp. p. 348–9, fig. 175). Should this conjecture of Schultze's prove to be true (whereby a parallel with the texture of the gray substance of the central organ results), and should the system of processes of a cone-fibre hereby enter into different ganglion-cells, complication of the nervous channels would indeed result which could not be mastered by our present accessories.

Probably an inwardly directed broad process of the ganglion-cells of the retina corresponds to the so-called axis-cylinder process of the central cell, and simply assumes the form of a primitive fibre of the nervous layer (*h'*).

It would lead us far beyond the narrow limits of this book, and the requirements of our circle of readers, were we to here make more complete mention of the most recent acquisitions in this domain. Thus, a problematical axis-filament has been ascribed to the rods (fig. 335, 2) which is certainly wanting in the outer portion (Schultze). At the inner portion of the rod, where it joins the outer portion, a singular lenticular body of semispherical or plano-parabolic form has been met with (fig. 336 *a*). Something of the kind also appears to occur in the cones (fig. 337 *b*). For the (certainly transitory) preservation of these structures the solution of the bichromate of potash may be tried.

Of interest is furthermore a disk-like structure of the rods (fig. 336, 5), which was incompletely seen many years ago, but which has been recently more accurately recognized and studied. On the fresh rod it is seldom that anything of it can be seen, only examinations with oblique light and a rotary stage show us a trace of the same, if one of the strongest immersion systems can be used. It becomes distinct only when distending reagents are resorted to, as, for example, dilute serum to which a little acetic acid may be added, dilute nitric acid, etc. Osmic acid

(1–2 per cent.) also affords good preparations, and, with careful picking in water, presents transverse sections of these plates. The same foliaceous disintegration also occurs on the outer portions of the cone-rods (fig. 337, 338, 2 *a*).

Finally, M. Schultze (we have thus far quoted from his works) found an infinitely delicate longitudinal fibrillary structure covering the rods and cones externally throughout their entire length (fig. 338). One may think of the primitive fibrillæ of the axis-cylinders and ascribe the latter signification to the rod- and cone-fibres.

The macula lutea and fovea centralis require no new methods. Their structure must be ascertained from the text-books on histology.

The usual hardening treatment with chromic acid and chromate of potash (and alcohol) also serves to show the relation of the blood-vessels to the retinal layers, and their advancement to

Fig. 339. Vessels of the human retina. *a*, arterial, *c*, venous branches; *b*, the capillary network.

near the intergranular layer; while the delicate vascular network in its extension (fig. 339) (for the demonstration of which we recommend the injection of the eye of the ox and the sheep) requires surface views.

For the investigation of the lymphatics of the eyeball, a still more uncertain domain (concerning which Schwalbe has recently made valuable contributions), use the puncturing method; for the posterior half, in the space between the sclerotic and choroid; for the anterior portion, the anterior chamber of the eye. Either a cold-flowing watery Prussian blue or a solution of gelatine serves as an injection mass. The practised hand will resort to the syringe; the constant pressure may possibly afford better results.

With regard to the pathological changes of the retina, we are at present acquainted with hypertrophies of the connective-tissue parts, with a corresponding degeneration of the nervous elements, proliferations of the granular layers, amyloid bodies, fatty degeneration of the nervous (but also of the connective-tissue) parts, embolia of the retinal vessels, likewise pigmentations coming partly from the extravasated blood, partly caused by the proliferated choroidal epithelium which has penetrated the retina, and which latter hereby frequently lies near the retinal blood-vessels, etc. Tumors of the retina are, as a rule, either sarcomata or glioma (p. 358), and only very rarely carcinomata.

Preparations from more hardened retinas (after the use of Müller's fluid) may be readily preserved in glycerine in the form of review specimens; for many views we would recommend tingeing them with carmine. The finest textural relations of the several layers and their elements could not, on the contrary, from the condition of the microscopical technique, be preserved for a long time. Attempts to mount them in their macerating fluids rapidly came to an end, as a rule, in the destruction of the preparation. M. Schultze has recently gladdened us with the important information that osmium preparations may be preserved for years in a solution of the acetate of potash (p. 212).

Fœtal eyes may be studied on very small embryos immersed fresh in chromic acid. With older fœtuses the eye is to be taken out and further treated in accordance with the directions given for the adult. The eyes of new-born kittens are to be recommended for injecting the magnificent vascular network of the membrana capsulo-pupillaris.

5. With regard to the organs of hearing, the external parts of the same, such as the auricle and the external auditory canal, require no special directions.

The ceruminous glands, with their coil-shaped bodies and short excretory ducts, are to be examined in the same manner as the related sudoriparous glands.

The membrana tympani is to be studied either in the fresh condition, with the aid of the knife and needles, and with the use of acetic acid as well as the alkaline solutions, or the previously dried organ is employed for fine sections, whereby we would also urgently recommend carmine tingeing.

The walls of the cavity of the tympanum and Eustachian tube, with the covering of ciliated cells, the ossicula auditus, with their porous bone substance and their transversely striated muscles, are to be examined by the methods customary for the tissues in question.

The investigation of the labyrinth is far more difficult. Even the opening by means of saw and chisel must be accomplished with caution; and from the delicacy of many structural conditions, only quite fresh, just previously killed animals are to be used. For the primary examination the labyrinths of larger mammalial animals, such as the calf and ox, are to be selected. If a certain practice in such procedures has once been acquired, the exposure also succeeds later with smaller creatures—the dog, the cat, and the rabbit. The great alterability of the elements also renders it necessary, as with the retina, to use fluid media which are as indifferent as possible; among these he would recommend blood- and iodine-serum, vitreous fluid, and dilute albumen. Dilute solutions of chromic acid may also be suitably applied to the fresh tissue.

The hardening and, in general, the immersion in solutions of chromic acid, bichromate of potash, and Müller's fluid also appears to be of the greatest importance here. The latter fluid, diluted with an equal volume of water, may be very highly recommended.

The aggregations of ear-stones or otoliths (as the polarizing microscope teaches, columns crystallized in the arragonite form) may be perceived in the sacculi of the vestibule as white spots, surrounded by an especial thin membrane. They are generally small and, according to many statements, possess an organic basis. Their appearance may be represented by fig. 340.

Fig. 340. Otoliths.

With regard to the distribution of the acoustic nerve on both sacculi of the vestibule, and on the membranous ampullæ of the semicircular canals, the coarser arrangement is not difficult to recognize. The nerve-fibres enter duplicatures of the walls, which are, especially in the ampullæ, distinctly recognized as prominences projecting into their cavity. This projection, called the septum nerveum, contains the termination. An earlier epoch, without having any presentiment of the difficulties of such investigations, would here convince itself of the presence of terminal loops.

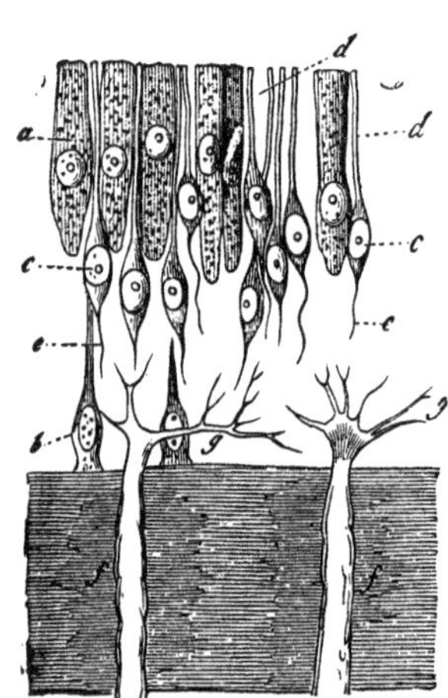

Fig. 341. From the crista acustica of the ampullæ of Raja clavata. *a*, cylinder cells; *b*, basal cells; *c*, fibre cells, with the upper rod-shaped processes *d*, and lower fine fibrillary ones *e*; *f*, nerve-fibres, becoming at *g* pale ramifying axis-cylinders.

That conditions also occur here which are quite similar to those we have mentioned in connection with other organs of sense; that there are auditory cells as well as olfactory and gustatory cells, was first demonstrated by Reich and M. Schultze; the former by the investigation of the lamprey, the latter by that of the ray and shark. Our fig. 341 may bring such conditions of the plagiostomy before the reader, and show the characteristic cells, *c*, with their rod-shaped projection *d*, and their lower fine terminal filament *e*. The latter is undoubtedly the terminal nerve-fibrilla, although even here a continuous connection could be recognized by M. Schultze with as little certainty as in the olfactory organ. Strange long hairs also occur on the peculiar epithelium of this locality.

Here also chromic acid, with those dilute solutions which we have had to mention so frequently for the higher organs of sense, plays at the present time the rôle of the most important accessory. Osmic acid and chloride of gold will be tried by some subsequent observer.

According to the few observations which have at present been published, similar textural conditions appear to occur in the vestibulum of the mammalial animal (ox). Although all of our knowledge of these subtile arrangements is still in embryo, a penetration of the nerve fibrillæ into a peculiar epithelium and the existence of specific cells is nevertheless quite certain.

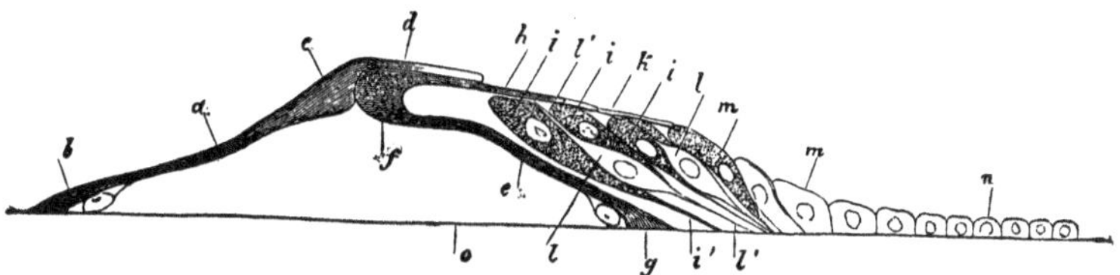

Fig. 342. Side view of the Cortian organ. *a*, inner fibre ; *b*, commencement of the same ; *c*, joint portion ; *d*, transparent laminal appendix (commencement of the lamina velamentosa) ; *e*, outer fibre ; *f*, its joint portion ; *g*, end at the membrana basilaris *o* ; *h*, rod-like process of the outer fibre ; *k*, outer portion of the lamina velamentosa ; *i*, Cortian cells (*i'*, their terminal filaments to the membrana basilaris) ; *l*, Deiters' cells (ciliated cells *D*) ; *l'*, their processes ; *m*, larger epithelial cells bounding the outer portion of the Cortian organ ; *n*, small-celled epithelium of the zona pectinata.

Infinitely more difficult, and leading into a true chaos of the most intricate structural conditions, is the termination of the nervus cochlearis in the Reissnerian cochlear canal, or the so-called scala media. Since Corti undertook the first successful excursion into this domain full of wonders, our knowledge has been enriched by each of the subsequent investigations by the discovery of new fragments. Within a period not long elapsed, Deiters especially has rendered the greatest service in connection with the structure of the cochlea ; and Kölliker, by the discovery and closer investigation of the almost forgotten Reissnerian cochlear canal, has established a new acceptation of the scala media. Among the successors, Hensen and Waldeyer deserve prominent mention.

It would lead us far beyond the limits of this work were we to mention here the structural conditions which have thus far been investigated, especially the structure of the so-called Cortian organs (fig. 342). Notwithstanding all previous efforts, the knowledge of the nerve terminations is not as far advanced as in the other organs of sense. Probably, however, the terminal nervous structures of the nervus cochlearis are present in certain cells which have recently been more clearly investigated by Deiters and Kölliker. For more accurate information, we would refer to the monographic statements of Deiters, Hensen, Kölliker, and Waldeyer.

The exposure of the parts in question may be accomplished and learned on the quite fresh auditory apparatus of one of our larger cattle (the ox). After some practice this may also be gradually achieved with smaller creatures. Fragments which are obtained in this manner from the scala media are to be examined by means of indifferent media or strongly diluted chromic acid. The latter, or chromate of potash, also serves for the immersion and hardening of the opened cochlea. For many cellular conditions of the Cortian organs, high degrees of dilution, similar to those introduced into histology by Schultze, are to be tried; and certainly also osmic acid, chloride of gold, and chloride of gold and potassium. To obtain transverse sections of the entire Reissnerian cochlear canal, the organ, previously hardened by means of chromic acid or Müller's fluid, is to be cautiously decalcified by adding a few drops of muriatic acid to these fluids and frequently changing the entire mixture. The lamina spiralis of the larger animals may be isolated and exposed to such a procedure. The nerve distribution in the zona ossea may likewise be brought to view in this manner. Several years ago Hensen, to obtain the membrane of Corti in its position, made use of a three months' immersion in Müller's fluid, and injected tolerably concentrated gelatine through a puncture in the tympanum secundarium, which then transuded into the cochlear canal. The external wall of the cochlea was afterwards broken open, and the scala media with the gelatine cast isolated. Hensen also found carmine imbibition appropriate here.

Waldeyer, the latest investigator in this difficult domain, already recommends, besides the review of fresh objects in humor aqueus, the osmic acid, to which he ascribes the same importance as for the retina of the eye. Degrees of concentration of 0.1–1 per cent. are to be used; the former for picking, the latter for hardening. A solution of chloride of sodium of 0.25–0.5 per cent. rendered him excellent service for the former preparations. Chromic acid of 0.05 per cent. chloride of gold, according to Cohnheim's directions, and a 1 per cent. solution of nitrate of silver, are likewise useful for many purposes.

To obtain good section preparations, remove as much bone as possible from large cochleæ, and make two or three small openings into the capsule. Smaller organs, on the contrary, are to be left intact. The cochleæ are then to be placed for a day in a liberal quantity of a solution of chloride of palladium (0.001 per cent.), or in one of osmic acid of 0.2 per cent. if the organs are small; while more voluminous ones require an increased concentration of 0.5–1 per cent. Such objects are then to be exposed to the action of absolute alcohol for 24 hours, and at once placed in the decalcifying fluid.

Waldeyer found chloride of palladium (0.001 per cent.) with $\frac{1}{10}$ part muriatic acid, or chromic acid of 0.25–1 per cent. most serviceable for the latter purpose. After the decalcification, the preparations are to be washed out with absolute alcohol; it may then (for the subsequent preparation of sections) be embedded in a piece of fresh spinal cord or fresh liver, and the whole then replaced in the absolute alcohol. During the hardening the enveloping piece of organ shrinks so firmly around the cochlea that the latter remains immovable and permits the desired sections to be made.

Before this embedding the cavity of the cochlea may be filled with gelatine-glycerine (1 : 1), or (not so good) with a mixture of wax and oil. This filling, however, is not absolutely necessary.

The primary views of the cochlear canal may be obtained

without great difficulty from embryos, treated in a similar manner, by means of suitable transverse sections through the petrous bone.

For cabinet preparations the same remarks apply which we have made (p. 608) concerning the retina. Preparations of the Cortian organ, however, I have preserved for years entirely unaltered in watery glycerine. Hensen recommends the watery solution of arsenious acid for mounting. Acetate of potash should also be tried.

INDEX.

B.

C.

H.

I.

J.

K.

L.

V.

W.

X.

Z.

PRICE-LISTS OF MICROSCOPE FIRMS.

No. 1.—*Price-List of the Achromatic Microscopes of* E. HARTNACK & Co., *Successors to* G. OBERHAUSER & Co., *Paris* (21 *Place Dauphine*) *and Potsdam* (1870).

Price in Francs.

Remarks.—All Microscopes are contained in a mahogany case, furnished with a lock and key.
2. The models, Nos. 1, 2, 3 and 3A are furnished with lens systems of older construction. The remainder have new lens systems, with large angles of aperture.
3. From its arrangement, the polarizing apparatus may be advantageously employed on the models from 4 to 8.
4. From the following tables other systems and eye-pieces, which are required in the place of the customary ones, as, indeed, any desired outfit, may be readily reckoned.

A. Prices of the Individual Lens Systems, and of the other Apparatuses.

LENS SYSTEMS OF OLDER CONSTRUCTIONS. MAGNIFYING POWER WITH THE EYE-PIECES.

System.	Eye-piece No. 1.	No. 2.	No. 3.	No. 4.	No. 5.	No. 6.	Price.
No. 1	12	15	25	—	—	—	12
2	20	30	40	—	—	—	20
3	30	40	50	—	—	—	20
4	40	50	65	100	—	—	20
5	75	100	150	200	—	—	30
6	110	150	220	300	—	—	35
7	150	220	300	450	—	—	35
8	250	300	400	600	800	—	40
9	360	430	550	850	1000	—	60

NEW LENS SYSTEMS WITH LARGE ANGLES OF APERTURE.

System.	Eye-piece. No. 1.	No. 2.	No. 3.	No. 4.	No. 5.	No. 6.	Price.	Equivalent focus in inches.
No. 1	15	20	25	—	—	—	15	2
2	25	32	45	—	—	—	20	1
3	50	60	80	120	—	—	25	¾
4	60	70	90	140	—	—	30	½
5	100	125	160	240	—	—	35	¼
6	150	180	240	350	—	—	35	¼
7	200	240	300	450	600	750	40	1-6
8	250	300	400	600	800	1000	50	1-9
9	350	440	550	860	1100	1400	75	1-11

NEW SYSTEMS WITH IMMERSION AND CORRECTION.

System.	Eye. piece. No. 1.	No. 2.	No. 3.	No. 4.	No. 5.	No. 6.	Price.	Equivalent focus in inches.
No. 9	410	480	630	950	1300	1500	150	1-12
10	520	600	750	1100	1500	1800	200	1-16
11	600	690	850	1250	1750	2500	250	1-18
12	710	820	1010	1490	2060	2800	300	1-21
13	820	950	1170	1730	2370	3100	350	1-25
14	930	1080	1340	2000	2680	3350	400	1-28
15	1040	1200	1500	2200	3000	3600	450	1-33
16	1200	1400	1750	2570	3500	4200	500	1-40
17	1400	1600	2000	2940	4000	4800	500	1-45
18	1560	1800	2250	3300	4500	5400	500	1-50

Plain eye-piece....10 fr
Holosteric eye-piece....15 fr.
Eye-piece with adjusting screw....25 fr.
Micrometer eye-piece....25 fr.
Binocular stereoscopic eye-piece....180 fr.
Movable stage....40 fr.
Simple compressorium....20 fr.
New compressorium....25 fr.
Stage micrometer mounted in brass, the millimetre divided into 100, 500 and 1000 parts....20 fr.
New movable micrometer (gives with great accuracy 0.0001 millimetre)....50 fr.
Polarizing apparatus....50 fr.
Improved patent polarizing apparatus, with a polarizing eye-piece, a prism with a large field of vision, and a graduated circle....60 fr.
Goniometer....60 fr.
Universal goniometer....150 fr.
Dujardin's illuminating apparatus (improved construction)....50 fr.
Camera lucida of Oberhäuser; also serving for the conversion of the vertical into the horizontal microscope....50 fr.
Camera lucida of Milne-Edwards and Doyère....35 fr.
Brücke's loup (improved construction)....20 fr.
Stand for this loup....30 fr.
Loups....5-15 fr.
Lamp for microscopic examinations, with a large illuminating lens for obtaining parallel rays of light....35 fr.
Object slides, first quality, per dozen....2 fr.
Object slides, second quality, per dozen....1 fr.
Fine covering glasses, per dozen, 1 fr.; per hundred....6 fr.

B. Prices of the several Microscopes.

No. I.—Small microscope (d'hospice), with a lens system No. 7, and an eye-piece No. 3; magnifying power, 300; with a dozen slides and thin covers, brass forceps, scalpel and preparing needles....65 fr.

No. II. Achromatic microscope (with a wider stage), furnished with the lens systems 4 and 7, eye-pièces 2 and 3, and an illuminating lens for opaque bodies. Magnifying power from 50, 65, 220 to 300 diameters, micrometer-screw on the stage, rotary diaphragm....115 fr.

The same instrument, with the addition of lens system 8, and eye-piece 4, magnifying power from 50–600....165 fr.

No. II.A.—Achromatic microscope, with firm stage; micrometer screw on the stem; mirror freely movable for oblique illumination; optical apparatus as in II....120 fr.

To obtain a magnifying power up to 600....170 fr.

No. III.—Achromatic microscope, small drum stand, with a firm, large stage, rotary diaphragm, and the same optical apparatus....140 fr.

To obtain a magnifying power up to 600....190 fr.

No. III. Achromatic microscope; stand, in its upper part, similar to the previous one; but having a horse-shoe foot; freely movable mirror for oblique illumination; optical apparatus the same....140 fr.

To obtain a magnifying power up to 600....190 fr.

No. III.A.—Achromatic microscope, similar to the previous one; the stem is, however, provided with a joint for obtaining an oblique position; optical apparatus the same....155 fr.

To obtain a magnifying power up to 600....205 fr.

No. IV.—Achromatic microscope with rotary stage, and a black glass plate on the same; cylindrical diaphragms; lens systems 4 and 7, new construction, and eye-pieces 2 and 3; magnifying power, 70, 90, 220, 300....300 fr.

To obtain a magnifying power of 650....360 fr.

No. V.—Large achromatic microscope; foot and stage the same, only larger and more firm; optical apparatus the same as with No. IV....340 fr.

To obtain a magnifying power of 650....400 fr.

No. VI. Achromatic dissecting microscope, with long focal distance and inversion of the image; magnifying power (without changing the lenses or eye-pieces) from 10–100. Rotary stage with glass plate....250 fr.

No. VII. New, large patent achromatic microscope, the mechanical and optical construction of which differ essentially from the older large (Oberhäuser's) stand. It consists of 5 lens, systems 2, 4, 5, 7, and the immersion system 9, and 5 eye-pieces (one of which has a micrometer); magnifying power from 25–1,300 (each succeeding enlargement twice as great as the previous one). Coarse movement by an adjusting screw, the fine one by a micrometer screw. Large illuminating lens for opaque objects; all the necessary accessory apparatus....750 fr.

The same stand with a joint for inclining....800 fr.

No. VII.A. Achromatic microscope of the same construction, but somewhat smaller and without the adjusting screw for the coarse movement; optical arrangement the same..650 fr.

The same stand with a joint for inclining........680 fr.

No. VIII. New small stand, the arrangement of which, with the exception of the rotation of the stage, presents the same advantages as the microscope No. VII.; with the lens, systems 4 7, 8, and eye-pieces 2, 3, 4; magnifying from 50–650 times........275 fr.

The same with the lens systems 4 and 7, as well as the immersion system 9, and 3 eye-pieces, one of which is provided with a micrometer; magnifying from 50–1,000 times........390 fr.

With a joint for inclining........405 fr.

No. 2.—Nachet & Son, 17 *Rue Saint Séverin, Paris* (1872).

Price in Francs.

1. Microscope, large model, improved, complete and binocular, suspended on the axis in such a manner that it may be inclined and remain fixed in any position between the horizontal and vertical. Coarse adjustment by rack work; two movements for fine adjustment, one acting on the column supporting the body, the other very delicate and belonging especially to the tube carrying the objectives, and so disposed as to establish a constant elasticity for the protection of the objectives in case of contact with the preparation. The stage is rotable and is furnished with a double plate and so arranged that the objects may be moved without touching them. The stage is furnished with a glass plate to resist the destructive effects of reagents. Illumination by a double mirror movable in all directions. A sliding arrangement, placed between the mirror and stage, permits of the removal of the diaphragms and of focussing the condensers with the greatest precision. Micrometric apparatus for introducing the micrometer into the eye-pieces without deranging them, and for accurately adjusting it to the focus of the eye, and for placing it in all parts of the field of vision. Eight objectives with correcting apparatus, from No. 0 to No. 7. Magnifying from 30–1,400 diameters; four eye-pieces, binocular apparatus, goniometer, camera lucida, polarizing apparatus with selenite plates, condenser, eye-piece and stage micrometers. Condensing lens on a stand. Accessories for preparing. Strong mahogany case with brass corners, lined with velvet; objectives in a separate case........1,400 fr.
2. Microscope, large model, mounted like No. 1. Rotary stage with a black glass plate for use with acids. Rock-work for coarse adjustment, slider for diaphragms and condenser, double mirror, 3 eye-pieces, 6 objectives, Nos. 0, 1, 2, 3, 5, 7, for immersion and correction, magnifying from 30–1,400 diameters. Camera lucida, eye-piece and stage micrometers, condensing lens, accessories for dissections, etc. Mahogany case, brass corners, etc.....680 fr.
3. Vertical, large model, rotary stage, coarse and fine adjustment, arrangement for introducing the diaphragms and condenser without deranging the object; eye-piece and stage micrometer; 5 objectives, Nos. 1, 2, 3, 5, 7, immersion and correction, magnifying from 30–1,400 diameters. 3 eye-pieces, camera lucida, condensing lens, accessories for dissecting, etc. Mahogany case, etc........550 fr.
4. "*Microscope à disposition particulière*," to reduce the height as much as possible, model of Prof. H. de Lacaze-Duthiers—same objectives as in the model No. 3........650 fr.
5. Microscope, medium model for inclining. Coarse and fine adjustment, rotary stage and glass plate, double mirror, apparatus for adjusting diaphragms beneath the object, etc., 5 objectives, Nos. 1, 2, 3, 5, 7, immersion and correction, magnifying, with 3 eye-pieces, from 30–1,400 diameters, condensing lens, accessories, eye-piece micrometer, etc. Mahogany case........500 fr.
6. Vertical, medium model, similar to No. 3; rotary stage, black glass plate, 5 objectives, Nos. 1, 2, 3, 5, 7, immersion and correction, 3 eye-pieces, eye-piece micrometer, condensing lens, accessories. Mahogany case........450 fr.
7. New model for inclining. Immovable stage, block glass plate, coarse and fine adjustment, diaphragm arrangement as in medium models; condensing lens, 3 eye-pieces, 4 objectives, Nos. 1, 3, 5, 7, immersion and correction, magnifying from 25–1,400 diameters, accessories. Mahogany case........430 fr.
8. Same microscope with 3 objectives, Nos. 1, 3, 5; magnifying 30–700 diameters........280 fr.
9. Small model for inclining; mirror freely movable, movable diaphragm, coarse and fine adjustment, draw tube, 2 objectives, Nos. 1, 3; 2 eye-pieces; magnifying from 30–500 diameters; condensing lens, accessories. Mahogany case........150 fr.

The same with 3 objectives, Nos. 1, 3, 5, and 3 eye-pieces; magnifying from 30–700 diameters........200 fr.

10. Small model, vertical. Mirror freely movable; 2 objectives, Nos. 1, 3; 2 eye-pieces, condensing lens, accessories. Mahogany case........125 fr.

With 3 objectives, Nos. 1, 3, 5, and 3 eye-pieces; magnifying from 30–700 diameters........175 fr.

11. More simple microscope. Cast iron base, 1 eye-piece, 1 objective, No. 3; maximum 380 diameters, accessories. Mahogany case........80 fr.

The same with objective replaced by No. 5, magnifying 500 diameters........90 fr.

12. Large model, binocular microscope. Constructed in such a manner as that the prismatic apparatus gives at pleasure stereoscopic or pseudoscopic images. The eye-pieces may be approximated or separated according to the distance between the eyes of the observer;

coarse and fine adjustment, may be inclined horizontally, 3 objectives, Nos. 0, 1, 3; movable stage, condensing lens. Mahogany case .... 500 fr.

13. Small model, binocular, for inclining; coarse and fine adjustment, 3 objectives, Nos. 0, 1, 3; 2 eye-pieces, condensing lens; case .... 250 fr.

14. Binocular apparatus, applicable to all microscopes, with arrangement for adjusting to distance between the eyes; 2 eye-pieces, no objectives .... 150 fr.

15. Binocular, stereoscopic and pseudoscopić apparatus, applicable to all microscopes .... 175 fr.

16. Microscope for two persons to observe, simultaneously, the same object. Objectives, Nos. 0, 1, 3; condensing lens, accessories, etc.; case .... 300 fr.

17. Double body to apply to ordinary instruments; 2 eye-pieces, no objectives .... 80 fr.

18. Microscope for three persons to observe, simultaneously; coarse and fine adjustment; each observer may adjust the focus separately; 3 objectives, Nos. 0, 1, 3; case .... 400 fr.

19. New large model, inverted, with a silvered mirror at the crossing of the rays. In this microscope the distance between the objective and the eye-piece may be increased to 90 centimetres or 1 metre without inconvenience. This combination requires a very delicate construction of the mounting. The strongest objectives may be used in this form of instrument, the loss of light produced by the silvered mirror (Foucault's method) being insignificant. Achromatic condenser, mirrors, 2 eye-pieces, no objectives .... 800 fr.

20. Inverted microscope for chemical studies. The objectives being placed under the object, there is no liability of clear vision being impaired by the accumulation of vapors. The stage is gilded; 4 objectives, Nos. 0, 1, 3, 5; 1 movable eye-piece, accessories, alcohol lamp on an articulated support, excavated slips of glass, thin covers. Mahogany case .... 350 fr.

21. New inverted microscope for the study of anatomical elements in gases at a steady temperature, with numerous accessories; 3 objectives, Nos. 1, 3, 5; 2 eye-pieces; case .... 500 fr.

22. Pocket microscope. The instrument is 90 millimetres long by 50 millimetres broad, can be used with high powers; objectives, Nos. 1, 3, 5, are usually added; one eye-piece, slides and covers, all in a compact leather case .... 200 fr.

23. New portable microscope, larger than the preceding one, enclosed in a case 14 centimetres long by 8 broad; movable mirror; all objectives may be used; with 3 objectives, Nos. 1, 3, 5, and 1 eye-piece .... 180 fr.

24. Dissecting microscope, model of Dr. Cosson. This instrument has on one side an arm for carrying the doublets for dissections, and on the other a column with a horizontal support for the body of the microscope. It may be used either as a simple or compound microscope at pleasure. Coarse adjustment for the doublets, fine for the compound microscope. 2 objectives, Nos. 1 and 3; eye-piece; 3 doublets of varying powers; condensing lens; case .... 140 fr.

25. Stage of this microscope alone, as a simple microscope with 3 doublets, and base, with articulations for supporting the doublets; accessories; case .... 50 fr.

26. Dissecting microscope for laboratories, model of Prof. Ch. Robin. May be used with glass dishes or plates of cork on which opaque objects are fixed. It gives erect images and magnifies from 8–70 diameters .... 120 fr.

A stage with a mirror may be added for dissecting very small objects .... 20 fr.

27. Hand microscope for demonstrations. The instrument may be placed on a stand till the object is permanently adjusted, and then passed among the audience. All kinds of illumination may be used; coarse and fine movement; without objective .... 80 fr.

28. Microscope on a support for aquaria, without objective .... 120 fr.

29. Photographic microscope, with accessories and a series of objectives .... 300 fr.

30. Dark chamber, etc .... 80 fr.

Moitessier's photographic frames .... 45 fr.

OBJECTIVES.

MOUNTED IMMOVABLE.

No. 0	15 fr.	No. 4	35 fr.
1	20 fr.	5	40 fr.
2	25 fr.	6	50 fr.
3	30 fr.	7	80 fr.

OBJECTIVES WITH CORRECTION.

No. 3	50 fr.	No. 6	100 fr.
4	60 fr.	7	125 fr.
5	75 fr.		

IMMERSION OBJECTIVES.

MOUNTED IMMOVABLE.

No. 6	70 fr.	No. 7	100 fr.

IMMERSION OBJECTIVES WITH CORRECTING APPARATUS.

No. 6 .... 120 fr.
7 .... 150 fr.
8 .... 200 fr.
9 .... 250 fr.
No. 10 .... 300 fr.
11 .... 350 fr.
12 .... 400 fr.

LINEAR MAGNIFYING POWER OBTAINED BY THE COMBINATION OF THE OBJECTIVES WITH THE EYE-PIECES.

ORDINARY OBJECTIVES.

	0	1	2	3	4	5
Eye-pieces 1	30	80	180	260	300	350
Eye-pieces 2	40	100	260	380	420	480
Eye-pieces 3	60	140	350	500	590	680
Corresponding focus in inches	2	1	1-2	1-4	1-5	1-8
Angle of aperture—degrees	10	15	40	90	90	130

IMMERSION AND CORRECTION OBJECTIVES.

	6	7	8	9	10	11	12
Eye-pieces 1	460	580	775	900	1150	1320	1700
Eye-pieces 2	600	900	1100	1300	1560	1800	2400
Eye-pieces 3	900	1400	1600	2000	2200	2680	3260
Eye-pieces 4	1200	1750	2000	2500	2750	3150	4500
Corresponding focus in inches	1-10	1-14	1-15	1-20	1-30	1-40	1-50
Angle of aperture—degrees	140	160	175	175	175	175	175

31. Simple microscope for dissections, with doublets, rack-work for coarse adjustment, two wings at the sides of the stage for the support of the hands in fine dissections, with two doublets and case .... 60 fr.
32. Binocular microscope for dissections, magnifying 10–150 diameters .... 150 fr.
33. Loupestand, movement for adjusting the focus, new system of camera lucida, and 2 doublets .... 80 fr.
34. Porte-loup, large model of Lacaze-Duthiers, with illuminating lens and multiple articulations for supporting the doublets under the rays of light; 2 doublets; mounted on a thick board which can contain all the apparatus, thus rendering it very portable .... 80 fr.
35. Articulated stand with rack-work, without loupe .... 15 fr.
36. The same without the rack-work .... 8 fr.
37. Brücke's loupe .... 15 fr.
38. Doublets for dissections, focal distance, 20–5 mm. each .... 6 fr.
39. do. 5–2 mm .... 10 fr.
40. Stage micrometer, mounted in copper, the millimetre in 100 lbs .... 10 fr.
41. do. millimetre in 500 lbs .... 20 fr.
42. do. " " 1000 lbs .... 30 fr.
43. Camera lucida .... 25 fr.
44. do. ordinary, old form .... 18 fr.
45. Erecting prism, may be applied to all instruments .... 25 fr.
46. do. perfected, combined with an eye-piece to give a larger field .... 35 fr.
47. Revolving object-bearer .... 25 fr.
48. Condenser, direct .... 25 fr.
49. do. oblique .... 15 fr.
50. Amici's illuminating prism, mounted on a separate base, with articulations .... 25 fr.
51. Black ground illumination .... 15 fr.
52. Polarizing apparatus, perfected, 2 Nicols', placed one beneath the object, the other over the eye-piece .... 40 fr.
53. Goniometer .... 30 fr.
54. Section-cutter .... 60 fr.
55. do. simplified .... 35 fr.

56. Microtome for holding the objects to make sections by hand, with glass plate, model of Dr. Hayem ....18 fr.
57. Compressor ....30 fr.
58. Eye-pieces, each ....10 fr.
 do. " very strong, achromatic ....20 fr.
59. Eye-piece micrometer ....15 fr.

No. 3.—*Price-List of Microscopes of* G. S. MERZ, *formerly* UTZSCHNEIDER & FRAUNHOFER, *in Munich* (1869).

Price in Guldens and Thalers.

A. Complete Microscopes.

Microscope No. 1, with stand No. 1, vertical, firm stand, horizontal rotary stage (English form), coarse and fine movement on the tube, double mirror and loup for opaque objects. The instrument is provided with 6 objectives, $\frac{1}{3}$, $\frac{1}{9}$, $\frac{1}{12}$, $\frac{1}{15}$, $\frac{1}{18}$, $\frac{1}{24}$″, and 6 eye-pieces, 1, 1½, 2, 2½, 3, 4; affords a magnifying power of 60–1,920 diameters. It has a screw micrometer with which even 0.0001 of a Paris line may be measured, a polarizing apparatus, a drawing prism and a compressorium. The whole in an elegant case..Price, 420 fl.=240 thlr.

Microscope No. 2, with stand No. 1; 3 objectives, $\frac{1}{3}$, $\frac{1}{12}$, $\frac{1}{18}$″, and 4 eye-pieces, 60–1,440 diameters; glass micrometer included ....Price, 175 fl.=100 thlr.

Microscope No. 3, with stand No. 2, vertical and horizontal, firm stage, coarse and fine adjustment on tube, double mirror, 2 objectives, $\frac{1}{3}$, $\frac{1}{12}$″, and 4 eye-pieces, 60–960 diameters ....Price, 87½ fl.=50 thlr.

Microscope No. 4, with stand No. 2, simple model, vertical and horizontal, firm stage, coarse and fine adjustment on tube, 2 objectives, $\frac{1}{3}$, $\frac{1}{12}$″ and 3 eye-pieces, 60–480 diameters ....Price, 70 fl.=40 thlr.

Microscope No. 5, with stand No. 3, coarse adjustment on tube, firm on stage, one objective, $\frac{1}{9}$″, and 2 eye-pieces, 180–360 diameters ....Price, 49 fl.=28 thlr.

Microscope No. 6, with stand No. 3, objective, $\frac{1}{6}$″; reduced aperture, 2 eye-pieces, 120–240 diameters ....Price, 31½ fl.=18 thlr.

Microscope No. 7 (dissecting microscope), stage with wings, rack adjustment. The instrument has 3 achromatic lenses, forming together a single system of ⅓″, focal length and a terrestrial eye-piece, magnifying power 8, 16, 24, 40, 200 diameters ....Price, 56 fl.=32 thlr.

Microscope, No. 7 (a simple dissecting microscope), same mechanical accessories, achromatic lenses, 8, 16, 24 diameters ....Price, 24½ fl.=14 thlr.

Microscope, No. 8 (Donders' model), stand like No. 2, rack for coarse adjustment, micrometer screw for fine, double mirror. The instrument has 2 objectives, $\frac{1}{3}$, $\frac{1}{12}$″, and 4 eye-pieces, and serves at same time as simple dissecting microscope, magnifies 8–720 diameters .... Price, 91fl.=52 thlr.

B. Microscopic Apparatus.

Stand No. 1, with case ....Price, 98 fl.=56 thlr.
" 2, " ....Price, 42 fl.=24 thlr.
" 3, " ....Price, 21 fl.=12 thlr.

Objectives.			Price.
1″ ½″ ⅓″		angle of aperture, 20–40°	14 fl.=8 thlr
1-6″		" 100°	21 fl.=12 thlr.
1-9″ 1-12″		" 120°	28 fl.=16 thlr.
1-15″	ordinary and immersion systems	" 140–150°	42 fl.=24 thlr.
1-18″	ordinary and immersion systems	" 140–150°	56 fl.=32 thlr.
1-21″	immersion	" 160–170°	70 fl.=40 thlr.
1-24″	immersion	" 160–170°	98 fl.=56 thlr.

Correcting apparatus increases price for each ....7 fl.=4 thlr.
Eye-pieces ....5¼ fl.=3 thlr.
Eye-piece with micrometer ....14 fl.=8 thlr.
Stage micrometer, millimetre in 100 parts ....10½ fl.=6 thlr.
Screw micrometer ....63 fl.=36 thlr.
Goniometer ....35 fl.=20 thlr.
Polarizing apparatus ....28 fl.=16 thlr.
Drawing prism, simple ....7 fl.=4 thlr.
Camera lucida, with double reflection ....28 fl.=16 thlr.
Compressorium ....17½ fl.=10 thlr.
Loups, doublets magnifying 5, 12, 17, 24, 32 times ....5¼ fl.=3 thlr.
Loupstand ....14 fl.=8 thlr.

No. 4.—*Price-List of the Microscopes and Accessories of* CARL ZEISS, *in Jena* (1869).

Price in Thalers.

1. Large compound microscope (stand 0), horse-shoe foot, rotary stage, concave and plane mirror, freely movable, slider for changing the cylindrical diaphragm without disturbing the object, micrometer screw for fine adjustment, objectives A, B, C, D, E, F; 4 eye pieces, magnifying 20–1,500 .....127 thlr.
2. Larger compound microscope (stand I*b*.), horse-shoe foot, rotary stage, arched rotary diaphragm, placed very near the object, concave and plane mirror freely movable, micrometer screw for fine adjustment, objectives, A, B, C, D, E, F; 4 eye-pieces; magnifying 20–1,500 diameters .....114 thlr.
3. The same instrument, objectives A, C, D, F; 4 eye-pieces, magnifying 20–1,500 diameters .....89 thlr.
4. Larger compound microscope (1), round ring-shaped foot, arched rotary diaphragm, double mirror freely movable, micrometer screw, objectives A, B, C, D, E, F; 4 eye-pieces; 20–1,500 diameters .....106 thlr.
5. The same instrument, objectives A, C, D, F; 4 eye-pieces; 20–1,500 diameters .....81 thlr.
6. The same instrument, objectives C, F; 4 eye-pieces; 80–1,500 diameters .....64 thlr.
7. The same instrument, objectives B, E; 4 eye-pieces; 75–900 diameters .....55 thlr.
8. Medium compound microscope (stand II.), round foot, rotary diaphragm, mirror, etc., as in I., objectives C, D, F; 4 eye-pieces; 80–1,500 diameters .....70 thlr.
9. The same instrument, objectives B, E; 4 eye-pieces; 75–900 diameters .....49 thlr.
10. The same instrument, objectives A, D; 3 eye-pieces .....40 thlr.
11. Smaller compound microscope (stand III*b*.), horse-shoe foot, the remainder as in I., only somewhat smaller; objectives A, C, D, F; 4 eye-pieces; 20–1,500 diameters .....72 thlr.
12. The same instrument, objectives C, F; 4 eye-pieces; 80–1,500 diameters .....55 thlr.
13. The same instrument, objectives B, E; 4 eye-pieces; 75–900 diameters .....46 thlr.
14. The same instrument, objectives A, D, 3 eye-pieces .....37 thlr.
15. Smaller compound microsocpe (stand III. *c*); square, heavy foot, rotary stage, otherwise like I., but smaller; objectives A, B, C, D, E, F: 4 eye-pieces; 20–1,500 diam...104 thlr.
16. The same instrument, objectives A, C, D, F; 4 eye-pieces; 20–1,500 diameters .....79 thlr.
17. The same instrument; objectives A, C, E; 4 eye-pieces; 20–to 900 diameters .....60 thlr.
18. Smallest compound microscope (Stand IV.); round foot; concave mirror; micrometer screw (size of III. *b*); objectives A, D; 3 eye-pieces; 30–740 diameters .....33 thlr.
19. The same instrument; objective C; 4 eye-pieces; 80–330 diameters .....26 thlr.
20. The same instrument; objective A; 2 eye-pieces; 30–115 diameters .....19 thlr.
21. Smallest compound microscope (Stand V.); objective A; 1 eye-piece, No. 2 or 3; 30–75 or 45–115 diameters .....12½ thlr.
22. Smallest compound microscope (Stand V. *b*); micrometer screw on stage; objectives A, D: 3 eye pieces, 36–740 diameters .....29 thlr.
23. The same instrument; objective C; 4 eye-pieces; 80–330 diameters .....22½ thlr.
24. The same instrument; system, A; 2 eye pieces; 30–115 diameter .....15½ thlr.
25. New preparing microscope, 40, 60, 100 and 150 diameters .....21 thlr.

Price of stand with cases.

Stand	Price	Stand	Price
Stand 0	45 thlr.	Stand III *c*	22 thlr.
" I	24 "	" IV	11 "
" I. *b*	32 "	" V	6 "
" II	18 "	" V. *b*	7½ "
" III. *b*	15 "		

Prices and magnifying power of the objectives.

With eye-piece	1	2	3	4	
System A, upper lens alone	20	30	45		
" A, entire system	50	75	115	180	5 thlr.
" B	75	105	150	180	7 "
" C	80	120	200	240	9 "
" D	160	250	450	740	12 "
" E	240	350	600	900	18 "
" F	300	500	950	1500	25 "

Eye-pieces at........1½ thlr.
Accessory apparatus for compound microscopes. Eye-piece micrometer........2–4 thlr.
Stage micrometer, 1mm. in 50 divisions, in case........2 thlr.
Stage micrometer 1 mm. in 100 " " ....3 thlr.
Stage micrometer ½ mm. in 100 " " ....5 thlr.
Apparatus for measuring thickness of covers, with nonius giving 1–10 mm., estimating accurately 0.05 mm., in case........2½ thlr.
Camera lucida, after Nachet, in case........5 thlr.
Camera lucida, after Nobert, in case........5 thlr.
Camera lucida of 2 prisms........7 thlr.
Compressorium, improved form, in case........5 thlr.
Illuminating lens on stand, 3″ in diameter, plano-convex........12 thlr.
do. other varieties........4–5–3½ thlr.
Polarizing microscope........10–15 thlr.
Simple microscope........9–19 thlr.
Loups........18 sgr., to 4½ thlr.

No. 5.—*Price-List of Microscopes, &c., of* F. W. Schiek, *Berlin, Halle'sche Strasse* 14 (1870).

Price in Thalers.

A.—Large Compound Microscope, Horse-Shoe Stand.

Constructed for inclining; the stage rotates around its axis; coarse adjustment by rack and pinion, fine by micrometer screw, cylindrical diaphragm for insertion, with 7 objectives, 1, 2, 4, 5, 7, 9 and 11 (immersion system with correcting apparatus); 4 eye-pieces, three of which are achromatic; illuminating lens, insect box, steel forceps, cylindrical loup, eye-piece and stage micrometer, slides, convex glasses, etc., linear enlargement; 15–2,500 times........200 thlr.

B.—Large Compound Microscope, Horse-Shoe Stand.

Constructed for inclining; the stage rotates on its axis; coarse adjustment by rack and pinion, fine by micrometer screw, cylindrical diaphragms, six objectives, 1, 3, 5, 7, 9 and 10 (immersion system); 4 eye-pieces, two of which are achromatic; illuminating lens, eye-piece and stage micrometer, etc.; magnifying from 15–1,800 times linear........150 thlr.

C.—Medium Compound Microscope, Horse-Shoe Stand.

a. Not constructed for inclining; stage rotates on its axis; coarse adjustment by rack and pinion, fine by micrometer screw; cylindrical diaphragms; 5 objectives, 1, 3, 5, 7 and 9; 4 eye-pieces, one of which is achromatic; illuminating lens, etc.; giving 20–1,500 times linear........100 thlr.

b. The same model, with 3 eye-pieces and 4 objectives, 3, 5, 7 and 9; without illuminating lens; giving 50–1,200 times linear........80 thlr.

D.—Medium Compound Microscope, Stand with Folding Tripod.

Coarse adjustment by rack and pinion, fine by micrometer screw, which tilts the stage; stage not rotary; cylindrical diaphragms; 4 objectives, 3, 5, 7 and 9; 4 eye-pieces; illuminating lens, etc., giving 50–1,500 times linear........100 thlr.

E.—Compound Microscope, Horse-Shoe Stand.

Constructed to incline; rotary stage, micrometer screw, rotary diaphragm; 4 objectives, 3, 5, 7 and 8; 4 eye-pieces; illuminating lens, 50–860 linear........75 thlr.

F.—Compound Microscope, Horse-Shoe Stand.

a. Firm stage, not rotary, micrometer screw, rotary diaphragm; 4 objectives, 1, 3, 5, 7; 3 eye-pieces; 30–600 linear........50 thlr.

b. The same model, with objectives 3, 5, 7, and 9........65 thlr.

If rotary stage is desired, price is increased........10 thlr.

G.—Small Compound Microscope, Horse-Shoe Stand.

Immovable stage, micrometer screw, rotary diaphragm; 3 objectives, 3, 5, 7; 2 eye-pieces; 50–450 linear........28 thlr.

H.—Small Compound Microscope, Drum Stand.

Micrometer screw on stage, rotary diaphragm, 3 achromatic objective lenses, 1+2+3 (system 4); and objective 7; 2 eye-pieces, etc.; 25–500 linear........28 thlr.

I.—Small Compound Microscope, for the Use of Schools.

a. Micrometer screw, rotary diaphragm, 3 achromatic objective lenses, 1+2+3 (system 4); and objective 7; two eye-pieces; forceps, slides and covers, 40–450 linear..............25 thlr.

b. The same model, with two achromatic objective lenses, 1+2, and objective 5; one eye-piece, etc., 40–300 linear..............20 thlr.

c. The same model with cast-iron foot; stage without diaphragm, 2 objectives, 1+2 and 5; one eye-piece, etc..............15 thlr.

K.—Smallest Compound Microscope, Travelling Pocket Microscope.

The stand is constructed to fold together, with one eye-piece and one objective, consisting of three achromatic lenses, magnifying 200 times linear..............10 thlr.

N.B.—This microscope, which is especially adapted for foot travelling, is enclosed in a case 4″ long by 2½″ wide and 1½″ deep.

L.—Simple Achromatic Microscope, Simplex.

In a mahogany case, with 6 achromatic lenses, in two systems, magnifying 40 times linear. 20 thlr.

M.—The several Objectives and their Average Magnifying Power with the Various Eye-Pieces.

OBJECTIVE.	0	1	2	3	Price.
No. 1	20	35	....	....	5 thlr.
" 2	45	65	....	....	6 "
" 3	70	100	180	....	7½ "
" 4	90	125	210	....	8 "
" 5	150	220	350	....	9 "
" 6	200	300	450	550	10 "
" 7	260	375	550	650	10 "
" 8	400	550	700	960	12 "
" 9	450	700	900	1,200	16 "
* " 9	500	750	1,100	1,500	25 "
* " 10	600	800	1,200	1,800	30 "
* " 11	750	1,000	1,400	2,500	40 "

* Nos. 9, 10, and 11 are immersion systems with simple correcting apparatus. If double correcting apparatus (an arrangement by which the position of all three lenses may be changed with screw) is required, the price is increased 10 thalers each.

No. 6.—*Price-List of* E. Gundlach, *Berlin*, 1870.

Price in Thalers.

Microscopes.

1. Large binocular stereoscopic microscope. Large brass foot with two massive arms on which the body rests. It may be inclined from vertical to horizontal. The body may also be rotated around the optical axis without moving the illuminating apparatus.

   The stereoscopic double tube may be removed and replaced by a simple tube. The distance between the two stereoscopic eye-pieces may be adapted by an adjusting mechanism to the distance between the eyes. The tube has three arrangements for adjustment: the rapid movement (coarse adjustment) by rack-work; the medium, for fine adjustment, for powers below 500 diam.; and the extremely slow adjustment for the highest powers. The last two movements are without friction, and are combined with new and desirable improvements.

   The diaphragm is placed on a slider, so that it may be entirely removed from the instrument if necessary (6 cylinder diaphragms). Large double mirror, freely movable. Stage inlaid with glass.

   To this instrument belong: a stage, which is movable by means of fine screws, and which may be removed and replaced at pleasure (No. 22); a revolving nose piece for five objectives (No. 21), fitting the ordinary as well as the double tube; eye-piece micrometer, adjusting with fine screws (No. 20); a polarizing apparatus with goniometer (No. 18); Oberhäuser's drawing apparatus (No. 16); large illuminating lens for opaque objects (No. 23); an achromatic condenser; compressorium (No. 26); stage micrometer (No. 27).

   The objectives Nos. I, II, IV, V, VI*b*, VII*b*, VIII, and IX; 3 eye-pieces (magnifying 30-2300-fold), test objects, slides, covers, etc.; all in a strong mahogany case, bound with brass, and arranged for convenience in carrying; lenses in a separate leather case....362 thlr.

2. Large microscope. Rotary stage with divisions, and adjusting screws for correcting the centring; joint for inclining; draw-tube; large brass foot; coarse adjustment by rack work, fine by micrometer screw; freely movable double mirror; 4 cylindrical diaphragms with slide-movement; nose-piece for 4 objectives; movable stage; movable eye-piece

micrometer; polarizing apparatus with goniometer; Oberhäuser's drawing apparatus: large illuminating lens; condenser, etc. Objectives No. I, II, IV, V, VI*b*, VII*b*, and IX; 3 eye-pieces; power 30–2300 fold; test objects, slides, covers, etc.; lenses in leather case. The whole in mahogany case....252 thlr.

The same instrument, with objectives Nos. I, II, IV, V, VI*b*, and VIII....215 thlr.

The same instrument with the following accessories: nose-piece for 4 objectives: polarizing apparatus with goniometer: Oberhäuser's drawing apparatus; valuable eye-piece micrometer: condenser, etc. Objectives Nos. I, II, IV, V, and VII*b*; 3 eye-pieces, magnifying 30–1150 times; test objects, slides, covers, etc....168 thlr.

The same instrument without accessories; objectives Nos. I, II, IV, V, and VII*b*; 3 eye-pieces, last with micrometer; condenser, test objects, slides, and covers....122 thlr.

3. Medium microscope, with joint for inclining like No. 2; rotary stage with centring screws; heavy brass foot: coarse adjustment by rack-work, fine screws (without friction); 3 cylinder diaphragms; sliding apparatus; freely movable double mirror; objectives Nos. I, II, IV, V, and VII*b*; 3 eye-pieces; one micrometer for insertion; 30–1500 diam.; condenser, etc. In strong mahogany case....100 thlr.

4. Medium microscope, with joint for inclining, heavy brass foot: 3 cylinder diaphragms, freely movable double mirror; immovable stage; fine adjustment; objectives Nos. I, II, IV, V, and VII*b*; 3 eye-pieces and micrometer; condenser, etc.; magnifying power 30–1500 fold. Mahogany case, etc....88 thlr.

The same instrument, with objectives Nos. I, III, V, VII*a*; 3 eye-pieces, and micrometer, condenser, etc.; magnifying 30–1150 fold....76 thlr.

5. Medium microscope, without joint for inclining; brass foot, filled with lead; 3 cylinder diaphragms without slider; double mirror: fine screw adjustment; objectives Nos. I, III, V, and VII*b*; 3 eye-pieces, and micrometer, condenser, test objects, etc.; magnifying 30–1150 fold. Mahogany case, and leather case for lenses....76 thlr.

The same instrument, with objectives Nos. I, III, V, and VII*a*; 3 eye-pieces, and micrometer, test objects, case, etc.: magnifying 30–1150 fold....69 thlr.

The same instrument, with objectives Nos. II, V, and VII*a*; 2 eye-pieces, test objects, case, etc.; magnifying 70–1150 times....61 thlr.

The same instrument, with objectives Nos. I, III, and V; 2 eye-pieces, and micrometer, test objects, etc.; magnifying power 30–500 fold....50 thlr.
Case, 1 thlr. more.

The same instrument, with objectives Nos. II and V; 2 eye-pieces and micrometer; 70–500 times....45 thlr.

6. Travelling microscope, with folding tripod and draw-tube under stage for increasing the length of the instrument; screw for fine adjustment; objectives Nos. II, V, and VII*a*; 2 eye-pieces and micrometer, test objects, etc.; magnifying 70–1150 times. The whole in a mahogany case 21 ctmetrs. long, 12 ctmetrs. wide, and 8 ctmetrs. high....60 thlr.

The same instruments with objectives Nos. I, III, and V; 2 eye-pieces; micrometer, test objects, etc.; magnifying 300–500 times....50 thlr.

7. Simple microscope, round brass foot, fine adjustment by screw on stage; rotary diaphragm, concave mirror; objectives No. II and V; 2 eye-pieces and diaphragm, test objects, etc.; magnifying 70–500 times; mahogany case....32 thlr.

8. Small microscope, round foot, fine adjustment on stage: concave mirror; objectives II and V; 2 eye-pieces; test objects, etc.; magnifies 70–500 times; mahogany case..26 thlr.

The same instrument, with 3 achromatic objective lenses, and 1 eye-piece; magnifying 60, 100, and 180 times....16 thlr.

9. Demonstrating microscope (when the preparation is fastened in may be passed from hand to hand): 3 achromatic objective lenses, and 1 eye-piece; magnifies 40, 80, and 120 times....12 thlr.

10. Micro-photographic apparatus, adjustable to any stand....36 thlr.

The same apparatus, with the micro-photographic objectives ¼, ½, and 1 in. focus....69 thlr.

11. Large micro-photographic apparatus, to be used without any microscope stand; size of image 15 ctmetres diameter, with the micro-photographic objectives ⅙, ¼, and 1 in. focus....116 thlr.

12. Large horizontal micro-photographic apparatus; may be drawn out to a length of 2 metres: size of image, 30 ctmetres: with the micro-photographic objectives ⅙, ¼, ½ and 1 in. focus, and the immersion objective No. VII*a*....192 thlr.

Accessory Apparatus.

13. Preparing microscope (simplex): firm stage; adjustment by rack-work, on a polished mahogany case, with rests for the hands, and 2 drawers; 2 achromatic triplets....15 thlr.

The same instrument, with 3 triplets....18 thlr.

The same instrument, with folding case (compact for travelling), with 3 triplets....20 thlr.

14. Large stand loup, with heavy brass foot and a long double-jointed arm; 2 lenses of 40 mm. diameter; magnifying 3 and 6 times....6 thlr.

15. Small stand loup; arrangement as in No. 14; 2 lenses of 25 mm. diameter; magnifying 5 and 10 times....3½ thlr.

16. Drawing apparatus, after Oberhäuser....10 thlr.

17. Goniometer....................20 thlr.
18. Polarizing apparatus, with goniometer, after Hartnach (improved)..................20 thlr.
19. Simple polarizing apparatus (analyzer over eye-piece)..................14 thlr.
20. Movable eye-piece micrometer, in a special eye-piece, with a fine screw for horizontal movement, and correction for sharp adjustment..................8 thlr.
Simple eye-piece micrometer..................2 thlr.
21. Nose-piece for 5 objectives..................10 thlr.
The same, smaller, for 4 objectives..................8 thlr.
22. Movable stage (with 5 screws)..................8 thlr.
23. Large illuminating doublet for opaque objects, on a separate stand, with heavy brass foot 8 thlr.
24. Simple illuminating lens, on stand..................4 thlr.
25. Max Schultze's warm stage..................10 thlr.
26. Compressorium..................5 thlr.
27. Stage micrometer, in case; 1 mm. in 100 parts..................3 thlr.

Objectives and Eye-Pieces.

OBJECTIVES.	FOCUS.	ANGLE OF APERTURE.		Thlr.
	Inches.	Degrees.		
No. I	1	13	....	5
II	1-2	38	....	5
III	1 3	50	....	5
IV	1-4	80	....	8
V	1-8	150	....	10
VI *a*	1-12	170	Without correcting screw..........	15
VI *b*	1-12	170	With " "	20
VII *a*	1-16	175	Immersion, without correcting screw...	15
VII *b*	1-16	175	" with " " ...	20
VIII	1-24	175	" " " " ...	30
IX	1-32	175	" " " " ...	45

Micro-photographic objective, 1 in. focus..................12 thlr.
Micro-photographic objective, ½ in. focus..................10 thlr.
Micro-photographic objective, ¼ in. focus..................10 thlr.
Micro photographic objective, ⅛ in. focus..................15 thlr.
Eye-pieces, No. 0, I, II, and III..................at 2½ thlr.
Eye-piece, No. III, with micrometer..................4 thlr.
Periscopic eye-piece (larger and more level field), No. 0..................7 thlr.
Periscopic eye-piece (larger and more level field), No. 1..................6 thlr.
Periscopic eye-piece (larger and more level field), No. 2..................5 thlr.
Periscopic eye-piece (larger and more level field), No. 3..................5 thlr.
New objective No. X.; focus, 1-50 in.; angle of aperture, 175°; magnifying power, 1,260, 1,800 2,500, and 3,400 fold..................100 thlr.

Magnifying Powers.

NO. OF OBJECTIVE.	I	II	III	IV	V	VI	VII	VIII	IX
Eye-piece 0..................	20	45	65	95	190	315	420	630	840
Eye-piece I..................	30	65	90	138	275	450	600	900	1200
Eye-piece II..................	45	90	125	185	375	625	835	1250	1670
Eye-piece III..................	65	120	170	250	500	830	1150	1700	2300

No. 7.—Price-List of L. Bénèche, Berlin, Grossbeeren Strasse, No. 17. (1870).

Price in Thalers.

Compound Microscopes.

A. Horse-shoe foot; slider for changing the diaphragms; large rotary stage; micrometer adjustment on the tube: objectives 2, 4, 7, 9, 10, 11, 12: eye-pieces 1 to 5: adjustable eye-piece micrometer; condenser; magnifying 3,000 diameters......................230 thlr.

B. Stand similar to that of A, but smaller: objectives 4, 7, 9, 10; eye-piece 1 to 4; eye-piece micrometer; condenser; magnifying 300 diameters...............................100 thlr.

C. Horse-shoe foot; large firm stage; slider for changing the diaphragms; micrometer adjustment on tube; objectives 4, 7, 9; eye-pieces 2, 3, 4; eye-piece micrometer; magnifying power 600..50 thlr.

The same microscope, with objectives 4, 7, 10; giving 800 diameters.................... 60 thlr.

D. Round foot, micrometer adjustment on tube; arched diaphragm under the stage; objectives, 4, 7; 2 eye-pieces; magnifying 400 diameters..................................25 thlr.

The same microscope, with objectives 4, 7, 8; magnifying 500 diameters.................30 thlr.

The same microscope, with objectives 4, 7, 9; magnifying 500 diameters..................37 thlr.

Smaller Microscopes.

d. Round foot; micrometer adjustment on stage; diaphragm under stage; eye-piece micrometer; 2 eye-pieces; 6 different magnifying powers to 400 diameters.... ....15 thlr.

E. Round foot; micrometer adjustment on stage; 1 eye-piece; 3 different magnifying powers to 300 diameters...

Price of the Objectives.

No. 1	4 thlr.	No. 8	10 thlr.
" 2	5 "	" 9	12 "
" 3	6 "	" 10..(immersion with correction)	25 "
" 4	7 "	" 11	15 "
" 5	8 "	" 11..(with correction)	20 "
" 6	10 "	" 12..(immersion with correction)	50 "
" 7	10 "		

No. 8.—*Price-List of the Achromatic Microscopes of the Institute founded in Wetzlar by* C. Kellner, *Successor* Ernst Leitz (*formerly* Belthle & Leitz, 1870.

Price in Thalers.

Microscopes.

1. Large microscope. Coarse adjustment by rack and pinion, fine by micrometer screw; sliding arrangement for cylinder diaphragms; polarizing apparatus; drawing apparatus; eye-piece micrometer; double mirror; rotation on the optical axis; orthoscopic eye-pieces I, II, III, IV, and objectives 1, 2, 3, 5, 6, 7, and immersion Nos. 8 and 9. 20–2000 diameters...150 thlr.

The same, to incline...160 thlr.

2. Medium microscope; stand as in No. 1; eye-piece micrometer; 4 orthoscopic eye-pieces; objectives 1, 3, 5, 7, and immersion 9; 20-1500 diam...............................100 thlr.

3. The same stand, but without rotation on optical axis; 3 eye-pieces; objectives 1, 3, 7 and 9; immersion, 25–1500 diam...80 thlr.

The same, without immersion system 9...65 thlr.

4. Small microscope. Horse-shoe foot: rotation on optical axis; fine adjustment, micrometer screw, double mirror; cylindrical diaphragms; 3 eye-pieces; objectives 1, 3, 7, and immersion 8; magnifying 25, 40, 60, 75, 100, 140, 220, 350, 500, 600, 800, 1000..............60 thlr.

5. Small microscope: No. 4 stand: double mirror; 3 eye-pieces; objectives 2, 5, 7; magnifying 35, 60, 100, 120, 200, 300, 320, 500, 800..50 thlr.

6. Small microscope; stand like No. 4, but smaller; rotation on optical axis, cylinder diaphragms; 3 eye-pieces; objectives 1, 3, 7; magnifying 20, 30, 50, 60, 100, 130, 300, 450, 600...40 thlr.

7. The same microscope, without rotation .... 35 thlr.

8. Small microscope; fine adjustment by micrometer screw on tube; eye-piece micrometer; 3 eye-pieces; objectives 1, 3, 7; magnifying 20, 30, 50, 60, 100, 130, 300, 450, 600.......30 thlr.

9. Same stand as No. 7; 2 eye-pieces; objectives 1, 3, 7; magnifying 20, 30, 60, 100, 300, 500..25 thlr.

10. Smallest microscope; 2 eye-pieces; objectives 3, 7; magnifying 60-500.................20 thlr.
The same, without objective No. 7..15 thlr.
11. Laboratory microscope, after A. Stuart. Coarse adjustment by sliding on a square metallic column; fixation of same by means of a screw; fine adjustment by micrometer screw on tube; eye-piece micrometer; 2 eye-pieces; objectives 1, 3, 7; magnifying from 25-500..30 thlr.
The same, with drawing and polarizing apparatus...42 thlr.
12. Laboratory microscope, with stand like No. 10; one eye-piece; objectives 3, 6.........23 thlr.
The same, with drawing and polarizing apparatuses, and eye-piece micrometer............37 thlr.
N.B.—All microscopes are enclosed in a compact mahogany case with lock. Magnifying powers of the microscopes are calculated at 8 in. = 216 mm.

NEW SYSTEMS, WITH LARGE ANGLES OF APERTURE.

OBJECTIVE.		EYE-PIECES.						Total distance in mms.	Price, Thalers.
		*	I.		II.		III.		
*	†	†0	I.	II.	III.	IV.	V.		
No. 0..........	No. 1.......	20	25	35	40	50	60	30.	3
" 0.a........	" 2.......	35	50	60	70	85	100	17.5	6
" 1..........	" 3.......	60	75	90	100	115	140	5.5	6
" 1a.........	" 4.......	85	115	140	150	170	220	3.5	8
" 2..........	" 5.......	120	150	175	200	220	300	1.8	10
" 2a.........	" 6.......	180	220	320	350	380	500	1.45	10
" 3.........	" 7.......	250	320	450	500	650	800	1.06	11
" 3a.........	" 8.......	350	500	600	650	750	970	0.98	12
" 4.........	" 9.......	450	650	750	850	1,000	1,450	0.8	15
" 5..........	" 10.......	500	700	900	1,150	1,400	1,800	0.6	18
NEW IMMERSION OBJECTIVES WITH CORRECTION.									
	" 8.......	350	450	550	650	800	1,000	0.93	15
	" 9.......	450	650	750	850	1,000	1,450	0.8	20
	" 10.......	500	700	900	1,200	1,500	2,000	0.6	25

No. 9.—*Price-List of the Microscopes of* Bruno Hasert, *in Eisenach* (1867).

Price in Thalers.

Large stand with rotary stage for direct and oblique illumination, with achromatic condensing lens for slightly oblique light, with three eye-pieces, and an illuminating lens for opaque objects..45-50 thlr.
Small stand with rotary stage, achromatic condensing lens, and illuminating lens for opaque objects, with 2 eye-pieces...25-27 thlr.
Small stand without rotary stage, cylindrical diaphragm, and oblique illumination, illuminating lens for opaque objects, and two eye-pieces................................15 to 17 thlr.
A. Objective of the first quality, of 1-16 in. focus, which, without immersion, thoroughly resolves all known test objects..45 thlr.
B. Objective, first quality, 1-12 in. focus, which also shows the hexagons on the Pleurosigma angulatum, and also the striations on the Grammatophora subtilissima, without immersion..35 thlr.
C. Objective, first quality, 1-5 in. focus, which also shows without immersion the hexagons on the Pleurosigma angulatum...20 thlr.
D. Objective, second quality, of shorter focus, which shows the transverse striations on the butterfly's scales, and the striations of the Pleurosigma attenuatum; by unscrewing the front lens a lower power is obtained..8-10 thlr.
The price of the several stands and objectives may be readily calculated from the above, as, for instance, microscope, first rank, with objectives A, C, and D, giving from 60-2,400 linear..125 thlr.
Microscope with rotary stage, small model, with objectives B and D, sufficing for the most difficult examinations, giving 100-1,500 linear..75 thlr.
The same, with objectives C and D..60-65 thlr.

---

* Old designation of objectives and eye-pieces.
† New designation of objectives and eye-pieces.

Small stand with rotary stage, with objectives C and D, sufficing for most examinations, 600–700 linear........36–50 thlr.

The small microscope with two eye-pieces and one objective, the front lens of which unscrews, and thus forms a second objective, giving power up to 400........25–27 thlr.

N.B.—The best objectives are to be used without immersion, give unclouded and clear images, and the strongest do not need correction for covers.

No. 10.—*Price-List of Optical Institute of* S. Plössl & Co., *in Vienna* (1871).

Factory, Wieden, Theresianumgasse, No. 12.

Depot, Stadt, Rauhenstein & Himmelpfortgasse, 7.

Price in Guldens.

Complete Microscopes.

1. Large compound microscope, tube moved by rack-work, micrometer screw for fine adjustment, stage covered with glass, brass pedestal, microscope moves on one arm on its axis, stage has two clamps for holding all sorts of slides, freely movable mirror on a double arm, sliding arrangement for cylinder diaphragms. The instrument is furnished with six achromatic objectives, so arranged as to be used with or without covers; three eye-pieces, magnifying from 25–600 times linear, or 625–360,000 superficial; spherical illuminating prism (after Selligue), with movement for illuminating opaque objects, large condensing lens (to use with lamp-light) on a separate stand, slides, a Wilson's loup, two micrometers with divisions of the Vienna duodecimal line into 30 and 60 parts, or the millimetre into 25 and 50 parts, mounted in brass, for insertion into the eye-piece No. 2, with screw for adjustment. Polished case, lined with velvet, with lock and key.......210 fl.
2. The same with arrangement for measuring objects to 0.00001 Paris lines, by means of Fraunhofer's screw micrometer........300 fl.
3. Medium compound microscope, presenting the same advantages as No. 1, but smaller...200 fl.
4. Medium compound microscope, rack-work to move tube, micrometer screw for fine adjustment, immovable stage covered with glass, two clamps on stage to hold slides, concave mirror on a double arm, freely movable; three achromatic objectives, so arranged as to be used with or without covers; three eye-pieces, magnifying 25–600 times linear; illuminating lens for opaque objects, slides on a Wilson's loup, two eye-piece micrometers with screws for adjusting, millimetre divided into 25 and 50 parts, mounted in brass. Polished case, lined with velvet, lock, etc.,........130 fl.
5. Small microscope, micrometer screw on stage, movable concave mirror, ordinary rotary diaphragm, new objective C, and 2 eye-pieces; magnifies 60–120 times. Elegant polished case........52 fl.
6. Small microscope, exactly like No. 5, objective E, and two eye-pieces; magnifies 180-360 times........64 fl.
7. Small microscope, like No. 5, coarse adjustment by rack-work, with objectives C and E, and two eye-pieces; magnifies 60, 120, 180 and 360 times linear........75 fl.
8. New small working microscope, on a round brass foot, the body resting on a horizontal arm; stage has two clamps for slides, and moves by rack-work towards the objectives; movable concave mirror, slides, etc., one eye-piece and three achromatic objectives; magnifies from 20–150 times linear, made gradual by elongating the tube. The microscope is so arranged that the objects do not appear inverted, is therefore a very convenient instrument for dissecting. Elegant polished case........54 fl.
9. New large working microscope, body moves by rack-work towards the immovable stage; otherwise like No. 8; magnifies to 300 linear........70 fl.
10. Small travelling microscope, with a stand to be screwed into the cover of the case, stage with clamps, to be moved by rack-work towards the lenses; concave mirror with double movement, slides, needles, forceps, etc., six mounted double lenses, after Wallaston, magnifying from 12–250 times linear. Polished case, lined with velvet........60 fl.
11. The same microscope with 3 double lenses........36 fl.
12. do. " " 6 simple " ....45 fl.
13. do. " " 3 " " ....30 fl.
14. Microscope, heavy foot, immovable stage, fine adjustment by micrometer screw, rotary diaphragm, lateral movement to mirror; one eye-piece and objective, magnifies 500 linear; excellently adapted for examining silk disease.

Sun, gas and photo-electric microscopes made to order.

Microscopic Accessories.

15. Ordinary eye-pieces, 1, 2, 3, 4, 5........@ 5 fl.
Orthoscopic eye-pieces, 1, 2, 3, 4, 5........@ 8 fl.
Aplanatic " ....10–50 fl.
Solid glass " ....8 fl.

16. Achromatic objectives wth large angles of aeprture.
   A. Combination of 3 achromatic lenses designated 1, 2, 3 .... 15 fl.
   B. do. " " " " 4, 5, 6 .... 18 fl.
   C. System, 1-3″ .... 10 fl.
   D. " 1-6″ .... 18 fl.
   E. " 1-9″ .... 24 fl.
   F. " 1-12″, correcting apparatus .... 35 fl.
   G. " 1-18″ " " .... 60 fl.
17. Stage screw micrometer .... 90 fl.
18. Eye-piece " .... 60 fl.
19. Micrometer eye-piece .... 8 fl.
20. Eye-piece micrometer .... 5 fl.
21. Stage " .... 6 fl.
22. Goniometer .... 28 fl.
23. " with polarizing apparatus .... 45 fl.
24. Polarizing apparatus, A, over objective; P. under stage .... 24 fl.
25. " " A, over eye-piece; P. " " .... 28 fl.
26. Wallaston's camera lucida .... 12–20 fl.
27. Sömmering's speculum apparatus .... 12 fl.
28. Warm stage .... 18 fl.
29. Compressorium, after Purkinje .... 15 fl.
30. " " Plössl .... 12 fl.
31. Prism for horizontal vision .... 18 fl.
32. Electrical discharger, after Plössl .... 9 fl.
33. Spherical illuminating prism, after Selligue .... 15 fl.
34. Illuminating lens on stand .... 12 fl.

No. 11.—*Price-List of the Microscopes, etc., of* THOMAS ROSS, *Optician*, 7 *Wigmore street, Cavendish Square, London, W.* (1872).

Price in £ s. d.

Compound Microscopes.

	£	s.	d.
No. 1. A large compound microscope stand, with graduated concentric rotating stage, having one inch of motion in rectangular directions; rack and fine screw movements to the optical part, clamping arm for fixing the instrument at any inclination, graduated sub-stage for holding and adjusting by universal motions all the illuminating and polarizing apparatus placed beneath the object; 2 eye-pieces, flat and concave mirrors; diaphragm plate; stage forceps, and two glass plates with ledges	30	0	0
No, 2. A smaller microscope stand, having three-quarters of an inch of motion, and ordinary rotating object plate to the stage	21	10	0
No. 2 A ditto, without sub-stage	17	0	0
No. 2. A microscope stand without sub-stage; fine screw adjustment, or stage movements; with 2 eye-pieces, a one-inch object glass of 25 degrees angular aperture, and a one-quarter inch object-glass of 100 degrees—this is the basis of a complete instrument	18	11	0
No. 2. Mechanical stage for do	4	10	0
No. 2. Fine-screw adjustment for ditto	2	14	0
No. 2. Sub-stage and vertical rack motion for ditto	4	10	0
No. 3. A smaller microscope stand, with mechanical stage and fine screw adjustment to the optical part, with 2 eye-pieces	13	10	0
No. 3. A microscope stand without stage movements or fine-screw adjustment with 2 eye-pieces, a one-inch object-glass of 15 degrees, and a one-quarter inch object-glass of 100 degrees—this is the basis of a complete instrument	14	15	0
No. 3. Mechanical stage for ditto	4	0	0
No. 3. Fine screw adjustment for ditto	2	0	0
No. 3. Sub-stage and vertical rack motion for ditto	4	0	0
A compound universal and tank microscope stand, with four rack-and-pinion adjustments and 2 eye-pieces	15	0	0
A ditto, with binocular arrangement, rack-and-pinion adjustment to draw tubes, and two pairs of eye-pieces	21	12	6
A ditto, ditto, with sliding adjustment to draw tubes, and one pair of eye-pieces	19	7	6

Microscope Cases.

		Brass bound.
No. 1. Spanish mahogany cabinet case, with box for the apparatus..from	£5 5 0	...... £7 0 0
No. 1. Portable mahogany case......from	3 0 0	...... 4 5 0
No. 2. Portable mahogany case......from	2 10 0	...... 3 15 0
No. 3. Portable mahogany case......from	2 2 0	...... 3 3 0
No. 3. Flat portable case......from	1 16 0	...... 2 10 0
No. 3. Cupboard case......from	1 10 0	...... 2 7 6

Achromatic Object Glasses for Microscopes.

T. Ross particularly invites attention to the great penetrating power and perfection of the central and marginal definition of these objectives. The screws are cut to the standard gauge of the Royal Microscopical Society, and may be applied to any instrument not having the Society's screw by means of an adapter (the price of which is 2*s.* 6*d.*) Those marked * have an adjustment for covered and uncovered objects.

OBJECT GLASSES.	Angular Aperture.	Magnifying powers with the various eye-pieces.						Price.	Lieber-kuhns.
		A	B	C	D	E	F	£ *s.* *d.*	*s.* *d.*
5 inches.	7 degs.	8	13	24	36	52	72	1 10 0	
4 "	9 "	10	16	30	45	65	90	1 10 0	
3 "	12 "	13	20	35	56	84	112	3 0 0	
2 "	15 "	20	32	55	90	135	180	3 0 0	17 6
1 1-2 "	20 "	25	40	70	112	168	224	3 0 0	17 6
1 "	15 "	37	60	105	170	255	340	2 0 0	15 0
1 "	25 "	37	60	105	170	255	340	3 10 0	15 0
2-3 "	35 "	60	100	145	270	405	540	3 10 0	10 6
1.2* "	90 "	95	153	265	420	630	840	5 5 0	10 6
4-10* "	110 "	140	220	370	650	975	1300	6 6 0	
1-4* "	100 "	195	310	540	850	1275	1700	5 5 0	
1-4* "	140 "	195	310	540	850	1275	1700	6 10 0	
1-6* "	140 "	320	510	700	910	1260	1820	7 7 0	
1-8* "	140 "	420	670	900	1200	1800	2400	8 8 0	
1-12 "	170 "	600	870	1200	2000	3000	4000	12 12 0	

Immersion Object Glasses.

☞ A second anterior arrangement may be had with the 1-8 in. and the 1 12 in. object glasses, which, when substituted for the ordinary front combinations, converts them into IMMERSION OBJECTIVES, giving most brilliant definition. This addition, which is strongly recommended, thus enables the microscopist to use these powers either with or without water.

Immersion arrangement for the 1-8 in......£2 0 0 extra.

Immersion arrangement for the 1-12 in...... 2 10 0 "

Apparatus for Compound Microscopes.

	£ *s.* *d.*
Wenham's binocular arrangement, with 1 eye-piece and sliding adjustment to draw tubes.	5 5 0
Wenham's binocular arrangement, with 3 eye-pieces, adapter with rotating analyzing prism, combined rack-and-pinion adjustment to draw tubes, and all alterations complete......	8 10 0
Brooke's double nose-piece, for rapidly changing the object-glasses......	1 10 0

A, B and C eye-pieces, each........ 17 6
D, E and F eye-pieces, each........ 1 0 0
A or B eye-piece, with Slack's adjustable diaphragm........ 1 12 6
Putting ditto to eye-piece........ 16 0
Kelner's orthoscopic achromatic eye-pieces, giving double the usual field, C and D, each... 1 4 0
Erecting glasses, for dissecting with the compound microscope........ 1 0 0
Micrometer eye-piece........ 1 4 0
Screw micrometer........ 5 5 0
Jackson's micrometer........ 1 0 0
Stage micrometer, on slips of glass........ 6 0
Camera lucida (Wollaston's)........ 1 14 0
Neutral tint, ditto........ 7 6
Condensing lens, on stand........£1 and 1 10 0
Side condensing lens........ 14 0
Side reflector, for illuminating opaque objects........ 1 5 0
Polarizing apparatus, with 2 selenites........ 2 10 0
Ditto, with Darker's revolving selenite stage, and set of three selenites, in box........ 4 4 6
Ditto, with extra large polarizing prism, additional........ 1 10 0
Darker's revolving selenite stage, and set of three selenites, in box........ 2 3 6
Darker's revolving selenites, in improved setting, adapted to sub-stage; may be used and rotated either singly or in combinations of two or three........ 3 10 0
Ditto, ditto, with polarizing apparatus........ 5 11 0
Double image prism and plate, for exhibiting the decomposition of light, by polarization.. 1 2 6
Ross's achromatic condenser, for transparent illumination with high and low powers...... 3 0 0
Gillett's achromatic condenser, with series of apertures and stops for illuminating transparent objects and developing the markings on diatoms........ 7 0 0
New four-tenths achromatic condenser, with two concentric diaphragm plates........ 7 10 0
Achromatic condenser, with flat place of diaphragms........ 6 0 0
Reade's improved double hemispherical condenser, with adjustable apertures........ 3 10 0
Paraboloid for dark ground illumination........ 1 13 6
Spotted lens for ditto and for test objects........*7s. 6d* and 10 6
Ross's centring glass, for examining the centring of the optical part of microscope........ 14 0
Amici's prism on stand, with jointed arms, for condensing an oblique pencil of light on transparent objects........ 2 2 0
Rainey's light modifier........ 5 0
Slack's ditto........ 3 6
Plate for fixing fish, frogs, etc., for exhibiting the circulation of the blood........ 14 0
Glass troughs for holding polyps, etc........ 4 6

No. 12.—*Price-List of Microscopes, etc., of* M. Pillischer, 88 *New Bond Street, London, W.*, 1872.

Pillischer's Improved Microscopes.

The construction of these microscopes is so arranged that any form of apparatus can be adapted without requiring the instrument for fitting.

Prices of Stands only.

1. Improved microscope stand, with coarse and fine adjustments and graduated draw-tube (moved by rack and pinion) to the body, rack, and pinion stage, with one and one-quarter inch motions, in rectangular directions, sliding and rotating object-holder, and sliding spring clamp, secondary stage for holding and centring achromatic condenser, polarizing and other apparatus, plane and concave mirrors, revolving diaphragm, and Nos. 1, 2, 3 eye-pieces........£29 0 0
2. Improved smaller microscope stand, with coarse and fine adjustments and graduated sliding draw-tube to the body, rack and pinion stage, with one-inch motion in rectangular directions; sliding and rotating object-holder, and sliding spring clamp, secondary stage, with centring motions, plane and concave mirrors, revolving diaphragm, and Nos. 1, 2 eye-pieces........£14 14 0
3. Improved microscope stand, with coarse and fine adjustments and graduated sliding draw-tube to the body, rack and pinion stage, three-quarter inch in rectangular directions, sliding and rotating object-holder and sliding spring clamp, revolving diaphragm, plane and concave mirrors, and Nos. 1, 2 eye-pieces........£12 12 0
4. Improved microscope stand, with a Pillischer's lever stage movement instead of a rack and pinion, all the rest the same as above........£7 10 0

Pillischer's Improved Medical Microscope.

Inclusive of Object Glasses, Apparatus, and Case, as manufactured for Her Majesty's Army Medical Department.

5. Improved microscope, with coarse and fine adjustments and graduated draw-tube to the body, rack and pinion stage, with three-quarter inch motion in rectangular directions, sliding and rotating object-holder, and sliding spring clamp, plane and concave mirrors, revolving diaphragm, Nos. 1 and 2 eye-pieces, one-inch object glass, 25 degrees angular aperture, and one-quarter inch 80 degrees, condenser for opaque objects, live box, and upright mahogany case, with drawer for objects, etc. £17 17 0

Achromatic Object Glasses.

OBJECT GLASSES.	ANGULAR APERTURE.	MAGNIFYING POWER WITH THE VARIOUS EYE-PIECES.				Price.	Lieber-kuhn's.
		A.	B.	C.	D.		
						£ s. d.	
4 inch.	9 degrees.	10	16	30	35	1 10 0	
3 "	12 "	14	22	37	60	1 10 0	
2 "	15 "	20	35	60	90	2 10 0	15 6
2 "	12 "	"	"	"	"	1 10 0	
1½ "	22 "	28	45	72	120	2 10 0	15 6
1 "	25 "	40	65	110	175	2 10 0	14 0
1 "	15 "	"	"	"	"	1 10 0	
½ "	80 "	95	155	270	430	5 0 0	10 6
4-10 "	55 "	142	230	375	655	4 0 0	
4-10 "	95 "	142	230	375	655	5 0 0	
¼ "	100 "	195	310	540	850	5 0 0	
¼ "	80 "	"	"	"	"	3 3 0	
1-6 "	130 "	320	510	700	910	6 0 0	
⅛ "	140 "	425	675	900	1,200	7 10 0	
1-20 "	150 "	950	1,620	2,800	3,500	15 10 0	

NOTE.—The object glasses are all fitted with the Microscopical Society's Standard Screw.

Prices of Apparatus. £ s. d.

6. Wenham's binocular arrangement, with rack and pinion adjustment to draw-tubes, and one eye-piece 6 6 6
7. Third eye-piece 0 15 0
8. Fourth ditto 0 17 6
9. Improved achromatic condenser, with series of apertures and stops for illuminating objects of delicate structure 5 0 0
10. Amicis prism on stand, with jointed arms, for condensing an oblique pencil of light .... £1 10 and 2 2 0
11. Parabolic condenser, for dark ground illumination, with adjustable stop 1 10 0
12. Large spotted lens for dark ground illumination, with low powers 0 12 6
13. Polarizing apparatus of two Nichol's prisms and selenite 2 2 0
14. Tourmalins, from 0 15 0
15. Large bull's-eye condenser on stand, with double motions 1 2 6
16. Stage condenser 0 18 0
17. Side reflector for illuminating opaque objects 1 1 0
18. Camera lucida, with prism 0 18 0
19. Ditto, neutral tinted 0 7 6
20. Compressorium 0 18 0
21. Live box 0 4 6
22. Frog plate for exhibiting the circulation of the blood 0 6 6
23. Brass pliers 0 2 6
24. Set of three dark wells and holder 0 10 6
25. Stage forceps 0 7 6
26. Erecting eye-piece for dissecting with the compound microscope 0 18 0
27. Darker's selenite stage 2 2 0
28. Brook's double nose-piece, for rapidly changing the object glasses 1 7 6
29. Double image prism, for exhibiting the decomposition of light by polarization 1 5 0

30. Indicator to eye-piece 0 4 6
31. Jackson's micrometer for the eye-piece 0 15 0
32. Ditto ditto for the stage 0 6 0
33. Screw micrometer 5 0 0
34. Glass troughs for holding polypi, etc 0 4 0
35. Set of animalcula tubes in case 0 3 0
36. Pocket spectroscope 1 15 0
37. Ditto ditto adapted to the microscope, extra 0 10 0
38. Micro-spectroscope, for showing double spectrum 4 10 0

Prices of Cases to Nos. 1, 2, 3 *and* 4 *Microscopes.*

39. Spanish mahogany or oak case, with two extra boxes for apparatus, etc., to No. 1 microscope 5 0 0
40. Spanish mahogany or walnut plate-glass case, with four drawers for holding apparatus, etc., to ditto 7 7 0
41. Spanish mahogany case, with two extra boxes for apparatus, etc., to No. 2 microscope. 3 5 0
42. Spanish mahogany or walnut plate-glass case, with four drawers for apparatus, etc., etc., to ditto 6 0 0
43. Spanish mahogany case, with one extra box for apparatus, etc., to No. 3 microscope. . 2 15 0
44. Upright mahogany case, with drawer for objects, etc., to No. 4 microscope 1 10 0

Best Student's Microscopes.

Inclusive of Object Glasses, Apparatus, and Case.

45. Improved compound microscope, having coarse and fine adjustments to the body, a best rack and pinion movable stage half-inch in rectangular directions, concave and plane mirrors, and revolving diaphragm; the apparatus supplied with this instrument consists of Nos. 1 and 2 eye-pieces, one inch and one-quarter inch object glasses of 15 and 80 degrees angular aperture, a live box, stage forceps, condenser for opaque objects, polarizing apparatus and selenite, and a best made Spanish mahogany or oak case, with drawer for objects, etc £15 15 0
46. Same size microscope, with a plain stage, or one with Pillischer's lever movement, and all the rest the same as above £10 10 0
47. The microscope stand, in every respect the same as No. 2, with one eye-piece, one inch and one-quarter inch object glasses, a condenser for opaque objects, and a plain mahogany case without drawer £7 7 0

Additional Apparatus.

Which can be applied to Pillischer's best Students' Microscopes.

48. Wenham's binocular arrangement, with rack and pinion adjustment to draw-tubes, and one eye-piece £4 10 0
49. Third eye-piece 0 15 0
50. Fourth ditto 0 17 6
51. Bull's-eye condenser for opaque objects 0 8 6
52. Live box 0 4 6
53. Stage forceps 0 5 6
54. Polarizing apparatus and selenite 1 10 0
55. Parabolic reflector for dark ground illumination 1 5 0
56. Glass stage with cavity 0 1 0
57. Brass pliers 0 2 0
58. Frog plate for exhibiting the circulation of the blood 0 5 0
59. Spotted lens for dark ground illumination with low powers 0 10 6
60. Camera lucida .... 7*s.* 6*d.* and 0 18 0
61. Micrometer for the eye-piece with adjusting screw 0 12 6
62. Ditto for stage, divided into 100ths and 1000ths of an inch 0 6 0
63. Compressorium 0 18 0
64. Brook's double nose-piece 1 5 0
65. Achromatic condenser, simple form for illuminating objects of delicate structure 2 10 0
66. Set of dark wells and holders 0 10 0

No. 13.—*Price-List of Microscopes of* C. Baker, 243 *and* 244 *High Holborn, London* (1872).

Achromatic Microscopes.

1. Highly-finished large compound microscope stand, with all the latest improvements, having double supports to prevent vibration, vertical rack adjustment for the approximate focus, and fine screw motion for the more delicate optical adjustment. A mechanical stage, with one-inch motion, in opposite directions; a sliding and rotating object holder, a supplementary stage, with vertical rack and centring adjustment for applying the diaphragm; polariscope, achromatic condenser, spot lens, etc; with plain and concave mirrors, and two eye-pieces........£21 0 0

1. A.—A large microscope stand, with mechanical stage, quick and slow motion, double mirror, two eye-pieces, etc., as above, but without the supplementary stage........£14 10 0

1. B.—A smaller ditto, and in every respect as No. 1. A........£11 10 0

1. B.—A ditto, without mechanical stage........£7 15 0

2. A smaller ditto, with mechanical stage, quick and slow motion, double mirror, and one eye-piece........£8 15 0

2. A ditto, without mechanical stage........£6 15 0

Supplementary stages applied to No. 1 A, No. 1 B, or No. 2........from £2 0 0

3. A superior finished binocular microscope stand, with two eye-pieces, double mirror, sliding stage, and quick and slow motions........£5 10 0

Ditto, ditto, with rack to eye-pieces as above........£6 0 0

Ditto, ditto, with improved circular rotating stage........£6 15 0

The improved binocular arrangement (Wenham's), which allows the removal of the prism with its necessary fittings, so that when the monocular body is used, an uninterrupted field is obtained. Applied to any of the preceding microscopes........from £5 0 0

To students, or smaller instruments........from £3 10 0

4. The student's microscope, a well-finished instrument, with quick and slow motion, sliding and revolving stage, a combination of three achromatic powers, live box, stage, and dissecting forceps, all packed in neat mahogany case........£4 4 0

Ditto, ditto, with superior rotating stage, having universal movements........£4 14 0

5. The educational microscope. This exceedingly cheap achromatic microscope, which is so strongly recommended by most of the *Professors at the various Colleges, Medical Schools, and Institutions for Public and Private Education*, has quick and slow motion, sliding stage, live box, stage forceps, dissecting ditto, with three achromatic object glasses in combination, all packed in a neat mahogany case........£3 3 0

A superior-finished dissecting microscope, with rack adjustment, three object glasses, etc., in neat mahogany case........£1 15 0

Achromatic Object Glasses.

	Angular Aperture.	£	s.	d.
Four-inch	8 degrees	1	5	0
Three-inch	10 "	1	15	0
Two-inch	12 "	1	10	0
Ditto	15 "	1	17	6
One-and-a-half-inch	20 "	1	17	6
One-inch	15 "	1	10	0
Ditto	23 "	1	17	6
Ditto	30 "	2	2	0
Two-thirds-inch	35 "	2	5	0
Half-inch, with adjustment	60 "	3	0	0
Ditto, without	40 "	2	10	0
Four-tenths, with adjustment	70 "	3	5	0
Ditto ditto	95 "	3	10	0
Quarter-inch, with adjustment	75 "	3	5	0
Ditto ditto	95 "	3	15	0
Ditto, without adjustment	75 "	2	10	0
One-eighth ditto, with adjustment	115 "	5	5	0
Ditto ditto	125 "	6	6	0

N. B.—*These Object Glasses are the finest that can be made, and for defining and penetrating power cannot be surpassed.*

A beautifully-defining combination of German powers, half-inch, dividing so as to form one-inch and one-and-a-half-inch, mounted in a superior form.......£1 2 6
Ditto, ditto, quarter-inch, and forming, if divided, one-third-inch.......1 5 0
The student's new English quarter-inch, of 70 degrees aperture.......1 10 0
Lieberkuhns fitted to any of the above object glasses.......from 6*s*. to 0 15 0
The new day-light reflectors to ditto.......0 15 0
Ditto, with universal movements adapted to instrument.......1 1 0

Magnifying Power of C. Baker's Achromatic Object Glasses.

Object Glasses	With the following Eye-pieces. A.	B.	C.	D.
Four-inch	10	16	30	45
Three-inch	17	28	41	50
Two-inch	25	35	58	70
One-and-a-half-inch	32	54	75	90
One-inch	56	76	128	158
Two-thirds-inch	66	91	132	183
Half-inch	120	168	280	387
Four-tenths-inch	172	230	393	480
Quarter-inch	248	345	575	700
Quarter-inch	225	312	445	615
One-eighth-inch	348	558	870	1050
Quarter-inch German	226	316	453	625
Half-inch German	105	147	210	290

Condensers for Opaque Illumination.

Large condenser for illuminating opaque objects, with universal joint and dividing stand.......0 15 6
Ditto, ditto, smaller size, not dividing.......0 12 6
Ditto, ditto, for student's microscopes.......0 9 6
Ditto, ditto, for educational microscopes.......0 7 6

Achromatic Condensers and Illuminators.

Gillet's achromatic condenser, with revolving diaphragm, having complete sets of stops for central rays, etc.......£5 5 0
The new improved achromatic condenser, which is also suitable for the binocular microscope, with revolving diaphragm.......from £1 10 0
Parabolic condenser, with central stops.......from £1 5s. to 1 10 0
Ditto ditto, with rack adjustment.......1 15 6
Cloud condenser for binoculars, fitted to stage.......0 12 6
Dark ground illuminators or spot lenses, of all kinds.......from each 0 5 0
Reade's diatom prism.......from 0 5 0

Apparatus for Achromatic Microscopes.

Polariscope, with extra large pair of prisms, fitted and attached complete.......£1 12 6
Ditto, with analyzer expressly mounted for the binocular microscope.......1 15 0
Ditto, with revolving analyzer.......1 17 6
Polariscopes for student's or educational microscope.......1 5 0
Dr. Beale's neutral tint glass reflector for drawing.......0 5 0
Brooke's double nose-piece, for carrying two object glasses to facilitate the change of focus.......£1 5 0
C. Baker's triple ditto, for three object glasses.......1 10 0
Extra eye-pieces.......from 6s. to 0 12 6
Kelner's orthoscopic achromatic eye-piece, giving a very large field.......1 0 0
Revolving selenite stages, with complete sets of selenites.......2 0 0
Erecting glasses for dissecting, applied to draw tubes.......0 15 0
Camera lucida, for drawing the magnified image.......from 15s. to 1 0 0
Stage forceps.......from 3s. 6d. to 0 5 6
Dissecting forceps.......1s. to 0 1 9
Micrometer for stage.......0 4 6
Ditto, for eye-piece, mounted in brass.......0 8 6
Ditto ditto, unmounted.......0 6 0

No. 14.—*Price-List of the Microscopes and Microscopic Apparatus of* Chas. A. Spencer & Sons, *Canastota, N. Y.* (1872).

Price in Dollars.

Student's Microscope, No. 1. Height, 15 inches; weight, 6 lbs. The arm, pillar, and base are of japanned iron; arm attached by cradle-joint, and taking any position from vertical to horizontal; coarse adjustment by sliding tube in velveted clip, and the fine by milled head upon the stage; by a curved arm the double mirror (plane and concave) has a lateral movement for oblique light, or may be swung above the stage for side illumination of opaque objects. It is furnished with B eye-piece; 1 in. objective of 20° angle of aperture, with a very flat and well-defined field, giving a power of 85 diameters; a ¼ in. of 60° angle of aperture, adjustable by front lens, giving a power of 325 diam. In neat cabinet....$60 00

Student's Microscope, Stand No. 2. Like No. 1, with addition of rack and pinion for coarse adjustment, and micrometer screw and lever fine adjustment to nose-piece; camera lucida, and animalcule cage; same objectives, eye-piece and cabinet........................$100 00

Student's Microscope, Stand No. 3. Stand of finished bronze and brass; complete as in No. 2; A and B eye-pieces; 2 in. objective of 12° ang. ap.; 1 in. objective of 20° ang. ap.; ¼ in. of 60°; camera lucida; cabinet...$125 00

Student's Microscope. Stand like No. 3; in cabinet; B eye-piece; 1 in. objective of 20° ang. ap.; 1-5 in. 100° ang. ap., adjustable for cover..$125 00

Student's Microscope. Stand like No. 3; with A and B eye-pieces, and ¼ in. objective of 60°; cabinet..$105 00

Standard Microscope, No. 1. Stand 17 inches high; weight, 11 lbs. This mounting, of the simplest and cheapest form, has the arm, pillar, and base of finished japanned iron; the arm attached by cradle joint and taking any position from vertical to horizontal; the coarse adjustment by rack and pinion, or by new friction pinion; the mirror (plane and concave) adjustable for oblique light or for side illumination of opaque objects; packed in cabinet $75 00

The friction movement mentioned is new in arrangement, and so perfect in action as to render a fine adjustment unnecessary for most observers; it can be changed almost instantly, from giving a light free motion to one so tense as almost to lock the main tube. When required, however, a fine adjustment is added at a cost of $10 additional, or..........$85 00

Standard Microscope, same as No. 1, with addition of fine adjustment; graduated draw-tube; sub-stage for accessories, movable with rack and pinion, and with screws for centring; cabinet..$125 00

Same mounting, complete as last mentioned, of finished brass and bronze.................$150 00

Other forms and sizes of stands made to order.

First-class Objectives. Large Angles of Aperture.

FOCAL LENGTH.	ANG. OF AP.	Price	FOCAL LENGTH	ANG. OF AP.	Price.
4 in. adjustable to 3 in.		$35	1-4 inch.	160°	$ 65
2 inch..........	27°	30	1-4 "	170° to 175° new form	75
1½ "	37°	30	1-8 "	175° " "	80
1 "	30°	30	1-12 "	175° " "	100
1 "	40° to 45°	40	1-15 or 1-16 "	175° " "	120
1-2 "	80°	40	1-20 "	175° " "	150
4-10 "	90°	45	1-50 "	175° " "	200
1-4 or 1-5 "	140°	50			

(75–200: Dry or Immers'n)

First-class Objectives. Medium and Smaller Angles of Aperture.

FOCAL LENGTH.	ANG. OF AP.	Price	FOCAL LENGTH	ANG. OF AP.	Price.
2 inch................	12°	$15	1-4 or 1-5 inch.	80°	$35
1 "	20°	20	1-4 "	60° adjst. by fr. [lens.	25
½ "	55° to 60°	35	1-5 or 1-6 "	100°	40
¼ "	40° non-adjst.	20	1-12 "	140°	70

Objectives are made with L. M. Society's screw, or our standard bayonet catch, as may be ordered.

No. 15.—*Price-List of Microscopes and Microscopic Apparatus of* J. Grunow, 410 *Fourth Avenue, N. Y.* (1872).

Student's Microscope. Horse-shoe brass base; 12 inches high, weighs 5 lbs.; coarse adjustment by sliding tube, fine by screw to compound body; plain stage with spring clips; diaphragm sunk into upper surface of stage, so as to be close to the object slide; plane and concave mirror, with hinge-movement; two eye-pieces; objectives ⅝ in. of 25° angle of aperture, and 1-5 in. 90° ang. ap.; giving a power of 70–500 diameters. In case........$85 00

Student's Microscope, same as above, with camera lucida, stage micrometer, and compressorium........$100 00

First-class Achromatic Objectives.

4 in.	12 degrees angular aperture	$18 00
3 in.	12 " " "	18 00
2 in.	20 " " "	18 00
1⅓ in.	22 " " "	18 00
1 in.	25–30 " " "	20 00
2-3 in.	25 " " "	22 00
4-10 in.	75 " " "	25 00
4-10 in.	100–110 " " "	35 00
1-5 in.	110 " " "	35 00
1-5 in.	135–140 " " "	40 00
1-8 in.	140–150 " " "	45 00
1-10 in.	160 " " "	50 00
1-15 in.	160 " " "	75 00

The 1-10 in. and 1-15 in. are immersion objectives.

Objectives for Student's Microscopes.

In neat morocco cases, the same as the Objectives of Nachet, Hartnack, and others.

1⅓ in.	15 degrees angular aperture	$12 00
2-3 in.	25 " " "	15 00
4-10 in.	50 " " "	18 00
1-5 in.	90 " " "	20 00
1-10 in.	130 (immersion) "	30 00

Accessories for the Microscope.

Achromatic condenser	$35 00
Polariscope "	$25–30 00
Camera lucida	6 00
" best (invented by J. Grunow)	15 00
Erector " " "	7 00
Extra eye-pieces " "	6 00
Orthoscopic or Kellner eye-piece	10 00
Amicis' prism "	12 00
Eye-piece micrometer	5 00

Microscope Stands, of the best American make, are furnished to order, at manufacturers' prices.

No. 16.—*Price-List of Microscopes and Microscopic Apparatus of* R. B. Tolles (Charles Stodder, *Agent*), 66 *Milk Street, Boston, Mass.* (1872).

Student's Microscope. 15 inches high; weight, 6 lbs.; the base, uprights, and curved arm are of iron handsomely japanned; on a trunnion joint, made on a new plan to wear well, by which the instrument can be inclined from vertical to horizontal, with a stop to prevent movement beyond these points. It is furnished with a B eye-piece, two second-quality objectives, 1 in. and ¼ in. focus; giving about 60 and 280 diameters; a plain stage with spring clips for holding slides; revolving diaphragm; concave mirror, with movement to give oblique light; for illumination of opaque objects mirror is removed to an upright stand; coarse adjustment by sliding compound body, which is held in place by a steel spring; fine, by a new construction efficient with high powers. Upright black walnut or mahogany case........$70 00

The same, with fine adjustment on stage........$50 00

Additions.—Extra eye-pieces A and C, $4 each; a superior camera lucida, $5; sub-stage for accessory apparatus, $5; sliding stage giving vertical and horizontal motions by the hand, and adapted for the use of Maltwood finder, $15; rack and pinion for coarse adjustment, $12; stand, all brass, $10.

Large Microscope B. This instrument is intended to meet the wants of the scientific investigator; to attain everything that the microscopist can accomplish, without sparing the cost, and to permit the use of all the modern accessory apparatus; constructed on the curved arm (Jackson) model; 18 inches high, weighs about 14 lbs.; the curved arm is supported on a steel trunnion between two strong brass pillars, made for durability and not liable to get out of order, and provided with a method of compensation for wear; has rack and pinion for coarse, and micrometer screw for fine adjustment of focus; graduated draw-tube; sub-stage, with rack and pinion, and centring screws for accessory apparatus; plane and concave mirrors on double-jointed arm; Tolle's thin stage, admitting light of great obliquity, with rectangular movements by screw, and rack and pinion; rotation on the optical axis of about 270°; all that is practically necessary........................$225 00

A modification of this size, with the stage carried by friction rollers, and entire rotation on the optical axis, can be made........................$275 00

Largest Microscope A. This instrument is one of the largest yet produced anywhere. It is similar in all respects of style and construction to the B instrument, but larger and heavier, weighing 20 lbs. The stage is six inches in diameter, and makes a complete revolution on the optical axis. The whole instrument rotates on a stout plate graduated to degrees. Price of stand........................$300 00

Can be furnished with radial arm to carry accessory apparatus at any angle, for...........$50 00

Pocket microscope, for clinical and field or sea-side use; is a simple tube 6 in. long, ¼-in. objective; B eye-piece; fine and coarse adjustments; stage with spring clips to hold the object, which can be removed when not in use, and the objective covered with a brass cap........................$25 00

With draw tube for increasing the power........................30 00

As the same eye-pieces and objectives are used for the student's microscope, those who want both instruments require but one set of the optical parts.

Accessories.

Huyghen's negative eye-piece ....2-in. $10, 1½-in. $9, 1-in. $8, ¾-in. $7

Ditto, " " for student's microscopes........................$4 each.

Tolles' patent solid eye-pieces........................½-in. $8, ¼-in. $8, ⅙-in. $7

Tolles' solid orthoscopic eye-pieces (this eye-piece makes an excellent achromatic condenser), 1-in. $15, ¾-in. and higher powers........................$12 00

Tolles' binocular eye-piece........................80 00

Tolles' amplifier, to double the power of any specified objective and eye-piece, without disturbing the "corrections,"........................$12 00

Higher power amplifiers by special arrangement.

Tolles' achromatic triplets for the pocket, in silver cases, ½-in. and ¾-in. $12, ⅓-in. $14, ¾-in. extra large mounting........................$14 00

1-inch do. do. with 610-inch field........................$16 00

Double nose-piece........................$8 to $12 00

Camera lucida........................$5 00

Objectives.

First quality of large angular aperture. These are made for the highest requirements of the microscopist and histologist, as tracing nerve fibre, cell formation, resolution of Nobert's lines, etc., etc.

1-2 inch ang. ap 60° to 80°........................$40 00
4-10 " " " 90° to 110°........................45 00
4-10 " " " 135° to 145°........................65 00

This objective may be used as an achromatic condenser, with special advantage.

1-4 &1-5 inch ang. ap. 110° to 130°........................$50 00
1-4 " " " up to 150°........................60 00
1-4 " " " " " 170°........................70 00
1-6 " $5 advance on 1-4th.
1-8 " ang. ap. under 140°........................$65 00
1-8 " " " to 160°........................75 00
1-8 " " " to 175°........................80 00
1-10 " $5 advance on price of 1–8th.
1-12 " ang. ap. under 140°........................$80 00
1-12 " " " up to 160°........................100 00
1-12 " " " " to 175°........................115 00
1-15 " " " " to 160°........................120 00
1-15 " " " " to 175°........................125 00
1-20 " " " " to 175°........................180 00

First quality objectives of lower angular aperture having properly more penetration and better adapted for some anatomical and botanical investigations.

1-2 inch	ang. ap.	60° or less	$35 00
4-10 "	" "	85° " "	40 00
1-4 & 1-5 "	" "	100° " "	40 00
1-6 "	" "	100° " "	45 00
1.8 "	" "	100° " "	45 00
1-10 "	" "	100° " "	45 00
1-12 "	" "	120° " "	60 00
1-15 "	" "	100° " "	70 00
1.20 "	" "	100° " "	100 00

All of 1-5th inch or higher powers will be made either dry or immersion at the same price; to work both ways with the same lens, from $5 to $15 extra. With an extra "front" lens $10 to 2-5ths extra. All the foregoing have Tolles' adjustment for covering glass, which does not move *the front lens.*

First quality objectives, without adjustment for cover.

4-inch, adjustable to 3-inch $35 00
4-inch, changeable to 2½-inch 18 00
2-inch 20 00
2-inch, higher ang. ap 23 00
1½-inch and 1-inch, in one 28 00
1-inch 17° 20 00
1-inch 25° 23 00
1-inch 30° 30 00
¾-inch, new formula, specially flat-field 30 00
½-inch 25° to 40° 23 00
½-inch, specially constructed for viewing opaque objects 28 00
¼-inch, 40° to 70°, adjusting by front lens 26 00
⅛-inch, to 70° " " " " 35 00
Second quality objectives .... 1-inch and 2-inch $8, 1-2-inch $10, 1-4 and 1-6 $15 to $20

All objectives are made with the "Society" screw, so as to fit all recent English or American stands—unless ordered otherwise.

No. 17.—*Price-List of Microscopes and Microscopic Accessories of* J. ZENTMAYER, 147 *South 4th Street, Philadelphia, Pa.* (1872).

1. U. S. army hospital microscope, brass body, 16 inches high, on brass stand, with joint to incline it to any angle, double-milled head, rack and pinion for coarse adjustment, micrometer screw for fine adjustment, movable glass stage; under the stage a tube is fitted for carrying the accessory illuminating apparatus; concave and plane mirrors, arranged for direct or oblique illumination; draw tube for increasing the power; two eye-pieces; one achromatic object glass, $\frac{8}{10}$ of an inch focus of 32 degrees angular aperture; one achromatic object glass, $\frac{1}{5}$ of an inch focus of 80 degrees angular aperture (not adjustable for glass cover), giving power of 50, 100, 250 and 450 diameters; camera lucida, stage micrometer ruled $\frac{1}{100}$ and $\frac{1}{1000}$ of an inch, and a condensing lens two inches diameter on separate stand. Securely packed in a neat walnut box with lock and key $135 00
2. Grand American microscope. The most complete and perfect instrument of the kind now made. It is 19 inches high on tripod base. The two brass pillars upon which the body and stage are swung rest upon a revolving plate with graduated edge, by which the angular aperture of the object glasses can be ascertained; the body is moved with a double-milled head; pinion and rack for the coarse adjustment, and a fine micrometer screw for the delicate adjustment. The mechanical stage has a screw adjustment with milled head for the horizontal motion, and a delicate chain and pinion with milled head for the vertical motion. On the centre of the upper side of the stage a circular plate with graduated edge is attached for measuring angles of crystals; the whole thickness of the stage is but $\frac{3}{10}$ of an inch, but at the same time perfectly solid and steady, and affording unusual facility for great obliquity of illumination when difficult tests are to be resolved. Under the stage a small tube with rack and pinion is attached; in this tube the accessory illuminating apparatus is carried when in use. The mirror has one side plane and the other concave; the bar which carries it is jointed, to give the required motion for oblique illumination. There is a graduated draw tube sliding into the main tube of the body for increasing the magnifying power, by lengthening the distance between the object glass and eye-piece. Three eye-pieces; one achromatic object glass, 1½ inch focus, 22 degrees angle of aperture, and one achromatic object glass $\frac{4}{10}$ of an inch focus, 80 degrees angle of aperture; power 50–500 diameters; large condensing lens on separate stand. All packed in a handsome walnut cabinet, with lock and key $262 00

No. 3. Grand American microscope, the same as No. 2, but with the following accessories:—3 eye-pieces; achromatic object-glass, 1½ inch focus, 22 degrees angle of aperture; 1 achromatic object-glass, 8-10 of an inch focus, 32 degrees angle of aperture; 1 achromatic object-glass, 4-10 of an inch focus, 80 degrees angle of aperture, with adjustment for thin glass cover; 1 achromatic object-glass, 1-5 of an inch focus, 120 degrees angle of aperture, with adjustment for thin glass cover; polarizing apparatus with selenite plate, parabola for dark field illumination, erector, large condensing lens on separate stand, camera lucida, stage micrometer ruled to 1-100th and 1-1000th of an inch, stage forceps, animalcule cage, zoophyte trough, blue glass cap. The whole packed in a highly polished mahogany cabinet.......$400 00

No. 4. Grand American microscope. Stand only, same as No. 2, with three eye-pieces. No object-glasses, no box.......$200 00

No. 5. Student's microscope, brass stand and body, bronzed, 16 inches high, with joint to incline to any angle; micrometer screw for fine adjustment; broad stage with clips for holding the object; under the stage a tube is fitted for carrying the accessory illuminating apparatus; concave and plane mirrors, arranged for direct or oblique illumination; two eye-pieces; 1 achromatic object-glass, 8-10 of an inch focus, 24 degrees angular aperture; 1 achromatic object-glass, 1-5 of an inch focus, 80 degrees angular aperture (not adjustable for glass cover), giving power of 50, 100, 250 and 450 diameters, and a condensing lens attached to the stand. Securely packed in a neat walnut box, with lock and key......$75 00

No. 6. Clinical microscope. This instrument is especially arranged for class demonstration, and is an elaboration of, and improvement upon the class microscope of Dr. Beale. It consists of full-size body with draw-tube; coarse and fine adjustments of focus; 2 eye-pieces; 1 achromatic object-glass, 8-10 of an inch focus, 24 degrees angular aperture; 1 achromatic object-glass, 1-5 of an inch focus, 80 degrees angular aperture (not adjustable for glass cover), giving powers from 50–450 diam.; lamp for direct illumination, arranged upon a sliding bar, and giving every facility for the accurate examination of any object by a large class, by being passed from hand to hand. The whole enclosed in a neat and very portable oiled walnut case.......$60 00

Achromatic condenser, with centring adjustment, revolving diaphragm plate, achromatic combination of ½ and 1-5 in.......$38 00

Polarizing apparatus	$25—35 00
Selenite plate	1 00
Parabolic illuminator, mounted,	14 00
Achromatic oblique prism, mounted	14 00
Amici's prism, mounted	10 00
Erector, mounted	6 00
Bull's-eye condenser, 3 inches diameter, on stand	10 00
Camera lucida, mounted	8 00
Stage micrometer, divided 1-100 and 1-1000 of an inch	2 00
do. do. do. 1-100, 1-1000, and 1-2000 of an inch	2 50
Eye piece micrometer, with micrometer screw	6 00
Stage forceps	4 00
Animalcule cage	3 50
Zoophyte trough	3 00
Blue glass cap, to place under the stage to modify the light	1 50
Eye-pieces, each	6 00
Mechanical finger	20 00
4 and 5 in. objective combined, a very low power to overlook a large slide at once	15 00
3 and 4 in. objective combined	15 00
1½ in. objective. Ang. ap. 22°	15 00
8-10 in. " " 30°	18 00
4-10 in. " " 85° (adjustable)	25 00
1-5 in. " " 120°	35 00
8-10 in. " " 24°	12 00
1-5 in. " " 85°	18 00

No. 18.—*Price-List of Objectives made by* WM. WALES, *Fort Lee, N. J.* (1872).

4 in.	Angle aperture	9°	$12 00
3 in.	" "	12°	15 00
1½ in.	" "	23°	17 00
2-3 in.	" "	30°	20 00
4-10 in.	" "	75°	40 00
4-10 in.	" "	90°	40 00
4.10 in.	" "	110°	40 00
1-4 in.	" "	90°	30 00
1-5 in.	" "	135°	40 00
1-5 in.	" "	165° (immersion)	50 00
1-8 in.	" "	145°	45 00
1-10 in.	" "	160° (immersion)	50 00
1-15 in.	" "	170° "	75 00
1-30 in.	" "	160° "	120 00

No. 19.—*Price-List of Microscopes and Accessories of* Miller Brothers, 69 *Nassau Street, and* 1223 *Broadway, New York* (1872).

Price in Dollars.

Microscope stand A. This binocular microscope is of first-class quality in every respect. The stand is firm and free from tremor under observation, even while the adjustment of apparatus is going on. The binocular mechanism is very superior, realizing both the stereoscopic and perspective views of the object with remarkable ease and perfection. In addition to a rectangular motion of one inch in each direction, and rotation by hand, the whole stage rotates concentrically and independently by means of a rack and pinion on a circular plate, graduated so as to form a goniometer or position micrometer. The secondary or sub-stage has adjusting screws for centring all the supplementary apparatus which it receives, and affords facilities for the manipulation and use in the most convenient and efficient manner, possessing also the means of rotation by rack and pinion, with graduated divisions at the circumference. Fine adjustment of the most delicate and perfect construction, the index reading off differences in the focal position of the objective to the 5-1000 of an inch, perceptible to the observer's eye. Including 4 eye-pieces................$275 00

If with single body, 2 eye-pieces...$230 00

Microscope stand B. Constructed on the same general plan as A, but smaller. Stage has usual rectangular motion, and one of rotation, without rack and pinion. Sub-stage has a rotating cylinder, with adjustments for centring the apparatus which it receives, and provides for their use and application with freedom. Flat and concave mirror on double arm. Binocular, with 4 eye-pieces... $220

Single body, 2 eye-pieces.. [illegible] 00

Microscope stand C; rather smaller than B, stage is remarkably thin, has a rotating object plate, and a rectangular motion of ¾-inch, both in vertical and horizontal directions. Polarizer and sub-stage appliances are fixed by means of a sliding dovetail plate, with a stop to insure their concentric position when in use; with two eye-pieces...........$135 00

If binocular, with four eye-pieces...180 00

Microscope stand D; formed on same general design as B, but smaller; has a rack and pinion, and a fine adjustment; rectangular screw and rack motion to stage; rotary diaphragm; two Huyghenian eye-pieces A and B; concave and flat mirror...............$70 00

If binocular, four eye-pieces, rack and pinion..$100 00

New educational microscope E (stand entirely of brass), has a fine adjustment, and the following accessories fitted with it in a polished mahogany case: Two Huyghenian eye-pieces A and B: 1-inch objective, 16° angular aperture; ¼-inch objective, 75° angular aperture; condensing lens on separate stand; pair of pliers; four stage plates and cells...$80 00

Student's microscope, 16 inches high, weighs 5 pounds; base upright and curved arm of iron handsomely japanned; trunnion joint, made on a new plan to wear well, by which the instrument can be inclined from vertical to horizontal, with stop to prevent movement beyond these points. Is furnished with A and B eye-pieces, a good achromatic objective, ½-inch, with magnifying power of 153 diameters; plain stage; spring clasp for holding slides; concave mirror, with movement for oblique light, removed to stage for opaque objects; coarse adjustment by sliding tube through cloth packing, fine by a new construction, efficient with high powers...$35 00

With case...40 00

First-Class Achromatic Objectives.

Focal Length.	Angular Aperture	Linear Magnifying Power with the respective Eye-pieces.				Price.	Lieberkuhn's Reflectors, Extra.
		A.	B.	C.	D.		
Inches.	Degrees.						
3	12	13	20	35	56	$18.00	$8.00
2	15	20	32	55	90	18.00	7.00
1½	20	25	40	70	112	22.00	7.00
1	25	37	60	105	170	22.00	5.50
⅔	30	60	100	145	270	22.00	4.50
½	75	95	153	265	420	36.00	
¼	100	195	310	540	850	40.00	
1-6	130	320	510	700	910	54.00	
⅛	140	420	670	900	1200	72.00	

All powers higher than two-thirds have the adjustment for covered and uncovered objects.

Achromatic Objectives with less Angular Aperture.

3	10	13	20	35	56	$13.50	None of these have the adjustment for covered and uncovered objects.
2	12	20	32	55	90	13.50	
1	16	37	60	105	170	13.50	
½	55	95	153	265	420	18.00	
¼	70	195	310	540	850	18.00	
¼	80	195	310	540	850	27.00	

All the objectives constructed by Miller Brothers are made with the Royal Microscopical Society's Universal Screw.

Microscopical Apparatus, Etc.

Extra eye-pieces, A, B, C and D....each, $ 5 00
Erecting eye-piece for dissecting....9 00
Improved micrometer eye-piece....10 00
Kelner's orthoscopic eye-piece, C or D, double-size field....10 00
Sorby's spectroscope....$30 00 to 50 00
Bull's-eye condenser, on stand....$3 50, and 5, 6, 7, 8 9 00
Webster's achromatic condenser....$15 00 and 40 00
Hall's universal condenser and appliances....$27 00 to 45 00
Gillet's achromatic condenser, with latest improvements....60 00
Eye-piece for centring do....5 00
Plain achromatic condenser, with ¼-inch objective, central and annular stops....20 00
Read's hemispherical or kettle-drum condenser....13 50
Set of Rainey's light modifiers....4 00
Camera lucida, Wollaston's....$8 00 and 10 00
Do. do. neutral tint glass....3 00
Polarizing apparatus....$20 00 and 25 00
Selenites, selected colors, $1 00 each; best, brass mounted....2 25
Set of Darker's do., revolving, brass mounted....20 00
Darker's new series of 3 selenites to rotate either singly or combined, showing 13 colors and complementary tints....31 00
Lister's set dark wells....5 00
Dark field condenser, with adjustment....4 00 to 5 00
Wenham's parabolic reflector....$9 50; best 14 00
Silver side reflector, for opaque objects....9 50
Silver parbolic stage reflector....9 50
Beck's illuminator, for opaque objects under high power....4 00
Amici's prism, for oblique illumination....20 00
Nachet's do. do. do.....9 50
Rectangular do., reflection of parallel rays....18 00
Brooke's arm, for two objectives....10 00
Do. do. three do....25 00
Stage micrometer, ruled 100 and 1,000....3 00
Maltwood's object finder, in case....3 00
Forceps, straight and curved....25c. to 2 00
Stage tweezers, on jointed arm....3 50
Glass polyp trough, complete....4 00
Live boxes, for insects, etc....3 50
Animalcule or dipping tubes....10
Frog and fish plate, complete in glass or meta....$2 50 to 4 50
Glass fish boxes....3 00
Glass stage plates, various....50c. and 1 25
Spring compressorium, $6 00; do for high power....4 00
Best lever do. $16 00; inverting do....10 00
Edward's prism, for oblique illumination, mounted in German silver....$40 00 and 50 00
Read's prism....12 00

No. 19.—*Price-List of Microscopes and Microscopic Apparatus of* T. H. McAllister, 49 *Nassau Street, New York* (1872).

Professional microscope. Somewhat more than 15 inches high; base of iron lackered, with uprights to receive the axis upon which the body inclines to any convenient angle; body of brass, finely finished with extension draw-tube. Coarse adjustment by a delicate watch-chain, controlled by a large milled head on each side of the tube; far more efficient and precise than the majority of rack movements; will readily adjust for all but very highest powers; fine adjustment by micrometer screw. Stage large and steady, but thin enough to allow extreme obliquity of illumination. To upper surface a plate glass stage is attached; can be freely moved in vertical and horizontal directions, can also be revolved; the motion is so delicate and simple that many experienced microscopists prefer it to elaborate screw stage; a movement as minute as 1-12,000 can be effected by it. A brass rest, with springs to hold the object, is clamped to it, but can be removed in a moment, leaving a clean glass plate for examination of recent anatomical preparations, chemicals or other substances which would injure the usual brass stage. Beneath the stage is a separable collar carrying the diaphragm, and also adapted to receive the polarizer, parabolic illumnator and other accessories. Concave and plane mirrors mounted with universal motion, and slide on a jointed bar for direct or oblique illumination. The fitting for the objectives is made with the "London Society" screw, so that objectives of all first-class makers can be used with the instrument. Stand with two eye-pieces, but without other accessories, in neat upright walnut case, with lock and handle.................$75 00

Student's Microscope; somewhat more than 12 in. high; base of iron lackered, with uprights to receive the axis upon which the body inclines; tube of brass, with extension draw-tube; coarse adjustment by delicate watch-chain, controlled by large milled heads on each side of tube; fine micrometer adjustment attached to stage; stage of brass, with brass springs to hold the object, diaphragm plate beneath; stage forceps attached to stage, which are convenient for holding an insect, flowers, etc., while being examined by low powers; concave and plane mirrors with universal motion on jointed bar. With following accessories: one eye-piece; 1 in. achromatic objective; ¼ in. do.; magnifying powers, 50, 75, 250 and 400 diam.; in neat upright walnut case, with lock and handle.....................$50 00

Popular microscope; somewhat more than 10 in. high; base of iron lackered, with uprights to receive axis on which the body inclines; tube of brass, with extension draw-tube, milled head, fine adjusting focus; stage of brass, with springs to hold object, and with diaphragm plate beneath; stage forceps attached to stage; mirror mounted with universal motion on jointed bar. With following accessories: one eye-piece; one achromatic objective, separable; magnifying powers, 50, 75, 150 and 200 diam.; in neat upright walnut case, with handle..............................$25 00

Objectives, etc., of first makers on hand, and for sale at manufacturer's prices.

Microscopic Accessories.

Bull's-eye condensing lens on stand, large size, fine finish—adapted for the "Grand American," or the "Professional" microscope..................$10 00
Bull's-eye condensing lens on stand, adapted for "The Student's" or the "Popular" microscope.................. 5 00
Bull's-eye condensing lens on stand, adapted for "The Household" microscope.............. 2 50
Polarizing apparatus, adapted for "The Grand American" microscope.................. 35 00
Polarizing apparatus, adapted for "The Professional" microscope.................. 25 00
Polarizing apparatus, adapted for "The Student's" microscope.................. 20 00
Polarizing apparatus, adapted for "The Popular" microscope.................. 15 00
Selenite plate.................. 1 00
Zentmayer's achromatic condenser.................. 38 00
Zentmayer's achromatic oblique prism.................. 14 00
Zentmayer's Amici prism.................. 10 00
Camera lucida.................. 8 00
Parabolic illuminator, adapted for "The Grand American" microscope.................. 14 00
Parabolic illuminator, adapted for "The Professional" microscope.................. 10 00
Erector, or reducing eye-piece, for erecting the image, reducing the power, and enlarging the field of view—adapted for "The Grand American," "The Professional" and "The Student's" microscope.................. 7 00
Concave amplifier, for increasing the power.................. 5 00
Stage forceps.................. 4 00
Animalcule cage.................. 4 00
Zoophyte trough of glass.................. 4 00
Stage micrometer, 1-100 and 1-1000.................. 2 00
Eye-piece micrometer, with micrometer screw.................. 6 00
Maltwood finder, for ascertaining and registering the position of any particular object in a slide.................. 5 00
German student's lamp for burning coal oil, affording the best light for microscopic investigations.................. 7 00

Materials for Mounting Objects.

Plate glass slides, 3 by 1 inch, ground edges	per dozen	$ 75
Thin glass circular covers, 1 inch diameter	do.	75
Do. do. ⅞ do.	do.	50
Do. do. ¾ do.	do.	40
Do. do. ⅝ do.	do.	40
Canada balsam	per bottle	50
Gold size	do.	50
Brunswick black	do.	50
Marine glue		25
Shadbolt's turn table		4 00
Steel forceps, 4 inch long, straight points		1 00
Do. do. curved points		1 00
Scissors, 4½ inch long, straight points		1 50
Do. do. curved points		1 50
Prof. Valentine's double-bladed knife, for making fine sections of soft tissues, etc		8 00
Case of dissecting instruments		10 25
Needle-holder, with binding screw		75
Needle in steel handle		25
Scalpel		75
Dropping and dipping tubes		10

Cabinets for Holding Objects.

To hold 6 dozen objects, mahogny case	$4 00
Do. 2 do. do. case covered with cloth, lined with velvet	75
Do. 2 do. do. plain case	25
Do. 10 objects, do.	15
Do. 6 do. do.	10
Do. 1 do. do. (for mailing, etc.) per doz	80

No. 20.—*Price-List of* B. Pike's Son, 518 *Broadway, N. Y.* (1872).

Student's microscope; 15 inches high, weighs 5 lbs.; iron base and arm; rack and pinion for coarse adjustment, fine by screw with lever to stage; stage of glass, sliding by hand; plane and concave mirror; A and B eye-pieces; 1 in. and ⅙ in. objectives; magnifies from 50 to 600 diam.; has a condensing lens on a stand; mahogany case.......$75 00

Physician's microscope; same height and weight as the preceding one; iron base; coarse adjustment by chain movement, fine by screw with lever to compound body; stage of glass, sliding by hand, double mirror; A and B eye-pieces; 1 in. and ⅙ in. objectives; magnifying power from 50–960 diameters; condensing lens on stand, and draw-tube, with case.......$100 00

No. 21.—*Price-List of Microscopes of* Jas. W. Queen & Co., 924 *Chestnut Street, Philadelphia, and* 535 *Broadway, New York* (1872).

Student's microscope; 14 in. high, weighs 5 lbs.; iron base and arm; coarse adjustment by sliding the tube with the hand, fine by screw with lever to nose-piece; stage of glass, sliding by hand with fitting for accessories; has a plane and concave mirror with hinge movement; one eye-piece; one objective of 1-5 in. (dividing); magnifies from 100–300 diam.; stage-micrometer and condensing lens on stand; with case.......$65 00

The same, with two eye-pieces; 1 in. objective of 18° and 1-5 in. of 80° ang. aperture; magnifying 50–500 diameters; stage micrometer, condensing lens on stand, camera lucida, and compressorium. In a case.......$100 00